Cannabis and the Developing Brain

Cannabis and the Developing Brain

Edited by

Miriam Melis
Department of Biomedical Sciences, Division of Neuroscience and Clinical Pharmacology, University of Cagliari, Cagliari, Italy

Olivier JJ. Manzoni
Mediterranean Neurobiology Institute, INSERM U1249, Aix-Marseille University, Marseille, France

Academic Press is an imprint of Elsevier
125 London Wall, London EC2Y 5AS, United Kingdom
525 B Street, Suite 1650, San Diego, CA 92101, United States
50 Hampshire Street, 5th Floor, Cambridge, MA 02139, United States
The Boulevard, Langford Lane, Kidlington, Oxford OX5 1GB, United Kingdom

Copyright © 2022 Elsevier Inc. All rights reserved.

No part of this publication may be reproduced or transmitted in any form or by any means, electronic or mechanical, including photocopying, recording, or any information storage and retrieval system, without permission in writing from the publisher. Details on how to seek permission, further information about the Publisher's permissions policies and our arrangements with organizations such as the Copyright Clearance Center and the Copyright Licensing Agency, can be found at our website: www.elsevier.com/permissions.

This book and the individual contributions contained in it are protected under copyright by the Publisher (other than as may be noted herein).

Notices
Knowledge and best practice in this field are constantly changing. As new research and experience broaden our understanding, changes in research methods, professional practices, or medical treatment may become necessary.

Practitioners and researchers must always rely on their own experience and knowledge in evaluating and using any information, methods, compounds, or experiments described herein. In using such information or methods they should be mindful of their own safety and the safety of others, including parties for whom they have a professional responsibility.

To the fullest extent of the law, neither the Publisher nor the authors, contributors, or editors, assume any liability for any injury and/or damage to persons or property as a matter of products liability, negligence or otherwise, or from any use or operation of any methods, products, instructions, or ideas contained in the material herein.

ISBN 978-0-12-823490-7

For information on all Academic Press publications
visit our website at https://www.elsevier.com/books-and-journals

Publisher: Nikki P. Levy
Acquisitions Editor: Joslyn T. Chaiprasert-Paguio
Editorial Project Manager: Pat Gonzalez
Production Project Manager: Kiruthika Govindaraju
Cover Designer: Mark Rogers

Typeset by STRAIVE, India

In memoriam

As this book entered the final stages of publication, Dr. Daniela Parolaro, who contributed to this book, passed away at the age of 72. We grieve her tragic loss in the knowledge that her important contributions to science will live on through her legacy of rigorous science, mentorship, generosity, and kindness. Daniela was a pioneer in the field of cannabinoid research: her first article on the impact of chronic cannabinoid exposure on CB1 receptor gene expression in the brain was published in 1994. Since then, she developed a spectacular research program and has made seminal contributions to the field of cannabinoid research, spanning from the first evidence of cannabidiol actions in cancer therapy to the potential of phytocannabinoids in the treatment of cognitive impairments associated with neurodevelopmental disorders and the protracted effects of adolescent cannabis. Her scientific leadership and charisma let her to become the first woman President of International Cannabinoid Research Society in 2005, and a chairwoman of International Association for Cannabis as Medicine in 2013. Author and collaborator on more than 170 scientific papers, Daniela was a recipient of the 2020 Mechoulam Award for her outstanding contribution to cannabinoid research. We will remember Daniela as an inspiring teacher, a generous colleague, a treasured friend, and an incredible woman.

Contents

Contributors		xv
Preface		xix

1. Pre-clinical models of neurodevelopmental cannabinoid exposure — 1
Ken Mackie

Introduction	1
Model considerations	2
Timing of drug administration	2
Pre-natal models	2
Adolescent models	4
Route of drug administration	4
Inhaled	4
Oral	5
Injection	6
Nature of drug administered	6
Dose of drug administered	7
Summary	8
References	8

2. Epigenetic imprint: An underlying link to developmental effects of prenatal cannabis exposure — 13
Anissa Bara and Yasmin L. Hurd

Introduction	13
Molecular reprogramming after prenatal cannabinoid exposure	14
Prenatal cannabis exposure and epigenetic mechanisms	15
DNA methylation	16
Histone modifications	16
Non-coding RNAs	16
Epigenetic mechanisms and developmenal cannabinoid exposure	16
Conclusion	19
References	19

3. **Impact of adolescent THC exposure on later adulthood: Focus on mesocorticolimbic function and behaviors**   23
Anthony English, Benjamin Land, and Nephi Stella

Introduction	23
Cannabis-based products and devices	26
Cannabinoids, endocannabinoid signaling, and brain development	28
Evidence from human studies	33
Cognitive aptitude and behavior of adults who used cannabis during adolescence	34
Disruption of brain anatomy, connectivity, and dopamine function in human adults that used cannabis during adolescence	35
Evidence from rodent studies	37
Molecular and cellular responses to THC treatment during adolescence in rodents	38
Impact on adult mesocorticolimbic systems and reward/addiction-like behaviors associated with THC use during adolescence in rodents	40
Additional systems-level and behavioral responses to THC treatment during adolescence in rodents	42
THC self-administration by adolescent rodents	44
CBD interactions with THC during adolescence	45
Conclusion	47
References	48

4. **Endocannabinoids and sex differences in the developing social behavior network**   59
Margaret M. McCarthy, Ashley E. Marquardt, and Jonathan W. VanRyzin

Introduction	59
The social behavior network orchestrates expression of innate social behaviors	60
Sexual differentiation of the developing brain	61
Endocannabinoids in developing brain	62
Adolescent playfulness is a foundational behavior for adult sociality	65
The extended amygdala is a key node of the social behavior network	66
Endocannabinoids developmentally program playfulness	66
Endocannabinoids regulate microglia phagocytic activity in the developing amygdala	67
The medial preoptic area is also a key node in the social behavior network	70
Conclusions and open questions	71
References	72

5. **Behavioral consequences of pre/peri-natal *Cannabis* exposure**   79
   Antonia Manduca and Viviana Trezza

Introduction	79
Clinical studies on the behavioral consequences of maternal *Cannabis* exposure	80
Rodent studies on the behavioral consequences of maternal cannabinoid exposure	84
Future perspectives	87
Conclusions	88
Acknowledgment	89
References	89

6. **Correlates and consequences of cannabinoid exposure on adolescent brain remodeling: Focus on glial cells and epigenetics**   95
   Zamberletti Erica, Manenti Cristina, Gabaglio Marina, Rubino Tiziana, and Parolaro Daniela

Introduction	95
Glial cells	95
The endocannabinoid system in glial cells	95
Consequences of adolescent THC exposure on glial cells	97
Conclusions	100
Epigenetics	100
Epigenetic alterations induced by adolescent THC exposure	101
Conclusions	103
References	103

7. **Effects of prenatal THC exposure on the mesolimbic dopamine system: Unveiling an endophenotype of sensory information processing deficits**   107
   Roberto Frau and Miriam Melis

Introduction	107
Pre-natal cannabis exposure as an environmental risk factor for neuropsychiatric disorders	109
Pre-natal cannabis exposure impacts on the offspring mesolimbic dopamine system	112
Sex differences in the effects of pre-natal cannabis exposure on mesolimbic dopamine system function	116
Measuring vulnerability to mental illness induced by pre-natal cannabis exposure: A path toward an early clinical staging	118
Concluding remarks	121
References	122

8. **Perinatal cannabis exposure and long-term consequences on synaptic programming** 129
   Gabriele Giua, Olivier JJ. Manzoni, and Andrew Scheyer

   Parental germline cannabis exposure could affect progeny through transgenerational epigenetic alterations 130
      Endocannabinoid system and epigenetic re-programming cohabitation in re-productive tissues 130
      Direct effects of cannabinoids exposure on parent's germline 131
   Cannabinoids exposure and peri-embryo implantation influences 132
      Role of the endocannabinoid system in early embryo development and implantation 132
      Maternal exocannabinoid exposure and repercussions on embryo implantation and development 133
   Perinatal cannabis exposure and long-term consequences on synaptic programming 133
      Role of the ECS in perinatal neurodevelopment 133
      Protracted effects of PCE on synaptic programming and the neurodevelopment trajectory 134
   Conclusions 136
   Acknowledgments 137
   References 138

9. **Molecular and cellular principles of endocannabinoid signaling and their sensitivity to cannabis in the developing brain** 149
   Erik Keimpema and Tibor Harkany

   Introduction 149
   Cannabis use during pregnancy 151
   The endocannabinoid system 152
      Endocannabinoid-sensing receptors 152
      Enzymatic production and degradation of endocannabinoids: Subcellular partitioning for intercellular communication 153
   Pre-natal endocannabinoid signaling and the adverse developmental outcomes of its disruption 156
      Embryo implantation and growth 156
      Neurulation and neurogenesis 156
      Neurite outgrowth and target innervation 157
      Pre-natal exogenous cannabinoids perturb adolescent synapse refinement 158
      Endocannabinoids in the reward circuitry 159

Post-natal endocannabinoid signaling and adolescent circuit
   disruption 159
    The perinatal period 159
    Neonatal findings 160
    Early adolescence toward adulthood 160
Conclusion 161
Acknowledgments 162
References 162

10. How adolescent cannabinoid exposure sets the stage for long-term emotional and cognitive dysregulation: Impacts on molecular and neuronal risk pathways   171
Steven R. Laviolette

Introduction 171
How cannabinoids control emotional processing: Convergent
   impacts on mesocorticolimbic dopamine states 173
Cannabinoid transmission in pre-frontal cortex regulates
   emotional processing through control of sub-cortical
   dopamine 174
Translational rodent models of adolescent cannabinoid
   exposure 178
Effects of neurodevelopmental cannabinoid exposure
   on emotional processing circuits: Addiction
   implications 179
Adolescent cannabinoid exposure alters dopaminergic,
   glutamatergic and GABAergic signaling and associated
   molecular pathways: Implication for schizophrenia
   vulnerability 180
Effects of adolescent cannabinoid exposure on long-term
   mesolimbic dopamine activity states 181
Effects of adolescent cannabinoid exposure on pre-frontal
   cortical regulation of sub-cortical dopamine activity:
   Impacts on molecular biomarkers for
   schizophrenia and other neuropsychiatric
   disorders 183
Effects of adolescent cannabinoid exposure on pre-frontal cortical
   GABAergic and glutamatergic functional balance: Implications
   for the pathophysiology of schizophrenia 188
Conclusions and future directions 190
References 191

11. Molecular mechanisms underlying cannabis-induced risk of psychosis — 197
Paula Unzueta-Larrinaga, Luis F. Callado, and Leyre Urigüen

Introduction	197
Schizophrenia	198
General aspects	198
Etiology and pathogenesis	200
Comorbidity in schizophrenia	208
Cannabis and schizophrenia	210
5-HT2A receptors	216
The Akt/mTOR pathway	218
Cannabis-induced 5-HT2AR pro-hallucinogenic signaling and Akt/mTOR pathway regulation	222
Conclusions	224
References	225

12. Synthetic cannabinoids: State-of-the-art with a focus on fertility and development — 243
A.-L. Pélissier-Alicot

The emergence of synthetic cannabinoids	243
Chemical structure and nomenclature	244
Colloquial and serial names	244
Systematic chemical names	245
Epidemiology, patterns of use, and legal status	245
Pharmacology	247
Binding affinity	247
Pharmacokinetics and metabolism	248
Clinical aspects	249
Synthetic cannabinoids and fertility	250
Synthetic cannabinoids and the developing brain	252
References	253

13. Prenatal THC exposure interferes with the neurodevelopmental role of endocannabinoid signaling — 259
Ismael Galve-Roperh, Adán de Salas-Quiroga, Samuel Simón Sánchez, and Manuel Guzmán

Introduction	260
The ECB system in neurogenic niches	260
Role of ECB signaling in pyramidal neuron development	263
Role of the ECB system in GABAergic interneuron development	266
Long-lasting consequences of prenatal THC exposure	266

	Projection neurons	267
	Dopaminergic neurons	268
	GABAergic interneurons	268
	Functional consequences of PCE	270
	Contribution of prenatal ECB signaling alterations to neurodevelopmental disorders	271
	Prenatal cannabinoid exposure in human-based models	272
	Conclusions	274
	References	274
14.	**Cannabis effects on the adolescent brain**	283
	Kateryna Murlanova, Yuto Hasegawa, Atsushi Kamiya, and Mikhail V. Pletnikov	
	Introduction	283
	Adolescence is a critical period of brain development	284
	Morphological and functional transformations in the brain during adolescence	284
	Vulnerability to environmental factors during adolescence	287
	Endogenous cannabinoid (eCB) system in the adolescent brain	288
	eCB system	288
	Maturation of the eCB system	289
	Cannabis and the adolescent brain	291
	Developmental effects of cannabis	292
	Behavioral effects of chronic cannabis use during adolescence	295
	Neurobiological mechanisms underlying cannabinoid-induced behavioral alterations: Insights from animal studies	299
	Genetic vulnerability to cannabis	306
	Conclusion and future directions	308
	References	308
Index		331

Contributors

Numbers in parenthesis indicate the pages on which the authors' contributions begin.

Anissa Bara (13), Departments of Psychiatry and Neuroscience, Icahn School of Medicine at Mount Sinai, Addiction Institute of Mount Sinai, Friedman Brain Institute, Mount Sinai, NY, United States

Luis F. Callado (197), Department of Pharmacology, School of Medicine, University of the Basque Country UPV-EHU and BioCruces Bizkaia Health Research Institute, Barakaldo, Bizkaia; Biomedical Research Networking Center for Mental Health Network (CIBERSAM), Madrid, Spain

Manenti Cristina (95), Deptartment of Biotechnology and Life Sciences and Neuroscience Center, University of Insubria, Busto Arsizio, Italy

Parolaro Daniela (95), Zardi-Gori Foundation, Milan, Italy

Adán de Salas-Quiroga (259), School of Biology and Instituto Universitario de Investigación Neuroquímica (IUIN), Complutense University; Centro de Investigación Biomédica en Red Enfermedades Neurodegenerativas (CIBERNED) and Instituto Ramón y Cajal de Investigaciones Sanitarias (IRYCIS), Madrid, Spain; Champalimaud Centre for the Unknown, Lisboa, Portugal

Anthony English (23), Department of Pharmacology, University of Washington School of Medicine, Seattle, WA, United States

Zamberletti Erica (95), Deptartment of Biotechnology and Life Sciences and Neuroscience Center, University of Insubria, Busto Arsizio, Italy

Roberto Frau (107), Department of Biomedical Sciences, Division of Neuroscience and Clinical Pharmacology; "Guy Everett" Laboratory, University of Cagliari, Cagliari, Italy

Ismael Galve-Roperh (259), School of Biology and Instituto Universitario de Investigación Neuroquímica (IUIN), Complutense University; Centro de Investigación Biomédica en Red Enfermedades Neurodegenerativas (CIBERNED) and Instituto Ramón y Cajal de Investigaciones Sanitarias (IRYCIS), Madrid, Spain

Gabriele Giua (129), Cannalab, Cannabinoids Neuroscience Research International Associated Laboratory, INSERM-Aix-Marseille University; Mediterranean Neurobiology Institute, INSERM U1249, Aix-Marseille University, Marseille, France

Manuel Guzmán (259), School of Biology and Instituto Universitario de Investigación Neuroquímica (IUIN), Complutense University; Centro de Investigación Biomédica en Red Enfermedades Neurodegenerativas (CIBERNED) and Instituto Ramón y Cajal de Investigaciones Sanitarias (IRYCIS), Madrid, Spain

Tibor Harkany (149), Department of Molecular Neurosciences, Center for Brain Research, Medical University of Vienna, Vienna, Austria; Department of Neuroscience, Biomedicum 7D, Karolinska Institutet, Solna, Sweden

Yuto Hasegawa (283), Department of Psychiatry and Behavioral Sciences, Johns Hopkins University School of Medicine, Baltimore, MD, United States

Yasmin L. Hurd (13), Departments of Psychiatry and Neuroscience, Icahn School of Medicine at Mount Sinai, Addiction Institute of Mount Sinai, Friedman Brain Institute, Mount Sinai, NY, United States

Atsushi Kamiya (283), Department of Psychiatry and Behavioral Sciences, Johns Hopkins University School of Medicine, Baltimore, MD, United States

Erik Keimpema (149), Department of Molecular Neurosciences, Center for Brain Research, Medical University of Vienna, Vienna, Austria

Benjamin Land (23), Department of Pharmacology, University of Washington School of Medicine, Seattle, WA, United States

Steven R. Laviolette (171), Addiction Research Group, Department of Anatomy & Cell Biology; Department of Psychiatry, Schulich School of Medicine & Dentistry, University of Western Ontario, London, ON, Canada

Ken Mackie (1), Gill Center and Department of Psychological and Brain Sciences, Indiana University Bloomington, Bloomington, IN, United States

Antonia Manduca (79), Roma Tre University, Rome, Italy; Fondazione Santa Lucia, Rome, Italy

Olivier JJ. Manzoni (129), Mediterranean Neurobiology Institute, INSERM U1249, Aix-Marseille University, Marseille, France

Gabaglio Marina (95), Deptartment of Biotechnology and Life Sciences and Neuroscience Center, University of Insubria, Busto Arsizio, Italy

Ashley E. Marquardt (59), Department of Pharmacology and Program in Neuroscience, University of Maryland School of Medicine, Baltimore, MD, Unites States

Margaret M. McCarthy (59), Department of Pharmacology and Program in Neuroscience, University of Maryland School of Medicine, Baltimore, MD, Unites States

Miriam Melis (107), Department of Biomedical Sciences, Division of Neuroscience and Clinical Pharmacology, University of Cagliari, Cagliari, Italy

Kateryna Murlanova (283), Department of Physiology and Biophysics, Jacobs School of Medicine and Biomedical Sciences, State University of New York at Buffalo, Buffalo, NY, United States

A.-L. Pélissier-Alicot (243), Cannalab, Cannabinoids Neuroscience Research International Associated Laboratory, INSERM-Aix-Marseille University France/Indiana University, Bloomington, IN, United States; INMED, INSERM U1249, Marseille, France; Aix-Marseille University, Service de Medecine Legale, AP-HM, Marseille, France

Mikhail V. Pletnikov (283), Department of Physiology and Biophysics, Jacobs School of Medicine and Biomedical Sciences, State University of New York at Buffalo, Buffalo, NY, United States

Samuel Simón Sánchez (259), School of Biology and Instituto Universitario de Investigación Neuroquímica (IUIN), Complutense University; Centro de Investigación Biomédica en Red Enfermedades Neurodegenerativas (CIBERNED) and Instituto Ramón y Cajal de Investigaciones Sanitarias (IRYCIS), Madrid, Spain

Andrew Scheyer (129), Cannalab, Cannabinoids Neuroscience Research International Associated Laboratory, INSERM-Aix-Marseille University, Marseille, France; Mediterranean Neurobiology Institute, INSERM U1249, Aix-Marseille University, Marseille, France

Nephi Stella (23), Department of Pharmacology; Department of Psychiatry and Behavioral Sciences, University of Washington School of Medicine, Seattle, WA, United States

Rubino Tiziana (95), Deptartment of Biotechnology and Life Sciences and Neuroscience Center, University of Insubria, Busto Arsizio, Italy

Viviana Trezza (79), Roma Tre University, Rome, Italy

Paula Unzueta-Larrinaga (197), Department of Pharmacology, School of Medicine, University of the Basque Country UPV-EHU and BioCruces Bizkaia Health Research Institute, Barakaldo, Bizkaia, Spain

Leyre Urigüen (197), Department of Pharmacology, School of Medicine, University of the Basque Country UPV-EHU and BioCruces Bizkaia Health Research Institute, Barakaldo, Bizkaia; Biomedical Research Networking Center for Mental Health Network (CIBERSAM), Madrid, Spain

Jonathan W. VanRyzin (59), Department of Pharmacology and Program in Neuroscience, University of Maryland School of Medicine, Baltimore, MD, Unites States

Preface

At a time when the legalization and decriminalization of cannabis is accelerating worldwide, thanks to the reduction of stigma associated with its use and the misperception that cannabis is a natural and safe therapeutic option, the evidence is sorely lacking.

Cannabis is the most widely used illicit drug among adolescents and pregnant women, and there is growing concern about its long-term impact on brain function and behavior.

Cannabis effects vary among individuals, and its consequences are particularly heightened in vulnerable groups. In particular, evidence suggests that cannabis exposure during neurodevelopment (i.e., prenatal, perinatal, and adolescent stages) results in persistent alterations in neuromodulation at molecular and circuit level that contribute to the pathophysiology of several neuropsychiatric disorders at some point during a lifetime. Notably, gene by environment interaction plays a fundamental role in these disorders, with a growing body of evidence demonstrating a significant link between developmental cannabis exposure and psychiatric vulnerability, including susceptibility to depressive states and substance use disorders.

The aim of this book is twofold: first, providing students and researchers in basic science and medicine an up-to-date overview of the impact of cannabis exposure during sensitive and critical developmental windows in the larger context of neuropsychiatric diseases; and second, helping actions taken to restricting access, reducing the potency of cannabis derivatives, and to raising awareness on the risks of recreational use, particularly during adolescence and at the age of child-bearing.

This book is the first of its kind to offer readers the opportunity to (i) learn about the in vivo neural circuitry and molecular and cellular mechanisms affected by cannabis exposure during three different periods of brain vulnerability, (ii) gain insight into the unique and common neurobiological features of cannabinoid exposure during different developmental periods, and (iii) determine the negative impact of developmental cannabinoid exposure on specific cognitive, emotional, and behavioral areas.

The book brings together interdisciplinary, translational, animal model, and human studies. Thus, the effects of perinatal exposure on different areas and circuits of the brain in both clinical and preclinical settings are covered in the chapters by Drs Mackie, Hurd, McCarthy, Harkany, Trezza, Galve-Roperh,

Melis, and their coworkers. In addition, the effects of the exposure to cannabis (including cannabis-derived drugs) during childhood and adolescence, comprising the interaction with known genetic and environmental risk factors, have been covered in the chapters by Drs Parolaro, Stella, Urigüen, Laviolette, Pletnikov, Pelissier, and Manzoni.

Chapter 1

Pre-clinical models of neurodevelopmental cannabinoid exposure

Ken Mackie
Gill Center and Department of Psychological and Brain Sciences, Indiana University Bloomington, Bloomington, IN, United States

Introduction

Endogenous cannabinoids have been implicated in many aspects of nervous system development occurring pre-natally and during adolescence[1] (Tibor's and Olivier's chapters). Thus it can be hypothesized that pre-natal or adolescent exposure to exogenous cannabinoids (i.e., phytocannabinoids) from cannabis may impact proper development of the nervous system, leading to impaired function. Several large epidemiological studies and meta-analyses have consistently confirmed this hypothesis following both pre-natal cannabis exposure[2] and early adolescent cannabis use[3] and as discussed in chapters X and Y (Viviana Trezza's and Daniela Parolaro's chapters). These human findings have prompted extensive efforts to determine the mechanism(s) underlying cannabis' ability to impair the developing human nervous system. As ethical constraints make detailed mechanistic studies impossible in humans, this research, particularly when behavioral outcomes will be evaluated, has utilized pre-clinical animal models, chiefly rodents.[4–7] Cerebral organoids utilizing inducible pluripotent human stem cells offer an alternative to animal models for certain types of questions[8] but would not be discussed in this chapter. Some studies on the developmental effects of phytocannabinoids have been done with non-human primates, particularly to evaluate the effects of adolescent phytocannabinoids[9,10]; nonetheless, the vast majority of studies have used rodents. While rodents offer many advantages in neurodevelopmental studies, they also poise unavoidable limitations. In this review, we will discuss the primary considerations in choosing appropriate rodent models of pre-natal and adolescent cannabis exposure. While our primary focus will be on the choice of model for cannabinoid administration, equally important is determining which outcomes will be evaluated. This latter topic is discussed at length in other chapters in this volume.

Model considerations

Pre-clinical models involve tradeoffs between efficiency, validity, scalability, and so on, that influence the development and choice of a model. The following are some of the important considerations in choosing an appropriate pre-clinical model of developmental cannabis exposure: timing of drug administration, route of drug administration, nature of drug administered, and dosing frequency of the drug. In the following sections, we will discuss each of these considerations in detail. These considerations and recommendations are summarized in Table 1.

Timing of drug administration

Pre-natal models

A key goal of pre-clinical models of cannabis use is to expose the developing rodent brain over the range of dates corresponding to the development of the relevant structures in the human brain. This goal is made more difficult by the fact that the later stages of fetal human brain development that may be affected by maternal cannabis use take place during the first two post-natal weeks in rodents. Another issue is that brain regions in the rodent often mature at different times and rates than in the human. Thus as will be a recurring theme in this review, there is no perfect model to perfectly mimic human brain development, and drug exposure must be timed to correspond to the development of the brain region being examined and it may not be possible to accurately model the influence of phytocannabinoids on circuits involving several brain regions. There are several good reviews and online tools that attempt to match the stages of rodent and human brain development.[11,12] As a generalization, rodent brain development at birth corresponds roughly to human brain development at the end of the second trimester. Thus experiments that treat rodents only until birth are mimicking human exposure through the second trimester and not exposure throughout a full-term pregnancy. One way to model human third trimester cannabis exposure is to continue to administer drug to the dam (e.g., Ref. 7) or the pup (e.g., Ref. 13) post-natally. However, this introduces the confounds of uncertain transfer of drug via lactation[14] and potential effects on mothering behaviors (if drug is given to the dam) or stress of administration and determination of appropriate dose (see later) (if drug is given to the pups).[15] Exposure to THC only during the early post-natal period (post-natal days 1–10) in rats (approximately corresponding to the third trimester in humans) can have enduring effects on synaptic plasticity and behaviors,[7,16] so treatment during this time

TABLE 1 Considerations for pre-clinical models of developmental cannabis exposure.

Age	Timing of administration	Route of administration	Dosing	Drugs used
Pre-natal	Only during pregnancy? Extend into post-natal period? Match timing of cannabinoid exposure to the development of the relevant brain region(s) Paternal treatment? Cross-fostering if pups treated?	**Pharmacological:** Is it necessary to mimic the plasma time course of the drug in humans? Route of administration—dependent metabolites—implications and similar to humans? **Behavioral:** Contingent vs non-contingent administration Effects on maternal and/or paternal care	Match rodent plasma exposure to human exposure? How frequent? Should dose be escalated (to circumvent tolerance)? Maternal pharmacokinetics may vary from non-pregnant female pharmacokinetics	THC ± CBD Cannabis extract (defined components?) Pharmacokinetic interactions between various cannabinoids Synthetic cannabinoids for modeling "spice" use
Adolescent	Duration of adolescence: Several days before onset of puberty until sexual maturity Treat during specific times of adolescence Consider adult administration as control	**Pharmacological:** Plasma/tissue profiles mirror human profiles? Route of administration—dependent metabolites—similar to humans? **Behavioral:** Contingent vs non-contingent administration Cage-mates treated or not? Enriched environment?	Mimic the time course of drug in humans? Match plasma exposure to what humans experience? How frequent? Sex differences in metabolism Should dose be escalated? (to circumvent tolerance)	THC +/− CBD Cannabis extract (defined components?) Pharmacokinetic interactions between various cannabinoids Synthetic cannabinoids for modeling "spice" use

period should be included if a study is aiming to determine the effect(s) of using cannabis throughout pregnancy. In addition, models of the impact of phytocannabinoid exposure only during the period corresponding to human lactation are worth exploring as at least one study found that a significant fraction of women did not use cannabis during pregnancy, but did use it while nursing.[17] Finally, pharmacokinetics of many drugs are altered during pregnancy[18]; thus caution is advised if pharmacokinetic parameters derived from studies using non-pregnant subjects are applied in developing models of cannabinoid exposure during pregnancy. In this regard, sophisticated human pharmacokinetic models (e.g., Ref. 19) might help to inform the design of pre-clinical models.

Adolescent models

Determining the timing of adolescent cannabis consumption is usually done by defining rodent adolescence by matching sexual maturation of the rodent and human. The precise timing of adolescence in rodents varies by measure but is generally considered to start a few days before the onset of puberty (denoted by vaginal opening in females (~PND35) and presence of sperm in males (~PND40)) and to extend until the males and females are fully reproductively competent, between PND 50 and 60.[20] These are only rough guidelines and are influenced by several factors—species, strain, housing conditions, and so on.[21,22] Since human data on adolescent cannabis use suggests that cannabis use during early adolescence is the most detrimental,[23] many rodent studies initiate phytocannabinoid administration early in rodent adolescence, often starting sometime during the fifth post-natal week and extending for a week or more, depending on the nature of the question(s) being addressed.

Route of drug administration

Ideally, the route of phytocannabinoid administration in pre-clinical models mimics human modes of consumption with the goal of achieving plasma and target organ profiles (i.e., levels over time) of the drug similar to those observed in humans. When designing and interpreting rodent models of developmental cannabinoid exposure, the potential differences between the pharmacokinetics of phytocannabinoids in rodents and primates need to be considered.[5,9]

Inhaled

For modeling cannabis use, the preferred routes of administration would mimic those used in the community, so would be voluntarily and inhaled (for most users) combusted cannabis, "vaped" phytocannabinoid, or orally consumed (for a small, but increasing number of users) product. Voluntary consumption, defined as the intentional, goal-directed consumption of a phytocannabinoid, is difficult to routinely achieve for phytocannabinoids in either perinatal or adolescent pre-clinical models for several reasons. In contrast to other abused drugs such as nicotine, opioids, and cocaine, rodents are reluctant to self-administer THC (and would not self-administer cannabidiol (CBD), which is generally non-rewarding) and weak patterns of reinforcement are generally observed.[24]

While "vaping" chambers as a viable means of exposing rodents to cannabis-related drugs have emerged over the last few years,[25–28] the technology and best practices are still evolving. The impacts of several aspects of inhalation exposure remain to be better understood, particularly for developmental studies. For example, administration of phytocannabinoid in a vaping chamber can be passive (taking advantage of the inhaled route by placing the subject(s) in a closed container and providing volatilized phytocannabinoid in a regulated fashion) or contingent (better models the human situation by having the subject perform a nose poke or lever press to receive a puff of volatilized phytocannabinoid), brain levels of drug achieved during the treatment, potential stress of being in the chamber, difference in human (inhale and hold) and rodent (very rapid respiratory rate) patterns of inhalation, and so on. Since exposure to the vapor chamber can be a significant stressor,[29] it is important to include control vapor and non-vapor controls when developing a vapor administration system. One important potential advantage of vaping is that it avoids the substantial first pass metabolism seen with intra-peritoneal (i.p.) injections of THC[30]; thus it may more faithfully mirror the ratios of THC, 11-OH-THC, and 9-COOH-THC seen in human consumers of cannabis compared to profiles obtained with the more common i.p. injections.

Oral

Oral administration has the advantages of being simple to conduct and mimicking the increasing use of "edibles," particularly by those who are using cannabis products to treat symptoms. Two major oral routes have been employed. The first is by oral gavage. Though frequently used, at least in the past, oral gavage will be stressful to the rodent, particularly when used for chronic administration, which is clearly undesirable. The second is by mixing the phytocannabinoid with a palatable food. Various palatable foods have been used including gelatin,[31] cookie dough,[32] and sweetened condensed milk (K. Mackie, unpublished). Oral consumption is, however, a poor model of inhaled consumption due to the much lower peak blood levels of THC achieved after oral consumption, the prolonged elevation of blood levels, and the substantial first pass metabolism that occurs.[33] In the case of THC, first pass metabolism will generate much more 11-OH-THC than inhaled THC,[30] which may have different effects due to its greater potency.[34] A similar situation will occur for CBD, with first pass metabolism giving rise to high levels of 7-OH-CBD.[35] Thus oral administration is appropriate to use to model oral consumption. If it is used, it is important to determine brain levels of the drug and its metabolites to ensure that they approximate those anticipated from human studies.

Injection

Injection has the advantage of being simple to perform and allowing for a precise amount of drug to be delivered. Injection is typically by the intravenous (i.v.), i.p., and subcutaneous (s.c.) routes. I.v. injection has the advantage of being able to most closely mimic the plasma phytocannabinoid concentration-time curves obtained when humans inhale combusted or vaped cannabis or cannabis product. Disadvantages of i.v. injection include restraint of the rodent during the injection and the stress this causes, higher degree of technical skill required, and possible thrombosis of the vein if injections are frequently repeated (often necessary for developmental experiments). I.v. injection via catheter (e.g., as might be used during an i.v. self-administration experiment) introduces the confound of single housing after catheter implantation, which is a model for social isolation stress.[36] I.p. injection is very commonly used as it is easy to perform. Disadvantages of this approach include substantial first pass metabolism due to absorption into the portal circulation,[30] intermediate levels of stress, and potential for sterile peritonitis and other complications with repeated injection. S.c. injection, like i.p., is simple to perform and the least stressful of the injection techniques as it requires the least restraint. The slowness of absorption from s.c. injection needs to be considered (this may be an advantage or disadvantage, depending on the questions being asked) when planning experiments.

Nature of drug administered

Cannabis consists of a complex mixture of cannabinoids, terpenes, flavonoids, and several other classes of compounds. Different cultivars will vary widely in their composition of these compounds. It is challenging to determine the cannabinoid content of cannabis consumed in the community, and the concentrations of other components are seldom even measured. Thus when modeling the pre-natal use of THC-rich cultivars compromises need to be made when deciding what compounds to administer. The first decision is to decide which cannabinoid(s) should be administered, THC or a potent synthetic cannabinoid. Since THC, and not potent synthetic cannabinoids, is present in cannabis, experiments seeking to understand the impact on brain development of the consumption of THC-rich cannabis (typically, cannabis used for its psychoactivity) should use THC. Synthetic cannabinoids (with their very different pharmacology[37,38]) are appropriate to use in developmental studies mimicking exposure to "spice" compounds and perhaps for addressing specific questions involving the endocannabinoid system on central nervous system (CNS) development. However, they are not a good model for the effects of THC on CNS development. A second decision is determining which other (if any) cannabinoids should be co-administered with THC. CBD is a variable constituent of cannabis cultivars used recreationally and has attracted considerable interest for use on its own for a variety of maladies, including those that might accompany pregnancy.

CBD has a variety of pharmacodynamic and pharmacokinetic interactions with THC,[39,40] which both raise interesting questions (i.e., if CBD-rich cannabis is used pre-natally will this attenuate THC's effects?) and can complicate experimental design (i.e., does CBD inhibit THC metabolism in humans to the same extent it does in rodents?[10,41]). An understudied question is the extent to which CBD impacts the developing nervous system, especially important given the widespread use of CBD. Emerging evidence suggests that CBD can affect the developing brain.[42] Almost nothing is known about the impact of other cannabinoids, e.g., cannabinol and cannabigerol, on the developing nervous system. Similarly, it is unknown how the acid forms of cannabinoids (i.e., the forms synthesized in the plant, THCA, CBDA, etc.), which are increasingly being considered for their possible therapeutic benefit,[43,44] affect the developing CNS. Finally, it is not known if the myriad of terpenoids and flavonoids present in cannabis affect nervous system development at the doses likely to be consumed by humans using cannabis. All of these are important avenues for future study given the great attention that these compounds are receiving in the popular press as over the counter remedies for a variety of conditions.

Dose of drug administered

In pre-clinical models, the low toxicity of cannabinoids presents the temptation to push the dose of drug until an effect is seen. However, this approach is of little use if one is trying to understand the effects of drugs on the developing human nervous system. It is possible to achieve CNS phytocannabinoid concentrations several orders of magnitude higher than will be encountered by the CNS of a human fetus or adolescent. In this case, it is more appropriate to choose a dosing regimen for the pre-clinical model that gives target organ concentrations that are similar to those achieved in humans. The drawback to this approach is that the levels of THC and other phytocannabinoids in the developing human brain after cannabis use are not known. A compromise would be to determine the levels of these compounds in human fetal blood and assume that the partitioning between blood and brain in the fetal rodent and human brains are similar. However, even for THC, human fetal blood levels are not known with any certainty. A few studies have looked at fetal cannabinoid levels in umbilical cord (blood) at delivery. However, most of these focus on THC metabolites (e.g., THC-COOH[45,46]) for identifying exposure to cannabis, rather than active components, such as THC or 11-OH-THC, making them less useful for determining the quantity of active component that the developing brain has been exposed to. A single study has looked at fetal levels of THC following a single 0.3 mg/kg i.v. dose of THC in pregnant rhesus monkeys.[47] This study found peak plasma fetal values of THC to be ~5% of maternal THC and substantial THC was detected in fetal brain and liver. Interestingly, almost no THC-9-COOH was detected in the fetuses, suggesting impaired fetal phase I metabolism of THC. While the acute dosing data are helpful, fetal levels after chronic dosing would be more helpful in designing models for THC exposure. Further confounds for human studies including unknown dose, unknown duration between last dose and sampling, an incomplete understanding of transport of phytocannabinoids across the placental and fetal blood-brain barrier, and so on, make these values only useful as a rough guide. It is likely that in US states with more liberal cannabis legislation, additional studies are ongoing and these data will be forthcoming.

Other considerations with dosing are frequency of dose and if the dose should be increased during the treatment period to mimic escalating use as could occur with humans if tolerance develops. Frequency of dosing should incorporate patterns of human cannabis use. However, this is quite variable and may range from several times daily to once every few days. Thus if the pre-clinical model is mimicking heavy use, daily or twice daily dosing may be appropriate. However, for occasional use, dosing every few days would be most appropriate. As mentioned previously, it is important to determine target organ phytocannabinoid concentrations with these different dosing schemes to better understand their translational relevance and to facilitate comparison between experiments. Certainly, experiments that vary dosing schedule, while maintaining other variables constant, can give important data on how varying exposure of the target organ to the phytocannabinoid may modify the outcome.

Summary

While there will never be an "ideal" model for developmental cannabis exposure, the validity and predictive validity of the models used can be improved by careful attention to some of the details discussed previously.

Conflict of interest

None.

References

1. Bara A, Ferland JN, Rompala G, Szutorisz H, Hurd YL. Cannabis and synaptic reprogramming of the developing brain. *Nat Rev Neurosci* 2021;**22**(7):423–38.
2. Paul SE, Hatoum AS, Fine JD, Johnson EC, Hansen I, Karcher NR, et al. Associations between prenatal Cannabis exposure and childhood outcomes: results from the ABCD study. *JAMA Psychiatry* 2021;**78**(1):64–76.
3. Godin SL, Shehata S. Adolescent cannabis use and later development of schizophrenia: an updated systematic review of longitudinal studies. *J Clin Psychol* 2022. https://doi.org/10.1002/jclp.23312.
4. Frau R, Miczan V, Traccis F, Aroni S, Pongor CI, Saba P, et al. Prenatal THC exposure produces a hyperdopaminergic phenotype rescued by pregnenolone. *Nat Neurosci* 2019;**22**(12):1975–85.

5. Murphy M, Mills S, Winstone J, Leishman E, Wager-Miller J, Bradshaw H, et al. Chronic adolescent Delta(9)-tetrahydrocannabinol treatment of male mice leads to long-term cognitive and behavioral dysfunction, which are prevented by concurrent Cannabidiol treatment. *Cannabis Cannabinoid Res* 2017;**2**(1):235–46.
6. Prini P, Zamberletti E, Manenti C, Gabaglio M, Parolaro D, Rubino T. Neurobiological mechanisms underlying cannabis-induced memory impairment. *Eur Neuropsychopharmacol* 2020;**36**:181–90.
7. Scheyer AF, Borsoi M, Wager-Miller J, Pelissier-Alicot AL, Murphy MN, Mackie K, et al. Cannabinoid exposure via lactation in rats disrupts perinatal programming of the gamma-aminobutyric acid trajectory and select early-life behaviors. *Biol Psychiatry* 2020;**87**(7):666–77.
8. Ao Z, Cai H, Havert DJ, Wu Z, Gong Z, Beggs JM, et al. One-stop microfluidic assembly of human brain organoids to model prenatal Cannabis exposure. *Anal Chem* 2020;**92**(6):4630–8.
9. Withey SL, Bergman J, Huestis MA, George SR, Madras BK. THC and CBD blood and brain concentrations following daily administration to adolescent primates. *Drug Alcohol Depend* 2020;**213**, 108129.
10. Withey SL, Kangas BD, Charles S, Gumbert AB, Eisold JE, George SR, et al. Effects of daily Delta(9)-tetrahydrocannabinol (THC) alone or combined with cannabidiol (CBD) on cognition-based behavior and activity in adolescent nonhuman primates. *Drug Alcohol Depend* 2021;**221**, 108629.
11. Charvet CJ. *Translating neurodevelopmental time across mammalian species*; 2021. Available from: https://www.translatingtime.org/.
12. Workman AD, Charvet CJ, Clancy B, Darlington RB, Finlay BL. Modeling transformations of neurodevelopmental sequences across mammalian species. *J Neurosci* 2013;**33**(17):7368–83.
13. Fride E, Ginzburg Y, Breuer A, Bisogno T, Di Marzo V, Mechoulam R. Critical role of the endocannabinoid system in mouse pup suckling and growth. *Eur J Pharmacol* 2001;**419**(2–3):207–14.
14. Johnson CT, Dias de Abreu GH, Mackie K, Lu H-C, Bradshaw HB. *Cannabinoids accumulate in mouse breast milk and differentially regulate lipid composition and lipid signaling molecules involved in infant development.* Biorxiv; 2021.
15. Turner PV, Brabb T, Pekow C, Vasbinder MA. Administration of substances to laboratory animals: routes of administration and factors to consider. *J Am Assoc Lab Anim Sci* 2011;**50**(5):600–13.
16. Scheyer AF, Borsoi M, Pelissier-Alicot AL, Manzoni OJJ. Perinatal THC exposure via lactation induces lasting alterations to social behavior and prefrontal cortex function in rats at adulthood. *Neuropsychopharmacology* 2020;**45**(11):1826–33.
17. Astley SJ, Little RE. Maternal marijuana use during lactation and infant development at one year. *Neurotoxicol Teratol* 1990;**12**(2):161–8.
18. Grant KS, Petroff R, Isoherranen N, Stella N, Burbacher TM. Cannabis use during pregnancy: pharmacokinetics and effects on child development. *Pharmacol Ther* 2018;**182**:133–51.
19. Patilea-Vrana GI, Unadkat JD. Development and verification of a linked Delta (9)-THC/11-OH-THC physiologically based pharmacokinetic model in healthy, nonpregnant population and extrapolation to pregnant women. *Drug Metab Dispos* 2021;**49**(7):509–20.
20. Schneider M. Puberty as a highly vulnerable developmental period for the consequences of cannabis exposure. *Addict Biol* 2008;**13**(2):253–63.
21. Laffan SB, Posobiec LM, Uhl JE, Vidal JD. Species comparison of postnatal development of the female reproductive system. *Birth Defects Res* 2018;**110**(3):163–89.
22. van Weissenbruch MM, Engelbregt MJT, Veening MA, Delemarre-van de Waal HA. Fetal nutrition and timing of puberty. *Endocr Dev* 2005;**8**:15–33.

23. Kiburi SK, Molebatsi K, Ntlantsana V, Lynskey MT. Cannabis use in adolescence and risk of psychosis: are there factors that moderate this relationship? A systematic review and meta-analysis. *Subst Abus* 2021;**42**(4):527–42.
24. Wakeford AGP, Wetzell BB, Pomfrey RL, Clasen MM, Taylor WW, Hempel BJ, et al. The effects of cannabidiol (CBD) on Delta(9)-tetrahydrocannabinol (THC) self-administration in male and female long-Evans rats. *Exp Clin Psychopharmacol* 2017;**25**(4):242–8.
25. Freels TG, Baxter-Potter LN, Lugo JM, Glodosky NC, Wright HR, Baglot SL, et al. Vaporized Cannabis extracts have reinforcing properties and support conditioned drug-seeking behavior in rats. *J Neurosci* 2020;**40**(9):1897–908.
26. Manwell LA, Ford B, Matthews BA, Heipel H, Mallet PE. A vapourized Delta(9)-tetrahydrocannabinol (Delta(9)-THC) delivery system part II: comparison of behavioural effects of pulmonary versus parenteral cannabinoid exposure in rodents. *J Pharmacol Toxicol Methods* 2014;**70**(1):112–9.
27. Nguyen JD, Aarde SM, Vandewater SA, Grant Y, Stouffer DG, Parsons LH, et al. Inhaled delivery of Delta(9)-tetrahydrocannabinol (THC) to rats by e-cigarette vapor technology. *Neuropharmacology* 2016;**109**:112–20.
28. Ruiz CM, Torrens A, Lallai V, Castillo E, Manca L, Martinez MX, et al. Pharmacokinetic and pharmacodynamic properties of aerosolized ("vaped") THC in adolescent male and female rats. *Psychopharmacology* 2021;**238**(12):3595–605.
29. Weimar HV, Wright HR, Warrick CR, Brown AM, Lugo JM, Freels TG, et al. Long-term effects of maternal cannabis vapor exposure on emotional reactivity, social behavior, and behavioral flexibility in offspring. *Neuropharmacology* 2020;**179**, 108288.
30. Baglot SL, Hume C, Petrie GN, Aukema RJ, Lightfoot SHM, Grace LM, et al. Pharmacokinetics and central accumulation of delta-9-tetrahydrocannabinol (THC) and its bioactive metabolites are influenced by route of administration and sex in rats. *Sci Rep* 2021;**11**(1):23990.
31. Kruse LC, Cao JK, Viray K, Stella N, Clark JJ. Voluntary oral consumption of Delta(9)-tetrahydrocannabinol by adolescent rats impairs reward-predictive cue behaviors in adulthood. *Neuropsychopharmacology* 2019;**44**(8):1406–14.
32. Smoker MP, Mackie K, Lapish CC, Boehm 2nd SL. Self-administration of edible Delta(9)-tetrahydrocannabinol and associated behavioral effects in mice. *Drug Alcohol Depend* 2019;**199**:106–15.
33. Spindle TR, Cone EJ, Schlienz NJ, Mitchell JM, Bigelow GE, Flegel R, et al. Acute pharmacokinetic profile of smoked and vaporized Cannabis in human blood and oral fluid. *J Anal Toxicol* 2019;**43**(4):233–58.
34. Howlett AC, Barth F, Bonner TI, Cabral G, Casellas P, Devane WA, et al. International Union of Pharmacology. XXVII. Classification of cannabinoid receptors. *Pharmacol Rev* 2002;**54**(2):161–202.
35. Lucas CJ, Galettis P, Schneider J. The pharmacokinetics and the pharmacodynamics of cannabinoids. *Br J Clin Pharmacol* 2018;**84**(11):2477–82.
36. Rivera-Irizarry JK, Skelly MJ, Pleil KE. Social isolation stress in adolescence, but not adulthood, produces hypersocial behavior in adult male and female C57BL/6J mice. *Front Behav Neurosci* 2020;**14**:129.
37. Howlett AC, Thomas BF, Huffman JW. The spicy story of cannabimimetic indoles. *Molecules* 2021;**26**(20):6190.
38. Atwood BK, Huffman J, Straiker A, Mackie K. JWH018, a common constituent of 'Spice' herbal blends, is a potent and efficacious cannabinoid CB receptor agonist. *Br J Pharmacol* 2010;**160**(3):585–93.

39. Pennypacker SD, Romero-Sandoval EA. CBD and THC: do they complement each other like yin and Yang? *Pharmacotherapy* 2020;**40**(11):1152–65.
40. Vazquez M, Garcia-Carnelli C, Maldonado C, Fagiolino P. Clinical pharmacokinetics of cannabinoids and potential drug-drug interactions. *Adv Exp Med Biol* 2021;**1297**:27–42.
41. Hlozek T, Uttl L, Kaderabek L, Balikova M, Lhotkova E, Horsley RR, et al. Pharmacokinetic and behavioural profile of THC, CBD, and THC+CBD combination after pulmonary, oral, and subcutaneous administration in rats and confirmation of conversion in vivo of CBD to THC. *Eur Neuropsychopharmacol* 2017;**27**(12):1223–37.
42. Maciel IS, de Abreu GHD, Johnson CT, Bonday R, Bradshaw HB, Mackie K, et al. Perinatal CBD or THC exposure results in lasting resistance to fluoxetine in the forced swim test: reversal by fatty acid amide hydrolase inhibition. *Cannabis Cannabinoid Res* 2021. https://doi.org/10.1089/can.2021.0015.
43. Formato M, Crescente G, Scognamiglio M, Fiorentino A, Pecoraro MT, Piccolella S, et al. (−)-Cannabidiolic acid, a still overlooked bioactive compound: an introductory review and preliminary research. *Molecules* 2020;**25**(11):2638.
44. Russo EB. Cannabis therapeutics and the future of neurology. *Front Integr Neurosci* 2018;**12**:51.
45. Kim J, de Castro A, Lendoiro E, Cruz-Landeira A, Lopez-Rivadulla M, Concheiro M. Detection of in utero cannabis exposure by umbilical cord analysis. *Drug Test Anal* 2018;**10**(4):636–43.
46. Metz TD, Allshouse AA, Hogue CJ, Goldenberg RL, Dudley DJ, Varner MW, et al. Maternal marijuana use, adverse pregnancy outcomes, and neonatal morbidity. *Am J Obstet Gynecol* 2017;**217**(4):478.e1–8.
47. Bailey JR, Cunny HC, Paule MG, Slikker Jr W. Fetal disposition of delta 9-tetrahydrocannabinol (THC) during late pregnancy in the rhesus monkey. *Toxicol Appl Pharmacol* 1987;**90**(2):315–21.

Chapter 2

Epigenetic imprint: An underlying link to developmental effects of prenatal cannabis exposure

Anissa Bara and Yasmin L. Hurd

Departments of Psychiatry and Neuroscience, Icahn School of Medicine at Mount Sinai, Addiction Institute of Mount Sinai, Friedman Brain Institute, Mount Sinai, NY, United States

Introduction

The past few years have been a watershed period in the United States and many other countries with great attention given to the medicinal and recreational impact of cannabis. While this has led to important societal changes, such as the decriminalization of cannabis use, the potential for harm particularly for certain vulnerable populations has raised concern as more pregnant women use cannabis today. This is also combined with an exponential increase over the years in the concentration of Δ^9-tetrahydrocannabinol (THC; the main psychoactive intoxicant) in the cannabis used recreationally and sold in dispensaries.[1] The effects of THC are mediated by the endocannabinoid system (ECS), which plays a critical regulatory role throughout all developmental stages from the neurogenesis and neuronal migration to neuronal hardwiring during prenatal ontogeny.[2–4] As such, the supraphysiological impact of cannabis on the ECS during prenatal development could change the normal trajectory of cellular processing and neurocircuitry critical for forming behaviors at later stages in life. In recent years, multiple human and pre-clinical studies have shown that in utero exposure to cannabis induces profound molecular, cellular, synaptic, and behavioral long-lasting changes in the offspring's brain.[5,6] In this chapter, we also highlight epigenetic mechanisms as a critical link underlying the molecular processes that maintain protracted effects of developmental cannabis exposure.

Molecular reprogramming after prenatal cannabinoid exposure

The principal targets of THC and other cannabinoids are the cannabinoid receptors (CBRs), including cannabinoid receptor 1 (CB$_1$R). Prenatal THC exposure is often associated with reduced mRNA expression of *Cnr1*, the gene coding for CB$_1$R, during fetal development,[7,8] which would have significant downstream consequences considering the major role of the ECS in cell proliferation, migration, differentiation, and maturation of neurons relevant to the development of various neural systems. Indeed, prenatal exposure of rodent models to CBR agonists has repeatedly shown to alter glutamatergic, GABAergic, dopaminergic, opioidergic, and serotonergic systems.[9] Importantly, a number of the molecular changes observed in the prenatal THC animal models are also evident in the brains of human fetuses with in utero cannabis exposure such as decreased mRNA expression of dopaminergic receptor D2 (*DRD2*) in the nucleus accumbens (NAc) and amygdala[10,11] as well as altered opioid receptors and opioid neuropeptides in the amygdala, mediodorsal thalamus, and striatum.[12–14] Interestingly, sex differences were evident in mesocorticolimbic areas in human midgestational fetuses exposed in utero to cannabis, where, for example, reduction of the *DRD2* gene expression was evident in the males, not females.[11]

In utero cannabis has also been shown to compromise cytoskeletal stability that can lead to erroneous neurite growth and corticofugal development. For instance, prenatal THC exposure reduces the hippocampal and striatal expression of SCG10/stathmin-2, a microtubule-binding protein in axons, an alteration also detected in cannabis-exposed human fetuses.[8] Disturbingly, the reduction of SCG10 is associated with re-organization of the fetal cortical circuitry, deficits that last into adulthood causing re-arrangements in the localization of CB$_1$Rs and reduced long-term depression in the hippocampus.[8] Other studies have demonstrated that THC administration to pregnant mice interfered with subcerebral projection neuron generation altering corticospinal connectivity and produced long-lasting alterations in the fine motor performance of the adult offspring.[7] In addition to biochemical and molecular evidence, morphological experiments have also confirmed that maternal exposure to cannabinoids affects proliferation, neurite branching,[15–17] and migration and differentiation of GABAergic and glutamatergic neurons.[18] Nanoscale insights are now also being revealed by the recent use of super-resolution imaging of the synapse, which demonstrates that THC exposure during gestation causes sex-specific molecular crowding at the pre-synaptic active zone of GABAergic terminals in the ventral tegmental area (VTA) of male rats, leading to decreased neurotransmitter release.[19] This further aligns with a decrease in the ratio VGCCs/CB$_1$Rs detected at inhibitory synaptic terminals after maternal THC exposure.[19]

Finally, most studies have focused on the glutamatergic system in relation to fetal cannabinoid exposure. Such investigations provide evidence that baseline and evoked levels of glutamate are reduced in vivo in the hippocampus[20] and frontal cortex[21] in young and adult rats prenatally exposed to a synthetic high-affinity CBR agonist WIN-55,212-2. Glutamate receptors expression and composition are also altered by in utero THC exposure. Indeed, recent studies reveal an increased AMPA/NMDA ratio associated with changes in receptor subunit composition in the VTA[19] of adult male rats but not female and reduced mRNA expression of perisynaptic mGlu5 receptors in the pre-frontal cortex (PFC) of both male and female adult rats.[22]

Prenatal cannabis exposure and epigenetic mechanisms

The long-term consequences of developmental cannabinoid exposure naturally raise questions about the underlying molecular mechanisms that maintain such protracted effects. The epigenome provides a molecular fingerprint of individual and environmental experiences, such as cannabinoid exposure, and is, thus, considered a highly pertinent biological candidate to explain the maintenance of persistent neuronal and phenotypic changes over time.[23–25] While "epigenetics" has gained significant attention in recent years in regard to its molecular machinery to regulate gene expression that control protein production and ultimately downstream phenotypes, the original definition by Conrad Waddington in the 1950s framed that "an epigenetic trait is a stably heritable phenotype resulting from changes in a chromosome without alterations in the DNA sequence".[26,27] In the biological era of recent years, "epigenetics" is typically used to describe specific molecular biological mechanisms that modulate gene expression that involves the physically "marking" of DNA or its associated proteins, without altering the genetic code. Since epigenetic changes are not genetic and, thus, do not change the DNA, they are reversible. Epigenetics is not only crucial for normal development but can also create the most well-characterized biological frameworks known to maintain aberrant neuronal processing and persistent abnormalities over long periods of time. Generally, the interaction between genes and regulatory genomic DNA elements, epigenetic modifiers, and transcriptional factors determine the expression state of genes. The network of these processes is tightly coordinated during development and within different organ systems, including the brain.[28,29] The ECS is highly regulated by epigenetics,[30,31] and these processes are critical in controlling various aspects of development in early life,[32] the adult brain,[33] and various neuropsychiatric disorders.[34]

Much has been learned in recent years about epigenetics as the molecular mechanism used for the formation and storage of cellular information in response to environmental signals such as drug exposure. Main epigenetic mechanisms implicated in the effects of cannabinoids and that can regulate gene expression include DNA methylation, histone modifications, and non-coding RNAs. Mechanistic implications of these specific epigenetic processes are briefly summarized later.

DNA methylation

Process by which methyl groups are covalently added directly to cytosine and adenine nucleobases of DNA, of which cytosine methylation has been the most characterized. The role of DNA methylation is dependent upon the genomic location, developmental stage, cell type, and disorder. Methylation in promoter regions and transcriptional regulatory sequences has frequently been associated with gene silencing, whereas methylation within the gene body is less understood and may act as either positive or negative effectors.[35,36]

Histone modifications

Histone proteins act as spools around which ~150 bp DNA winds to create structural units that form the nucleosome, the basic unit of chromatin. Core histone octamers (composed of two copies each of the histone proteins H2A, H2B, H3, and H4) are characterized by the presence of N-terminal tails of various lengths that are subject to extensive post-translational covalent modifications including acetylation, methylation, phosphorylation, ubiquitination, and sumoylation[37] and a variety of more recently discovered structures in the brain.[38] The modifications can influence both the accessibility of genomic regions and the binding of transcription factors to the DNA. On the protein level, the main epigenetic mechanism that has been involved in neurobiological alterations linked to drug abuse is post-translational modifications of nucleosomal histones.

Non-coding RNAs

These abundant and functional RNA molecules are transcribed from DNA and regulate gene expression at the transcriptional and post-transcriptional level.[39] Non-coding RNAs include transfer RNAs and ribosomal RNAs, as well as small RNAs such as microRNAs, siRNAs, piRNAs, snoRNAs, snRNAs, exRNAs, scaRNAs, and the long ncRNAs such as Xist and HOTAIR. While the exact genomic targets of specific non-coding RNAs remain to be characterized, the multiple non-coding RNAs are mechanistically intriguing given the variety of tissue-specific cellular and developmental processes that can be influenced by them.[40] Some non-coding RNAs even persist during the maturation of germ cells and in early embryo development and, thus, are interesting candidates for the propagation of cannabinoid effects across generations.[41]

Epigenetic mechanisms and developmenal cannabinoid exposure

Despite the role of epigenetic factors in neurodevelopmental, there remains a dearth of studies regarding epigenetic mechanisms in relation to developmental cannabinoid exposure. Most information relevant to the epigenetic mechanisms implicated in the neurodevelopmental effects of prenatal cannabinoid exposure has focused on histone modifications. For example, our group revealed distinct alterations in the histone modification profile of the NAc in adult male rats with prenatal THC exposure.[10,42] Of particular relevance was the epigenetic relationship to the *Drd2* gene, which is highly implicated in several psychiatric disorders. Epigenetic perturbations were characterized by decreased levels of the trimethylation of lysine 4 on histone H3 (H3K4me3, a transcriptionally permissive mark), increased levels of dimethylation of lysine 9 on histone H3 (H3K9me2, a repressive mark), as well as decreased RNA polymerase II association with the promoter and coding regions of the *Drd2* gene. Importantly, these THC-related epigenetic modifications, which would be predictive of reduced transcription, were linked to reduced *Drd2* mRNA expression in both humans and rodents and persisted into adulthood. These findings are interesting since reduction of the dopamine D2 receptor is characteristic of individuals with substance use disorders,[43] suggesting that early cannabinoid exposure potentially enhances addiction vulnerability through altering the epigenetic processes at the *Drd2* gene locus. Consistent with these findings suggesting enhanced reward sensitivity, Frau and collaborators show sex-specific hyper-excitability of dopaminergic neurons in juvenile males prenatally exposed to THC.[19] Such changes in mesolimbic dopaminergic neurons that project from the VTA to the NAc could explain the increased vulnerability for enhanced drug intake seen in adult male rats exposed to THC during gestation.[10,12] Moreover, our group recently demonstrated that prenatal THC not only increases drug intake adulthood but also enhances motivation for natural reward in a food self-administration paradigm.[44] This motivational disturbance was *causally* linked to an upregulation of the epigenetic factor Ktm2a (a H3K4 histone methyltransferase) in the NAc of adult male rats.[44] Importantly, the epigenetic alteration of increased striatal expression of Kmt2a was maintained throughout lifetime of the prenatally exposed male rats and this specific epigenetic dysregulation strongly correlated with perturbations to synaptic plasticity-related genes.[44]

Reprogramming of the *Drd2* gene has also been demonstrated in animals with developmental THC exposure that present phenotypes relevant to schizophrenia. Adult male rats with perinatal THC exposure (gestational day 5 to postnatal day 9) were shown to elicit cognitive and social deficits, evident by impaired short-term recognition memory and reduced social interaction (i.e., increased social withdrawal), which are often considered as schizophrenia-like symptom.[45] This was associated with a reduction in DNA methylation selectively at *Drd2* gene promoter regulatory region in the adult PFC. Interestingly, there was also a significant correlation between *Drd2* gene expression and cognitive and social impairments in the perinatal THC-exposed rats. Moreover, *DRD2* DNA methylation was reduced in peripheral blood mononuclear cells of individuals with schizophrenia compared to controls. Whether timing of THC exposure during development and resulting epigenetic perturbations leads to different outcomes relevant to particular psychiatric and addiction vulnerability is unknown.

TABLE 1 Molecular and epigenetic dysregulation of the brain associated with prenatal cannabinoid exposure.

Cannabinoid	Subject and brain region studied	Molecular and/or epigenetic alterations	References
Cannabis	Human fetal NAc and amygdala	↓ DRD2 mRNA levels	10
Cannabis	Human fetal hippocampus and striatum	↓ SCG10/stathmin2 mRNA levels	8
Cannabis	Human fetal cerebrum	↓ CNR1, ↓ DAGLA, ↑ MGLL mRNA levels	8
THC	Fetal male rat striatum	↓ Drd2 mRNA levels	10
THC	Mouse juvenile hippocampus	↓ CB_1R density in hippocampus	7
THC	Adolescent male rat VTA	↓ GABA release, ↑ AMPAR/NMDAR ratio	19
THC or WIN-55,212-2	Adolescent PFC and hippocampus	↓ Glutamate levels	15,20,21,46
THC	Adult male rat NAc	H3K4me2, H3K9me3 ↓ Drd2 mRNA levels	10
THC	Adult male rat NAc	H3K4me3 ↑ Kmt2a mRNA and protein expressions	44
THC or WIN-55,212-2	Adult rat PFC	↓ Grm5, Trpv1, Dagla mRNA levels	22
THC	Adult female rat VTA and NAc	↑ DOPA/DA ratio	47
THC	Adult male rat Cortex and hippocampus	Δ CB_1R density	48
THC	Adult male rat hippocampus	↓ GABA levels	49
THC	Adult male rat PFC	↓ DNA methylation at Drd2 gene promoter	45

Increased (↑), decreased (↓) or change (Δ) of specific measure are noted. SCG10, superior cervical ganglion 10; CB_1R, cannabinoid receptor 1; CNR1, cannabinoid receptor 1 gene; DAGLA, diacylglycerol lipase alpha gene; MGLL, monoacylglycerol lipase gene; GABA, gamma-aminobutyric acid; NMDAR, N-methyl-D-aspartic acid receptor; AMPAR, alpha-amino-3-hydroxy-5-methyl-4-isoxazolepropionic acid receptor; VTA, ventral tegmental area; NAc, nucleus accumbens; PFC, prefrontal cortex; H3K4me2, dimethylation of Lysine 4 on Histone 3; H3K9me3, trimethylation of Lysine 9 on Histone 3; H3K4me3, trimethylation of Lysine 4 on Histone 3; Kmt2a, lysine methyltransferase 2A; Grm5, metabotropic glutamate receptor 5 gene; Trpv1, transient receptor potential vanilloid 1 gene; DOPA, dihydroxyphenylalanine; DA, dopamine; DNA, deoxyribonucleic acid; mRNA, messenger ribonucleic acid; THC, tetrahydrocannabinol.

Conclusion

The literature has repeatedly confirmed long-term molecular and behavioral effects of prenatal cannabinoid exposure. Though not as extensive to date, accumulating research would suggest that cannabis exposure during early stages of development can alter epigenetic mechanisms to change the trajectory of individual vulnerability to neurobiological and phenotypic dysregulation later in life. The marked epigenetic reprogramming induced by THC emphasizes, however, highlights that the observed neurobiological disturbances and behaviors may be reversible given the reversible nature of epigenetic processes. The dynamic nature of epigenetic mechanisms in contrast to static genetic inheritance could, thus, provide opportunities to counter negative consequences of developmental cannabis exposure (Table 1).

References

1. Chandra S, et al. New trends in cannabis potency in USA and Europe during the last decade (2008-2017). *Eur Arch Psychiatry Clin Neurosci* 2019;**269**:5–15. https://doi.org/10.1007/s00406-019-00983-5.
2. Berghuis P, et al. Hardwiring the brain: endocannabinoids shape neuronal connectivity. *Science* 2007;**316**:1212–6. https://doi.org/10.1126/science.1137406.
3. Maccarrone M, Guzmán M, Mackie K, Doherty P, Harkany T. Programming of neural cells by (endo)cannabinoids: from physiological rules to emerging therapies. *Nat Rev Neurosci* 2014;**15**:786–801. https://doi.org/10.1038/nrn3846.
4. Mulder J, et al. Endocannabinoid signaling controls pyramidal cell specification and long-range axon patterning. *Proc Natl Acad Sci U S A* 2008;**105**:8760–5. https://doi.org/10.1073/pnas.0803545105.
5. Grant KS, Conover E, Chambers CD. Update on the developmental consequences of cannabis use during pregnancy and lactation. *Birth Defects Res* 2020. https://doi.org/10.1002/bdr2.1766.
6. Bara A, Ferland JN, Rompala G, Szutorisz H, Hurd YL. Cannabis and synaptic reprogramming of the developing brain. *Nat Rev Neurosci* 2021. https://doi.org/10.1038/s41583-021-00465-5.
7. de Salas-Quiroga A, et al. Prenatal exposure to cannabinoids evokes long-lasting functional alterations by targeting CB1 receptors on developing cortical neurons. *Proc Natl Acad Sci U S A* 2015;**112**:13693–8. https://doi.org/10.1073/pnas.1514962112.
8. Tortoriello G, et al. Miswiring the brain: Δ9-tetrahydrocannabinol disrupts cortical development by inducing an SCG10/stathmin-2 degradation pathway. *EMBO J* 2014;**33**:668–85. https://doi.org/10.1002/embj.201386035.
9. Higuera-Matas A, Ucha M, Ambrosio E. Long-term consequences of perinatal and adolescent cannabinoid exposure on neural and psychological processes. *Neurosci Biobehav Rev* 2015;**55**:119–46. https://doi.org/10.1016/j.neubiorev.2015.04.020.
10. DiNieri JA, et al. Maternal cannabis use alters ventral striatal dopamine D2 gene regulation in the offspring. *Biol Psychiatry* 2011;**70**:763–9. https://doi.org/10.1016/j.biopsych.2011.06.027.
11. Wang X, Dow-Edwards D, Anderson V, Minkoff H, Hurd YL. In utero marijuana exposure associated with abnormal amygdala dopamine D2 gene expression in the human fetus. *Biol Psychiatry* 2004;**56**:909–15. https://doi.org/10.1016/j.biopsych.2004.10.015.
12. Ellgren M, Spano SM, Hurd YL. Adolescent cannabis exposure alters opiate intake and opioid limbic neuronal populations in adult rats. *Neuropsychopharmacology* 2007;**32**:607–15. https://doi.org/10.1038/sj.npp.1301127.

13. Spano MS, Ellgren M, Wang X, Hurd YL. Prenatal cannabis exposure increases heroin seeking with allostatic changes in limbic enkephalin systems in adulthood. *Biol Psychiatry* 2007;**61**:554–63. https://doi.org/10.1016/j.biopsych.2006.03.073.
14. Wang X, Dow-Edwards D, Anderson V, Minkoff H, Hurd YL. Discrete opioid gene expression impairment in the human fetal brain associated with maternal marijuana use. *Pharmacogenomics J* 2006;**6**:255–64. https://doi.org/10.1038/sj.tpj.6500375.
15. Castaldo P, et al. Altered regulation of glutamate release and decreased functional activity and expression of GLT1 and GLAST glutamate transporters in the hippocampus of adolescent rats perinatally exposed to Delta(9)-THC. *Pharmacol Res* 2010;**61**:334–41. https://doi.org/10.1016/j.phrs.2009.11.008.
16. Ferraro L, et al. Short- and long-term consequences of prenatal exposure to the cannabinoid agonist WIN55,212-2 on rat glutamate transmission and cognitive functions. *J Neural Transm (Vienna)* 2009;**116**:1017–27. https://doi.org/10.1007/s00702-009-0230-0.
17. Antonelli T, et al. Prenatal exposure to the CB1 receptor agonist WIN 55,212-2 causes learning disruption associated with impaired cortical NMDA receptor function and emotional reactivity changes in rat offspring. *Cereb Cortex* 2005;**15**:2013–20. https://doi.org/10.1093/cercor/bhi076.
18. Saez TM, Aronne MP, Caltana L, Brusco AH. Prenatal exposure to the CB1 and CB2 cannabinoid receptor agonist WIN 55,212-2 alters migration of early-born glutamatergic neurons and GABAergic interneurons in the rat cerebral cortex. *J Neurochem* 2014;**129**:637–48. https://doi.org/10.1111/jnc.12636.
19. Frau R, et al. Prenatal THC exposure produces a hyperdopaminergic phenotype rescued by pregnenolone. *Nat Neurosci* 2019;**22**:1975–85. https://doi.org/10.1038/s41593-019-0512-2.
20. Mereu G, et al. Prenatal exposure to a cannabinoid agonist produces memory deficits linked to dysfunction in hippocampal long-term potentiation and glutamate release. *Proc Natl Acad Sci U S A* 2003;**100**:4915–20. https://doi.org/10.1073/pnas.0537849100.
21. Castaldo P, et al. Prenatal exposure to the cannabinoid receptor agonist WIN 55,212-2 increases glutamate uptake through overexpression of GLT1 and EAAC1 glutamate transporter subtypes in rat frontal cerebral cortex. *Neuropharmacology* 2007;**53**:369–78. https://doi.org/10.1016/j.neuropharm.2007.05.019.
22. Bara A, et al. Sex-dependent effects of in utero cannabinoid exposure on cortical function. *elife* 2018;**7**. https://doi.org/10.7554/eLife.36234.
23. Szutorisz H, Egervari G, Sperry J, Carter JM, Hurd YL. Cross-generational THC exposure alters the developmental sensitivity of ventral and dorsal striatal gene expression in male and female offspring. *Neurotoxicol Teratol* 2016;**58**:107–14. https://doi.org/10.1016/j.ntt.2016.05.005.
24. Szutorisz H, Hurd YL. Epigenetic effects of Cannabis exposure. *Biol Psychiatry* 2016;**79**:586–94. https://doi.org/10.1016/j.biopsych.2015.09.014.
25. Szutorisz H, Hurd YL. High times for cannabis: epigenetic imprint and its legacy on brain and behavior. *Neurosci Biobehav Rev* 2017. https://doi.org/10.1016/j.neubiorev.2017.05.011.
26. Baedke J. The epigenetic landscape in the course of time: Conrad Hal Waddington's methodological impact on the life sciences. *Stud Hist Phil Biol Biomed Sci* 2013;**44**:756–73. https://doi.org/10.1016/j.shpsc.2013.06.001.
27. Van Speybroeck L. From epigenesis to epigenetics: the case of C. H. Waddington. *Ann N Y Acad Sci* 2002;**981**:61–81.
28. Dambacher S, de Almeida GP, Schotta G. Dynamic changes of the epigenetic landscape during cellular differentiation. *Epigenomics* 2013;**5**:701–13. https://doi.org/10.2217/epi.13.67.
29. Dillon N. Factor mediated gene priming in pluripotent stem cells sets the stage for lineage specification. *Bioessays* 2012;**34**:194–204. https://doi.org/10.1002/bies.201100137.

30. D'Addario C, Di Francesco A, Pucci M, Finazzi Agro A, Maccarrone M. Epigenetic mechanisms and endocannabinoid signalling. *FEBS J* 2013;**280**:1905–17. https://doi.org/10.1111/febs.12125.
31. Meccariello R, et al. The epigenetics of the endocannabinoid system. *Int J Mol Sci* 2020;**21**. https://doi.org/10.3390/ijms21031113.
32. Weaver IC. Integrating early life experience, gene expression, brain development, and emergent phenotypes: unraveling the thread of nature via nurture. *Adv Genet* 2014;**86**:277–307. https://doi.org/10.1016/B978-0-12-800222-3.00011-5.
33. Bayraktar G, Kreutz MR. Neuronal DNA methyltransferases: epigenetic mediators between synaptic activity and gene expression? *Neuroscientist* 2018;**24**:171–85. https://doi.org/10.1177/1073858417707457.
34. Batool S, et al. Synapse formation: from cellular and molecular mechanisms to neurodevelopmental and neurodegenerative disorders. *J Neurophysiol* 2019;**121**:1381–97. https://doi.org/10.1152/jn.00833.2018.
35. Armstrong MJ, Jin Y, Allen EG, Jin P. Diverse and dynamic DNA modifications in brain and diseases. *Hum Mol Genet* 2019;**28**:R241–53. https://doi.org/10.1093/hmg/ddz179.
36. Greenberg MVC, Bourc'his D. The diverse roles of DNA methylation in mammalian development and disease. *Nat Rev Mol Cell Biol* 2019;**20**:590–607. https://doi.org/10.1038/s41580-019-0159-6.
37. Bhaumik SR, Smith E, Shilatifard A. Covalent modifications of histones during development and disease pathogenesis. *Nat Struct Mol Biol* 2007;**14**:1008–16. https://doi.org/10.1038/nsmb1337.
38. Chan JC, Maze I. Nothing is yet set in (hi)stone: novel post-translational modifications regulating chromatin function. *Trends Biochem Sci* 2020;**45**:829–44. https://doi.org/10.1016/j.tibs.2020.05.009.
39. Iyengar BR, et al. Non-coding RNA interact to regulate neuronal development and function. *Front Cell Neurosci* 2014;**8**:47. https://doi.org/10.3389/fncel.2014.00047.
40. Yoshino Y, Dwivedi Y. Non-coding RNAs in psychiatric disorders and suicidal behavior. *Front Psychiatry* 2020;**11**. https://doi.org/10.3389/fpsyt.2020.543893, 543893.
41. Legoff L, D'Cruz SC, Tevosian S, Primig M, Smagulova F. Transgenerational inheritance of environmentally induced epigenetic alterations during mammalian development. *Cells* 2019;**8**. https://doi.org/10.3390/cells8121559.
42. Morris CV, DiNieri JA, Szutorisz H, Hurd YL. Molecular mechanisms of maternal cannabis and cigarette use on human neurodevelopment. *Eur J Neurosci* 2011;**34**:1574–83. https://doi.org/10.1111/j.1460-9568.2011.07884.x.
43. Volkow ND, Morales M. The brain on drugs: from reward to addiction. *Cell* 2015;**162**:712–25. https://doi.org/10.1016/j.cell.2015.07.046.
44. Ellis RJ, et al. Prenatal Δ(9)-tetrahydrocannabinol exposure in males leads to motivational disturbances related to striatal epigenetic dysregulation. *Biol Psychiatry* 2021. https://doi.org/10.1016/j.biopsych.2021.09.017.
45. Di Bartolomeo M, et al. Crosstalk between the transcriptional regulation of dopamine D2 and cannabinoid CB1 receptors in schizophrenia: analyses in patients and in perinatal Δ9-tetrahydrocannabinol-exposed rats. *Pharmacol Res* 2021;**164**. https://doi.org/10.1016/j.phrs.2020.105357, 105357.
46. Campolongo P, et al. Perinatal exposure to delta-9-tetrahydrocannabinol causes enduring cognitive deficits associated with alteration of cortical gene expression and neurotransmission in rats. *Addict Biol* 2007;**12**:485–95. https://doi.org/10.1111/j.1369-1600.2007.00074.x.

47. González B, et al. Effects of perinatal exposure to delta 9-tetrahydrocannabinol on operant morphine-reinforced behavior. *Pharmacol Biochem Behav* 2003;**75**:577–84. https://doi.org/10.1016/s0091-3057(03)00115-1.
48. García-Gil L, Romero J, Ramos JA, Fernández-Ruiz JJ. Cannabinoid receptor binding and mRNA levels in several brain regions of adult male and female rats perinatally exposed to delta9-tetrahydrocannabinol. *Drug Alcohol Depend* 1999;**55**:127–36. https://doi.org/10.1016/s0376-8716(98)00189-6.
49. Beggiato S, et al. Long-lasting alterations of hippocampal GABAergic neurotransmission in adult rats following perinatal Δ(9)-THC exposure. *Neurobiol Learn Mem* 2017;**139**:135–43. https://doi.org/10.1016/j.nlm.2016.12.023.

Chapter 3

Impact of adolescent THC exposure on later adulthood: Focus on mesocorticolimbic function and behaviors

Anthony English[a], Benjamin Land[a], and Nephi Stella[a,b]
[a]Department of Pharmacology, University of Washington School of Medicine, Seattle, WA, United States, [b]Department of Psychiatry and Behavioral Sciences, University of Washington School of Medicine, Seattle, WA, United States

Introduction

The legal use of the *Cannabis* plant and *Cannabis*-based products for medical and recreational purposes is rapidly evolving and thus there is an urgent need to understand how this might impact human health and society. THC and CBD are the two most abundant phyto-cannabinoids (phyto-CB) produced by the *Cannabis* plant (Fig. 1A). THC is often referred to as the primary psychotropic compound and CBD as the primary non-psychotropic compound (Fig. 1B). The recent recreational legalization of *Cannabis*-based product use in several states of the US, with Washington and Colorado having led the way, has implemented a *Cannabis* market. Thus, the bioactivity of cannabinoids occurs along a continuum, from beneficial effects to harm reduction properties and potential harm in vulnerable population, a continuum that also occurs along the market-oriented legalization of the use of cannabis-based products (Fig. 1C). The recently developed cultivation methods of the *Cannabis* plant either boost THC concentrations or boost CBD concentrations while reducing THC concentrations below 0.3%, a plant now referred to as *Hemp*. Specifically, the THC:CBD ratio has increased from 14-fold in 1995 to 80-fold in 2014, and more recent *Cannabis*-based products using *Cannabis* extracts can contain up to 90%–95% THC.[1] Concomitantly, *Cannabis* use seems to be increasingly accepted as a safe recreational drug, as indicated by 16.4% of individuals ages 12–17 years and 51.9% of individuals ages 18–25 years in the US reporting the use of *Cannabis* in their lifetime (2021 NIDA). Furthermore, the age of onset of

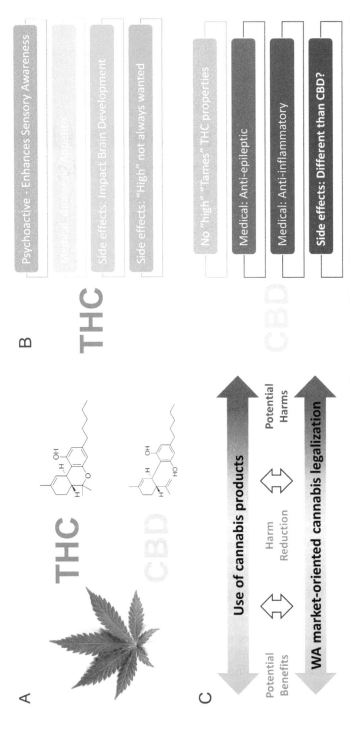

FIG. 1 Phyto-cannabinoid bioactivity: (A) The *Cannabis* plant produces different ratios of THC and CBD depending on its genetic make-up and growing conditions. (B) Main bioactivity associate with the use of THC and CBD by humans. (C) The bioactivity of cannabinoids occurs along a continuum, from beneficial effects, to harm reduction and potential harm in vulnerable population. A continuum that also occurs in the market-oriented legalization of the use of *Cannabis-based* products.

Impact of adolescence THC exposure on later adulthood **Chapter | 3 25**

daily use of *Cannabis*-based products is rapidly shifting towards younger ages (2021 NIDA). Together, these alarming statistics indicate that there is a shift towards increased use of high potency *Cannabis*-based products by adolescents, a scenario that may have important consequences on adolescent brain development and subsequent adulthood behavior.

The relatively young field of Cannabis research has provided a strong understanding of the mechanism of action and bioactivity of phyto-CBs at the molecular, cellular, and systems levels. THC and CBD modulate neuronal activity by interacting with distinct receptors: THC acting principally on cannabinoid 1 and 2 receptors (CB_1R and CB_2R) and the glycine receptor (GLRA3), and CBD acting on G protein-coupled receptor 55 (GPR55) and transient receptor potential cation channel sub-family V member 1 (TRPV1), at least as evidenced by preclinical in vivo studies that validated the involvement of these receptors using genetic knockout mice (Fig. 2A and B). Thus, both THC and CBD modulate the activity of G protein-coupled receptors (GPCRs) and ligand-gated ion channels that are normally modulated by endogenous ligands. This includes the eCBs anandamide (AEA) and 2-arachidonoyl glycerol (2-AG) at CB_1R and CB_2R,

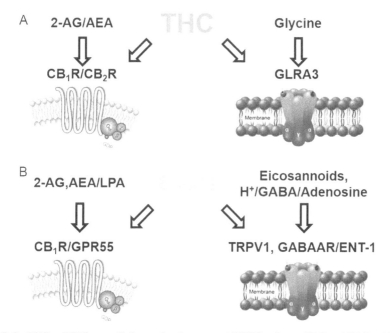

FIG. 2 THC and CBD act at distinct molecular targets: (A) THC activates CB_1R and CB_2R, which are endogenously modulated by 2-AG and AEA, and activates GLRA3, which is endogenously modulated by glycine). (B) CBD is a negative allosteric modulator at CB_1R, which is endogenously modulated by 2-AG and AEA, is an antagonist at GPR55, which is endogenously modulated by LPA, is an agonist at TRPV1 and $GABA_AR$, which are endogenously medullated by eicosanoids, protons, and GABA, respectively, and blocks ENT-1, an adenosine transporter.

lysophosphatidic acid (LPA) at GPR55, eicosanoids and protons at TRPV1, and the amino acids glycine at GLRA3 and GABA at $GABA_AR$ (Fig. 2A and B).

More than two decades of peer-reviewed studies have demonstrated that the adolescent brain exhibits significant vulnerability to continued exposure of THC.[2] Functional neuronal connectivity between the ventral tegmental area (VTA), prefrontal cortex (PFC), and nucleus accumbens (NAc) (i.e., mesocorticolimbic circuitry) underlie executive function, reward processing, and appropriate decision-making, and are precisely defined during critical stages of brain development: childhood, adolescence, and young adulthood.[3,4] This places *Cannabis*-based products in the bull's-eye of public health concerns and suggests that thousands, if not millions, of adolescents will be exposed to these chemically complex products over the coming decades. Specifically, THC can have acute and durable impacts on the mental health of adolescents and can influence mental health much later in adulthood. For example, daily use of *Cannabis*-based products with high THC by adolescents increases the risk of developing a psychotic disorder, including schizophrenia, and is related to an earlier onset of symptoms compared to people who do not use cannabis.[5–8] Relevant to this chapter, the dopaminergic system encompasses neurons that release dopamine, originating in the midbrain and sending their axons to select areas of the forebrain, a system well known to regulate motivated behaviors. Dysregulation of the dopaminergic system is associated with several neurological and cognitive diseases, including drug addiction and schizophrenia.

The marriage of pharmacological PK/PD approaches with molecular, genetic, and behavioral approaches has greatly increased our understanding of the bioactivity of THC on brain development; particularly on the development of mesocorticolimbic brain areas. This chapter was written to bridge our current understanding of the available evidence on the relationship between THC content and health outcomes, and adverse events associated with consuming highly concentrated, manufactured *Cannabis*-based products. We focused our discussion on peer-reviewed surveys and controlled studies (controlled intake and readouts). The body of work reviewed in this chapter has provided an interdisciplinary framework to generate data-driven messages about the impact of THC use on adolescent brain and highlight directions for future research objectives.

Cannabis-based products and devices

Cannabis is a dioecious plant that grows around the world and is one of the world's oldest crops whose use dates back about 12,000 years.[9] THC produces psychotropic effects, typically described as enhanced sensory perception, distorted perception of space and time, and altered interpersonal relationships and thought processes.[10,11] CBD, often referred to as a non-psychotropic phyto-CB because it does not induce the euphoria and intoxication triggered by THC.[12] CBD does greatly influence specific brain functions and behaviors such as neuronal activity and seizure incidence through a different mechanism than THC.[13,14] Additional phyto-CBs that exhibit a certain level of bioactivity include cannabinol, cannabigerol, and cannabichromine.[15,16] Thus, while hundreds of structurally related phyto-CBs are synthesized by *Cannabis* plant expressing enzymes, only a handful of them appear to exhibit significant bioactivity when taken alone or in combinations.

A key question is the potential difference in the bioactivity of whole extract *Cannabis* that include terpenes versus the bioactivity of phyto-CB isolates.[17,18] Terpenoids (terpenes) are another major bioactive constituent of the *Cannabis* plant, although present in lower quantities than phyto-CBs.[19] These 5-carbon compound isoprenes generally add fragrance to a strain of *Cannabis*, as limonene, pinene, myrcene, and caryophyllene all have distinctive odors. They bind to odorant receptors[20] and there is growing evidence that terpenoids can also engage CB_1R and CB_2R,[21] here providing plausible support to the idea that terpenoids might modulate phyto-CB bioactivity. Thus, *Cannabis*-based products generated by extracting the plant material and adding the principal bioactive compounds, THC, CBD, and terpenes at set ratios might produce comparable biological effects.

The legalization of *Cannabis* use bolstered the development of novel *Cannabis*-based products and devices. The most common are flower cigarettes (joints) and edibles (cookies and candies) (Fig. 3A). The newly developed

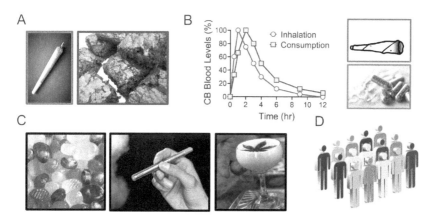

FIG. 3 Cannabis-based products, devices, and pharmacokinetics: (A) The most common *Cannabis*-based products were flower cigarettes (joints) and edibles (cookies). (B) The predominant route of administration in human is through the lungs and digestive system, which produces different PK profiles. The newly developed devices greatly increase THC intake and provide both faster and higher quantities of phyto-CBs delivered per use. (C) The legalization of *Cannabis* use led to the development of new *Cannabis*-based products and devices that delivery higher amounts of THC with faster onset and are appealing adolescents, including vape pens, dabbing devices, colorful candies, and soft drinks, all of which are appealing to adolescents. (D) There is a need to identify and better understand the human vulnerable population, which might depend on health, sex, race, and age.

devices greatly increase THC intake and provide both faster and higher quantities of phyto-CBs delivered per use (Fig. 3B). Examples include electronic vaping devices that deliver 60%–80% THC and dabbing devices that delivers high dose of concentrated cannabis resin containing 90%–95% THC in one hit exposure[22] (Fig. 3C). Unlike *Cannabis* flower cigarettes and edibles, vaping and dabbing devices, as well as colorful candies and soft drinks, tend to be more common among adolescents who use more *Cannabis* overall (Fig. 3B). The ever-increasing landscape of devices that deliver high-dose THC further promotes potentially dangerous consummatory behaviors. For example, these riskier methods of consuming THC does not allow effective titration of their dose.[23] In Washington state, epidemiological data available since legalizing Cannabis use in 2012 show that *Cannabis* users are now more likely to use concentrated forms through dabbing, eating, or vaping than prior to legalization.[24] Further, it is important to better understand the diversity in THC's impact on human health, whether it varies depending on health, sex, ethnicity and age (Fig. 3D). To summarize, the legalization of *Cannabis*-based products resulted in the production of new devices that rapidly deliver THC concentrates of high potency and thus increase their psychotropic response, which raise serious concerns of the impact of such products on vulnerable population, in particular adolescents.

Cannabinoids, endocannabinoid signaling, and brain development

THC and its primary metabolite, 11-hydroxy-Δ^9-THC, activate CB_1R, GPCRs expressed by select types of neurons and glial cell.[25,26] CB_1R activation modulates neuronal functions by reducing the release of many neurotransmitters, including glutamate and GABA, and adjusting synaptic circuits and energy metabolism[27,28] (Fig. 4A). Recent evidence also emphasizes a prominent role for CB_1R expressed by astrocytes in the modulation of neuronal functions[27] (Fig. 4B). At the molecular level, THC modulated that activity of **CB_1R**, **CB_2R** and **GLRA3**, receptors normally modulated by eCBs and glycine, respectively (see Fig. 2A). Thus, the psychoactive responses produced by THC exposure are mediated through various receptors expressed by a complex network of neuronal and glial cells in the central nervous system. The pharmacological activity of CBD appears to involve multiple receptors (see Fig. 2B). For example, evidence suggests that CBD might modulate CB_1R as a negative allosteric modulator,[29] a positive allosteric modulator at $GABA_A$ receptors,[30] antagonizes GPR55, activate TRPV1 and inhibit ENT-1 that mediated adenosine uptake.[31–34] As mentioned above, much less is known about the bioactivity and mechanism of action (MOA) of other phyto-CBs, including cannabinol, cannabigerol, and cannabichromine.

Significant medicinal chemistry efforts were dedicated to the synthesis of artificial cannabinoids (often referred to as "synthetic cannabinoids") that are not produced by the *Cannabis* plant (or in fact any biological organism) and

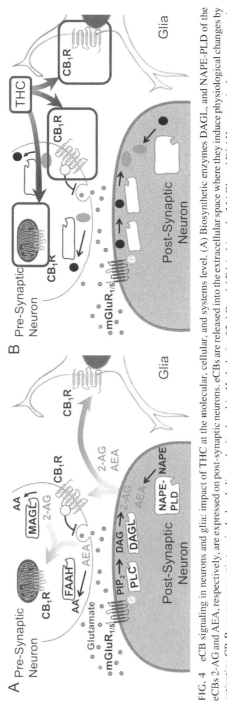

FIG. 4 eCB signaling in neurons and glia: impact of THC at the molecular, cellular, and systems level. (A) Biosynthetic enzymes DAGL, and NAPE-PLD of the eCBs 2-AG and AEA, respectively, are expressed on post-synaptic neurons. eCBs are released into the extracellular space where they induce physiological changes by activating CB_1Rs on pre-synaptic terminals, local glia, and mitochondria. Hydrolysis of 2-AG and AEA is driven by MAGL and FAAH, respectively on pre-synaptic neurons. CB_1R. (B) Exogenous introduction of THC impairs CB_1R signaling by competing with eCBs, affecting neurotransmission, metabolism, and glial cell function (highlighted in red). *Black circles*: eCB precursors and metabolites, *green circles*: 2-AG and AEA.

activate CB_1R. Some of the most notorious artificial cannabinoids include JWH-018, the bioactive ingredient in Spice and K2.[35,36] Spice has been made illegal in many countries in the world; however, the development of potent analogues, such as JWH-019, JWH-073, JWH-081, and other synthetic cannabinoids have circumvented conventional drug laws before their novel structures are made illegal. Additional artificial cannabinoids such as WIN55,212–2 (WIN) and CP55,940 (CP) are used primarily for preclinical study, as referenced below. Artificial cannabinoids exhibit a pronounced toxicity profile, and adolescents are particularly vulnerable to this class of compounds.[13,37]

Cannabinoid receptors are endogenously activated by eCBs, including anandamide (AEA) and 2-arachidonoyl glycerol (2-AG), signaling lipids produced and inactivated by distinct enzymes expressed by neurons and glial cells, resulting in functionally parallel eCB signaling systems[27,38,39] (Fig. 4A and B). Specifically, eCBs are produced on-demand by neurons and glial cells in response to select stimuli and increased cellular activity (typically associated with increases in intracellular calcium) via lipase activation. N-acetylphosphatidylethanolamine-hydrolysing phospholipase D (NAPE-PLD) and diacyl glycerol lipase (DAGL) release eCBs from their plasma membrane precursors to generate AEA and 2-AG, respectively.[40–45] Released eCB's act in a retrograde manner from the post-synaptic production site to the pre-synaptically expressed CB_1R, a concept that stems from the post-synaptic expression of eCB synthesizing enzymes and pre-synaptic expression of CB_1R.[46] Furthermore, glial cells produce eCB and activate cannabinoid receptors expressed on other neural cells, here emphasizing the paracellular signaling mechanism of eCBs. 2-AG (functioning as a full agonist) and AEA (functioning as a partial agonist) stabilize CB_1R in conformations that activate $G_{i/o}$-proteins, induce β-arrestin signaling, and inhibit neurotransmitter release.[47] 2-AG and AEA also activate the CB_2R, primarily expressed on non-neuronal cell types, to modulate immunological activity.[39] Inactivation of eCBs occurs in two steps: a rapid uptake across cell plasma membranes via active transport followed by intracellular hydrolysis: fatty acid amide hydrolase (FAAH) for AEA[48–50] and monoacylglycerol lipase (MAGL) and α/β-hydrolase domain 6 (ABHD6) for 2-AG.[38] 2-AG and AEA can additionally activate CB_2R, a CB_1R homolog that is primarily expressed in immune cells to modulate anti-inflammatory activity.[51] Thus, eCB signaling in the CNS encompasses CB_1R, AEA and 2-AG, and the enzymes that produce and inactivate these two main eCBs, which both mediate paracellular signaling between cells and intracellular signaling in mitochondria (Fig. 4A and B).

GPR55 is a newly de-orphanized Class A GPCR that has been implicated in neuronal development and regulation of neurotransmission in the adult brain.[52,53] Several endogenous ligands for GPR55 have been identified, including the signaling lipid LPI, PACAP-27 and several newly discovered small peptides.[54,55] CBD acts as an antagonist at GPR55 and modulates neurotransmission under both physiological and pathological conditions and exhibits clearly promising anti-epileptic properties for the treatment of juvenile epilepsy.[56–59]

Relevant to this chapter, eCB signaling plays an overarching regulatory role in development, starting during the initial stages of embryonic development, ensuing prenatal development and differentiation.[60,61] This signaling system then undergoes a switch in function from the determination of cell fate during adolescence to the homeostatic regulation of metabolic pathways and transmission in the mature CNS.[62,63] Thus, the developing adolescent brain undergoes substantial structural remodeling that makes it particularly vulnerable to the harmful effects of exogenous bioactive agents, such as THC.[64] For example, excessive activation of CB_1R during brain development influences multiple fundamental cellular functions, including cell proliferation, migration, and differentiation through control of select signaling pathways and changes in the expression of morphogenetic factors[65–67] (Fig. 5A). Thus, the molecular diversification into neuronal and glial progenies during brain development is regulated by morphogenetic signaling molecules, including eCB signaling that contribute to the building of complex tissues. An interesting example is provided by eCB-mediated activation of CB_1R expressed by neural stem cells, which enhances differentiation into neurons without affecting astrocytes and oligodendrocytes, as evidenced by increased neurite outgrowth and expression in neuronal markers.[66] By contrast, activation of CB_1R expressed by post-natal radial and neuronal stem cells controls differentiation in the adult brain by promoting astroglial differentiation of newly born cells.[68,69] Such remodeling occurs in many brain areas involved in vital neuronal function, including sensory inputs and higher-order cognitive processes such as learning, memory, and decision-making.[70] Given that CB_1R expressed by neural progenitor cells in the adolescent developing forebrain regulates the ratio of neurons to astrocytes in areas such as the hippocampus and cerebral cortex, THC triggered activation of CB_1R during this critical period is likely to influence the connectivity of these brain regions[71,72] (Fig. 5B and C). In addition, evidence suggests that focal eCB gradients are generated to guide the direction of cell migration[73] (Fig. 5A). Accordingly, the downregulation of DAGLα and DAGLβ expression (enzymes involved in 2-AG release as well as production from membranes) following neuronal specification represents an essential step to increase the reliance of post-mitotic neurons on extracellular 2-AG.[73] Relevant to this chapter, cortical plate pyramidal neurons produce and release enough quantities of 2-AG to provide positional cues.[73] Additional molecular mechanisms involving eCB signaling that regulate directed migration include the transactivation of specific receptor tyrosine kinases following activation of CB_1R.[67,70,74,75] Together, these studies provide a detailed molecular picture depicting the role eCB signaling in brain development and the vulnerabilities to exogenous cannabinoid compounds such as THC.

The PFC is considered one of the most evolved brain regions enabling evolutionary cognition and the emotional and cognitive capabilities of the

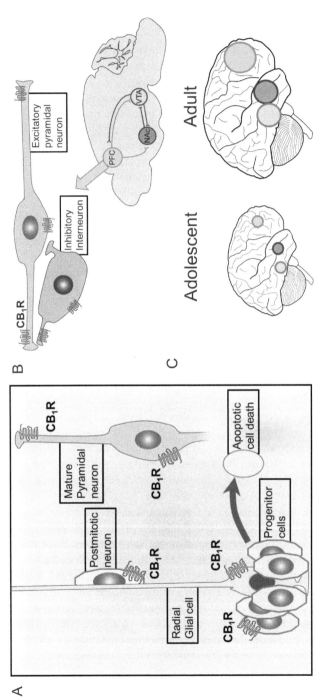

FIG. 5 Endocannabinoid signaling and brain development: (A) CB_1R are involved in brain development by their ability to control cell viability, proliferation, migration and fate, as well as neurotransmission, metabolism and connectivity. (B) CB_1R are abundantly expressed by excitatory pyramidal neurons and inhibitory interneurons in mesocorticolimbic structures of the developing adolescent brain, a circuitry that connects the prefrontal context (PFC, green), nucleus accumbens (NAc, red) and ventral tegmentum area (VTA, blue) in rodent brain. (C) Mesocorticolimbic brain areas are actively developing during human adolescence.

developed, matured brain, making it particularly vulnerable during its developmental stages.[76,77] To develop the PFC and establish its complex network with other brain structures, this tissue undergoes vast number of molecular and cellular changes during adolescence that are crucial to properly form and connect brain structures, a process that is highly sensitive to bioactive agents known to impair development.[78] Specifically, the PFC undergoes significant synaptic pruning, enhanced myelination, and interneuron migration that contributes to proper adolescent maturation.[79] The complex maturation that the PFC must undergo during adolescence leaves it as one of the final brain regions to fully develop. Structural maturation of brain regions commonly occurs concurrently to functional maturation and therefore the proper maturation of the PFC is critical for developed executive functioning[80] (Fig. 5C). Given the importance of proper PFC development, several studies have linked impaired development with future psychiatric disorders.[81–83] More recently, dysfunction of microRNA-mediated maturation of the PFC leading to psychiatric disorders has been reported.[83] In line with these findings, studies have linked *Cannabis* use during adolescence to neuro-psychiatric disorders in adulthood.[82]

To conclude this section, eCB signaling plays a pivotal role for appropriate neuronal maturation during adolescence and neuronal maintenance during adulthood. Exogenous tampering from phyto-CB's impacts the CB_1R-driven proliferation, differentiation and maturation and pruning processes during adolescence. Further, the PFC appears to be a highly vulnerable brain region to the deleterious effects of *Cannabis* used during adolescent given that its development occurs through adolescence and that this intricate process depends on eCB signaling. The following two sections detail human clinical evidence and preclinical evidence showing that adolescent cannabinoid intake disrupts brain physiology and behavior, with particular attention paid to the PFC and mesocorticolimbic network, and its impact on adulthood anatomy and behavior.

Evidence from human studies

Several studies have addressed the health impact of *Cannabis* use during human adolescence, however only few studies offer interpretable results by providing key parameters on the population that was studied, the product used, and the amount and regimen of THC use. The gold standard of research is Randomized Control Trials (**RCTs**) in which study participants are randomized to two or more conditions. However, RCTs are not always possible or ethical to conduct in humans. Research on the impact of *Cannabis* use during human adolescence must therefore rely on alternative study designs and use sophisticated statistical methods to mimic RCTs to draw conclusions. Here we selected peer-reviewed human studies that were rigorously designed as emphasized by (1) the recruitment protocol of subjects with a specific health outcome and (2) a documented dose and frequency of *Cannabis* use. We also relied on surveys and secondary data analysis, where people report their cannabis use patterns, health behaviors, and health outcomes.

Because measuring THC content in *Cannabis*-based products is not feasible for large population-based studies, a more common approach is to collect detailed information on the modes of *Cannabis* administration that typically use concentrated *Cannabis*-based products. For example, using this approach, a recent study show that adults who regularly consume *Cannabis* can self-titrate their use and adjust their intake to compensate for potency.[84] While for adults this might mitigate a potential increase in the detrimental effects of high potency products, this might not be the same for adolescents. Specifically, adverse effects linked to the use of manufactured products are especially high among adolescents, and exposure to vaping *Cannabis* products is more likely to need medical intervention.[85] Many current *Cannabis*-based products use formulations with higher levels of THC content compared to *Cannabis* flower, and these modes of use have been associated with adverse health events like acute toxicity, emergency department visits, and poison center calls.[1,86,87] Thus, the increase in high potency *Cannabis*-based product is now associated with various acute toxicities, a scenario that contrasts with the previously common belief that *Cannabis* use is safe and not be associated with acute toxicity.

An additional alarming change is the recent identification of a "new" psychiatric disorder introduced in the DSM IV, the development of *Cannabis* use disorder (CUD) or addiction to *Cannabis*, particularly among adolescents.[88] It should be emphasized that use of high potency *Cannabis*-based products increases the risk of developing CUD, a condition that impairs social functioning, memory, decision-making, school/work performance, and is more likely to be expressed by adolescent long-term users of high-dose THC compared to adults using similar amounts.[89–92] Furthermore, in a population of adolescents experimenting with *Cannabis*-based products, the use of highly concentrated THC is associated with the progression and persistence of further use compared to adolescents that use less potent products. This suggests that THC potency may represent a contributing factor to experimental use as adolescents transition to frequent use.[90] For example, a study performed in the U.S. monitored THC potency and CUD symptom onset and found that high potency cannabis products used at *Cannabis* initiation is associated with over four times the risk of CUD symptom onset within the first year of initiation.[89] Further, adolescents that frequently used high potency products report a higher risk of *Cannabis* dependence.[92] Other distressing studies indicate that daily use of high-potency *Cannabis* during adolescence is associated with an earlier onset of psychotic symptoms (6 years earlier) than non-cannabis users.[6] Together, this evidence suggests an increased risk of experiencing CUD and other mental disorders when adolescents frequently use high THC *Cannabis* products.

Cognitive aptitude and behavior of adults who used cannabis during adolescence

Several studies explored how cognitive aptitude and behavior during adolescence might influence *Cannabis* use behavior in adulthood. For example: "Does cognitive aptitude during adolescence influence the pattern of *Cannabis* use during adolescence and later in adulthood?" An analysis of a nationally representative longitudinal cohort identified five latent trajectories of *Cannabis* use frequency between ages 16 and 26 years: abstainers, dabblers, early heavy quitters, consistent users, and persistent heavy users.[93] When examining how cognitive aptitude in early adolescence is associated with heterogeneous pathways of *Cannabis* use, there was a statistical relationship between adolescents with a higher rating of cognitive aptitude who start using *Cannabis* products in early adolescence and the likelihood to enter consistent patterns of use (i.e., without extreme trajectories of *Cannabis* use as they age into young adulthood).[93]

A systematic review and meta-analysis of 11 studies encompassing 23,317 individuals younger than 18 years of age indicated that adolescent *Cannabis* consumption is likely associated with increased risk of developing mental health disorders, specifically depression and suicidal behavior later in life, even in the absence of a premorbid condition.[94] Notably, this study did not find any association with anxiety incidence. This study measured *Cannabis* use in the last year or the last 6 months using self-reported questionnaires, and distinguished weekly users, daily users, and occasional users. Considering the high prevalence of adolescents consuming *Cannabis*-based products, this study suggests that the large number of young adults who develop depression and suicidality might be partially attributable to *Cannabis* use. Another recent study explored associations between *Cannabis* potency, substance use and mental health outcomes, while accounting for preceding mental health and frequency of *Cannabis* use; finding that use of high-potency *Cannabis* was associated with both increased frequency of *Cannabis* use and increased likelihood of anxiety disorder onset.[95] This result was based on analyzing outcomes and exposures collected from 1087 participants with an average age of 24 years who self-reported *Cannabis* use during adolescence (average onset of use: 16.7 years of age). Of note, this study found no evidence of association between the use of high-potency *Cannabis* on either alcohol use disorder or depression.[95]

Thus, the few reports on *Cannabis* use by human adolescents suggest that cognitive aptitude and behaviors during adolescence influences the pattern of *Cannabis* use, and that increased use of high THC products is associated with increased risk of developing mental health disorders, including CUD and possibly depression. Together, these studies provide a clear message of caution: pre-adolescents and adolescents should avoid using *Cannabis*-based products as it might be associated with a significant increased risk of developing mental health disorders in adulthood.

Disruption of brain anatomy, connectivity, and dopamine function in human adults that used cannabis during adolescence

Several laboratories studied whether *Cannabis* use during adolescence affects brain anatomy and connectivity in adulthood by combining self-reports of *Cannabis* use and brain imaging technologies. A 2014 study used high-resolution

MRI scans of adolescent *Cannabis* users and compared them to non-using controls.[96] They measured several morphometry readouts, including gray matter density, brain and regional volumes and shapes, and found greater gray matter density and brain structure shape differences in *Cannabis* users compared to non-using participants, particularly in the left nucleus accumbens (NAc) that extended to subcallosal cortex, hypothalamus, sublenticular extended amygdala.[96] Participants in this study were 20 young adults (age 18–25 years) that currently used *Cannabis* and 20 non-using controls. *Cannabis* participants used it at least once a week but were not dependent according to a Structured Clinical Interview for the DSM-IV. Thus, this study suggested that *Cannabis* exposure during young adulthood might be associated with exposure-dependent alterations of the neural matrix of core brain reward structures.

The first longitudinal study that compared resting functional connectivity of frontally mediated networks (cingulate cortex and frontal gyrus) that mediate cognition and executive function using fMRI scan compared 43 healthy controls and 22 treatment-seeking adolescents with CUD.[97] This study found the expected increase in resting functional connectivity measured in healthy controls did not occur in adolescents with CUD, and that high amounts of *Cannabis* use during the 18-month interval predicted lower intelligence quotient and slower cognitive function as measured by full-scale IQ and reaction time. Here, the average age of cannabis use onset in the CUD cohort was 17.6 years of age and all were lifetime *Cannabis* users, representing 1000–1200 days of *Cannabis* use/individual. This study suggests that repeated exposure to *Cannabis*-based products during adolescence may have detrimental effects on brain resting functional connectivity, intelligence, and cognitive function, notably in the PFC.[97] In line with this report, a study following 799 adolescent participants found increased thinning of the PFC associated with dose-dependent experimental *Cannabis* use.[98] Participants were *Cannabis*-naïve (mean age: 14.4 years) at initial magnetic resonance (MR) image collection and 5-year follow-up MR images evaluated based on *Cannabis* use throughout the adolescent window. They also determined, by a self-report questionnaire, that *Cannabis* use and cortical thinning throughout adolescence was associated with increased impulsive behaviors.[98]

A 2019 study measured the gray matter volume (GMV) by voxel-based morphometry of 46 individuals that reported just one or two instances of cannabis use at 14-year-old human adolescents (males and females). The results indicated greater GMV in *Cannabis* users of the bilateral medial temporal lobes and posterior cingulate, and of the lingual gyri and cerebellum, compared to carefully matched THC-naive controls.[99] The authors noted that the GMV differences were unlikely to precede *Cannabis* use and were more likely linked to generalized anxiety symptoms in the *Cannabis* users.[99] This study outlines the provocative idea that structural brain and cognitive effects might occur after just one or two instances of *Cannabis* use in adolescence.

A 2018 study leveraged two positron emission tomography scans to measure striatal dopamine release and found a lower dopamine release 30%–50% in the associative striatum of adults that use cannabis during adolescents qualify as severely cannabis-dependent participants (onset *Cannabis* use at 16.3 ± 3.2 years of age, used *Cannabis* for an average of 11.3 ± 3.6 years). The authors also reported a correlation between inattention and negative symptoms in severely *Cannabis*-dependent participants such as poorer working memory and probabilistic category learning performance.[100]

Together, these studies provide independent evidence that *Cannabis* use during human adolescence may result in disrupted brain anatomy and connectivity in adulthood. To address these detrimental effects of adolescent *Cannabis* use, we urgently need additional studies to confirm and extend the current pool of knowledge. More specifically, the necessity for studies that report PK data in humans using different products, routes of administrations, and quantitative measures of any impact on brain anatomy, connectivity, and neurotransmitter functions (e.g., the dopaminergic system) are increasingly prevalent. An additional approach is to study the impact of THC use on adolescent non-human primates. A recent study in squirrel monkeys treated daily for 4 months with escalating dose of THC (0.1–1 mg/kg) showed that initial low doses of THC impair the performance of adolescent monkeys in a cognitive test designed to study repeated acquisition and discrimination reversal using a touchscreen-based cognitive test.[101] THC treatment also reduced motor activity and increased sedentary behavior during the initial week of treatment, and progressive tolerance to treatment developed, starting the second week.[101] This study provides an example of a highly translational model system to study the impact of THC use during adolescence.

Evidence from rodent studies

There has been a recent increase in the number of studies reporting the impact of THC in pre-clinical adolescent rodent models. Provided the well-known neurodevelopmental stages that occur during rodent adolescence, conclusions can be accurately drawn when comparing THC-treated rodents to controls. An additional advantage of such studies is the ability to precisely deliver set regimens of THC, and measure changes in molecular, cellular, and behavioral parameters as a proxy of the acute impact in adolescence and subsequent consequences in adulthood. On the other hand, rodents do not model all aspects of human physiology and behavior, and many studies deliver THC using i.p. injections, which do not accurately recapitulate human use. Additionally, exposure to THC, CBD, other phyto-CBs, and terpenes as single agents does not encapsulate the full bioactive profile of the *Cannabis* plant. These important limitations must be considered when interpretating the results of pre-clinical studies and their translational values.

Considering an average THC content of herbal *Cannabis* is approximately 10% in Europe, and using the transformation of human-equivalent doses proposed by the Food and Drug Administration (**FDA**), one can extrapolate that i.p. injections of 2.5, 5, and 10 mg/kg THC corresponds to half, one, and two joints, respectively.[102,103] Several studies reported the PK profile of THC injected i.p. and delivered orally to mice and rats,[104] and it is known that THC is mainly metabolized by select P450 enzymes expressed in the liver.[105] Meanwhile, the ABC transporters P-glycoprotein (P-gp, Abcb1) and breast cancer resistance protein (Bcrp, Abcg2) regulate the brain disposition of THC.[106,107] Relevant to this chapter, there are multiple differences in the distribution and metabolism of THC between adolescent and adult mice, which might influence the pharmacological response to the drug.[108] For example, i.p. injections of THC (5 mg/kg) reaches 50% higher circulating concentration in adolescent male mice (PND 37) compared to adult mice (PND 70). Conversely, THC brain-to-plasma ratios measured in adolescent mice brain relative to adult mice brain indicate 40%–60% lower brain concentrations in adolescents, most likely due to higher expression of P-gp by endothelial cells of the adolescent blood brain barrier.[108] Accordingly, i.p. injections of THC (5 mg/kg) reduces spontaneous locomotor activity in adult, but not adolescent, mice.[108] Further, adolescent female rats (PND 27–45) show a stronger metabolism compared to males. Here, the 0.5 mg/kg dose of THC i.p. was selected based upon those that typically produce rewarding and anxiolytic effects (<1 mg/kg), and the 5 mg/kg dose i.p. was selected because it typically produces anxiogenic and aversive effects (>5 mg/kg) in rodents.[109] Thus, this study reports dose-dependent and sex-dependent effects on behavior, neural activity, and functional connectivity across multiple nodes of brain stress and reward networks as measured by the elevated plus maze (EPM), novel environment activity, conditioned place preference (CPP), as well as changes in cFOS expression using network analysis.[109] In summary, THC metabolism in rodents varies depending on the sex and species of rodents.

Treating rodents with cannabinoid agonists triggers characteristic cannabimimetic behaviors, including four hallmark responses referred to as the "tetrad response": hypothermia, hypolocomotion, analgesia, and catalepsy[110,111] (Fig. 6A). However, acute treatment with cannabinoids and pharmacological agents that target eCB signaling influence additional mouse behaviors, including ambulation, motor coordination, stress, short term memory, spatial memory, and acoustic startle.[112–115] Thus, it is important to study the impact of THC in adolescent mice using multiple behavioral readouts.

Molecular and cellular responses to THC treatment during adolescence in rodents

How does THC treatment impact CB_1R function, gene expression, and neuronal functions in adolescent rodents? One answer is provided by studies showing that

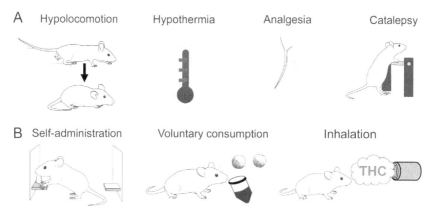

FIG. 6 Measuring cannabimimetic activities: devices and self-administration studies in rodents: (A) Tetrad behaviors classically used for measuring cannabimimetic responses in rodents. (B) Recent developed methodological and experimental approaches in rodents to study voluntary self-administration: e.g., lever pressing and consumption of gelatin or dough. (C) Alternative routes of THC administration in rodents: vaporized THC for inhalation.

CB_1Rs activation on the axonal surface induces repulsive growth cone turning and eventual collapse.[72,116,117] The molecular mechanism of CB_1R-mediated cytoskeletal instability in growth cones involves RHO-family GTPases, RAS, and PI3K-AKT-β-catenin signaling.[72,118,119] Accordingly, THC exposure leads to ectopic formation of filipodia and alterations in axon morphology, together limiting the computational power of neuronal circuits involved in higher cognitive function, such as the mPFC, in affected individuals.[41,73] Another molecular and cellular response linked to repeated activation of CB_1R by THC during the sensitive period of adolescent brain development are the pronounced changes in the expression and functionality of signaling proteins, including dopamine receptors critically involved in higher cognitive functions.[120,121]

While it is known that the hippocampus is particularly sensitive to THC treatment,[122] increasing evidence indicates additional brain structures are also sensitive, including the PFC. Specifically, as a vital cortical region with a high glutamatergic to GABAergic anatomical distribution, the balance of excitatory and inhibitory function within the PFC is critical to proper development and adult cognitive function.[123] Increased eCB production and treatments with THC or synthetic cannabinoids activate CB_1Rs expressed by different neuronal sub-populations and fine-tuned excitatory and inhibitory neurotransmission and neuronal function.[27,124] Distinct neuronal sub-populations in the PFC express CB_1R at different levels: higher levels by GABAergic neurons and parvalbumin (PV)-expressing interneurons and lower levels by cortical neurons that project to the striatum.[125] In fact the levels of *Cnr1* mRNA (which encodes CB_1R) in PFC are reduced during adolescence.[126] Given the sensitive balance of glutamatergic and GABAergic transmission with the PFC, decreased *Cnr1* mRNA suggests a shift in eCB-mediated control of synaptic development that may be uniquely vulnerable to exogenous treatment.[127] In line with this notion, activation of CB_1R shifted the balance between excitation and inhibition towards excitation,[128] and results in a remarkable up-regulation in CB_1R expression and down-regulation of BNDF in the PFC, as expected by down-regulation of CB_1R that controls BDNF expression.[67,129]

Additional studies show that THC treatment of adolescent rodents modifies the excitatory/inhibitory balance in the mPFC. Adolescent female rats (PND 35–45) treated with increasing doses of THC twice a day (i.p., 2.5 mg/kg 35–37 PND; 5 mg/kg 38–41 PND; 10 mg/kg 42–45 PND) show decreased expression in components of eCB signaling: CB_1R, MAGL and FAAH.[130] Accordingly, the authors found reduced eCB-mediated LTD in the adult PFC as indicated by electrophysiological slices recordings in the cortical layer V (L-V) upon stimulation of layer II (L-II). Females presented with increased GluN2B in adulthood and AMPA GluA1 with no changes in GluA2 sub-units. Adolescent exposure to the CB_1R agonist WIN impairs neuronal development and downregulates local inhibitory GABA transmission in the mPFC.[121,131] By contrast, chronic THC exposure in adolescent mice decreases NMDA current and plasma membrane expression of GluN1 in adulthood.[132] Together, this evidence suggests that THC exposure may delay adolescent maturation of the glutamatergic system, thus resulting in a less functional adult mPFC. Of note, fluctuations and dynamic state of eCB levels (2-AG and AEA) in the PFC do not appear impacted by chronic THC exposure during adolescence, thought these levels were measured by LC-MS which has clear limitations.[133] Specifically, our understanding of eCB dynamics in vivo has been limited by the low spatiotemporal resolution of common analytical chemistry approaches (rapid freezing of tissue follow by LC-MS quantification). While well suited for measuring changes in eCB tone, quantification of bulk brain eCBs does not resolve their fluctuations occurring in *sec* to min triggered by transient changes in brain activity. To overcome this limitation, novel genetically encoded eCB sensors, e.g., GRABeCB2.0, are paving the way to study localized changes in low micromolar eCB concentrations within seconds.[134,135] Together, this significant body of works convincingly shows that THC treatment of adolescent rodents results in profound changes in the molecular and cellular components involved in healthy neuronal maturation and development.

Impact on adult mesocorticolimbic systems and reward/addiction-like behaviors associated with THC use during adolescence in rodents

The THC-triggered changes in the expression of select proteins and neuronal functions in adolescent rodent brain has a pronounced impact at the system's level of adult brain function. Subsequent rodent behaviors are commonly studied as a proxy of behaviors, cognitive tasks, and pathological behavioral impairments in humans. Early studies from the laboratory of Dr. Yasmin Hurd addressed this important question in the context of the mesocorticolimbic systems and reward/addiction-like behaviors, setting a solid foundation for future studies.

One persistent claim is that *Cannabis* acts as a gateway drug; that its use may predispose individuals to want to take 'harder' drugs like opiates or psychostimulants. Because adolescence represents a time of marked plasticity, it is possible that *Cannabis* use during this period may change the underlying neurophysiology to make already reinforcing drugs more rewarding. A handful of pre-clinical studies have attempted to behaviorally model this 'gateway' hypothesis. Those that have typically treated adolescent rats with THC between the ages of PND 21–45 and use operant self-administration (SA) of the secondary drug in adulthood. Note that most studies are almost exclusively performed in rats, and it is noted when mice were used.

Opiates. In one of the earliest attempts to model the gateway hypothesis, the Hurd lab reported that adolescent THC exposure increases heroin SA in adult rats.[136] Rats with THC exposure (1.5 mg/kg, i.p., 8 injections over 21 days) pressed the heroin-associated lever more and received more total heroin after 15 days of heroin exposure. Analysis of changes in mRNA levels showed increases in pro-enkephalin mRNA in the NAc as a function of ado-THC exposure that correlated with increases in heroin SA.[136] A follow-up study showed that this pro-enkephalin was critical for expression of heroin SA after ado-THC.[137] Studies from other groups using opiates generally support this initial behavioral finding, including identifying reinstatement of heroin seeking as influenced by ado-THC.[138] Note that the genetic makeup contributed to specific aspects of opiate-seeking behavior.[139] Using two rat strains, Lecca et al. showed that the Lewis rat strain (a more addiction prone strain) exhibit increased heroin SA after ado-THC, while the effect was largely absent in the Fischer 344 strain.[139] This study suggests that individuals who may already be addiction-prone may further increase risk of subsequent secondary drug use when using THC in adolescence. Thus, results from pre-clinical rodent models point to a direct relationship between adolescent *Cannabis* use and heightened opiate consummatory behavior in adulthood.

Cocaine. In contrast to the consistent results reported on opiates, three studies investigating adolescent cannabinoid exposure on cocaine-triggered behaviors in adulthood and provided opposing results. In CP treated adolescent rats (0.4 mg/kg, i.p., 11 days), adult female but not male rats increased acquisition of cocaine SA.[140] In a study published decade later, this effect was repeated with THC in adolescents (1 mg/kg, i.p., 18 days) and importantly, only using lower cocaine doses as adults.[141] Finally, using mice, Gobira et al. (2021) found that adolescent exposure to WIN (3 mg/kg, 8 injections over 21 days) blunted CPP to cocaine (15 mg/kg, i.p.), a dose that is typically rewarding (Gobira et al., 2021). The authors also report an increase in PFC expression of a methyltransferase protein important in epigenetic regulation. Interestingly, this parallels another report which shows WIN-induced reduction in HDAC6,[142] suggesting both methylation and acetylation changes as a result adolescent WIN exposure. Thus, while there appear to be effects of cannabinoids in adolescence on later cocaine sensitivity/intake, it may be both sex and dose dependent.

Other drugs of abuse. Similar to cocaine, the effect of cannabinoid use during adolescence on other non-opiate drug reward behaviors in adulthood are less conclusive. When adolescent THC (escalating dose up to 12 mg/kg) was paired with adult nicotine self-administration, no effects were found on adult acquisition of nicotine SA, extinction, or reinstatement.[143] When adolescent mice were instead injected with WIN (0.2 mg/kg, 12 injections), this did enhance subsequent adult nicotine SA at lower doses of nicotine,[144] but this effect was seen only in males. The impact of THC use during adolescence on its use in adulthood also appears to be variable. One study indicated that adolescent THC exposure produced an increase in adult WIN SA,[142] and showed that this THC exposure lowered DA cell activity in the VTA as well as DA release in the NAc. However, another study published the same year found that THC exposure during adolescence did not alter subsequent place preference to THC itself.[145]

Taken together, two main conclusions may be drawn from this limited set of studies. The first is that adolescent exposure appears to enhance adult opioid intake, while its influence on other drug effects are unclear. However, the type of behavioral test may also be important, as CPP studies tended to produce more negative results compared to SA within each drug category. A second conclusion is that synthetic cannabinoid agonists (CP and WIN) have distinct pharmacological activity from THC, including their ability to both produce long-term changes in abuse liability and be abused themselves. This has extremely important ramifications considering the increase in artificial cannabinoid agonists that are illegally sold, such as *Spice*. If these potent CB_1R agonists are uniquely detrimental during adolescence because of their ability to alter future drug taking, further consideration of legal status is warranted.

Additional systems-level and behavioral responses to THC treatment during adolescence in rodents

Adolescent use of THC in rodents also produces significant changes in non-addiction models of anxiety and locomotor behavior suggesting potentially dose-dependent modifications to the systems level neural connectivity of the adult brain. Adolescent chronic exposure of male mice to THC (daily 3 mg/kg i.p. for 3 weeks) increases repetitive and compulsive-like behaviors, as measured by the Nestlet shredding task.[146] Chronic administration of THC, either during adolescence or during adulthood, leads to a delayed increase in anxiety as measured by the EPM. Adolescent male rats (PND 35–37) treated with escalating low-doses of THC for 10 days (twice daily, 0.3–3 mg/kg) show increased spontaneous open-field activity without affecting pre-pulse inhibition (PPI) and attentional set-shifting performance in adulthood (PND 75, i.e., 30-day interval).[129] Specifically, a study investigating acute THC use in adolescence report that adolescent mice responding to THC do not show significant locomotor impairments or anxiogenic behaviors in adulthood.[147] One study showed that a combination of formalin-induced chronic pain and adolescent THC exposure impaired sociability in mice, a behavior not observed from either stimulant alone.[148] Locomotion and sociability activity of adolescent male and female rats was also not affected 24 h after a single dose of WIN given that sex-dependent differences were noticeable in female adult rats.[149] Here it is important to emphasize that their THC exposure paradigm was shorter (single dose of WIN) than studies showing negative consequences such as impaired recognition memory in adult mice after 3 weeks of 8 mg/kg THC exposure per day in adolescence.[150] Another set of studies with a longer THC exposure paradigm (3 weeks of daily 3 mg/kg exposure) showed cognitive and behavioral dysfunction in adulthood as measured by deficits in working memory and novel object recognition behavioral tasks.[146,151]

At the mechanistic level, a recent study shows that the behavioral changes measured in adulthood resulting from THC treatment of adolescent mice depends on the expression of reelin, a signaling protein cardinally implicated in brain development implicated in psychiatric disorders when its function is impaired. Specifically, mice (PND 28) injected daily with THC (10 mg/kg, i.p.) for 3 weeks and studied 2 weeks after the last injection exhibited sex-dependent and reelin expression dependent impaired social behaviors as determined by measuring working memory, social interaction, locomotor activity.[152] THC treated adolescent mice heterozygote for reelin also exhibited elevated disinhibitory phenotypes as measured by anxiety-like responses and stress reactivity, and increased reactivity to aversive situations as measuring using PPI of acoustic startle.[152]

Long-term 'two hit' behavioral effects of chronic young-adult treatment with CP (0.2 mg/kg, 8–10 weeks of age) in combination with maternal separation (3 h every day from post-natal days 2–14) was studied in male and female rats.[153] In male rats, the combination of maternal separation and CP exposure decreased sucrose preference and time spent on the open arms of the EPM. By contrast, no effect was detected on PPI, memory performance in the Y-maze and novel object recognition. This difference in behavioral consequences due to varied THC exposure paradigms may point to a unique vulnerability of the mPFC during this adolescent period due to its complexity and developmental complexity and timeline.

Together, these studies paint a picture of diverse behavioral modifications in anxiety, sociability, locomotion, and sucrose preference that are initiated by adolescent exposure of THC. The diversity of behaviors impacted in adulthood suggests that THC's bioactivity during neurodevelopment is modifying a larger systems-level connectivity throughout the nervous system. Further investigations to understand the molecular, cellular, and systems-level effects of adolescent THC exposure will contribute to deciphering the alterations in region-to-region connectivity.

THC self-administration by adolescent rodents

Few pre-clinical approaches are available to study the health impact of voluntary consumption *Cannabis*-based products in rodents. The primary route of administration of such products by humans is inhalation through smoking and more recently vaporizing and dabbing; however, consumption of edibles infused with *Cannabis* extracts and cannabinoids is rapidly increasing[154] (Fig. 3). Of particular concern, human adolescents classified as frequent users of THC-infused edibles report both a younger age of first use compared to users who never consumed edibles and increased likelihood to have recently used *Cannabis*-based products.[155] Adolescents are particularly drawn to such products over traditional methods of using *Cannabis* given their strong appealing cues and packaging (e.g., brownies, lollipops, gummies, etc.), inconspicuous appearance, and perceived lower risk.[156,157] Given the increased prevalence of adolescent being exposed and consuming THC-containing edibles, research on the short-term and long-term consequences of this behavior is urgently needed.

The bioactivity of THC in pre-clinical rodent models has been studied using various routes of administration, including i.p., intravenous, subcutaneous, inhalation, solution drinking, and oral gavage.[146,154,158] However, a limitation of these approaches is that they do not study the effects of THC self-administration at doses that trigger cannabimimetic responses, and often require food and/or water-deprived conditions. In a recent effort to translationally represent oral self-administration of phyto-CBs by rodents, both gelatin- and dough-based formulations have been designed for behavioral analysis (Fig. 6). For example, voluntary oral consumption of THC-containing gelatin by rodents can be measured second-by-second using high-resolution, piezoelectric scales, and the amount of THC consumed by rodents over 1–2 h trigger reliable cannabimimetic responses.[159,160] Specifically, we found that adolescent rats self-administer 2–3 mg/kg of THC-gelatin over 1 h that results in 2–3 ng/mL THC in plasma THC levels measured at the end of the intake period, and significant cannabimimetic responses.[160] THC consumption by adolescent male rats and not female rats leads to impaired Pavlovian reward-predictive cue behaviors in adulthood consistent with a male-specific reduction in CB_1R expression by vGlut-1 synaptic terminals in the VTA.[160] Thus, THC-gelatin consumption by adolescent rats (PND 25–58) is associated with sex-dependent behavioral impairment in adulthood.[160] Consumption of THC-dough by adolescent mice (PND 26–38) impairs drug-free rotarod performance and reduces THC-triggered hypothermic responses at the end of consummatory exposure.[161] An important observation made with both of the gelatin and dough paradigms is that rodents consume less THC at higher doses, raising the possibility that there might be an avoidance to the taste/odor of THC and/or to the unwanted cannabimimetic behaviors experienced by the animal (suggesting self-titration).[160,161] Furthermore, adolescent rodents learn to consume THC gelatin faster if access is reduced over time, thus reaching similar consumption in less time and suggesting that they learn to eat faster.[160]

It is important to acknowledge that inhaled THC methods have also been implemented pre-clinically, providing a second ingestion method with human relevance (Fig. 6C). Typically, this involves either burning *Cannabis* cigarettes or vaporizing THC extract in an airtight environment similar to an operant chamber.[162–164] Several labs have used involuntary THC vapor exposure to elicit cannabimimetic responses in adult rodents, including effects on anxiety and locomotor behaviors.[165] Through continued exposure, THC dependence has also been shown.[166] More relevant to this section, reliable self-administration of THC vapor has also been demonstrated.[162] Rats were trained to nose poke for THC vapor and showed high rates of responding, as well as motivation to take vapor by measuring progressive ratio and extinction responses. In contrast to i.p. injection, inhaled cannabinoids show behavioral responses more consistent with human behavior, such as increased food intake.[163] Thus, pre-clinical models are beginning to both represent human consumption patterns more accurately and provide relevant PK and PD profiles of cannabis exposure and consumption patterns, and how this triggers cannabimimetic effects.

In summary, while most studies published thus far have established our understanding of the consequences of acute and prolonged THC exposure administered by the experimenter, only few studies exist on the effect of freely accessible *Cannabis*-containing and THC-containing products to adolescent rodents and its impact on behavior and brain function in adulthood, a paradigm that better recapitulates an increasingly predominant form of *Cannabis* use by human adolescents. Such studies offer a novel experimental paradigm that provide more relevant pre-clinical results when studying the biological effects at the molecular, cellular, systems levels, and the behavioral consequences in adolescence and later in adulthood resulting from THC self-administration. This work will provide a solid foundation for the development of therapeutics for the treatment of THC's impact on adolescent brain and its consequences in adulthood.

CBD interactions with THC during adolescence

Several studies have shown that the bioactivity of CBD is often biphasic, producing one set of behavioral changes at low doses and exhibiting a distinct bioactivity at high doses, including reducing seizures.[166] Specifically, high dose of CBD (approximately 300 mg/kg in both humans and rodents) proved to exhibit groundbreaking anti-seizure activity for the treatment of early-onset epilepsies that develop in adolescent patients with Dravet syndrome and Lennox-Gaustat; and this treatment regimen exhibits a promising safety profile characterized by mild side-effects (somnolence, decreased appetite, diarrhea, and fatigue).[57,58,167] These studies emphasize the promising safety profile of CBD even at relatively high doses. One also need to consider non-medicinal preparations often referred to as "cannabis light" that contain CBD (e.g., 1–30 mg/kg) and low levels of THC (e.g., 0.2%–1%), products that are legally available in select countries.[103,168] Thus, one approach to study the mitigating effect of CBD on THC's bioactivity is to consider THC/CBD ratios currently used by human adolescents. For example, THC/CBD ratios of 3 and 0.33 are reminiscent of THC-rich/CBD-poor and CBD-rich/THC-poor cannabis-based product that were confiscated in the illegal market.[103,169]

To our knowledge, the first study reporting the mitigating effect of CBD on THC use in adolescent rodents was published in 2017 by the laboratory of Dr. Kenneth Mackie. Adolescent mice were treated for 15 days (PND 28–48) with daily i.p. injections of either THC (3 mg/kg), CBD (3 mg/kg) or both.[146] THC triggered immediate and long-term impairments in working memory measured by the novel object recognition task, as well as in increased adulthood anxiety measured on the EPM, and increased repetitive and compulsive-like behaviors measured with the Nestlet shredding task and marble burying.[146] All THC-induced behavioral abnormalities were prevented by the coadministration of CBD+THC, and CBD alone did not influence behavioral outcomes.[146] Additionally, a study investigated the potential effects of CBD to mitigate the negative anxiogenic effects experienced from THC, finding that a single co-administration of CBD (3 mg/kg) alleviates the anxiogenic behavior produced from THC (1 mg/kg).[170]

A recent study in adolescent female rats (PND 35–45) explored the impact of 15 days treatment with twice daily i.p. injections of increasing doses of THC (2.5, 5, and 10 mg/kg), CBD (0.8, 1.7, and 3.3 mg/kg), or of THC:CBD (1:1).[103] This paradigm results in adulthood impaired emotional behaviors measured by the swim test, sucrose intake, palatable food intake, and EPM; as well as deviations in social interaction behaviors measured with short-term recognition memory, and impaired memory measured with novel object recognition.[103] These behavioral changes correlate with molecular and cellular changes in the PFC, including reduction in CB_1R and Glutamic Acid Decarboxylase-67 (GAD67) expression measured by Western blot, increased CD11b expression measured by Western blot, and microglial cell activation measured by changes in cell morphology analyzed by IHC.[103] These results suggest that PFC activity in adulthood is impacted by changing the balance between excitatory and inhibitory neurotransmission and concomitant neuroinflammation. Here too, CBD mitigates some of the long-term behavioral alterations induced by adolescent THC exposure as well as the changes in long-term changes in PFC molecular components.[103]

In summary, these studies provide convincing pre-clinical evidence that the impact associated with repeated administration of THC during adolescence is lessened by CBD. The molecular mechanism by which CBD lessens THC's impact on adolescent brain is unknown and needs to be urgently studied as it might offer promising molecular targets to develop therapeutics for the treatment of, for example, CUD, that exhibit good safety profiles.

Conclusion

The significant number of recent peer-reviewed studies in humans and pre-clinical rodent models has provided unequivocal evidence that the use of high-content THC products by adolescences may have a detrimental impact on adolescent brain function and development, and that this impact may persist into adulthood. This new evidence stems from the large increase in THC amounts delivered by recently developed devices and products. An important question to address here is whether some adolescents are more vulnerable to THC than others, and if this is the case, what are the molecular bases of such vulnerability.

Our knowledge of molecular and cellular mechanism involved in brain development indicates that developing mesocortolimbic brain structures are particularly vulnerable to THC, most likely because CB_1R are involved in this process, including by regulating neurotransmission, establishing neuronal connections, regulating metabolism, and establishing neuronal phenotypes. Accordingly, THC use during adolescence was found to trigger irregular dendrite growth in medium spiny neurons and premature cortical thinning in adolescent humans.[98,171] Another key element to consider when studying vulnerability is that there are fundamental sex-differences in the response to THC, as emphasized by pre-clinical studies showing distinct behavioral impairments occurring in either males or females.[172,173] Thus, pre-clinical findings should be explored and interpreted in the context of human population as much as possible, especially to identify populations that may be uniquely at risk from THC impact of adolescent brain.

What is currently missing in our field of research? Considering the rapid development of new devices and diverse products, we need to continue developing innovative methods and experimental paradigms to study voluntary self-administration consumption of high-THC products, including voluntary vaporization chambers, dabbing devices, and oral consumption. This will provide a greater degree of face validity and allow us to make better inferences to the human population. Furthermore, the diversity in content of *Cannabis*-based products that are now available illustrates that we need to study the effect of: *Cannabis* extracts, phytoCBs combinations, combination of phytoCB and terpenes, artificial cannabinoids, and recently rediscovered potent cannabinoids, such as Δ^8-THC and Δ^{10}-THC.[174,175] An exciting area of research that is germinating is the study of the potential mitigating effect of CBD on the bioactivity of THC. Thus, we need to better understand the pharmacology of this phytoCB and its analogues, including the molecular targets that it modulates, the endogenous signaling systems it modulates, and the cellular and system level response on the developing brain. For example, recent CBD formulation have been developed to ovoid first pass liver metabolism and enhance dosing and bioactivity onset.[176] Another example is the limited understanding of the enzymes that produce and inactivate LPI, the endogenous ligand at GPR55, and whether pharmacologically modulating the action of LPI at GPR55 might mimic the therapeutic properties of CBD. Such endeavor would not only help the development of innovative and safe therapeutic approaches for the treatment of, for example, CUD, but also guide rationale policy decisions regarding percentages of THC versus CBD in *Cannabis*-based products that are sold commercially.

Considering the recent legalization of selling and using *Cannabis*-based products by several countries, there is an urgent need to gather ongoing data on the THC:CBD contents in such products, the frequency of their use, and the behavioral and psychiatric outcomes resulting from acute and chronic use, and the population using these products. Thus, there is a need to implement a monitory system (for example, inspired by the monitoring systems implemented for alcohol and nicotine use) that will provide accurate data to both analyze these questions, empower scientific research aimed at studying the health impact of this change in human behavior, foster the development of possible groundbreaking therapeutic approaches and educate human adolescents on the epidemiology of the use of *Cannabis*-based products.

In conclusion, the evidence discussed above convincing show that the use of high THC-containing products during adolescence impacts brain development and behaviors in adulthood. To mitigate persistence of individuals developing CUD and other mental health disorders, adolescents should be limited in their access to such products until their full side-effects are better understood.

References

1. Smart R, Caulkins JP, Kilmer B, Davenport S, Midgette G. Variation in cannabis potency and prices in a newly legal market: evidence from 30 million cannabis sales in Washington state. *Addiction* 2017;**112**:2167–77.
2. Hurd YL, Manzoni OJ, Pletnikov MV, Lee FS, Bhattacharyya S, Melis M. Cannabis and the developing brain: insights into its long-lasting effects. *J Neurosci* 2019;**39**:8250–8.
3. Le Merre P, Ährlund-Richter S, Carlén M. The mouse prefrontal cortex: unity in diversity. *Neuron* 2021;**109**:1925–44.
4. White FJ. Synaptic regulation of mesocorticolimbic dopamine neurons. *Annu Rev Neurosci* 1996;**19**:405–36.
5. Di Forti M, Quattrone D, Freeman TP, Tripoli G, Gayer-Anderson C, Quigley H, et al. The contribution of cannabis use to variation in the incidence of psychotic disorder across Europe (EU-GEI): a multicentre case-control study. *Lancet Psychiatry* 2019;**6**:427–36.
6. Di Forti M, Sallis H, Allegri F, Trotta A, Ferraro L, Stilo SA, et al. Daily use, especially of high-potency cannabis, drives the earlier onset of psychosis in cannabis users. *Schizophr Bull* 2014;**40**:1509–17.
7. Pierre JM, Gandal M, Son M. Cannabis-induced psychosis associated with high potency "wax dabs". *Schizophr Res* 2016;**172**:211–2.

8. Van der Steur SJ, Batalla A, Bossong MG. Factors moderating the association between cannabis use and psychosis risk: a systematic review. *Brain Sci* 2020;**10**:97.
9. Ryz NR, Remillard DJ, Russo EB. Cannabis roots: a traditional therapy with future potential for treating inflammation and pain. *Cannabis Cannabinoid Res* 2017;**2**:210–6.
10. Hollister LE, Gillespie H. Delta-8-and delta-9-tetrahydrocannabinol; Comparison in man by oral and intravenous administration. *Clin Pharmacol Ther* 1973;**14**:353–7.
11. Tart CT. Marijuana intoxication: common experiences. *Nature* 1970;**226**:701–4.
12. Grotenhermen F, Russo E, Zuardi AW. Even high doses of oral cannabidiol do not cause THC-like effects in humans: comment on Merrick et al. *Cannabis Cannabinoid Res* 2017;**1**(1):102–12. https://doi.org/10.1089/can.2015.0004 [Cannabis and Cannabinoid Research 2, 1–4].
13. Renard J, Rosen LG, Loureiro M, De Oliveira C, Schmid S, Rushlow WJ, et al. Adolescent cannabinoid exposure induces a persistent sub-cortical hyper-dopaminergic state and associated molecular adaptations in the prefrontal cortex. *Cereb Cortex* 2017;**27**:1297–310.
14. Todd S, Arnold J. Neural correlates of interactions between cannabidiol and Δ^9-tetrahydrocannabinol in mice: implications for medical cannabis. *Br J Pharmacol* 2016;**173**:53–65.
15. Rosenthaler S, Pöhn B, Kolmanz C, Huu CN, Krewenka C, Huber A, et al. Differences in receptor binding affinity of several phytocannabinoids do not explain their effects on neural cell cultures. *Neurotoxicol Teratol* 2014;**46**:49–56.
16. Turner SE, Williams CM, Iversen L, Whalley BJ. Molecular pharmacology of phytocannabinoids. *Prog Chem Org Nat Prod* 2017;**61**–101.
17. Karniol I, Carlini E. The content of (−) Δ^9-trans-tetrahydrocannabinol (Δ^9-THC) does not explain all biological activity of some Brazilian marihuana samples. *J Pharm Pharmacol* 1972;**24**:833–5.
18. Sexton M, Shelton K, Haley P, West M. Evaluation of cannabinoid and terpenoid content: cannabis flower compared to supercritical CO2 concentrate. *Planta Med* 2018;**84**:234–41.
19. Russo EB, Marcu J. Cannabis pharmacology: the usual suspects and a few promising leads. *Adv Pharmacol* 2017;**80**:67–134.
20. Geithe C, Noe F, Kreissl J, Krautwurst D. The broadly tuned odorant receptor OR1A1 is highly selective for 3-methyl-2, 4-nonanedione, a key food odorant in aged wines, tea, and other foods. *Chem Senses* 2017;**42**:181–93.
21. LaVigne JE, Hecksel R, Keresztes A, Streicher JM. Cannabis sativa terpenes are cannabimimetic and selectively enhance cannabinoid activity. *Sci Rep* 2021;**11**:1–15.
22. Sagar KA, Lambros AM, Dahlgren MK, Smith RT, Gruber SA. Made from concentrate? A national web survey assessing dab use in the United States. *Drug Alcohol Depend* 2018;**190**:133–42.
23. Sagar KA, Gruber SA. Marijuana matters: reviewing the impact of marijuana on cognition, brain structure and function, & exploring policy implications and barriers to research. *Int Rev Psychiatry* 2018;**30**:251–67.
24. Firth CL, Davenport S, Smart R, Dilley JA. How high: differences in the developments of cannabis markets in two legalized states. *Int J Drug Policy* 2020;**75**, 102611.
25. Devane WA, Dysarz FR, Johnson MR, Melvin LS, Howlett AC. Determination and characterization of a cannabinoid receptor in rat brain. *Mol Pharmacol* 1988;**34**:605–13.
26. Lutz B. Neurobiology of cannabinoid receptor signaling. *Dialogues Clin Neurosci* 2020;**22**:207.
27. Busquets-Garcia A, Bains J, Marsicano G. CB 1 receptor signaling in the brain: extracting specificity from ubiquity. *Neuropsychopharmacology* 2018;**43**:4–20.
28. Katona I, Freund TF. Endocannabinoid signaling as a synaptic circuit breaker in neurological disease. *Nat Med* 2008;**14**:923–30.

29. Laprairie R, Bagher A, Kelly M, Denovan-Wright E. Cannabidiol is a negative allosteric modulator of the cannabinoid CB1 receptor. *Br J Pharmacol* 2015;**172**:4790–805.
30. Bakas T, Van Nieuwenhuijzen P, Devenish S, McGregor I, Arnold J, Chebib M. The direct actions of cannabidiol and 2-arachidonoyl glycerol at GABAA receptors. *Pharmacol Res* 2017;**119**:358–70.
31. Iannotti FA, Hill CL, Leo A, Alhusaini A, Soubrane C, Mazzarella E, et al. Nonpsychotropic plant cannabinoids, cannabidivarin (CBDV) and cannabidiol (CBD), activate and desensitize transient receptor potential vanilloid 1 (TRPV1) channels in vitro: potential for the treatment of neuronal hyperexcitability. *ACS Chem Nerosci* 2014;**5**:1131–41.
32. Ross RA. The enigmatic pharmacology of GPR55. *Trends Pharmacol Sci* 2009;**30**:156–63.
33. Sharir H, Console-Bram L, Mundy C, Popoff SN, Kapur A, Abood ME. The endocannabinoids anandamide and virodhamine modulate the activity of the candidate cannabinoid receptor GPR55. *J Neuroimmune Pharmacol* 2012;**7**:856–65.
34. Stollenwerk TM, Pollock S, Hillard CJ. Contribution of the adenosine 2A receptor to behavioral effects of tetrahydrocannabinol, cannabidiol and PECS-101. *Molecules* 2021;**26**:5354.
35. Atwood BK, Huffman J, Straiker A, Mackie K. JWH018, a common constituent of 'Spice' herbal blends, is a potent and efficacious cannabinoid CB1 receptor agonist. *Br J Pharmacol* 2010;**160**:585–93.
36. Theunissen EL, Hutten NR, Mason NL, Toennes SW, Kuypers KP, Ramaekers JG. Neurocognition and subjective experience following acute doses of the synthetic cannabinoid JWH-018: responders versus nonresponders. *Cannabis Cannabinoid Res* 2019;**4**:51–61.
37. Dalton VS, Zavitsanou K. Cannabinoid effects on CB1 receptor density in the adolescent brain: an autoradiographic study using the synthetic cannabinoid HU210. *Synapse* 2010;**64**:845–54.
38. Cao JK, Kaplan J, Stella N. ABHD6: its place in endocannabinoid signaling and beyond. *Trends Pharmacol Sci* 2019;**40**:267–77.
39. Stella N. Cannabinoid and cannabinoid-like receptors in microglia, astrocytes, and astrocytomas. *Glia* 2010;**58**:1017–30.
40. Bisogno T, Howell F, Williams G, Minassi A, Cascio MG, Ligresti A, et al. Cloning of the first sn1-DAG lipases points to the spatial and temporal regulation of endocannabinoid signaling in the brain. *J Cell Biol* 2003;**163**:463–8.
41. Di Marzo V, Fontana A, Cadas H, Schinelli S, Cimino G, Schwartz J-C, et al. Formation and inactivation of endogenous cannabinoid anandamide in central neurons. *Nature* 1994;**372**:686–91.
42. Gao Y, Vasilyev DV, Goncalves MB, Howell FV, Hobbs C, Reisenberg M, et al. Loss of retrograde endocannabinoid signaling and reduced adult neurogenesis in diacylglycerol lipase knock-out mice. *J Neurosci* 2010;**30**:2017–24.
43. Leishman E, Mackie K, Luquet S, Bradshaw HB. Lipidomics profile of a NAPE-PLD KO mouse provides evidence of a broader role of this enzyme in lipid metabolism in the brain. *Biochim Biophys Acta (BBA) Mol Cell Biol Lipids* 2016;**1861**:491–500.
44. Stella N, Piomelli D. Receptor-dependent formation of endogenous cannabinoids in cortical neurons. *Eur J Pharmacol* 2001;**425**:189–96.
45. Stella N, Schweitzer P, Piomelli D. A second endogenous cannabinoid that modulates long-term potentiation. *Nature* 1997;**388**:773–8.
46. Ohno-Shosaku T, Kano M. Endocannabinoid-mediated retrograde modulation of synaptic transmission. *Curr Opin Neurobiol* 2014;**29**:1–8.
47. Guo J, Ikeda SR. Endocannabinoids modulate N-type calcium channels and G-protein-coupled inwardly rectifying potassium channels via CB1 cannabinoid receptors heterologously expressed in mammalian neurons. *Mol Pharmacol* 2004;**65**:665–74.

48. Cravatt BF, Demarest K, Patricelli MP, Bracey MH, Giang DK, Martin BR, et al. Supersensitivity to anandamide and enhanced endogenous cannabinoid signaling in mice lacking fatty acid amide hydrolase. *Proc Natl Acad Sci U S A* 2001;**98**:9371–6.
49. Fu J, Bottegoni G, Sasso O, Bertorelli R, Rocchia W, Masetti M, et al. A catalytically silent FAAH-1 variant drives anandamide transport in neurons. *Nat Neurosci* 2012;**15**:64–9.
50. McKinney MK, Cravatt BF. Structure and function of fatty acid amide hydrolase. *Annu Rev Biochem* 2005;**74**:411–32.
51. Van Sickle MD, Duncan M, Kingsley PJ, Mouihate A, Urbani P, Mackie K, et al. Identification and functional characterization of brainstem cannabinoid CB2 receptors. *Science* 2005;**310**:329–32.
52. Lauckner JE, Jensen JB, Chen H-Y, Lu H-C, Hille B, Mackie K. GPR55 is a cannabinoid receptor that increases intracellular calcium and inhibits M current. *Proc Natl Acad Sci U S A* 2008;**105**:2699–704.
53. Wu CS, Zhu J, Wager-Miller J, Wang S, O'Leary D, Monory K, et al. Requirement of cannabinoid CB1 receptors in cortical pyramidal neurons for appropriate development of corticothalamic and thalamocortical projections. *Eur J Neurosci* 2010;**32**:693–706.
54. Foster SR, Hauser AS, Vedel L, Strachan RT, Huang X-P, Gavin AC, et al. Discovery of human signaling systems: pairing peptides to G protein-coupled receptors. *Cell* 2019;**179**. 895–908.e821.
55. Lingerfelt MA, Zhao P, Sharir HP, Hurst DP, Reggio PH, Abood ME. Identification of crucial amino acid residues involved in agonist signaling at the GPR55 receptor. *Biochemistry* 2017;**56**:473–86.
56. Chuang S-H, Westenbroek RE, Stella N, Catterall WA. Combined antiseizure efficacy of cannabidiol and clonazepam in a conditional mouse model of Dravet syndrome. *J Exp Neurol* 2021;**2**:81.
57. Devinsky O, Cross JH, Laux L, Marsh E, Miller I, Nabbout R, et al. Trial of cannabidiol for drug-resistant seizures in the Dravet syndrome. *N Engl J Med* 2017;**376**:2011–20.
58. Kaplan JS, Stella N, Catterall WA, Westenbroek RE. Cannabidiol attenuates seizures and social deficits in a mouse model of Dravet syndrome. *Proc Natl Acad Sci U S A* 2017;**114**:11229–34.
59. Sylantyev S, Jensen TP, Ross RA, Rusakov DA. Cannabinoid- and lysophosphatidylinositol-sensitive receptor GPR55 boosts neurotransmitter release at central synapses. *Proc Natl Acad Sci U S A* 2013;**110**:5193–8.
60. Grant KS, Petroff R, Isoherranen N, Stella N, Burbacher TM. Cannabis use during pregnancy: pharmacokinetics and effects on child development. *Pharmacol Ther* 2018;**182**:133–51.
61. Wu C-S, Jew CP, Lu H-C. Lasting impacts of prenatal cannabis exposure and the role of endogenous cannabinoids in the developing brain. *Future Neurol* 2011;**6**:459–80.
62. Anavi-Goffer S, Mulder J. The polarised life of the endocannabinoid system in CNS development. *Chembiochem* 2009;**10**:1591–8.
63. Harkany T, Guzman M, Galve-Roperh I, Berghuis P, Devi LA, Mackie K. The emerging functions of endocannabinoid signaling during CNS development. *Trends Pharmacol Sci* 2007;**28**:83–92.
64. Crews DE. Senescence, aging, and disease. *J Physiol Anthropol* 2007;**26**:365–72.
65. Calvigioni D, Hurd YL, Harkany T, Keimpema E. Neuronal substrates and functional consequences of prenatal cannabis exposure. *Eur Child Adolesc Psychiatry* 2014;**23**:931–41.
66. de Salas-Quiroga A, Díaz-Alonso J, García-Rincón D, Remmers F, Vega D, Gómez-Cañas M, et al. Prenatal exposure to cannabinoids evokes long-lasting functional alterations by targeting CB1 receptors on developing cortical neurons. *Proc Natl Acad Sci U S A* 2015;**112**:13693–8.

67. Marsicano G, Goodenough S, Monory K, Hermann H, Eder M, Cannich A, et al. CB1 cannabinoid receptors and on-demand defense against excitotoxicity. *Science* 2003;**302**:84–8.
68. Aguado T, Monory K, Palazuelos J, Stella N, Cravatt B, Lutz B, et al. The endocannabinoid system drives neural progenitor proliferation. *FASEB J* 2005;**19**:1704–6.
69. Aguado T, Palazuelos J, Monory K, Stella N, Cravatt B, Lutz B, et al. The endocannabinoid system promotes astroglial differentiation by acting on neural progenitor cells. *J Neurosci* 2006;**26**:1551–61.
70. Galve-Roperh I, Chiurchiù V, Díaz-Alonso J, Bari M, Guzmán M, Maccarrone M. Cannabinoid receptor signaling in progenitor/stem cell proliferation and differentiation. *Prog Lipid Res* 2013;**52**:633–50.
71. Berghuis P, Rajnicek AM, Morozov YM, Ross RA, Mulder J, Urbán GM, et al. Hardwiring the brain: endocannabinoids shape neuronal connectivity. *Science* 2007;**316**:1212–6.
72. Maccarrone M, Guzmán M, Mackie K, Doherty P, Harkany T. Programming of neural cells by (endo) cannabinoids: from physiological rules to emerging therapies. *Nat Rev Neurosci* 2014;**15**:786–801.
73. Tortoriello G, Morris CV, Alpar A, Fuzik J, Shirran SL, Calvigioni D, et al. Miswiring the brain: Δ^9-tetrahydrocannabinol disrupts cortical development by inducing an SCG 10/stathmin-2 degradation pathway. *EMBO J* 2014;**33**:668–85.
74. Cudaback E, Marrs W, Moeller T, Stella N. The expression level of CB1 and CB2 receptors determines their efficacy at inducing apoptosis in astrocytomas. *PLoS One* 2010;**5**, e8702.
75. Priestley R, Glass M, Kendall D. Functional selectivity at cannabinoid receptors. *Adv Pharmacol* 2017;**80**:207–21.
76. Passingham RE, Smaers JB. Is the prefrontal cortex especially enlarged in the human brain? Allometric relations and remapping factors. *Brain Behav Evol* 2014;**84**:156–66.
77. Smaers JB, Gómez-Robles A, Parks AN, Sherwood CC. Exceptional evolutionary expansion of prefrontal cortex in great apes and humans. *Curr Biol* 2017;**27**:714–20.
78. Archer T. Effects of exogenous agents on brain development: stress, abuse and therapeutic compounds. *CNS Neurosci Ther* 2011;**17**:470–89.
79. Kolb B, Mychasiuk R, Muhammad A, Li Y, Frost DO, Gibb R. Experience and the developing prefrontal cortex. *Proc Natl Acad Sci U S A* 2012;**109**:17186–93.
80. Macht VA. Neuro-immune interactions across development: a look at glutamate in the prefrontal cortex. *Neurosci Biobehav Rev* 2016;**71**:267–80.
81. Bristot G, De Bastiani MA, Pfaffenseller B, Kapczinski F, Kauer-Sant'Anna M. Gene regulatory network of dorsolateral prefrontal cortex: a master regulator analysis of major psychiatric disorders. *Mol Neurobiol* 2020;**57**:1305–16.
82. Caspi A, Moffitt TE, Cannon M, McClay J, Murray R, Harrington H, et al. Moderation of the effect of adolescent-onset cannabis use on adult psychosis by a functional polymorphism in the catechol-O-methyltransferase gene: longitudinal evidence of a gene X environment interaction. *Biol Psychiatry* 2005;**57**:1117–27.
83. Morgunova A, Flores C. MicroRNA regulation of prefrontal cortex development and psychiatric risk in adolescence. In: *Paper Presented at: Seminars in Cell & Developmental Biology*. Elsevier; 2021.
84. Bidwell LC, Ellingson JM, Karoly HC, YorkWilliams SL, Hitchcock LN, Tracy BL, et al. Association of naturalistic administration of cannabis flower and concentrates with intoxication and impairment. *JAMA Psychiat* 2020;**77**:787–96.
85. Whitehill JM, Dilley JA, Brooks-Russell A, Terpak L, Graves JM. Edible cannabis exposures among children: 2017–2019. *Pediatrics* 2021;**147**.

86. Bidwell LC, YorkWilliams SL, Mueller RL, Bryan AD, Hutchison KE. Exploring cannabis concentrates on the legal market: user profiles, product strength, and health-related outcomes. *Addict Behav Rep* 2018;**8**:102–6.
87. Davenport S. Price and product variation in Washington's recreational cannabis market. *Int J Drug Policy* 2019;, 102547.
88. Mennis J, Stahler GJ, McKeon TP. Young adult cannabis use disorder treatment admissions declined as past month cannabis use increased in the US: an analysis of states by year, 2008–2017. *Addict Behav* 2021;**123**, 107049.
89. Arterberry BJ, Padovano HT, Foster KT, Zucker RA, Hicks BM. Higher average potency across the United States is associated with progression to first cannabis use disorder symptom. *Drug Alcohol Depend* 2019;**195**:186–92.
90. Barrington-Trimis JL, Cho J, Ewusi-Boisvert E, Hasin D, Unger JB, Miech RA, et al. Risk of persistence and progression of use of 5 cannabis products after experimentation among adolescents. *JAMA Netw Open* 2020;**3**, e1919792.
91. Freeman T, Winstock A. Examining the profile of high-potency cannabis and its association with severity of cannabis dependence. *Psychol Med* 2015;**45**:3181–9.
92. Gunn RL, Aston ER, Sokolovsky AW, White HR, Jackson KM. Complex cannabis use patterns: associations with cannabis consequences and cannabis use disorder symptomatology. *Addict Behav* 2020;**105**, 106329.
93. Kelly BC, Vuolo M. Cognitive aptitude, peers, and trajectories of marijuana use from adolescence through young adulthood. *PLoS One* 2019;**14**, e0223152.
94. Gobbi G, Atkin T, Zytynski T, Wang S, Askari S, Boruff J, et al. Association of cannabis use in adolescence and risk of depression, anxiety, and suicidality in young adulthood: a systematic review and meta-analysis. *JAMA Psychiat* 2019;**76**:426–34.
95. Hines LA, Freeman TP, Gage SH, Zammit S, Hickman M, Cannon M, et al. Association of high-potency cannabis use with mental health and substance use in adolescence. *JAMA Psychiat* 2020;**77**:1044–51.
96. Gilman JM, Kuster JK, Lee S, Lee MJ, Kim BW, Makris N, et al. Cannabis use is quantitatively associated with nucleus accumbens and amygdala abnormalities in young adult recreational users. *J Neurosci* 2014;**34**:5529–38.
97. Camchong J, Lim KO, Kumra S. Adverse effects of cannabis on adolescent brain development: a longitudinal study. *Cereb Cortex* 2017;**27**:1922–30.
98. Albaugh MD, Ottino-Gonzalez J, Sidwell A, Lepage C, Juliano A, Owens MM, et al. Association of cannabis use during adolescence with neurodevelopment. *JAMA Psychiat* 2021;**78**:1–11.
99. Orr C, Spechler P, Cao Z, Albaugh M, Chaarani B, Mackey S, et al. Grey matter volume differences associated with extremely low levels of cannabis use in adolescence. *J Neurosci* 2019;**39**:1817–27.
100. Weinstein JJ, van de Giessen E, Rosengard RJ, Xu X, Ojeil N, Brucato G, et al. PET imaging of dopamine-D2 receptor internalization in schizophrenia. *Mol Psychiatry* 2018;**23**:1506–11.
101. Withey SL, Kangas BD, Charles S, Gumbert AB, Eisold JE, George SR, et al. Effects of daily Δ^9-tetrahydrocannabinol (THC) alone or combined with cannabidiol (CBD) on cognition-based behavior and activity in adolescent nonhuman primates. *Drug Alcohol Depend* 2021;**221**, 108629.
102. Freeman TP, Groshkova T, Cunningham A, Sedefov R, Griffiths P, Lynskey MT. Increasing potency and price of cannabis in Europe, 2006-16. *Addiction* 2019;**114**:1015–23.
103. Gabaglio M, Zamberletti E, Manenti C, Parolaro D, Rubino T. Long-term consequences of adolescent exposure to THC-rich/CBD-poor and CBD-rich/THC-poor combinations: a comparison with pure THC treatment in female rats. *Int J Mol Sci* 2021;**22**:8899.

104. Wiley JL, Burston JJ. Sex differences in Δ^9-tetrahydrocannabinol metabolism and in vivo pharmacology following acute and repeated dosing in adolescent rats. *Neurosci Lett* 2014;**576**:51–5.
105. Narimatsu S, Watanabe K, Yamamoto I, Yoshimura H. Sex difference in the oxidative metabolism of Δ^9-tetrahydrocannabinol in the rat. *Biochem Pharmacol* 1991;**41**:1187–94.
106. Bonhomme-Faivre L, Benyamina A, Reynaud M, Farinotti R, Abbara C. PRECLINICAL STUDY: disposition of Δ^9 tetrahydrocannabinol in CF1 mice deficient in mdr1a P-glycoprotein. *Addict Biol* 2008;**13**:295–300.
107. Spiro AS, Wong A, Boucher AA, Arnold JC. Enhanced brain disposition and effects of Δ^9-tetrahydrocannabinol in P-glycoprotein and breast cancer resistance protein knockout mice. *PLoS One* 2012;**7**, e35937.
108. Torrens A, Vozella V, Huff H, McNeil B, Ahmed F, Ghidini A, et al. Comparative pharmacokinetics of Δ^9-tetrahydrocannabinol in adolescent and adult male mice. *J Pharmacol Exp Ther* 2020;**374**:151–60.
109. Ruiz CM, Torrens A, Castillo E, Perrone CR, Cevallos J, Inshishian VC, et al. Pharmacokinetic, behavioral, and brain activity effects of Δ^9-tetrahydrocannabinol in adolescent male and female rats. *Neuropsychopharmacology* 2021;**46**:959–69.
110. Metna-Laurent M, Mondésir M, Grel A, Vallée M, Piazza PV. Cannabinoid-induced tetrad in mice. *Curr Protoc Neurosci* 2017;**80**. 9.59. 51-59.59. 10.
111. Wiley JL, Martin BR. Cannabinoid pharmacological properties common to other centrally acting drugs. *Eur J Pharmacol* 2003;**471**:185–93.
112. Calabrese EJ, Rubio-Casillas A. Biphasic effects of THC in memory and cognition. *Eur J Clin Invest* 2018;**48**, e12920.
113. Dar MS. Cerebellar CB1 receptor mediation of Δ9-THC-induced motor incoordination and its potentiation by ethanol and modulation by the cerebellar adenosinergic A1 receptor in the mouse. *Brain Res* 2000;**864**:186–94.
114. Saravia R, Ten-Blanco M, Julià-Hernández M, Gagliano H, Andero R, Armario A, et al. Concomitant THC and stress adolescent exposure induces impaired fear extinction and related neurobiological changes in adulthood. *Neuropharmacology* 2019;**144**:345–57.
115. Tournier BB, Ginovart N. Repeated but not acute treatment with Δ^9-tetrahydrocannabinol disrupts prepulse inhibition of the acoustic startle: reversal by the dopamine D2/3 receptor antagonist haloperidol. *Eur Neuropsychopharmacol* 2014;**24**:1415–23.
116. Harkany T, Keimpema E, Barabás K, Mulder J. Endocannabinoid functions controlling neuronal specification during brain development. *Mol Cell Endocrinol* 2008;**286**:S84–90.
117. Harkany T, Mackie K, Doherty P. Wiring and firing neuronal networks: endocannabinoids take center stage. *Curr Opin Neurobiol* 2008;**18**:338–45.
118. Alpár A, Tortoriello G, Calvigioni D, Niphakis MJ, Milenkovic I, Bakker J, et al. Endocannabinoids modulate cortical development by configuring Slit2/Robo1 signalling. *Nat Commun* 2014;**5**:1–13.
119. Díaz-Alonso J, Aguado T, Wu C-S, Palazuelos J, Hofmann C, Garcez P, et al. The CB1 cannabinoid receptor drives corticospinal motor neuron differentiation through the Ctip2/Satb2 transcriptional regulation axis. *J Neurosci* 2012;**32**:16651–65.
120. Renard J, Norris C, Rushlow W, Laviolette SR. Neuronal and molecular effects of cannabidiol on the mesolimbic dopamine system: implications for novel schizophrenia treatments. *Neurosci Biobehav Rev* 2017;**75**:157–65.
121. Renard J, Szkudlarek HJ, Kramar CP, Jobson CE, Moura K, Rushlow WJ, et al. Adolescent THC exposure causes enduring prefrontal cortical disruption of GABAergic inhibition and dysregulation of sub-cortical dopamine function. *Sci Rep* 2017;**7**:1–14.

122. Chen R, Zhang J, Fan N, Teng Z-Q, Wu Y, Yang H, et al. Δ9-THC-caused synaptic and memory impairments are mediated through COX-2 signaling. *Cell* 2013;**155**:1154–65.
123. Ferguson BR, Gao W-J. PV interneurons: critical regulators of E/I balance for prefrontal cortex-dependent behavior and psychiatric disorders. *Front Neural Circuits* 2018;**12**:37.
124. De Giacomo V, Ruehle S, Lutz B, Häring M, Remmers F. Cell type-specific genetic reconstitution of CB1 receptor subsets to assess their role in exploratory behaviour, sociability, and memory. *Eur J Neurosci* 2020;**55**.
125. Fortin DA, Levine ES. Differential effects of endocannabinoids on glutamatergic and GABAergic inputs to layer 5 pyramidal neurons. *Cereb Cortex* 2007;**17**:163–74.
126. Heng L, Beverley JA, Steiner H, Tseng KY. Differential developmental trajectories for CB1 cannabinoid receptor expression in limbic/associative and sensorimotor cortical areas. *Synapse* 2011;**65**:278–86.
127. Caballero A, Granberg R, Tseng KY. Mechanisms contributing to prefrontal cortex maturation during adolescence. *Neurosci Biobehav Rev* 2016;**70**:4–12.
128. Den Boon FS, Werkman TR, Schaafsma-Zhao Q, Houthuijs K, Vitalis T, Kruse CG, et al. Activation of type-1 cannabinoid receptor shifts the balance between excitation and inhibition towards excitation in layer II/III pyramidal neurons of the rat prelimbic cortex. *Pflugers Arch* 2015;**467**:1551–64.
129. Poulia N, Delis F, Brakatselos C, Polissidis A, Koutmani Y, Kokras N, et al. Detrimental effects of adolescent escalating low-dose Δ^9-tetrahydrocannabinol leads to a specific bio-behavioural profile in adult male rats. *Br J Pharmacol* 2021;**178**:1722–36.
130. Rubino T, Prini P, Piscitelli F, Zamberletti E, Trusel M, Melis M, et al. Adolescent exposure to THC in female rats disrupts developmental changes in the prefrontal cortex. *Neurobiol Dis* 2015;**73**:60–9.
131. Cass DK, Flores-Barrera E, Thomases DR, Vital WF, Caballero A, Tseng KY. CB1 cannabinoid receptor stimulation during adolescence impairs the maturation of GABA function in the adult rat prefrontal cortex. *Mol Psychiatry* 2014;**19**:536–43.
132. Pickel VM, Bourie F, Chan J, Mackie K, Lane DA, Wang G. Chronic adolescent exposure to Δ^9-tetrahydrocannabinol decreases NMDA current and extrasynaptic plasmalemmal density of NMDA GluN1 subunits in the prelimbic cortex of adult male mice. *Neuropsychopharmacology* 2020;**45**:374–83.
133. Ellgren M, Artmann A, Tkalych O, Gupta A, Hansen H, Hansen S, et al. Dynamic changes of the endogenous cannabinoid and opioid mesocorticolimbic systems during adolescence: THC effects. *Eur Neuropsychopharmacol* 2008;**18**:826–34.
134. Dong A, He K, Dudok B, Farrell JS, Guan W, Liput DJ, et al. A fluorescent sensor for spatiotemporally resolved endocannabinoid dynamics in vitro and in vivo. *bioRxiv* 2020. 2020. 2010.2008.329169.
135. Farrell JS, Colangeli R, Dong A, George AG, Addo-Osafo K, Kingsley PJ, et al. In vivo endocannabinoid dynamics at the timescale of physiological and pathological neural activity. *Neuron* 2021;**109**. 2398–2403.e2394.
136. Ellgren M, Spano SM, Hurd YL. Adolescent cannabis exposure alters opiate intake and opioid limbic neuronal populations in adult rats. *Neuropsychopharmacology* 2007;**32**:607–15.
137. Tomasiewicz HC, Jacobs MM, Wilkinson MB, Wilson SP, Nestler EJ, Hurd YL. Proenkephalin mediates the enduring effects of adolescent cannabis exposure associated with adult opiate vulnerability. *Biol Psychiatry* 2012;**72**:803–10.
138. Stopponi S, Soverchia L, Ubaldi M, Cippitelli A, Serpelloni G, Ciccocioppo R. Chronic THC during adolescence increases the vulnerability to stress-induced relapse to heroin seeking in adult rats. *Eur Neuropsychopharmacol* 2014;**24**:1037–45.

139. Lecca D, Scifo A, Pisanu A, Valentini V, Piras G, Sil A, et al. Adolescent cannabis exposure increases heroin reinforcement in rats genetically vulnerable to addiction. *Neuropharmacology* 2020;**166**, 107974.
140. Higuera-Matas A, Soto-Montenegro ML, Del Olmo N, Miguéns M, Torres I, Vaquero JJ, et al. Augmented acquisition of cocaine self-administration and altered brain glucose metabolism in adult female but not male rats exposed to a cannabinoid agonist during adolescence. *Neuropsychopharmacology* 2008;**33**:806–13.
141. Friedman AL, Meurice C, Jutkiewicz EM. Effects of adolescent Δ^9-tetrahydrocannabinol exposure on the behavioral effects of cocaine in adult Sprague-Dawley rats. *Exp Clin Psychopharmacol* 2019;**27**:326.
142. Scherma M, Dessì C, Muntoni AL, Lecca S, Satta V, Luchicchi A, et al. Adolescent Δ^9-tetrahydrocannabinol exposure alters WIN55, 212-2 self-administration in adult rats. *Neuropsychopharmacology* 2016;**41**:1416–26.
143. Flores Á, Maldonado R, Berrendero F. THC exposure during adolescence does not modify nicotine reinforcing effects and relapse in adult male mice. *Psychopharmacology (Berl)* 2020;**237**:801–9.
144. Dukes AJ, Fowler JP, Lallai V, Pushkin AN, Fowler CD. Adolescent cannabinoid and nicotine exposure differentially alters adult nicotine self-administration in males and females. *Nicotine Tob Res* 2020;**22**:1364–73.
145. Hempel BJ, Wakeford AG, Clasen MM, Friar MA, Riley AL. Delta-9-tetrahydrocannabinol (THC) history fails to affect THC's ability to induce place preferences in rats. *Pharmacol Biochem Behav* 2016;**144**:1–6.
146. Murphy M, Mills S, Winstone J, Leishman E, Wager-Miller J, Bradshaw H, et al. Chronic adolescent Δ^9-tetrahydrocannabinol treatment of male mice leads to long-term cognitive and behavioral dysfunction, which are prevented by concurrent cannabidiol treatment. *Cannabis Cannabinoid Res* 2017;**2**:235–46.
147. Kasten CR, Zhang Y, Boehm SL. Acute cannabinoids produce robust anxiety-like and locomotor effects in mice, but long-term consequences are age-and sex-dependent. *Front Behav Neurosci* 2019;**13**:32.
148. Tagne AM, Fotio Y, Springs ZA, Su S, Piomelli D. Frequent Δ^9-tetrahydrocannabinol exposure during adolescence impairs sociability in adult mice exposed to an aversive painful stimulus. *Eur Neuropsychopharmacol* 2021;**53**:19–24.
149. Borsoi M, Manduca A, Bara A, Lassalle O, Pelissier-Alicot A-L, Manzoni OJ. Sex differences in the Behavioral and synaptic consequences of a single in vivo exposure to the synthetic cannabimimetic WIN55, 212-2 at puberty and adulthood. *Front Behav Neurosci* 2019;**13**:23.
150. Jouroukhin Y, Zhu X, Shevelkin AV, Hasegawa Y, Abazyan B, Saito A, et al. Adolescent Δ^9-tetrahydrocannabinol exposure and astrocyte-specific genetic vulnerability converge on nuclear factor-κB-cyclooxygenase-2 signaling to impair memory in adulthood. *Biol Psychiatry* 2019;**85**:891–903.
151. Chen H-T, Mackie K. Adolescent Δ^9-tetrahydrocannabinol exposure selectively impairs working memory but not several other mPFC-mediated behaviors. *Front Psych* 2020;**11**.
152. Iemolo A, Montilla-Perez P, Nguyen J, Risbrough VB, Taffe MA, Telese F. Reelin deficiency contributes to long-term behavioral abnormalities induced by chronic adolescent exposure to Δ^9-tetrahydrocannabinol in mice. *Neuropharmacology* 2021;**187**, 108495.
153. Klug M, van den Buuse M. Chronic cannabinoid treatment during young adulthood induces sex-specific behavioural deficits in maternally separated rats. *Behav Brain Res* 2012;**233**:305–13.

154. Barrus DG, Lefever TW, Wiley JL. Evaluation of reinforcing and aversive effects of voluntary Δ^9-tetrahydrocannabinol ingestion in rats. *Neuropharmacology* 2018;**137**:133–40.
155. Peters EN, Bae D, Barrington-Trimis JL, Jarvis BP, Leventhal AM. Prevalence and sociodemographic correlates of adolescent use and polyuse of combustible, vaporized, and edible cannabis products. *JAMA Netw Open* 2018;**1**, e182765.
156. Hammond CJ, Chaney A, Hendrickson B, Sharma P. Cannabis use among US adolescents in the era of marijuana legalization: a review of changing use patterns, comorbidity, and health correlates. *Int Rev Psychiatry* 2020;**32**:221–34.
157. Knapp AA, Lee DC, Borodovsky JT, Auty SG, Gabrielli J, Budney AJ. Emerging trends in cannabis administration among adolescent cannabis users. *J Adolesc Health* 2019;**64**:487–93.
158. Dow-Edwards D, Zhao N. Oral THC produces minimal behavioral alterations in preadolescent rats. *Neurotoxicol Teratol* 2008;**30**:385–9.
159. Abraham AD, Leung EJ, Wong BA, Rivera ZM, Kruse LC, Clark JJ, et al. Orally consumed cannabinoids provide long-lasting relief of allodynia in a mouse model of chronic neuropathic pain. *Neuropsychopharmacology* 2020;**45**:1105–14.
160. Kruse LC, Cao JK, Viray K, Stella N, Clark JJ. Voluntary oral consumption of Δ^9-tetrahydrocannabinol by adolescent rats impairs reward-predictive cue behaviors in adulthood. *Neuropsychopharmacology* 2019;**44**:1406–14.
161. Smoker MP, Hernandez M, Zhang Y, Boehm 2nd SL. Assessment of acute motor effects and tolerance following self-administration of alcohol and edible (9)-tetrahydrocannabinol in adolescent male mice. *Alcohol Clin Exp Res* 2019;**43**:2446–57.
162. Freels TG, Baxter-Potter LN, Lugo JM, Glodosky NC, Wright HR, Baglot SL, et al. Vaporized cannabis extracts have reinforcing properties and support conditioned drug-seeking behavior in rats. *J Neurosci* 2020;**40**:1897–908.
163. Manwell LA, Charchoglyan A, Brewer D, Matthews BA, Heipel H, Mallet PE. A vaporized Δ^9-tetrahydrocannabinol (Δ^9-THC) delivery system part I: development and validation of a pulmonary cannabinoid route of exposure for experimental pharmacology studies in rodents. *J Pharmacol Toxicol Methods* 2014;**70**:120–7.
164. Taffe MA, Creehan KM, Vandewater SA, Kerr TM, Cole M. Effects of Δ^9-tetrahydrocannabinol (THC) vapor inhalation in Sprague-Dawley and Wistar rats. *Exp Clin Psychopharmacol* 2020;**29**:1–13.
165. Bruijnzeel AW, Qi X, Guzhva LV, Wall S, Deng JV, Gold MS, et al. Behavioral characterization of the effects of cannabis smoke and anandamide in rats. *PLoS One* 2016;**11**, e0153327.
166. Huizenga MN, Fureman BE, Soltesz I, Stella N. Proceedings of the Epilepsy Foundation's 2017 cannabinoids in epilepsy therapy workshop. *Epilepsy Behav* 2018;**85**:237–42.
167. Mandelbaum DE. Cannabidiol in patients with treatment-resistant epilepsy. *Lancet Neurol* 2016;**15**:544–5.
168. Small E. Evolution and classification of Cannabis sativa (marijuana, hemp) in relation to human utilization. *Bot Rev* 2015;**81**:189–294.
169. Hädener M, Gelmi TJ, Martin-Fabritius M, Weinmann W, Pfäffli M. Cannabinoid concentrations in confiscated cannabis samples and in whole blood and urine after smoking CBD-rich cannabis as a "tobacco substitute". *Int J Leg Med* 2019;**133**:821–32.
170. Salviato BZ, Raymundi AM, da Silva TR, Salemme BW, Sohn JMB, Araújo FS, et al. Female but not male rats show biphasic effects of low doses of Δ^9-tetrahydrocannabinol on anxiety: can cannabidiol interfere with these effects? *Neuropharmacology* 2021;**196**, 108684.
171. Fernandez-Cabrera MR, Higuera-Matas A, Fernaud-Espinosa I, DeFelipe J, Ambrosio E, Miguens M. Selective effects of Δ^9-tetrahydrocannabinol on medium spiny neurons in the striatum. *PLoS One* 2018;**13**, e0200950.

172. Cha YM, Jones KH, Kuhn CM, Wilson WA, Swartzwelder HS. Sex differences in the effects of Δ^9-tetrahydrocannabinol on spatial learning in adolescent and adult rats. *Behav Pharmacol* 2007;**18**:563–9.
173. Tseng AH, Harding JW, Craft RM. Pharmacokinetic factors in sex differences in Δ^9-tetrahydrocannabinol-induced behavioral effects in rats. *Behav Brain Res* 2004;**154**:77–83.
174. Christensen H, Freudenthal R, Gidley J, Rosenfeld R, Boegli G, Testino L, et al. Activity of $\Delta 8$- and Δ^9-tetrahydrocannabinol and related compounds in the mouse. *Science* 1971;**172**:165–7.
175. Thapa D, Cairns EA, Szczesniak A-M, Toguri JT, Caldwell MD, Kelly ME. The cannabinoids Δ8THC, CBD, and HU-308 act via distinct receptors to reduce corneal pain and inflammation. *Cannabis Cannabinoid Res* 2018;**3**:11–20.
176. Devinsky O, Kraft K, Rusch L, Fein M, Leone-Bay A. Improved bioavailability with dry powder cannabidiol inhalation: a phase 1 clinical study. *J Pharm Sci* 2021;**110**:3946–52.

Chapter 4

Endocannabinoids and sex differences in the developing social behavior network

Margaret M. McCarthy, Ashley E. Marquardt, and Jonathan W. VanRyzin

Department of Pharmacology and Program in Neuroscience, University of Maryland School of Medicine, Baltimore, MD, Unites States

Introduction

The concept of a social neural network that coordinates a range of behaviors, from mating to parenting to aggressing on your neighbor, was first introduced in the 1990s based on behavioral and anatomical assays in rodents that confirmed convergence of olfactory and hormonal stimuli in a few nodes within a distributed network.[1] This concept was subsequently confirmed and expanded to encompass all vertebrate species examined to-date.[2] Since then we have made enormous inroads into identifying the essential nodes and critical connections that drive social behaviors in real time,[3–6] and in some cases how those features differ in males versus females.[7–9] Where we have lagged is in characterizing how the social network develops, how it develops differently as a function of sex, and how it goes awry in response to challenge.

As membrane-derived fast acting small molecules, endocannabinoids (EDCs) are at first pass a surprising purveyor of social network formation. Derived from lipid membrane precursors, EDCs are among the most evolutionarily ancient and earliest expressed signaling molecules in the brain, detectable as early as 5-weeks of gestation in humans.[10] When placed in the context of ancient evolutionary origins, pan species importance and early appearance in brain development, the centrality of EDCs seems self-evident and is probably underestimated. The goal of this review is to highlight what we know to-date and identify fruitful avenues for further investigation. Studying how sex differences in the brain are established and undergird later sex differences in social behavior is compelling in its own right; at the same, it can also serve as a valuable investigative wedge for prying out specific aspects of circuit formation and function.

The social behavior network orchestrates expression of innate social behaviors

Discussion on the neural machinery underlying key social behaviors reached a turning point in 1999 when Sarah Newman proposed the existence of a core "social behavior network," or SBN, in mammals.[1] As proposed by Newman, this network consists of six nodes: (1) the medial extended amygdala (medial amygdala, MeA, and medial bed nucleus of the stria terminalis, BNST), (2) medial preoptic area (POA), (3) lateral septum (LH), (4) anterior hypothalamus (AH), (5) ventromedial hypothalamus (VMH), and (6) midbrain (periaqueductal gray, PAG, and tegmentum) (Fig. 1). Studies on the neural circuitry underlying various social behaviors including male and female sex behavior, aggression, maternal behavior, and territorial marking revealed a surprising amount of overlap, leading to the identification of this critical set of limbic regions. All reciprocally connected, each of the six regions was independently implicated in the expression of one, or often more, of these social behaviors. Newman envisioned a system in

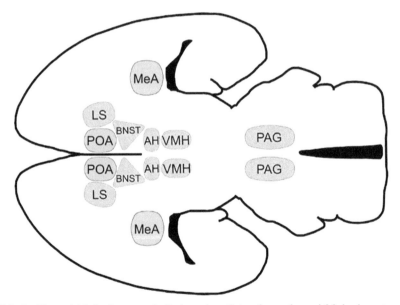

FIG. 1 The social behavior network. Brain regions that make up the social behavior network (SBN) as originally proposed by Newman (1999) are shown on an example horizontal section of rodent brain. All regions in the SBN have a known role in social behavior and express steroid hormone receptors (purple). However, a role for endocannabinoid signaling in social behavior has only been identified for two nodes (MeA and POA; green) thus far. Further research is needed to identify the dynamics and potential significance of endocannabinoids in the other regions in the network in both development and adulthood. AH, anterior hypothalamus; BNST, bed nucleus of the stria terminalis; LS, lateral septum; MeA, medial amygdala; PAG, periaqueductal gray; POA, preoptic area; VMH, ventromedial hypothalamus.

which each node of the SBN responded to a variety of social stimuli, whereby the relative activation of each of these nodes led to the expression of the particular behavior, e.g., parental behavior as opposed to aggression. As such, while the nodes and connections of the SBN are similar in both sexes, the pattern of activation differs in males and females when displaying distinct sex-typical social behaviors.

Importantly, all six regions of the SBN are also hormone-sensitive, showing enriched expression of gonadal hormone receptors. Consequently, Newman suggested that the SBN is both shaped and modified by changes in the levels of steroid hormones across development and in adulthood. Early in life, differences in steroid hormone milieu (most critically, estradiol) shape the sex-typical development of the nodes of the SBN; later in life, these steroid-responsive nodes exhibit plastic changes in response to differences in circulating hormone levels or to changes in the social environment.

Sexual differentiation of the developing brain

The singular event in the formation of any vertebrate fetus is the determination of sex. In mammals, the presence of a single gene, Sry, on the Y chromosome will direct the formation of the testis from the bipotential gonadal anlage. If Sry is absent or disabled, an ovary forms instead.[11] This process occurs stunningly early in development, by day 13 post-fertilization in rodents and starting in weeks 4–5 in humans, and thus sets in motion all the secondary events tied to success as a particular sex. This includes formation of the reproductive tract and genitalia, body size and adult adornments needed for attracting the opposite sex or fending off competitors of the same sex. None of these physical features matter, however, unless the brain is in register with the body. This requires both the appropriate physiology, meaning the brain directs the control of the anterior pituitary to regulate continuous production of sperm versus periodic shedding of ova, and the appropriate behavior, starting with mating. In the majority of mammals, sex behavior is fundamentally different in males and females, both in frequency (continuous in males, periodic in females) and in motor patterns (mounting and intromitting in males, lordosis or other sexually receptive postures in females). Not surprisingly then, fundamental aspects of the nodes within the SBN that control mating and reproductive physiology are also programmed developmentally to match gonadal status.[12]

Sex-specific programming of the developing brain occurs during a critical period. It is called "critical" because if it does not happen then, it never will, and "period" because there is a restricted time window in which it is critical that the programming happen. The critical period is operationally defined by the onset of steroidogenesis by the fetal male testis, providing the brain with high levels of androgens in only this sex, and the offset being the point at which exogenously supplying females with androgen levels equivalent to that of males fails to masculinize their brains.[13] This has all been known for a very long time, over half a century, and was originally encoded as the Organizational/Activational Hypothesis which refers to the developmental programming of the brain by steroids and the subsequent activation of the brain after puberty.[14] Using sex behavior as an example, an adult male needs to have been exposed to testosterone developmentally and again post-puberty in order to mate like a proper male. The absence of testosterone at either time results in a failure to execute. Females, on the other hand, need to have not been exposed to testosterone developmentally and as adults require temporally constrained exposure to estradiol and progesterone in order to become sexually receptive (with humans being a notable exception). The requirement of a temporal pattern of hormonal exposure is, not surprisingly, tied to ovulation. While a useful framework for understanding many sex differences in the SBN, particularly those directly associated with reproduction, not all features of complex mammalian behaviors are so nicely constrained, which challenges the ability to interrogate the mechanistic origins of developmental programming.

One overarching principle evident from the mechanistic studies conducted to-date is that what is true for one node in the SBN is not true for another, and even within a single node there can be multiple mechanisms at play. For instance, the medial preoptic nucleus is central to male sexual behavior. The sexual differentiation of neurons in this region involves a complex interplay between multiple cell types that include mast cells, microglia and astrocytes which direct the synaptic patterning on the local neurons that is required to promote male mating.[15–17] Conversely, the ventromedial nucleus of the hypothalamus is central to female sexual behavior. The sexual differentiation of this region includes no role for mast cells or astrocytes, but instead involves neuron-to-neuron signaling via membrane bound estrogen receptors and glutamate receptor signaling.[18,19] On the one hand, this is frustrating as it means there is no "unified field theory" for sexual differentiation of the brain; however, it does not mean there are no commonalities across regions. The more we learn, the more specific patterns begin to emerge. Among them is an unexpected role for membrane derived signaling molecules, in particular prostaglandins and EDCs, which are themselves connected via overlapping synthetic pathways involving arachidonic acid (Fig. 2). Indeed, the two systems are so interconnected that it is unclear if it is possible to perturb one selectively from the other.[20–22] Notably, both of these lipid-derived signaling systems have been implicated in developmental programming of nodes in the SBN[23,24] and acutely modifying adult social behavior expression.[25–27]

Endocannabinoids in developing brain

The developmental importance of EDCs precedes the formation of the brain as the system plays an unexpectedly important role in reproduction writ large. The proper implantation of the blastocyst in the uterine endometrium depends upon appropriately balanced EDC levels. Once the placenta is formed, EDCs can

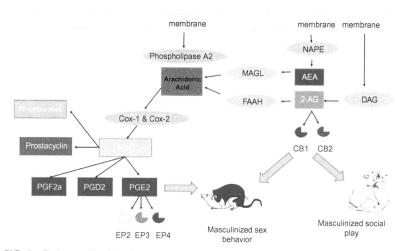

FIG. 2 Endocannabinoid and prostaglandin synthesis are interlocked. The endocannabinoids, 2-acylglycerol (2-AG) and anandamide (AEA) and the prostaglandins (PG's), prostacycline and thromboxanes are membrane derived signaling molecules which converge at arachidonic acid as either a precursor or metabolic product. Thus disturbing one system is likely to disturb the other in as of yet poorly understood ways. Both endocannabinoids and prostaglandins are essential signaling molecules for the sexual differentiation of the social behavior network during a developmental critical period.

readily traffic into the developing fetus. In addition to direct effects on the reproductive tract, the complex immune profile of pregnancy appears to be at least in part mediated by the anti-inflammatory actions of EDCs (reviewed in Ref. 28). The importance of the placenta to the health of the developing brain is at once inherently obvious and only recently verified.[29,30] Likewise, critical roles for the immune system in both healthy and dysregulated brain development are also emerging, including formation of circuitries controlling social behaviors.[31–35]

There are numerous members of the EDC family but dominant are 2-arachidonoylglycerol (2-AG), which is most plentiful, and anandamide (AEA) which shares many properties in regards to signal transduction but is also distinct from 2-AG. Two receptors, CB1 and CB2, also dominate although others exist.[36] The degradative enzyme, fatty acid amide hydrolase (FAAH), is expressed at the 2-cell stage of embryogenesis and CB1 and CB2 receptors detected shortly thereafter. The central importance of EDCs to adult brain functioning is well established, with the canonical view being one of post-synaptic release of 2-AG binding to pre-synaptic CB1 receptors which inhibit voltage-gated-calcium channels and dampen release of neurotransmitters, mostly GABA and glutamate.[37,38] In this scenario, EDCs are synthesized on demand and released from the cell to act in a paracrine fashion. However, the developing brain does not have established mature and regulated synapses but is instead a combination of differing degrees of cell genesis, axonal and dendritic growth and the formation of synapses, many of which will be later pruned away. As a result, we need a different lens by which to understand how EDCs sculpt neural circuits in an enduring way.

The importance of EDCs to brain development is revealed in two ways. First is detection of the components of the EDC system (enzymes, receptors, ligand), in the brain as it matures. For instance, CB1 receptor expression peaks in the early post-natal period and then declines to adult levels (reviewed in Ref. 39). Second is the clear impact of perturbing that system with exogenous cannabis exposure either in utero or postnatally, demonstrating the functionality of the existing receptors. Because of the ubiquity of CB1 distribution in the brain, it is not surprising that the roles attributed to the EDC system are widespread. Broad ranging effects of CB1, CB2, and TRPV1 receptors on neuronal proliferation, neurite outgrowth and axonal directionality have been thoroughly reviewed.[40] Yet, we remain primitive in our understanding of the cellular and molecular mechanisms by which either the endogenous ligands or those found in cannabis manifest changes in so many fundamental parameters of brain development. Despite this, rates of cannabis use during pregnancy are high and increasing. The 2016 report of the United States National Survey on Drug Use and Health finds 5% of all pregnant women reported use of cannabis in the past month and in women 18–25 yrs. of age the percent increases to 8.5% (https://www.samhsa.gov/data/data-we-collect/nsduh-national-survey-drug-use-and-health). The rate of use has been steadily increasing in recent years, particularly among young women, with greater than 70% of all women considering weekly cannabis use of little risk.[41] Detection of THC in the meconium of newborns confirms rates of exposure in the 10%–20% range and self-reports in some demographics find rates as high as 25%.[42] Similar values are reported for women in Canada[43] and parts of Europe,[42] reflecting the impact of increased legalization. The most common reported reasons for cannabis use are treatment of nausea and perceptions that cannabis is safer than tobacco or alcohol ingestion during pregnancy. Cannabis use is often comorbid with other substances and is frequently combined with substantial levels of psychological stress.[42] Children exposed to cannabis during pregnancy exhibit anomalies beginning around age 4 with the dominant effects being deficits in executive functioning and visual problem-solving.[42] Reading and spelling deficits are found in some studies,[44] as well as increased symptoms associated with neuropsychiatric disorders.[45,46] Extended use of cannabis during the adolescent period is an established risk factor for early adult onset schizophrenia, and results of animal studies suggest pre-natal cannabis exposure increases drug seeking behavior in adolescence (reviewed in Ref. 40) and adulthood,[45,47] creating a entrapping web of abuse. Based on animal models, the later propensity for drug abuse may be manifest via dysregulation of the dopamine[48] and/or opioid systems[49–51] by pre-natal cannabis exposure. Dysregulation to these systems would also impair the rewarding aspects of a variety of social behaviors. Many of these changes appear to be enduring due to epigenetic modifications,[52] but there is still much to be learned.

In an extraordinarily effective use of pre-clinical findings to investigation, Hurd, Harkany and colleagues used human fetal cerebrum tissue from electively aborted fetuses to confirm that THC use by pregnant women during the 2nd trimester modified the expression of a gene key to growth cone formation and axonal growth.[53] The reorganization of axonal morphology appears to be due at least in part to mis-localization of CB1 receptor in the growth cones, providing just one example of how THC exposure to the developing brain may exert enduring effects. The same group further demonstrated tandem roles for CB1 and CB2 receptors in axonal guidance and myelination. Proper cortical development requires a tightly orchestrated multi-step process involving axonal projections across the midline to the contralateral hemisphere while simultaneously myelinating those axons by oligodendrocytes. A combination of attraction and repulsion via the SLIT/Robo signaling complex assures appropriate directionality and termination. 2-AG activation of CB1 receptors on axonal growth cones and CB2 receptors on neighboring oligodendrocytes orchestrates coordinated expression of Robo1 and Slit1, respectively, and thereby guiding formatting of major axonal tracts.[54] This example illustrates the powerful impact of a single ligand, 2-AG, that simultaneously engages distinct cell specific receptors with divergent signal transduction pathways during the most fundamental stages of brain construction. While this study focused on the formation of the corticofugal pathway, the authors speculate the same phenomenon could be widespread throughout the brain. Similar widespread effects of developmental THC exposure could result from the documented impact on GABAergic[55] and glutamatergic[56] transmission, both of which have been tied to cognitive impairment.

Adolescent playfulness is a foundational behavior for adult sociality

The bulk of emphasis on developmental cannabis exposure has been on cognitive-associated outcomes with relatively little attention to social behaviors. Yet, impairments in sociability are associated with further impairments in cognition. For instance, isolating adolescent animals so that they are deprived of social interactions results in poorer performance on cognitive tasks as adults.[57,58] The impact of the lack of sociability as a juvenile is compounding in that adult males are also less adept at mating as adults.[59] Thus, understanding how EDCs and exogenous cannabis sculpt the Social Behavior Network developmentally has broad significance for elucidating the enduring impacts of drug exposure.

During the juvenile age, one social behavior that predominates across most mammalian species is social play. Often referred to as rough-and-tumble play, this evolutionarily conserved, tightly orchestrated behavior involves cooperation between two or more juveniles with rules of engagement which, when followed, make participation rewarding for both parties.[57] The ultimate function or benefit of play is debated by child psychologists, anthropologists, behavioral neuroscientists, evolutionary biologists, teachers, and exhausted parents. While there is no agreement as to precisely its purpose, there is consensus that play is essential to appropriate adult social behavior, cognition and reproductive success.[60,61] Play between males is more robust than that between females across nearly all species that play, including humans, but the mechanisms establishing this sex difference are elusive.[62–66] Children across the globe have been largely deprived of physical play with peers for the past 2 years, perhaps further contributing to the devasting setbacks in cognitive growth and development from the COVID-19 pandemic, the magnitude of which we have yet to fully realize.

The extended amygdala is a key node of the social behavior network

The amygdala draws its name from the Latin word for almond and consists of a collection of sub-nuclei which sub-serve separate but related functions, most of which are related to social behavior and those things which modify it such as fear and anxiety.[67–72] All of the amygdaloid nuclei are components of the SBN,[1] and they are among the most strongly expressing of estrogen and androgen receptors (ARs) in the brain,[73–75] as well as the steroidogenic enzyme, aromatase,[76–80] which converts androgens into estrogens. As a result, the amygdala is also a region rich with neuroanatomical and other sex differences.[9,69,81–87]

CB1 receptors are found preferentially in the amygdala and hippocampus of developing humans as early as mid-gestation. More intriguingly, the amygdala appears differentially sensitive to maternal cannabis use with an inverse correlation in dopamine receptor (D2R) with increasing use of cannabis selectively in the basal amygdala, and also selectively only in males.[88] Adult rats prenatally exposed to THC show elevated preproenkephalin in the MeA (as well as the basolateral amygdala and nucleus accumbens), a change associated with increased heroin seeking in adulthood.[51] However, whether the amygdala plays a role in the myriad of impacts of cannabis exposure during development in humans has been impossible to determine and so we turn to animal models.

The Sprague-Dawley laboratory rat has been a particularly robust model system for elucidating the biological origins of sex differences in the brain. Many effects are the result of testosterone aromatization into estradiol, but the MeA is a notable exception for the direct effect of androgens acting on AR to generate sex differences and masculinize social play.[64,65,89] Therefore, it is not surprising that AR are richly expressed in both neurons and astrocytes of the MeA, based on immunohistochemical analysis,[90,91] but little else is known about their phenotype.

Endocannabinoids developmentally program playfulness

For purely serendipitous reasons, we discovered that levels of 2-AG in the developing amygdala were significantly higher in males compared to females,[92] and that this sex difference is programed by the higher androgen levels normally experienced by males.[93] The sex difference is gone by 2-weeks of age, thus only present during the critical period for sexual differentiation of the brain. As a first step towards understanding how the sex difference in playfulness is established and what role EDCs might play in that process, we treated newborn rat pups with agonists and antagonists for both CB1 and CB2 receptors over the first 4-days of life. The treated animals were then assessed for their playfulness when adolescents, about 3 weeks later. Females in which both the CB1 and CB2 receptor was activated developmentally increased their level of play to that of males, while males treated with antagonists to both receptors were less playful, instead resembling females.[24] The requirement for both receptors to be either agonized or antagonized is unusual and remains unexplained but highlights the importance of not generalizing findings from the adult to the neonate. Nonetheless, the results demonstrate that it is the overall "tone" of the EDC system that is the driving force during the critical period.

The potential significance of EDC tone is intriguing given recent evidence that much of EDC signaling may actually be autocrine, which is achieved by the membrane derived EDCs never being released from the cell but instead activating receptors via lateral diffusion through the lipid bilayer. This autocrine activation lends itself naturally to the maintenance of an EDC "tone" in which the EDCs are continuously produced. More importantly, they can also continuously activate the GPCR CB1 and CB2 receptors.[94] Elucidating the importance of "tone" has been a challenge but in the developing amygdala we have a natural experiment in which, based on 2-AG level, the tone is significantly higher in neonatal males as a result of androgen action.[92,93] How androgens regulate the tone, where the EDCs are being synthesized and what cells they are acting on (i.e., where are the receptors) are questions of paramount importance but are challenged by the expression of the synthetic enzymes in a wide range of cell types. Given every cell has a membrane, all that is required is a few short enzymatic reactions to make a potent local fast-acting signaling molecule. Understanding how EDCs are locally released and regulated are central questions in all of neuroscience and the answers likely differ for varying functions, with what is true in the adult not necessarily being true in the developing brain and vice versa.

Endocannabinoids regulate microglia phagocytic activity in the developing amygdala

In parallel to the divergence in EDC actions in the developing versus matured brain are the dichotomous actions of the neuroimmune system. Microglia are the innate immune cells of the brain with myriad unique features, including sex differences in maturation rates and immune reactivity,[95,96] as well as originally unanticipated roles in both brain development[97,98] and aging.[99] The developmental role is dramatically highlighted by the ability of microglia to prune synapses in the lateral geniculate nucleus, thereby achieving the high-fidelity innervation required for visual acuity.[100] The targeting of particular synapses is mediated by the complement system and the specificity is achieved by activity-dependent expression. Complement is a system first discovered in the 1890's to "complement" or assist in the destruction of bacteria. Complement proteins are either soluble in the blood or bound to proteins in plasma membranes. Once bound to membranes, a process called opsonization, a series of enzymatic reactions produce anaphylatoxins which exert a variety of effects ranging from chemoattraction to apoptosis.[101] As with much in immunology, the varied effects of complement are still being discovered, including unique roles in brain development,[102] as well as neurodegenerative diseases.[103] We have found an important role for both microglia and complement in controlling cell number and doing so to a different degree in males versus females.[104] The classic complement pathway is initiated with C1q's and through a series of convertase reactions ends with the opsonin C3b.[101] This is the arm of the complement system central to both synaptic pruning and the phagocytosis of live cells we observe in the developing MeA.[93,105] This feature has led to the moniker "eat me signals" which attract the attention of phagocytic microglia if not balanced or overridden by a separate class of membrane bound "don't eat me signals."[106]

We discovered a brief span just after birth when the MeA undergoes extensive remodeling by highly phagocytic microglia. These microglia preferentially consume newly born cells.[93] A central question was, what cells are being consumed? The three main cell types of the brain are neurons, astrocytes and oligodendrocytes, which arise from multi-potent neural stem cells that then generate progenitors for each of the lineages and which take on unique fundamental properties at the earliest stages.[107] Neural progenitors quickly differentiate and the potential for further division is ultimately lost with the exception of a few neurogenic niches (SVZ, dentate gyrus), whereas astrocytes retain the ability to divide later in development and although largely non-proliferative in the adult brain, can be reactivated following injury, so called "reactive gliosis".[108,109] We have found astrocyte progenitors are broadly distributed throughout the developing amygdala and are actively proliferative at birth, with a gradual decline across the first week of post-natal life. The adjustment is achieved via the active elimination of recently born astrocytic progenitors (APs) by phagocytic microglia. Over the span of a 6–7 day critical window beginning at birth, the AP population undergoes a loss of approximately 100–200,000 cells in this one tiny brain region. Several unique features distinguish this extraordinary developmental sensitive period. First is that APs are alive and well at the time they are consumed and destroyed by microglia. Second is that targeting of cells for consumption is dependent upon the complement system. Third is that more cells are consumed in the amygdala of newborn males than females, and lastly, EDCs are the key driver of the sex difference in phagocytosis. This dramatic culling of cells is not without purpose as the reduction in astrocyte number in the adolescent MeA is permissive to higher levels of neuronal activity and an increase in adolescent playfulness[93] (Fig. 3).

This series of observations also led to the question, which cells are making the EDCs which are driving up the EDC tone which in in turn promotes microglia phagocytosis. In the adult brain, microglia are EDC synthesizers[110] but we found in the developing MeA microglia do NOT synthesize EDCs. We depleted microglia by >90% with liposomal clodronate infusions into the developing amygdala and measured no change in 2-AG levels in either males or females, and the previously established higher EDC tone in males was still evident. This leaves either the neurons expressing ARs and neighboring the targeted astrocyte precursors, or the precursors themselves, essentially signaling their own demise. These important questions are the topic of ongoing investigations.

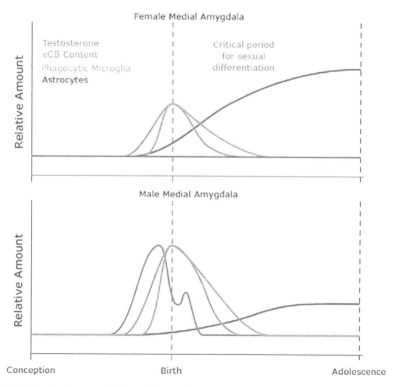

FIG. 3 The development of the medial amygdala amygdala (MeA) differs in magnitude between males and females. In females (top) the absence of perinatal testosterone leads to a smaller peak in endocannabinoid (EDC) content, which in turn generates a lesser phagocytic microglia population during the critical period for sexual differentiation. The presence of perinatal testosterone in males (bottom), in comparison, generates a larger peak in EDC content which produces a more phagocytic microglia population during this time. By the end of the critical period, males and females will differ in the number of astrocytes within the MeA.

The medial preoptic area is also a key node in the social behavior network

Copulation is arguably the most important, most intensely evolutionarily regulated social behavior; without mating, nothing else matters. The neural control of mating is distributed throughout the SBN as it involves a complex interplay of appropriately interpreting olfactory and other cues, soliciting the interest of the opposite sex through courtship or territorial defense, and the act of copulation. The relative weight of nodes in the network differs for males and females, with the medial preoptic area (mPOA), being the most critical for male sexual behavior across all species examined to date, including fish, reptiles, amphibians, birds, and mammals.

In adults, EDCs or cannabis have been found to both promote,[111] impair[112] or have no effect[113] on male sexual performance. There do not appear to be any studies of the effect of pre- or perinatal exposure to cannabis in animals on sexual behavior, and quantifying sexual performance in men is fraught with difficulties. However, as noted at the beginning, the EDC and prostaglandin synthetic pathways are closely intertwined, with 2-AG metabolism being a source of arachidonic acid,[22] which is the substrate used by cyclooxygenase enzymes (COX-1, COX-2) to produce an array of prostanoids, most important being prostaglandin-E2 or PGE2.[114]

The capacity for adult male sexual behavior is programmed developmentally by androgens (aromatized to estrogens in rodents) in the same manner as rough-and-tumble play described above. However, rather than regulating the population of local astrocytes in the MeA to presumably indirectly control neuronal activity, in this instance it is the mPOA that is programmed and neuronal excitation is programmed directly by altering the number of excitatory synapses. Meaning, in response to gonadal steroid action during the critical period, males form twice the density of excitatory dendritic spine synapses on mPOA neurons as found in females and this pattern endures across the life span, positively correlating with the robustness of adult male sexual behavior.[23] While trying to understand how steroids (estradiol in particular in the rat) promote the developmental formation of excitatory synapses, we made the surprising discovery that PGE2 was the key signaling molecule that through a series of steps involving glutamate AMPA receptors led to the increase in synaptogenesis. Agents which interfered with PGE2 production during the critical period, including aspirin, resulted in life-long deficits in male copulatory ability.[15] Thus similar to the MeA, increased production of membrane-bound lipid signaling molecules are essential to the masculinization of nodes in the SBN.

A common aspect of the two observations is that EDCs and PGE2 appear to either substitute for or enhance androgen action. Moreover, we find that inhibitory neurons of the MeA are central to the sex difference in playfulness while others have found that inhibitory neurons of the mPOA are central to male mating, which in adulthood is an androgen dependent behavior.[115] These convergences provide the beginning of a framework for determining if effects of EDCs on the developing SBN share the common feature of modifications to inhibitory neurons which are normally the site for influence by androgens. We are not suggesting that EDCs or THC have androgenic action, but rather may converge on the same cellular populations and endpoints. If so, this could have implications for other brain regions which highly express ARs, such as the developing hippocampus which exhibits sex differences associated with EDCs.[116]

Conclusions and open questions

There are vast gaps in our knowledge of how EDCs and cannabis act in the developing brain. As we highlight here, just focusing on one component, the SBN, there are more nodes in the network about which we know little to nothing than there are those in which we know anything. Establishing patterns of impact will require substantially more inquiry into the basic mechanisms by which EDCs act in the immature brain, and this will require a region by region assessment. What we do know is that the developing brain is not just a smaller version of the adult brain. The cellular location and signal transduction pathways of the CB1 and CB2 receptors seem to bear little resemblance to that seen in the adult. Our observation that activation of either CB1 or CB2 has the same impact on the neural circuitry of social play seems to have no parallel in the adult. Moreover, pre-natal exposure to cannabis impacts how the EDC system functions in adulthood and the effects are sex-specific both in terms of signal transduction and functional impact.[117]

The establishment of the SBN involves differential gonadal steroid exposure during the critical period for sexual differentiation and invokes the EDC system. However, how that occurs precisely is unknown. It has been established for some time that EDC signaling in the hippocampus is modified by sex,[118] but connecting developmental exposure to cannabis and/or gonadal steroids to these adult endpoints has not been achieved.

The target of steroid action could range from the synthetic and degradative enzymes to the receptors and associated signal transduction pathways. It is reasonable to assume that the primary cell type involved expresses ARs, but again as demonstrated in the MeA, the primary cell may release EDCs to act on other cells, so for instance neurons may be modulating the activity of microglia by releasing 2-AG. How the specificity in this cell-to-cell communication would be achieved given the diffuse and short-lived nature of these membrane derived signaling molecules is entirely unclear.

Lastly, the developing brain is characterized by distinct epochs of critical events which are tightly constrained in a temporal manner. In other words, timing is everything and we understand very little about whether there are more or less sensitive periods for cannabis exposure and EDC action for the myriad of delicate and complicated events that must occur in sequence. Given the earliness with which EDC signaling appears, the ubiquity of the receptors and the universality of the cell membrane as the source of the ligand, it will be extraordinarily challenging. But these same features also highlight why it is extraordinarily important to know.

References

1. Newman SW. The medial extended amygdala in male reproductive behavior. A node in the mammalian social behavior network. *Ann N Y Acad Sci* 1999;**877**:242–57. https://doi.org/10.1111/j.1749-6632.1999.tb09271.x.
2. Goodson JL. The vertebrate social behavior network: evolutionary themes and variations. *Horm Behav* 2005;**48**:11–22. https://doi.org/10.1016/j.yhbeh.2005.02.003.
3. Lee H, et al. Scalable control of mounting and attack by Esr1+ neurons in the ventromedial hypothalamus. *Nature* 2014;**509**:627–32. https://doi.org/10.1038/nature13169.
4. Hong W, Kim DW, Anderson DJ. Antagonistic control of social versus repetitive self-grooming behaviors by separable amygdala neuronal subsets. *Cell* 2014;**158**:1348–61. https://doi.org/10.1016/j.cell.2014.07.049.
5. Yang T, et al. Social control of hypothalamus-mediated male aggression. *Neuron* 2017;**95**. https://doi.org/10.1016/j.neuron.2017.06.046. 955–970.e954.
6. Wei D, Talwar V, Lin D. Neural circuits of social behaviors: innate yet flexible. *Neuron* 2021;**109**:1600–20. https://doi.org/10.1016/j.neuron.2021.02.012.
7. Kohl J, et al. Functional circuit architecture underlying parental behaviour. *Nature* 2018;**556**:326–31. https://doi.org/10.1038/s41586-018-0027-0.
8. Dulac C, O'Connell LA, Wu Z. Neural control of maternal and paternal behaviors. *Science* 2014;**345**:765–70. https://doi.org/10.1126/science.1253291.
9. Bergan JF, Ben-Shaul Y, Dulac C. Sex-specific processing of social cues in the medial amygdala. *eLife* 2014;**3**. https://doi.org/10.7554/eLife.02743, e02743.
10. Schneider M. Cannabis use in pregnancy and early life and its consequences: animal models. *Eur Arch Psychiatry Clin Neurosci* 2009;**259**:383–93. https://doi.org/10.1007/s00406-009-0026-0.
11. Goodfellow PN, Lovell-Badge R. SRY and sex determination in mammals. *Annu Rev Genet* 1993;**27**:71–92. https://doi.org/10.1146/annurev.ge.27.120193.000443.
12. McCarthy MM, De Vries GJ, Forger NG. In: Pfaff DW, Joels Pfaff M, editors. *Hormones, brain and behavior*. vol. 3. Elsevier; 2017. p. 3–32 [Chapter 5].
13. McCarthy MM, Herold K, Stockman SL. Fast, furious and enduring: sensitive versus critical periods in sexual differentiation of the brain. *Physiol Behav* 2018;**187**:13–9. https://doi.org/10.1016/j.physbeh.2017.10.030.
14. Phoenix CH, Goy RW, Gerall AA, Young WC. Organizing action of prenatally administered testosterone proprionate on the tissues mediating mating behavior in the female Guinea pig. *Endocrinology* 1959;**65**:369–82.
15. Amateau SK, McCarthy MM. Induction of PGE(2) by estradiol mediates developmental masculinization of sex behavior. *Nat Neurosci* 2004;**7**:643–50.
16. Lenz KM, et al. Mast cells in the developing brain determine adult sexual behavior. *J Neurosci* 2018;**38**:8044–59. https://doi.org/10.1523/JNEUROSCI.1176-18.2018.
17. Amateau SK, McCarthy MM. Sexual differentiation of astrocyte morphology in the developing rat preoptic area. *J Neuroendocrinol* 2002;**14**:904–10.

18. Schwarz JM, McCarthy MM. The role of neonatal NMDA receptor activation in defeminization and masculinization of sex behavior in the rat. *Horm Behav* 2008;**54**:662–8.
19. Schwarz JM, Liang S-L, Thompson SM, McCarthy MM. Estradiol induces hypothalamic dendritic spines by enhancing glutamate release: a mechanism for organizational sex differences. *Neuron* 2008;**58**:584–98.
20. Holt S, et al. Inhibition of fatty acid amide hydrolase, a key endocannabinoid metabolizing enzyme, by analogues of ibuprofen and indomethacin. *Eur J Pharmacol* 2007;**565**:26–36. https://doi.org/10.1016/j.ejphar.2007.02.051.
21. Fowler CJ. The contribution of cyclooxygenase-2 to endocannabinoid metabolism and action. *Br J Pharmacol* 2007;**152**:594–601. https://doi.org/10.1038/sj.bjp.0707379.
22. Nomura DK, et al. Endocannabinoid hydrolysis generates brain prostaglandins that promote neuroinflammation. *Science* 2011;**334**:809–13. https://doi.org/10.1126/science.1209200.
23. Wright CL, Burks SR, McCarthy MM. Identification of prostaglandin E2 receptors mediating perinatal masculinization of adult sex behavior and neuroanatomical correlates. *Dev Neurobiol* 2008;**68**.
24. Argue KJ, et al. Activation of Both CB1 and CB2 endocannabinoid receptors is critical for masculinization of the developing medial amygdala and juvenile social play behavior. *eNeuro* 2017;**4**. https://doi.org/10.1523/ENEURO.0344-16.2017.
25. Stuart JM, Paris JJ, Frye C, Bradshaw HB. Brain levels of prostaglandins, endocannabinoids, and related lipids are affected by mating strategies. *Int J Endocrinol* 2013;**2013**. https://doi.org/10.1155/2013/436252, 436252.
26. Juntti SA, et al. A neural basis for control of cichlid female reproductive behavior by prostaglandin F2alpha. *Curr Biol* 2016;**26**:943–9. https://doi.org/10.1016/j.cub.2016.01.067.
27. Wei D, Allsop S, Tye K, Piomelli D. Endocannabinoid Signaling in the control of social behavior. *Trends Neurosci* 2017;**40**:385–96. https://doi.org/10.1016/j.tins.2017.04.005.
28. Taylor AH, et al. Endocannabinoids and pregnancy. *Clin Chim Acta* 2010;**411**:921–30. https://doi.org/10.1016/j.cca.2010.03.012.
29. Ursini G, et al. Convergence of placenta biology and genetic risk for schizophrenia. *Nat Med* 2018. https://doi.org/10.1038/s41591-018-0021-y.
30. Howerton CL, Bale TL. Targeted placental deletion of OGT recapitulates the prenatal stress phenotype including hypothalamic mitochondrial dysfunction. *Proc Natl Acad Sci U S A* 2014;**111**:9639–44. https://doi.org/10.1073/pnas.1401203111.
31. Filiano AJ, et al. Unexpected role of interferon-gamma in regulating neuronal connectivity and social behaviour. *Nature* 2016;**535**:425–9. https://doi.org/10.1038/nature18626.
32. Estes ML, McAllister AK. Maternal immune activation: implications for neuropsychiatric disorders. *Science* 2016;**353**:772–7. https://doi.org/10.1126/science.aag3194.
33. McCarthy MM. Sex differences in neuroimmunity as an inherent risk factor. *Neuropsychopharmacology* 2019;**44**:38–44. https://doi.org/10.1038/s41386-018-0138-1.
34. McCarthy MM, Wright CL. Convergence of sex differences and the neuroimmune system in autism spectrum disorder. *Biol Psychiatr* 2017;**81**:402–10. https://doi.org/10.1016/j.biopsych.2016.10.004.
35. McCarthy MM, Nugent BM, Lenz KM. Neuroimmunology and neuroepigenetics in the establishment of sex differences in the brain. *Nat Rev Neurosci* 2017;**18**:471–84. https://doi.org/10.1038/nrn.2017.61.
36. Mackie K, Stella N. Cannabinoid receptors and endocannabinoids: evidence for new players. *AAPS J* 2006;**8**:E298–306. https://doi.org/10.1007/BF02854900.
37. Alger BE. Retrograde signaling in the regulation of synaptic transmission: focus on endocannabinoids. *Prog Neurobiol* 2002;**68**:247–86.

38. Lovinger DM. Presynaptic modulation by endocannabinoids. *Handb Exp Pharmacol* 2008; **435-477**. https://doi.org/10.1007/978-3-540-74805-2_14.
39. Trezza V, Cuomo V, Vanderschuren LJ. Cannabis and the developing brain: insights from behavior. *Eur J Pharmacol* 2008;**585**:441–52. https://doi.org/10.1016/j.ejphar.2008.01.058.
40. Alpar A, Di Marzo V, Harkany T. At the tip of an iceberg: prenatal marijuana and its possible relation to neuropsychiatric outcome in the offspring. *Biol Psychiatry* 2016;**79**:e33–45. https://doi.org/10.1016/j.biopsych.2015.09.009.
41. Scheyer AF, Melis M, Trezza V, Manzoni OJJ. Consequences of perinatal cannabis exposure. *Trends Neurosci* 2019;**42**:871–84. https://doi.org/10.1016/j.tins.2019.08.010.
42. Ryan SA, et al. Marijuana use during pregnancy and breastfeeding: implications for neonatal and childhood outcomes. *Pediatrics* 2018;**142**. https://doi.org/10.1542/peds.2018-1889.
43. Ordean A, Kim G. Cannabis use during lactation: literature review and clinical recommendations. *J Obstet Gynaecol Can* 2020. https://doi.org/10.1016/j.jogc.2019.11.003.
44. Goldschmidt L, Richardson GA, Cornelius MD, Day NL. Prenatal marijuana and alcohol exposure and academic achievement at age 10. *Neurotoxicol Teratol* 2004;**26**:521–32. https://doi.org/10.1016/j.ntt.2004.04.003.
45. Day NL, Goldschmidt L, Day R, Larkby C, Richardson GA. Prenatal marijuana exposure, age of marijuana initiation, and the development of psychotic symptoms in young adults. *Psychol Med* 2015;**45**:1779–87. https://doi.org/10.1017/S0033291714002906.
46. Day NL, Leech SL, Goldschmidt L. The effects of prenatal marijuana exposure on delinquent behaviors are mediated by measures of neurocognitive functioning. *Neurotoxicol Teratol* 2011;**33**:129–36. https://doi.org/10.1016/j.ntt.2010.07.006.
47. Sonon KE, Richardson GA, Cornelius JR, Kim KH, Day NL. Prenatal marijuana exposure predicts marijuana use in young adulthood. *Neurotoxicol Teratol* 2015;**47**:10–5. https://doi.org/10.1016/j.ntt.2014.11.003.
48. DiNieri JA, et al. Maternal cannabis use alters ventral striatal dopamine D2 gene regulation in the offspring. *Biol Psychiatr* 2011;**70**:763–9. https://doi.org/10.1016/j.biopsych.2011.06.027.
49. Jutras-Aswad D, DiNieri JA, Harkany T, Hurd YL. Neurobiological consequences of maternal cannabis on human fetal development and its neuropsychiatric outcome. *Eur Arch Psychiatr Clin Neurosci* 2009;**259**:395–412. https://doi.org/10.1007/s00406-009-0027-z.
50. Ellgren M, Spano SM, Hurd YL. Adolescent cannabis exposure alters opiate intake and opioid limbic neuronal populations in adult rats. *Neuropsychopharmacology* 2007;**32**:607–15. https://doi.org/10.1038/sj.npp.1301127.
51. Spano MS, Ellgren M, Wang X, Hurd YL. Prenatal cannabis exposure increases heroin seeking with allostatic changes in limbic enkephalin systems in adulthood. *Biol Psychiatr* 2007;**61**:554–63. https://doi.org/10.1016/j.biopsych.2006.03.073.
52. Morris CV, DiNieri JA, Szutorisz H, Hurd YL. Molecular mechanisms of maternal cannabis and cigarette use on human neurodevelopment. *Eur J Neurosci* 2011;**34**:1574–83. https://doi.org/10.1111/j.1460-9568.2011.07884.x.
53. Tortoriello G, et al. Miswiring the brain: Delta9-tetrahydrocannabinol disrupts cortical development by inducing an SCG10/stathmin-2 degradation pathway. *EMBO J* 2014;**33**:668–85. https://doi.org/10.1002/embj.201386035.
54. Alpar A, et al. Endocannabinoids modulate cortical development by configuring Slit2/Robo1 signalling. *Nat Commun* 2014;**5**:4421. https://doi.org/10.1038/ncomms5421.
55. Beggiato S, et al. Long-lasting alterations of hippocampal GABAergic neurotransmission in adult rats following perinatal Delta(9)-THC exposure. *Neurobiol Learn Mem* 2017;**139**:135–43. https://doi.org/10.1016/j.nlm.2016.12.023.

56. Antonelli T, et al. Long-term effects on cortical glutamate release induced by prenatal exposure to the cannabinoid receptor agonist (R)-(+)-[2,3-dihydro-5-methyl-3-(4-morpholinyl-methyl)pyrrolo[1,2,3-de]-1,4-benzo xazin-6-yl]-1-naphthalenylmethanone: an in vivo microdialysis study in the awake rat. *Neuroscience* 2004;**124**:367–75. https://doi.org/10.1016/j.neuroscience.2003.10.034.
57. Vanderschuren LJ, Trezza V. What the laboratory rat has taught us about social play behavior: role in behavioral development and neural mechanisms. *Curr Top Behav Neurosci* 2014;**16**:189–212. https://doi.org/10.1007/7854_2013_268.
58. Baarendse PJ, Counotte DS, O'Donnell P, Vanderschuren LJ. Early social experience is critical for the development of cognitive control and dopamine modulation of prefrontal cortex function. *Neuropsychopharmacology* 2013;**38**:1485–94. https://doi.org/10.1038/npp.2013.47.
59. Gerall HD, Ward IL, Gerall AA. Disruption of the male rat's sexual behaviour induced by social isolation. *Anim Behav* 1967;**15**:54–8. https://doi.org/10.1016/s0003-3472(67)80010-1.
60. Pellis SM, Pellis VC, Pelletier A, Leca JB. Is play a behavior system, and, if so, what kind? *Behav Process* 2019;**160**:1–9. https://doi.org/10.1016/j.beproc.2018.12.011.
61. Blake BE, McCoy KA. Hormonal programming of rat social play behavior: standardized techniques will aid synthesis and translation to human health. *Neurosci Biobehav Rev* 2015;**55**:184–97. https://doi.org/10.1016/j.neubiorev.2015.04.021.
62. Argue KJ, McCarthy MM. Characterization of juvenile play in rats: importance of sex of self and sex of partner. *Biol Sex Differ* 2015;**6**:16. https://doi.org/10.1186/s13293-015-0034-x.
63. Auger AP, Olesen KM. Brain sex differences and the organisation of juvenile social play behaviour. *J Neuroendocrinol* 2009;**21**:519–25. https://doi.org/10.1111/j.1365-2826.2009.01871.x. JNE1871 [pii].
64. Meaney MJ, Stewart J. Neonatal androgens influence the social play of prepubescent rats. *Horm Behav* 1981;**15**:197–213.
65. Meaney MJ, Stewart J, Poulin P, McEwen BS. Sexual differentiation of social play in rat pups in mediated by the neonatal androgen-receptor system. *Neuroendocrinology* 1983;**37**:85–90.
66. Olioff M, Stewart J. Sex differences in the play behavior of prepubescent rats. *Physiol Behav* 1978;**20**:113–5.
67. Raam T, Hong W. Organization of neural circuits underlying social behavior: a consideration of the medial amygdala. *Curr Opin Neurobiol* 2021;**68**:124–36. https://doi.org/10.1016/j.conb.2021.02.008.
68. Yamaguchi T, et al. Posterior amygdala regulates sexual and aggressive behaviors in male mice. *Nat Neurosci* 2020;**23**:1111–24. https://doi.org/10.1038/s41593-020-0675-x.
69. Walker DM, et al. Adolescent social isolation reprograms the medial amygdala: transcriptome and sex differences in reward. *bioRxiv* 2020. https://doi.org/10.1101/2020.02.18.955187.
70. Chen PB, et al. Sexually dimorphic control of parenting behavior by the medial amygdala. *Cell* 2019;**176**:1206–1221 e1218. https://doi.org/10.1016/j.cell.2019.01.024.
71. Li Y, et al. Neuronal representation of social information in the medial amygdala of awake behaving mice. *Cell* 2017;**171**:1176–1190 e1117. https://doi.org/10.1016/j.cell.2017.10.015.
72. Wolff SB, et al. Amygdala interneuron subtypes control fear learning through disinhibition. *Nature* 2014;**509**:453–8. https://doi.org/10.1038/nature13258.
73. MacLusky NJ, Lieberburg I, McEwen BS. The development of estrogen receptor systems in the rat brain: perinatal development. *Brain Res* 1979;**178**:129–42.
74. Kashon ML, Arbogast JA, Sisk CL. Distribution and hormonal regulation of androgen receptor immunoreactivity in the forebrain of the male European ferret. *J Comp Neurol* 1996;**376**:567–86. https://doi.org/10.1002/(SICI)1096-9861(19961223)376:4<567::AID-CNE6>3.0.CO;2-#.

75. Simerly RB, Chang C, Muramatsu M, Swanson LW. Distribution of androgen and estrogen receptor mRNA-containing cells in the rat brain: an in situ hybridization study. *J Comp Neurol* 1990;**294**:76–95.
76. Cisternas CD, Cabrera Zapata LE, Arevalo MA, Garcia-Segura LM, Cambiasso MJ. Regulation of aromatase expression in the anterior amygdala of the developing mouse brain depends on ERbeta and sex chromosome complement. *Sci Rep* 2017;**7**:5320. https://doi.org/10.1038/s41598-017-05658-6.
77. Unger EK, et al. Medial amygdalar aromatase neurons regulate aggression in both sexes. *Cell Rep* 2015;**10**:453–62. https://doi.org/10.1016/j.celrep.2014.12.040.
78. Roselli CE, Resko JA. Aromatase activity in the rat brain: hormonal regulation and sex differences. *J Steroid Biochem Mol Biol* 1993;**44**:499–508.
79. Roselli CE, Ellinwood WE, Resko JA. Regulation of brain aromatase activity in rats. *Endocrinology* 1984;**114**:192–200.
80. Billing A, Henrique Correia M, Kelly DA, Li GL, Bergan JF. Synaptic connections of aromatase circuits in the medial amygdala are sex specific. *eNeuro* 2020;**7**. https://doi.org/10.1523/ENEURO.0489-19.2020.
81. Kolodkin MH, Auger AP. Sex difference in the expression of DNA methyltransferase 3a in the rat amygdala during development. *J Neuroendocrinol* 2011;**23**:577–83. https://doi.org/10.1111/j.1365-2826.2011.02147.x.
82. Blanton RE, Chaplin TM, Sinha R. Sex differences in the correlation of emotional control and amygdala volumes in adolescents. *Neuroreport* 2010;**21**:953–7. https://doi.org/10.1097/WNR.0b013e32833e7866.
83. Kilpatrick LA, Zald DH, Pardo JV, Cahill LF. Sex-related differences in amygdala functional connectivity during resting conditions. *NeuroImage* 2006;**30**:452–61. https://doi.org/10.1016/j.neuroimage.2005.09.065. S1053-8119(05)00764-0 [pii].
84. Cooke BM, Woolley CS. Sexually dimorphic synaptic organization of the medial amygdala. *J Neurosci* 2005;**25**:10759–67. https://doi.org/10.1523/JNEUROSCI.2919-05.2005.
85. Kerchner M, Malsbury CW, Ward OB, Ward IL. Sexually dimorphic areas in the rat medial amygdala: resistance to the demasculinizing effects of prenatal stress. *Brain Res* 1995;**672**:251–60.
86. Hines M, Allen LS, Gorski RA. Sex differences in subregions of the medial nucleus of the amygdala and the bed nucleus of the stria terminalis of the rat. *Brain Res* 1992;**579**:321–6.
87. Mizukami S, Nishizuka M, Arai Y. Sexual difference in nuclear volume and its ontogeny in the rat amygdala. *Exp Neurol* 1983;**79**:569–75.
88. Wang X, Dow-Edwards D, Keller E, Hurd YL. Preferential limbic expression of the cannabinoid receptor mRNA in the human fetal brain. *Neuroscience* 2003;**118**:681–94.
89. Meaney MJ, McEwen BS. Testosterone implants into the amygdala during the neonatal period masculinize the social play of juvenile female rats. *Brain Res* 1986;**398**:324–8.
90. Romeo RD, Diedrich SL, Sisk CL. Effects of gonadal steroids during pubertal development on androgen and estrogen receptor-alpha immunoreactivity in the hypothalamus and amygdala. *J Neurobiol* 2000;**44**:361–8. https://doi.org/10.1002/1097-4695(20000905)44:3<361::AID-NEU6>3.0.CO;2-P.
91. Pfau DR, Hobbs NJ, Breedlove SM, Jordan CL. Sex and laterality differences in medial amygdala neurons and astrocytes of adult mice. *J Comp Neurol* 2016;**524**:2492–502. https://doi.org/10.1002/cne.23964.
92. Krebs-Kraft DL, Hill MN, Hillard CJ, McCarthy MM. Sex difference in cell proliferation in developing rat amygdala mediated by endocannabinoids has implications for social behavior. *Proc Natl Acad Sci U S A* 2010;**107**:20535–40.

93. VanRyzin JW, et al. Microglial phagocytosis of newborn cells is induced by endocannabinoids and sculpts sex differences in juvenile rat social play. *Neuron* 2019. https://doi.org/10.1016/j.neuron.2019.02.006.
94. Howlett AC, et al. Endocannabinoid tone versus constitutive activity of cannabinoid receptors. *Br J Pharmacol* 2011;**163**:1329–43. https://doi.org/10.1111/j.1476-5381.2011.01364.x.
95. Hanamsagar R, et al. Generation of a microglial developmental index in mice and in humans reveals a sex difference in maturation and immune reactivity. *Glia* 2017. https://doi.org/10.1002/glia.23176.
96. Lenz KM, Nugent BM, Haliyur R, McCarthy MM. Microglia are essential to masculinization of brain and behavior. *J Neurosci* 2013;**33**:2761–72.
97. Bilbo SD, Block CL, Bolton JL, Hanamsagar R, Tran PK. Beyond infection—maternal immune activation by environmental factors, microglial development, and relevance for autism spectrum disorders. *Exp Neurol* 2018;**299**:241–51. https://doi.org/10.1016/j.expneurol.2017.07.002.
98. Bilbo SD, Schwarz JM. The immune system and developmental programming of brain and behavior. *Front Neuroendocrinol* 2012;**33**:267–86. https://doi.org/10.1016/j.yfrne.2012.08.006.
99. Hemonnot AL, Hua J, Ulmann L, Hirbec H. Microglia in Alzheimer disease: well-known targets and new opportunities. *Front Aging Neurosci* 2019;**11**:233. https://doi.org/10.3389/fnagi.2019.00233.
100. Hong S, Dissing-Olesen L, Stevens B. New insights on the role of microglia in synaptic pruning in health and disease. *Curr Opin Neurobiol* 2016;**36**:128–34. https://doi.org/10.1016/j.conb.2015.12.004.
101. Sarma JV, Ward PA. The complement system. *Cell Tissue Res* 2011;**343**:227–35. https://doi.org/10.1007/s00441-010-1034-0.
102. Kopec AM, Smith CJ, Ayre NR, Sweat SC, Bilbo SD. Microglial dopamine receptor elimination defines sex-specific nucleus accumbens development and social behavior in adolescent rats. *Nat Commun* 2018;**9**:3769. https://doi.org/10.1038/s41467-018-06118-z.
103. Presumey J, Bialas AR, Carroll MC. Complement system in neural synapse elimination in development and disease. *Adv Immunol* 2017;**135**:53–79. https://doi.org/10.1016/bs.ai.2017.06.004.
104. VanRyzin JW, Pickett LA, McCarthy MM. Microglia: driving critical periods and sexual differentiation of the brain. *Dev Neurobiol* 2018;**78**:580–92. https://doi.org/10.1002/dneu.22569.
105. Schafer DP, et al. Microglia sculpt postnatal neural circuits in an activity and complement-dependent manner. *Neuron* 2012;**74**:691–705. https://doi.org/10.1016/j.neuron.2012.03.026.
106. Brown GC, Neher JJ. Eaten alive! Cell death by primary phagocytosis: 'phagoptosis'. *Trends Biochem Sci* 2012;**37**:325–32. https://doi.org/10.1016/j.tibs.2012.05.002.
107. Nourse JL, et al. Membrane biophysics define neuron and astrocyte progenitors in the neural lineage. *Stem Cells* 2014;**32**:706–16. https://doi.org/10.1002/stem.1535.
108. Eng LF, Ghirnikar RS. GFAP and astrogliosis. *Brain Pathol* 1994;**4**:229–37.
109. Lindsay R. In: Feodoroff S, Vernadakis A, editors. *Astrocytes*. vol. 3. Academic Press; 1986. p. 231–62.
110. Witting A, Walter L, Wacker J, Moller T, Stella N. P2X7 receptors control 2-arachidonoylglycerol production by microglial cells. *Proc Natl Acad Sci U S A* 2004;**101**:3214–9. https://doi.org/10.1073/pnas.0306707101.
111. Canseco-Alba A, Rodriguez-Manzo G. Low anandamide doses facilitate male rat sexual behaviour through the activation of CB1 receptors. *Psychopharmacology* 2014;**231**:4071–80. https://doi.org/10.1007/s00213-014-3547-9.

112. Carvalho RK, et al. Chronic cannabidiol exposure promotes functional impairment in sexual behavior and fertility of male mice. *Reprod Toxicol* 2018;**81**:34–40. https://doi.org/10.1016/j.reprotox.2018.06.013.
113. Shiff B, et al. The impact of cannabis use on male sexual function: a 10-year, single-center experience. *Can Urol Assoc J* 2021. https://doi.org/10.5489/cuaj.7185.
114. Kaufmann W, Andreasson K, Isakson P, Worley P. Cyclooxygenases and the central nervous system. *Prostaglandins* 1997;**54**:601–24.
115. Wu MV, Tollkuhn J. Estrogen receptor alpha is required in GABAergic, but not glutamatergic, neurons to masculinize behavior. *Horm Behav* 2017;**95**:3–12. https://doi.org/10.1016/j.yhbeh.2017.07.001.
116. Oberlander JG, Woolley CS. 17beta-Estradiol acutely potentiates glutamatergic synaptic transmission in the hippocampus through distinct mechanisms in males and females. *J Neurosci* 2016;**36**:2677–90. https://doi.org/10.1523/JNEUROSCI.4437-15.2016.
117. Bara A, et al. Sex-dependent effects of in utero cannabinoid exposure on cortical function. *eLife* 2018;**7**. https://doi.org/10.7554/eLife.36234.
118. Huang GZ, Woolley CS. Estradiol acutely suppresses inhibition in the hippocampus through a sex-specific endocannabinoid and mGluR-dependent mechanism. *Neuron* 2012;**74**:801–8. https://doi.org/10.1016/j.neuron.2012.03.035.

Chapter 5

Behavioral consequences of pre/peri-natal *Cannabis* exposure

Antonia Manduca[a,b] and Viviana Trezza[a]
[a]Roma Tre University, Rome, Italy, [b]Fondazione Santa Lucia, Rome, Italy

Introduction

Cannabis is the world's most widely used (illicit) drug among pregnant and breastfeeding women.[1–3] The main psychoactive component of *Cannabis* Δ[9]-tetrahydrocannabinol (THC) enters the maternal circulation and easily crosses the placenta, exerting deleterious effects on the developing fetus.[4–6] Furthermore, being a lipid molecule, THC can easily be transferred from the mother to the infant through the maternal milk during breastfeeding.[7,8] There is concern that legalization of non-medical *Cannabis* use which is rapidly expanding worldwide will lead to low perceived risk of harm, particularly for vulnerable populations, as more pregnant and breastfeeding women use *Cannabis* today and conceivably large numbers of children will be prenatally (i.e., during pregnancy) or perinatally (i.e., during pregnancy and lactation) exposed to marijuana ingredients over the next decades. Considering that the legalization/decriminalization of *Cannabis* is rapidly expanding worldwide and given that breastfeeding is strongly encouraged in many countries, rigorous scientific research about the impact of maternal *Cannabis* use on health and well-being becomes paramount.

Several factors are related to the decision to consume *Cannabis* during pregnancy: maternal self-reports highlight that the drug is predominantly used to treat depression, anxiety, stress, pain, nausea, and vomiting.[9–11] A statewide cross-sectional study in the United States (U.S.) reported that 70%–80% of medically licensed *Cannabis* dispensaries in Colorado recommended *Cannabis*-based products to pregnant women during the first trimester to alleviate their morning sickness, without encouraging a previous discussion with a health care.[12] This suggests that unlike the use of other illicit substances during pregnancy (i.e., tobacco, alcohol), there is a lack of perceived risk for the use of maternal *Cannabis* among medically licensed dispensaries and healthcare providers which may result in a perceived medicinal incentive and facilitate increasing rates of *Cannabis* use during pregnancy. This is in strong contrast with the available scientific evidence suggesting that *Cannabis* exposure at key developmental periods (i.e., pregnancy, lactation, adolescence) can trigger psychopathology later in life, and in line with the fact that, to date, there are no *Cannabis*-derived medications that are approved by the U.S. Food and Drug Administration (FDA) and/or by the European Medicines Agency (EMA) for treating conditions associated with pregnancy, including nausea. In this context, additional research is needed on the health effects associated with *Cannabis* exposure during pregnancy and breastfeeding, particularly studies examining the effects of the *Cannabis* products used by women today, the patterns by which they are used, and their co-use with other substances to encourage pregnant women who use *Cannabis* to discontinue its use.

In this scenario, investigating the neurobehavioral consequences of maternal *Cannabis* use on the human progeny poses some challenges, mainly due to (1) under-reporting of frequency of use (e.g., number of joints/per day) by pregnant women because of the fear of legal consequences, the possible loss of the children's custody, and the feelings of guilt caused by the potential effects on the baby; (2) wide variation in *Cannabis* potency due to the sharp spike in THC concentrations observed over the past two decades together with individual smoking habits (the number, depth, and volume of smoking or vaping); (3) selection bias (i.e., error in selecting the study participants and/or from factors affecting the study participation including environmental and genetic factors) and information bias (i.e., misclassification and/or inaccuracy of some diagnostic tests). This highlights the advantage of using animal models that allow to control over the possible confounding factors that characterize human studies to evaluate the contribution of pre/peri-natal *Cannabis* to adverse, even subtle neurodevelopmental consequences in the offspring (for exhaustive reviews see Refs. 5,7,13–17) and to understand the mechanisms through which *Cannabis* might affect the brain of the fetus and cause neurochemical and neuroanatomical changes along development.

In this chapter, we summarize available clinical and pre-clinical data on the behavioral consequences of maternal *Cannabis* exposure along development (from infancy through adolescence till adulthood), discuss the limitations of the available studies and highlight future perspectives in this research field.

Clinical studies on the behavioral consequences of maternal *Cannabis* exposure

Despite decreasing rates of alcohol and tobacco consumption during pregnancy, the use of maternal *Cannabis* is on the rise.[18] According to data collected from the National Survey on Drug Use and Health (NSDUH), the use of *Cannabis* in pregnant women aged 18–44 years rose from 2.37% in 2002 to 3.85% in 2014 in the U.S., with 21.1% of *Cannabis* users reporting that they used the drug daily during pregnancy. Considering that *Cannabis* legalization policies move forward worldwide and that conceivably large numbers of children will be prenatally exposed to its ingredients over the next decades, unrevealing the neurobiological processes that mediate the offspring vulnerability to the detrimental effects of maternal *Cannabis* use is clinically and scientifically relevant.

Human studies have provided invaluable information on the behavioral effects of pre-natal *Cannabis* exposure in the offspring from the neonatal period through to early adulthood. To date, much of what is known about the consequences of maternal *Cannabis* exposure on child neurodevelopment is based on data collected from three large prospective longitudinal cohorts studies: 1. The Ottawa Prenatal Prospective Study (OPPS), initiated in 1978 by Fried and colleagues, included a low-risk, European-American, middle-class population of pregnant women and followed the offspring till the age of 18–22 years.[19–26] 2. The Maternal Health Practices and Child Development Study (MHPCD), which began in 1982, focused on high-risk pregnant women of mixed ethnicity and low socioeconomic status, with follow-up of the offspring till the age of 14.[27–32] 3. The Generation R study, initiated in 2001, enrolled multi-ethnic pregnant women of middle-high socioeconomic status, with follow-up for most measures until the age of 6 years.[33–37]

There are plenty of reports based on results obtained from the OPPS, MHPCD, and Generation R longitudinal studies and we here resume some of the most relevant data. As early consequence of prenatal *Cannabis* exposure, the MHPCD study found lower mental development scores in 9-month-old babies exposed to *Cannabis*, which disappeared at the age of 19 months.[38] Both the OPPS and MHPCD studies reported impairment in verbal reasoning scores and short-term memory at the age of 3–4 years.[23,28] Interestingly, the Generation R study reported an increase only in girls' aggression and inattention at the age of 18 months[39] that lost statistically significance at 36 months.[39,40] This raises the important question of whether maternal *Cannabis* might affect the male and female progeny differently.[41] Some effects of in utero *Cannabis* exposure are indeed sexually divergent: in 2004, the group of Yasmin Hurd discovered that the decrease in dopamine receptor mRNA in the amygdalae of *Cannabis*-exposed fetuses (all between 18 and 22 weeks of gestation and donated by women who underwent voluntary abortions) was statistically significant in the male but not female offspring.[42] Recent pre-clinical studies have confirmed a certain degree of sexual divergence in the detrimental neurobehavioral consequences of fetal cannabinoids in the rat offspring at different developmental ages,[43–47] which warrants further investigations to decipher the mechanisms determining sex-related effects following maternal *Cannabis* exposure.

Beyond infancy, both the OPPS and MHPCD studies reported an increase in impulsivity and hyperactivity in *Cannabis*-exposed children at 6 and 10 years.[25,29,48] More recently, the Generation R study found that maternal *Cannabis* use during pregnancy was associated with externalizing symptoms (i.e., aggressive and rule-breaking behaviors) at 7–10 years[49] and increased risk of psychotic-like experiences at the age of 10.[50] This study also revealed that paternal *Cannabis* use was predictive of psychotic-like experiences and behavioral deficits in the offspring at ages 7–10 years,[50] even though this study presents the limitation that paternal *Cannabis* use was derived from maternal reports, and was only determined for the pregnancy period, not prior to pregnancy. Reports from the MHPCD cohort showed that heavy pre-natal *Cannabis* consumption (~1 or more joints/day) during the first trimester predicted levels of self-reported anxiety and depressive symptoms in children aged 10 years[30,51,52]; however the more recent Generation R study failed to confirm these findings.[49] This discrepancy could depend on the different demographic characteristics of the women enrolled in the two studies (e.g., low versus slightly higher socioeconomic status in the MHPCD and Generation R cohorts, respectively) or concurrent environmental events.

Beyond affective and psychosis-like conditions, there is evidence that a history of pre-natal *Cannabis* use might be associated with enhanced sensitivity to drugs of abuse later in life. Both the MHPCD and OPPS studies found that the male offspring of mothers who reported using *Cannabis* during pregnancy were at increased risk for subsequent initiation of *Cannabis* use when adolescent/young adult as compared to the offspring of mothers who did not report using the drug while pregnant.[53–55] While intriguing, a positive relationship between pre-natal *Cannabis* exposure and the early onset of *Cannabis* use was not found in another study that utilized both maternal self-report and infant meconium to measure levels of gestational exposure.[56] Data from the OPPS study also indicate that use of *Cannabis* during pregnancy impairs cognitive flexibility, sustained and focused attention, planning and working memory and goal-directed behavior in the offspring: functional magnetic resonance imaging (fMRI) experiments showed that 18–22 year-old young adults exposed to *Cannabis* during pregnancy had altered neuronal functioning in several brain areas during both visuospatial working memory processing[57] and response inhibition.[58] Collectively, the OPPS, MHPCD, and Generation R studies demonstrated a subtle rather than gross effect of *Cannabis* upon later functioning including specific cognitive deficits especially in visuospatial function, impulsivity, inattention and hyperactivity, depressive symptoms, and substance use disorders.

In 2015 the National Institute of Health (NIH) launched the Adolescent Brain Cognitive Development (ABCD) study, the largest ongoing long-term study of brain development and child health ever conducted in the U.S., to investigate the effects of substance use (including *Cannabis*), as well as environmental, social, genetic, and other biological factors on the developing adolescent brain. The study recruited around 11,750 children between the ages of 9 and 10 years and exposure to *Cannabis* during pregnancy was identified through retrospective reports from parents and/or caregivers. Internalizing (i.e., anxious/depressed, withdrawn-depressed, and somatic complaints) and externalizing (i.e., rule-breaking and aggressive behavior) problems in children were assessed once or twice a year to verify for psychotic symptoms; behavioral assessments, together with physiological measures of cardiovascular health (e.g., blood pressure, cholesterol) and neuroimaging of brain structure and function, were performed every 2 years. The first results showed that, compared to children whose mothers did not use *Cannabis*, children whose mothers consumed the drug exhibited worse outcomes across multiple variables: exposed children were more likely to have higher levels of psychotic symptoms, as well as internalizing and externalizing behaviors, more frequent sleep disturbances, social problems, and lower scores on tests of attention and cognition.[59] This evidence should encourage pregnant women who use Cannabis to discontinue its use for reducing child's risk for psychopathology later in life.

Together with *Cannabis* use, there has been a dramatic increase in the acceptance and use of prescription opioids to treat moderate-to-severe pain especially in the U.S., posing threats to society. According to data collected from the Centers for Disease Control and Prevention (CDC), the number of women with opioid-related diagnoses documented at delivery increased by 131% from 2010 to 2017; among them, there is a high rate of polypharmacy, including (but not limited to) the concomitant use of alcohol, cocaine, benzodiazepines, and *Cannabis*.[60] Studies evaluating the consequences of concomitant opioid and marijuana use during pregnancy still remain limited. A recent evidence reported that neonates born to marijuana-positive women in opioid-exposed pregnancy were more likely to be born pre-term, small for gestational age, have low birth weight, and be admitted to neonatal intensive care unit.[61] Given the increasing number of states legalizing *Cannabis*, there is a need to understand infant health outcomes among mothers with opioid-related diagnoses and who also present concomitant marijuana use to educate about the risks of concurrent marijuana use during pregnancy.

A retrospective analysis of all live births in Ontario, Canada, between April 1, 2007 and March 31, 2012 linked pregnancy and birth data to provincial health administrative databases to ascertain child neurodevelopmental outcomes.[62] This analysis found an association between maternal *Cannabis* use in pregnancy and the incidence of autism spectrum disorder (ASD) in the offspring. The incidence of ASD diagnosis after age 18 months was 4.00 per 1000 person-years among children with maternal exposure to *Cannabis* compared to 2.42 among unexposed children; moreover, the incidence of attention deficit hyperactivity disorder (ADHD), intellectual disability and learning disorders after age 4 was higher among offspring of mothers who use *Cannabis* in pregnancy, although less statistically robust. Even though these findings need a cautious interpretation (for a commentary see Ref. 63), expectant mothers should avoid using *Cannabis* for recreational or medical purposes until further research is available given potential harms to infants beyond ASD.

Cannabinoids can easily be transferred through the maternal milk during breastfeeding.[7,64] Considering that the legalization/decriminalization of *Cannabis* is rapidly expanding worldwide and given that breastfeeding is strongly encouraged in many countries, it is concerning that neonates may be exposed to *Cannabis* during early infancy when breastfed. Very few studies have analyzed the harms of *Cannabis* exposure through lactation in the progeny (for a recent review see Refs. 7,65,66). The only studies investigating the consequences of *Cannabis* use during breastfeeding were conducted more than 30 years ago when the concentrations of THC in available *Cannabis* products were likely much lower than today and reported conflicting results on motor and mental performance in 1-year-old children.[67,68] A more recent cross-sectional survey conducted in Colorado in 2014 and 2015 found that both pre-natal and postnatal *Cannabis* use were associated with a shorter duration of breastfeeding.[69] Moreover, women who reported using the drug were more likely to smoke cigarettes, experience postpartum depressive symptoms, and breastfeed for less than 8 weeks.[70] Given the evident challenges and confounding factors that characterize human studies, particularly at this developmental epoch, animal models are critical to clarify the impact of developmental exposure to cannabinoids, and to shed light on the neurobiological substrates underlying such effects.

Rodent studies on the behavioral consequences of maternal cannabinoid exposure

A key strategy to overcome the intrinsic limitations of the clinical studies is to use animal models, which allow (1) to control over the possible confounding factors that characterize human studies, and also (2) to dissect the role of specific brain areas and neural pathways on *Cannabis*-induced developmental insult. The endocannabinoid system mediates the actions of THC and plays a critical regulatory role throughout all developmental stages[71]; moreover *Cannabis* exposure during critical periods (e.g., pregnancy, lactation) can impair brain maturation and pre-dispose the offspring to behavioral disturbances later in life.[5,14,72,73] A recent review manuscript from Bara and colleagues provided an elegant overview of the neurobiological effects of cannabinoid exposure during pre-natal/peri-natal and adolescent periods by highlighting that impaired synaptic plasticity induced in different cell types and neurotransmitter systems is relevant to the behavioral outcomes associated with in utero cannabinoid exposure. The authors also highlighted the important contribution of epigenetic factors in maintaining the enduring phenotypic changes which persist at adulthood and across generation following pre-natal/peri-natal *Cannabis* exposure.[7]

Different protocols are used to model maternal cannabinoid exposure in rodents, depending on the cannabinoid agonist used (i.e., either the active ingredient of *Cannabis* THC or synthetic cannabinoid receptor agonists, mainly WIN55,212-2), the doses and routes of administration, the time window of exposure and the behavioral outcomes measured in the offspring. As for the clinical relevance of the pre-clinical studies, it is important to estimate, by extrapolation, whether the dose of THC or other synthetic cannabinoid receptor agonists administered to the experimental animals is comparable to the dose of THC absorbed by

Cannabis users. It has been estimated that an oral dose of 5 mg/kg of THC in rats given daily through pregnancy and/or lactation corresponds to a moderate exposure to the drug in humans, correcting for the differences in route of administration and body weight surface area.[74,75] This dose is devoid of overt signs of toxicity and/or gross malformations and for this reason it has been widely used in pre-clinical studies. The synthetic cannabinoid agonist WIN55,212-2 has been found to be 3–10 times more potent than THC, depending on the administration route and the behavioral endpoints considered.[76,77] Based on these considerations, the dose of WIN55,212-2 normally used in maternal exposure protocols is 0.5 mg/kg via a subcutaneous injection.[44,78] As for the time window of exposure, most protocols are based upon either pre-natal [most often from gestational day (GD) 0 till GD 20] or peri-natal (i.e., through pregnancy and lactation) cannabinoid exposure, with the pre-natal period being approximately equivalent to the first and second trimesters of human pregnancy and the first 10 postnatal days (PNDs) in rodents approximately equivalent to the third trimester.[79] Based on this evidence, pioneering animal studies have demonstrated specific deficits in rodents prenatally exposed to cannabinoids: for instance, a wide range of cognitive deficits have been reported in rats exposed to cannabinoid agonists (e.g., WIN55,212-2) during the pre/peri-natal period, from early deficits in olfaction-based social discrimination[80] to disrupted memory retention in the inhibitory avoidance task both at adolescence and adulthood[78,81,82] and impaired discrimination abilities in both the social discrimination[81] and object recognition[83] tests. These cognitive deficits have been related to changes in brain glutamatergic neurotransmission.[81,84–87] As for the impact of maternal cannabinoid exposure on emotional-related behaviors in rodents, both increased[88] and decreased[44,80] rates of ultrasonic vocalizations (USVs) have been found in infant rats exposed to cannabinoid drugs (peri-natal THC vs. pre-natal WIN55,212-2). In this scenario, our group has recently revealed a previously undisclosed sexual divergence in the consequences of fetal cannabinoids on newborns at early developmental ages, which is dependent on mGlu5 receptor signaling. Prenatally WIN-exposed male infant pups emitted less isolation-induced USVs when separated from the dam and siblings at PND 10 compared with male control pups, and showed increased locomotor activity at PND 13, while females were spared. These effects were normalized when male pups were treated with the positive allosteric modulator of mGlu5 receptor CDPPB, confirming a key role of early brain glutamatergic neurotransmission in mediating the detrimental effects of maternal cannabinoid exposure. When tested at the pre-pubertal and pubertal periods, WIN-prenatally exposed rats of both sexes did not show any difference in social play behavior, anxiety and temporal order memory.[44] At later ages, maternal cannabinoid exposure also impacts the social repertoire of the offspring in a sex-dependent manner: thus, adult male but not female rats prenatally exposed to cannabinoid drugs showed social deficits associated with long-term depression (LTD) and heightened excitability of prefrontal cortex (PFC) pyramidal neurons.[43] Interestingly, positive allosteric modulation of mGlu5 receptors and enhancement of anandamide levels restored LTD and social interaction in cannabinoid-exposed males.[43] Overall, these studies provide new impetus for the urgent need to investigate the functional and behavioral substrates of maternal cannabinoid exposure in both the male and female progeny.

Previous findings showed sexual divergence in the long-term functional and behavioral consequences of maternal cannabinoid exposure when the progeny was tested for drug self-administration later in life: indeed, female but not male rats perinatally exposed to THC showed an increase in the rate of acquisition of intravenous morphine self-administration.[89] Moreover, male rats prenatally exposed to THC showed enhanced heroin-seeking only during mild stress and extinction.[90] Concerning the effects of maternal cannabinoid exposure on adult sensitivity to other drugs of abuse, peri-natal THC exposure did not influence ethanol self-administration in the adult male offspring,[91] while THC-exposed rats of both sexes showed a blunted locomotor response to amphetamine.[82] Recent research demonstrated that the male but not female rat offspring prenatally exposed to THC exhibit extensive molecular, cellular, and synaptic changes in dopamine neurons of the ventral tegmental area (VTA), resulting in a susceptible mesolimbic dopaminergic system associated with a psychotic-like endophenotype that is unmasked by a single exposure to THC later in life.[45,46] Moreover, the male rat offspring prenatally exposed to THC displayed a reduced population activity of VTA dopaminergic neurons in vivo and also exhibited enhanced sensitivity to dopamine D2 receptor activation and a vulnerability to acute stress, associated with compromised sensorimotor gating functions. This may lead to hypothesize that maternal cannabinoid exposure renders a neural substrate highly susceptible to subsequent challenges that may trigger psychotic-like outcomes.[92] As suggested by the authors, their findings indicate that maternal cannabinoid exposure impacts mesolimbic dopamine function and its related behavioral domains in a sex-dependent manner and warrant further investigations to decipher the mechanisms determining this sex-related protective effect from intrauterine THC exposure.

Aberrant dopaminergic neurotransmission represents a neuropathological feature in developmental THC model.[45,93] Recent evidence showed that male rats perinatally exposed to THC (i.e., from GD 15 to PND 9) exhibited social withdrawal and cognitive impairments at adulthood, together with brain alterations of dopamine D2 and cannabinoid CB1 receptors[94] which resemble those described in schizophrenic patients.[95] These adult abnormalities in rats were preceded at neonatal age by delayed appearance of neonatal reflexes, higher dopamine D2 mRNA and lower 2-arachidonoylglycerol (2-AG) brain levels, which persisted till adulthood. Interestingly, the detrimental effects induced by peri-natal THC were counteracted by peripubertal injection of the non-psychoactive phytocannabinoid cannabidiol (CBD),[94] further supporting the therapeutic potential of this cannabinoid for the treatment of neuropsychiatric disorders.[96,97]

Despite advances, challenges remain to understand the mechanisms underlying the consequences of pre/peri-natal *Cannabis* exposure. Among others, cannabinoids are inhibitors of the essential Hedgehog (HH) signaling pathway which has diverse functions in animal tissue development (i.e., it is involved in the growth and morphogenesis of several body structures including limbs, brain, heart, and craniofacial structure) and homeostasis.[98] It is thus possible that in utero exposure to THC that perturbs HH signaling at specific times during embryogenesis could represent a risk factor for developmental disorders, perhaps acting alongside pre-disposing genetic variants.[99]

The pre-clinical studies examining the effects of pre/peri-natal cannabinoid exposure typically employ injections of synthetic cannabinoids or isolated *Cannabis* constituents that may not accurately model *Cannabis* use in the human populations. In this context, a novel e-cigarette technology-based system to deliver vaporized *Cannabis* extracts to pregnant rats has been recently developed, that has translational relevance since in humans *Cannabis* is typically inhaled (through smoking or vaping).[100] The inhalation procedure induces rapid drug delivery and onset of physiological effects, together with a faster and higher peak of THC concentration. The authors tested the effects of pre-natal cannabinoid exposure on emotional, social, and cognitive endpoints in the male and female offspring from early development into adulthood. Dams were exposed to vaporized THC-enriched vapor (CAN_{THC}), vehicle vapor (VEH), or no vapor (AIR) twice daily during mating and gestation. The offspring exposed to CAN_{THC} emitted more USVs at PND 6 relative to VEH-exposed offspring, suggesting an altered emotional reactivity at infancy. At adolescence (PND 26), male CAN_{THC} offspring engaged in fewer social investigation behaviors than the VEH-exposed male offspring. At adulthood, CAN_{THC}-exposed offspring spent less time exploring the open arms of the elevated plus-maze and exhibited dose-dependent deficits in behavioral flexibility in an attentional set-shifting task relative to AIR controls. These data collectively indicate that pre-natal cannabinoid exposure may cause enduring emotional disturbances in the male offspring.

A few studies in rodents have begun to examine the consequences of cannabinoid exposure during lactation.[101–103] THC exposure via lactation (from PND 1 to PND 10) is associated with impaired early social communication (i.e., USVs) in rat pups,[102] in accordance with the results following in utero cannabinoid exposure.[44,80] Exposure to THC during lactation also induces lasting deficits in behavior and synaptic function which persist into adulthood in both the male and female progeny.[101,103] While a large gap in knowledge remains regarding the specific impact of cannabinoid exposure during lactation, this evidence suggests potential harm which warrants further investigations.

Future perspectives

In our opinion, there are many challenges to overcome in the understanding of the consequences of pre/peri-natal *Cannabis* exposure: (1) *Cannabis* and *Cannabis*-derived products have become increasingly available in recent years, with new and different types of products, especially those based on the non-psychoactive phytocannabinoid CBD. These products raise questions and concerns, especially because there is no comprehensive research studying the effects of CBD on the pregnant mother and the developing fetus. While THC has been widely studied in developmental and reproductive toxicology, there is only one study published in 1986 showing that high doses of CBD in pregnant mice negatively impacted the reproductive system of developing male fetuses.[104] This represents an important line of future research especially because CBD, due to its lipophilic nature, might readily cross the placenta and influence fetal development via endocannabinoid- and non-endocannabinoid-mediated mechanisms. (2) Prospective investigations delineating the behavioral outcomes in the offspring following paternal-only and/or combined maternal and paternal exposure to *Cannabis* are lacking. As previously discussed in this chapter, analysis from the Generation R cohort revealed that paternal *Cannabis* use was predictive of psychotic-like experiences and behavioral deficits in the offspring at ages 7–10 years.[50] In both humans and rats, it has been demonstrated that THC exposure alters DNA methylation in sperm cells[105] and there is pre-clinical evidence that in rats THC exposure during adolescence, prior to mating, may influence neurodevelopmental and behavioral outcomes in subsequent generations.[106,107] Investigating how paternal exposure to *Cannabis* might influence genetic expression, and therefore development, in the offspring becomes paramount. (3) There are insufficient data to evaluate the effects of *Cannabis* use on infants during lactation and breastfeeding, and in the absence of such data, the use of the drugs is discouraged. Different windows of maternal exposure including lactation should be thoroughly investigated since they represent important periods for postnatal development of the offspring. (4) There is a dearth of literature on the potential intergenerational effects of maternal cannabinoid exposure. Few studies suggested that developmental cannabinoid exposure may increase the intergenerational risk of psychiatric disease through epigenetic mechanisms.[108,109] This represents an appealing line of research whose data might be used by regulators and health communicators to inform consumers of potential risks associated with *Cannabis* use during specific time points in the life course.

Conclusions

In this chapter, we summarized the main available clinical and pre-clinical data on the behavioral effects of pre/peri-natal cannabinoid exposure. Since side effects affecting the fetus cannot be detected early enough to prevent potentially life-long damages, studying the consequences of maternal *Cannabis* exposure remains an important goal of current neuroscience research. Considering the increased popularity and legalization of *Cannabis*, more studies are warranted to assess *Cannabis* safety to aid clinicians and policymakers in evidence-informed decision-making. Moreover, with further understanding of the underlying mechanisms involved, safe interventions could be employed to ameliorate the detrimental outcomes for children exposed to cannabinoids in utero.

Acknowledgment

This work was supported by Autism Speaks grant #11690 (AM and VT), PRIN 2017 SXEXT5 (VT), Regione Lazio Progetti di Gruppi Ricerca 2020 (VT).

References

1. Azofeifa A, Mattson ME, Grant A. Monitoring marijuana use in the United States: challenges in an evolving environment. *JAMA* 2016;**316**:1765–6.
2. Metz TD, Stickrath EH. Marijuana use in pregnancy and lactation: a review of the evidence. *Am J Obstet Gynecol* 2015;**213**:761–78.
3. Young-Wolff KC, Tucker LY, Alexeeff S, et al. Trends in self-reported and biochemically tested marijuana use among pregnant females in California from 2009–2016. *JAMA* 2017;**318**:2490–1.
4. Hurd YL, Wang X, Anderson V, Beck O, Minkoff H, Dow-Edwards D. Marijuana impairs growth in mid-gestation fetuses. *Neurotoxicol Teratol* 2005;**27**:221–9.
5. Scheyer AF, Melis M, Trezza V, Manzoni OJJ. Consequences of perinatal *Cannabis* exposure. *Trends Neurosci* 2019;**42**:871–84.
6. Kim J, de Castro A, Lendoiro E, Cruz-Landeira A, Lopez-Rivadulla M, Concheiro M. Detection of in utero *Cannabis* exposure by umbilical cord analysis. *Drug Test Anal* 2018;**10**:636–43.
7. Bara A, Ferland JN, Rompala G, Szutorisz H, Hurd YL. *Cannabis* and synaptic reprogramming of the developing brain. *Nat Rev Neurosci* 2021;**22**:423–38.
8. Bertrand KA, Hanan NJ, Honerkamp-Smith G, Best BM, Chambers CD. Marijuana use by breastfeeding mothers and cannabinoid concentrations in breast milk. *Pediatrics* 2018;**142**.
9. Brown QL, Sarvet AL, Shmulewitz D, Martins SS, Wall MM, Hasin DS. Trends in marijuana use among pregnant and nonpregnant reproductive-aged women, 2002–2014. *JAMA* 2017;**317**:207–9.
10. Mark K, Gryczynski J, Axenfeld E, Schwartz RP, Terplan M. Pregnant Women's current and intended *Cannabis* use in relation to their views toward legalization and knowledge of potential harm. *J Addict Med* 2017;**11**:211–6.
11. Volkow ND, Han B, Compton WM, Blanco C. Marijuana use during stages of pregnancy in the United States. *Ann Intern Med* 2017;**166**:763–4.
12. Dickson B, Mansfield C, Guiahi M, et al. Recommendations from *Cannabis* dispensaries about first-trimester *Cannabis* use. *Obstet Gynecol* 2018;**131**:1031–8.
13. Hurd YL, Manzoni OJ, Pletnikov MV, Lee FS, Bhattacharyya S, Melis M. *Cannabis* and the developing brain: insights into its long-lasting effects. *J Neurosci* 2019;**39**:8250–8.
14. Trezza V, Campolongo P, Manduca A, et al. Altering endocannabinoid neurotransmission at critical developmental ages: impact on rodent emotionality and cognitive performance. *Front Behav Neurosci* 2012;**6**:2.
15. Tirado-Munoz J, Lopez-Rodriguez AB, Fonseca F, Farre M, Torrens M, Viveros MP. Effects of *Cannabis* exposure in the prenatal and adolescent periods: preclinical and clinical studies in both sexes. *Front Neuroendocrinol* 2020;**57**, 100841.
16. Calvigioni D, Hurd YL, Harkany T, Keimpema E. Neuronal substrates and functional consequences of prenatal *Cannabis* exposure. *Eur Child Adolesc Psychiatry* 2014;**23**:931–41.
17. Szutorisz H, Hurd YL. High times for *Cannabis*: epigenetic imprint and its legacy on brain and behavior. *Neurosci Biobehav Rev* 2018;**85**:93–101.
18. Agrawal A, Rogers CE, Lessov-Schlaggar CN, Carter EB, Lenze SN, Grucza RA. Alcohol, cigarette, and *Cannabis* use between 2002 and 2016 in pregnant women from a nationally representative sample. *JAMA Pediatr* 2019;**173**:95–6.

19. Fried PA. Postnatal consequences of maternal marijuana use in humans. *Ann N Y Acad Sci* 1989;**562**:123–32.
20. Fried PA. Adolescents prenatally exposed to marijuana: examination of facets of complex behaviors and comparisons with the influence of in utero cigarettes. *J Clin Pharmacol* 2002;**42**:97S–102S.
21. Fried PA, O'Connell CM, Watkinson B. 60- and 72-month follow-up of children prenatally exposed to marijuana, cigarettes, and alcohol: cognitive and language assessment. *J Dev Behav Pediatr* 1992;**13**:383–91.
22. Fried PA, Smith AM. A literature review of the consequences of prenatal marihuana exposure. An emerging theme of a deficiency in aspects of executive function. *Neurotoxicol Teratol* 2001;**23**:1–11.
23. Fried PA, Watkinson B. 36- and 48-month neurobehavioral follow-up of children prenatally exposed to marijuana, cigarettes, and alcohol. *J Dev Behav Pediatr* 1990;**11**:49–58.
24. Fried PA, Watkinson B. Visuoperceptual functioning differs in 9- to 12-year olds prenatally exposed to cigarettes and marihuana. *Neurotoxicol Teratol* 2000;**22**:11–20.
25. Fried PA, Watkinson B, Gray R. A follow-up study of attentional behavior in 6-year-old children exposed prenatally to marihuana, cigarettes, and alcohol. *Neurotoxicol Teratol* 1992;**14**:299–311.
26. Fried PA, Watkinson B, Gray R. Differential effects on cognitive functioning in 13- to 16-year-olds prenatally exposed to cigarettes and marihuana. *Neurotoxicol Teratol* 2003;**25**:427–36.
27. Day N, Cornelius M, Goldschmidt L, Richardson G, Robles N, Taylor P. The effects of prenatal tobacco and marijuana use on offspring growth from birth through 3 years of age. *Neurotoxicol Teratol* 1992;**14**:407–14.
28. Day NL, Richardson GA, Goldschmidt L, et al. Effect of prenatal marijuana exposure on the cognitive development of offspring at age three. *Neurotoxicol Teratol* 1994;**16**:169–75.
29. Goldschmidt L, Day NL, Richardson GA. Effects of prenatal marijuana exposure on child behavior problems at age 10. *Neurotoxicol Teratol* 2000;**22**:325–36.
30. Goldschmidt L, Richardson GA, Cornelius MD, Day NL. Prenatal marijuana and alcohol exposure and academic achievement at age 10. *Neurotoxicol Teratol* 2004;**26**:521–32.
31. Goldschmidt L, Richardson GA, Willford J, Day NL. Prenatal marijuana exposure and intelligence test performance at age 6. *J Am Acad Child Adolesc Psychiatry* 2008;**47**:254–63.
32. Goldschmidt L, Richardson GA, Willford JA, Severtson SG, Day NL. School achievement in 14-year-old youths prenatally exposed to marijuana. *Neurotoxicol Teratol* 2012;**34**:161–7.
33. Hofman A, Jaddoe VW, Mackenbach JP, et al. Growth, development and health from early fetal life until young adulthood: the Generation R Study. *Paediatr Perinat Epidemiol* 2004;**18**:61–72.
34. El Marroun H, Tiemeier H, Jaddoe VW, et al. Demographic, emotional and social determinants of *Cannabis* use in early pregnancy: the Generation R study. *Drug Alcohol Depend* 2008;**98**:218–26.
35. Jaddoe VW, van Duijn CM, van der Heijden AJ, et al. The Generation R Study: design and cohort update 2010. *Eur J Epidemiol* 2010;**25**:823–41.
36. Kruithof CJ, Kooijman MN, van Duijn CM, et al. The Generation R Study: biobank update 2015. *Eur J Epidemiol* 2014;**29**:911–27.
37. Kooijman MN, Kruithof CJ, van Duijn CM, et al. The Generation R Study: design and cohort update 2017. *Eur J Epidemiol* 2016;**31**:1243–64.
38. Richardson GA, Day NL, Goldschmidt L. Prenatal alcohol, marijuana, and tobacco use: infant mental and motor development. *Neurotoxicol Teratol* 1995;**17**:479–87.
39. El Marroun H, Hudziak JJ, Tiemeier H, et al. Intrauterine *Cannabis* exposure leads to more aggressive behavior and attention problems in 18-month-old girls. *Drug Alcohol Depend* 2011;**118**:470–4.

40. Jaddoe VW, van Duijn CM, Franco OH, et al. The Generation R Study: design and cohort update 2012. *Eur J Epidemiol* 2012;**27**:739–56.
41. Traccis F, Frau R, Melis M. Gender differences in the outcome of offspring prenatally exposed to drugs of abuse. *Front Behav Neurosci* 2020;**14**:72.
42. Wang X, Dow-Edwards D, Anderson V, Minkoff H, Hurd YL. In utero marijuana exposure associated with abnormal amygdala dopamine D2 gene expression in the human fetus. *Biol Psychiatry* 2004;**56**:909–15.
43. Bara A, Manduca A, Bernabeu A, et al. Sex-dependent effects of in utero cannabinoid exposure on cortical function. *Elife* 2018;**7**.
44. Manduca A, Servadio M, Melancia F, Schiavi S, Manzoni OJ, Trezza V. Sex-specific behavioural deficits induced at early life by prenatal exposure to the cannabinoid receptor agonist WIN55,212-2 depend on mGlu5 receptor signalling. *Br J Pharmacol* 2020;**177**:449–63.
45. Frau R, Miczan V, Traccis F, et al. Prenatal THC exposure produces a hyperdopaminergic phenotype rescued by pregnenolone. *Nat Neurosci* 2019;**22**:1975–85.
46. Traccis F, Serra V, Sagheddu C, et al. Prenatal THC does not affect female mesolimbic dopaminergic system in preadolescent rats. *Int J Mol Sci* 2021;**22**.
47. de Salas-Quiroga A, Garcia-Rincon D, Gomez-Dominguez D, et al. Long-term hippocampal interneuronopathy drives sex-dimorphic spatial memory impairment induced by prenatal THC exposure. *Neuropsychopharmacology* 2020;**45**:877–86.
48. Leech SL, Richardson GA, Goldschmidt L, Day NL. Prenatal substance exposure: effects on attention and impulsivity of 6-year-olds. *Neurotoxicol Teratol* 1999;**21**:109–18.
49. El Marroun H, Bolhuis K, Franken IHA, et al. Preconception and prenatal *Cannabis* use and the risk of behavioural and emotional problems in the offspring; a multi-informant prospective longitudinal study. *Int J Epidemiol* 2019;**48**:287–96.
50. Bolhuis K, Kushner SA, Yalniz S, et al. Maternal and paternal *Cannabis* use during pregnancy and the risk of psychotic-like experiences in the offspring. *Schizophr Res* 2018;**202**:322–7.
51. Gray KA, Day NL, Leech S, Richardson GA. Prenatal marijuana exposure: effect on child depressive symptoms at ten years of age. *Neurotoxicol Teratol* 2005;**27**:439–48.
52. Leech SL, Larkby CA, Day R, Day NL. Predictors and correlates of high levels of depression and anxiety symptoms among children at age 10. *J Am Acad Child Adolesc Psychiatry* 2006;**45**:223–30.
53. Day NL, Goldschmidt L, Thomas CA. Prenatal marijuana exposure contributes to the prediction of marijuana use at age 14. *Addiction* 2006;**101**:1313–22.
54. Porath AJ, Fried PA. Effects of prenatal cigarette and marijuana exposure on drug use among offspring. *Neurotoxicol Teratol* 2005;**27**:267–77.
55. Sonon KE, Richardson GA, Cornelius JR, Kim KH, Day NL. Prenatal marijuana exposure predicts marijuana use in young adulthood. *Neurotoxicol Teratol* 2015;**47**:10–5.
56. Frank DA, Kuranz S, Appugliese D, et al. Problematic substance use in urban adolescents: role of intrauterine exposures to cocaine and marijuana and post-natal environment. *Drug Alcohol Depend* 2014;**142**:181–90.
57. Smith AM, Fried PA, Hogan MJ, Cameron I. Effects of prenatal marijuana on visuospatial working memory: an fMRI study in young adults. *Neurotoxicol Teratol* 2006;**28**:286–95.
58. Smith AM, Fried PA, Hogan MJ, Cameron I. Effects of prenatal marijuana on response inhibition: an fMRI study of young adults. *Neurotoxicol Teratol* 2004;**26**:533–42.
59. Paul SE, Hatoum AS, Fine JD, et al. Associations between prenatal *Cannabis* exposure and childhood outcomes: results from the ABCD study. *JAMA Psychiat* 2021;**78**:64–76.
60. Brogly SB, Saia KE, Werler MM, Regan E, Hernandez-Diaz S. Prenatal treatment and outcomes of women with opioid use disorder. *Obstet Gynecol* 2018;**132**:916–22.

61. Shah DS, Turner EL, Chroust AJ, Duvall KL, Wood DL, Bailey BA. Marijuana use in opioid exposed pregnancy increases risk of preterm birth. *J Matern Fetal Neonatal Med* 2021;1–6.
62. Corsi DJ, Donelle J, Sucha E, et al. Maternal *Cannabis* use in pregnancy and child neurodevelopmental outcomes. *Nat Med* 2020;**26**:1536–40.
63. Sajdeya R, Brown JD, Goodin AJ. Evidence in context—commentary: perinatal *Cannabis* exposures and autism spectrum disorders. *Med Cannabis Cannabinoids* 2021;**4**:67–71.
64. Baker T, Datta P, Rewers-Felkins K, Thompson H, Kallem RR, Hale TW. Transfer of inhaled *Cannabis* into human breast milk. *Obstet Gynecol* 2018;**131**:783–8.
65. Badowski S, Smith G. *Cannabis* use during pregnancy and postpartum. *Can Fam Physician* 2020;**66**:98–103.
66. Ordean A, Kim G. *Cannabis* use during lactation: literature review and clinical recommendations. *J Obstet Gynaecol Can* 2020;**42**:1248–53.
67. Tennes K, Avitable N, Blackard C, et al. Marijuana: prenatal and postnatal exposure in the human. *NIDA Res Monogr* 1985;**59**:48–60.
68. Astley SJ, Little RE. Maternal marijuana use during lactation and infant development at one year. *Neurotoxicol Teratol* 1990;**12**:161–8.
69. Crume TL, Juhl AL, Brooks-Russell A, Hall KE, Wymore E, Borgelt LM. *Cannabis* use during the perinatal period in a state with legalized recreational and medical marijuana: the association between maternal characteristics, breastfeeding patterns, and neonatal outcomes. *J Pediatr* 2018;**197**:90–6.
70. Ko JY, Tong VT, Bombard JM, Hayes DK, Davy J, Perham-Hester KA. Marijuana use during and after pregnancy and association of prenatal use on birth outcomes: a population-based study. *Drug Alcohol Depend* 2018;**187**:72–8.
71. Harkany T, Keimpema E, Barabas K, Mulder J. Endocannabinoid functions controlling neuronal specification during brain development. *Mol Cell Endocrinol* 2008;**286**:S84–90.
72. Alpar A, Di Marzo V, Harkany T. At the tip of an iceberg: prenatal marijuana and its possible relation to neuropsychiatric outcome in the offspring. *Biol Psychiatry* 2016;**79**:e33–45.
73. Higuera-Matas A, Ucha M, Ambrosio E. Long-term consequences of perinatal and adolescent cannabinoid exposure on neural and psychological processes. *Neurosci Biobehav Rev* 2015;**55**:119–46.
74. Garcia-Gil L, De Miguel R, Munoz RM, et al. Perinatal delta(9)-tetrahydrocannabinol exposure alters the responsiveness of hypothalamic dopaminergic neurons to dopamine-acting drugs in adult rats. *Neurotoxicol Teratol* 1997;**19**:477–87.
75. Molina-Holgado F, Amaro A, Gonzalez MI, Alvarez FJ, Leret ML. Effect of maternal delta-9-tetrahydrocannabinol on developing serotonergic system. *Eur J Pharmacol* 1996;**316**:39–42.
76. Compton DR, Gold LH, Ward SJ, Balster RL, Martin BR. Aminoalkylindole analogs: cannabimimetic activity of a class of compounds structurally distinct from delta-9-tetrahydrocannabinol. *J Pharmacol Exp Ther* 1992;**263**:1118–26.
77. French ED, Dillon K, Wu X. Cannabinoids excite dopamine neurons in the ventral tegmentum and substantia nigra. *Neuroreport* 1997;**8**:649–52.
78. Mereu G, Fa M, Ferraro L, et al. Prenatal exposure to a cannabinoid agonist produces memory deficits linked to dysfunction in hippocampal long-term potentiation and glutamate release. *Proc Natl Acad Sci U S A* 2003;**100**:4915–20.
79. Spear LP, File SE. Methodological considerations in neurobehavioral teratology. *Pharmacol Biochem Behav* 1996;**55**:455–7.
80. Antonelli T, Tomasini MC, Tattoli M, et al. Prenatal exposure to the CB1 receptor agonist WIN 55,212-2 causes learning disruption associated with impaired cortical NMDA receptor function and emotional reactivity changes in rat offspring. *Cereb Cortex* 2005;**15**:2013–20.

81. Campolongo P, Trezza V, Cassano T, et al. Perinatal exposure to delta-9-tetrahydrocannabinol causes enduring cognitive deficits associated with alteration of cortical gene expression and neurotransmission in rats. *Addict Biol* 2007;**12**:485–95.
82. Silva L, Zhao N, Popp S, Dow-Edwards D. Prenatal tetrahydrocannabinol (THC) alters cognitive function and amphetamine response from weaning to adulthood in the rat. *Neurotoxicol Teratol* 2012;**34**:63–71.
83. O'Shea M, McGregor IS, Mallet PE. Repeated cannabinoid exposure during perinatal, adolescent or early adult ages produces similar longlasting deficits in object recognition and reduced social interaction in rats. *J Psychopharmacol* 2006;**20**:611–21.
84. Antonelli T, Tanganelli S, Tomasini MC, et al. Long-term effects on cortical glutamate release induced by prenatal exposure to the cannabinoid receptor agonist (R)-(+)-[2,3-dihydro-5-methyl-3-(4-morpholinyl-methyl)pyrrolo[1,2,3-de]-1, 4-benzoxazin-6-yl]-1-naphthalenylmethanone: an in vivo microdialysis study in the awake rat. *Neuroscience* 2004;**124**:367–75.
85. Castaldo P, Magi S, Gaetani S, et al. Prenatal exposure to the cannabinoid receptor agonist WIN 55,212-2 increases glutamate uptake through overexpression of GLT1 and EAAC1 glutamate transporter subtypes in rat frontal cerebral cortex. *Neuropharmacology* 2007;**53**:369–78.
86. Ferraro L, Tomasini MC, Beggiato S, et al. Short- and long-term consequences of prenatal exposure to the cannabinoid agonist WIN55,212-2 on rat glutamate transmission and cognitive functions. *J Neural Transm* 2009;**116**:1017–27.
87. Castaldo P, Magi S, Cataldi M, et al. Altered regulation of glutamate release and decreased functional activity and expression of GLT1 and GLAST glutamate transporters in the hippocampus of adolescent rats perinatally exposed to Delta(9)-THC. *Pharmacol Res* 2010;**61**:334–41.
88. Trezza V, Campolongo P, Cassano T, et al. Effects of perinatal exposure to delta-9-tetrahydrocannabinol on the emotional reactivity of the offspring: a longitudinal behavioral study in Wistar rats. *Psychopharmacology (Berlin)* 2008;**198**:529–37.
89. Vela G, Martin S, Garcia-Gil L, et al. Maternal exposure to delta-9-tetrahydrocannabinol facilitates morphine self-administration behavior and changes regional binding to central mu opioid receptors in adult offspring female rats. *Brain Res* 1998;**807**:101–9.
90. Spano MS, Ellgren M, Wang X, Hurd YL. Prenatal *Cannabis* exposure increases heroin seeking with allostatic changes in limbic enkephalin systems in adulthood. *Biol Psychiatry* 2007;**61**:554–63.
91. Economidou D, Mattioli L, Ubaldi M, et al. Role of cannabinoidergic mechanisms in ethanol self-administration and ethanol seeking in rat adult offspring following perinatal exposure to delta-9-tetrahydrocannabinol. *Toxicol Appl Pharmacol* 2007;**223**:73–85.
92. Sagheddu C, Traccis F, Serra V, et al. Mesolimbic dopamine dysregulation as a signature of information processing deficits imposed by prenatal THC exposure. *Prog Neuropsychopharmacol Biol Psychiatry* 2021;**105**, 110128.
93. Renard J, Rosen LG, Loureiro M, et al. Adolescent cannabinoid exposure induces a persistent sub-cortical hyper-dopaminergic state and associated molecular adaptations in the prefrontal cortex. *Cereb Cortex* 2017;**27**:1297–310.
94. Di Bartolomeo M, Stark T, Maurel OM, et al. Crosstalk between the transcriptional regulation of dopamine D2 and cannabinoid CB1 receptors in schizophrenia: analyses in patients and in perinatal delta-9-tetrahydrocannabinol-exposed rats. *Pharmacol Res* 2021;**164**, 105357.
95. D'Addario C, Micale V, Di Bartolomeo M, et al. A preliminary study of endocannabinoid system regulation in psychosis: distinct alterations of CNR1 promoter DNA methylation in patients with schizophrenia. *Schizophr Res* 2017;**188**:132–40.

96. Premoli M, Aria F, Bonini SA, et al. Cannabidiol: recent advances and new insights for neuropsychiatric disorders treatment. *Life Sci* 2019;**224**:120–7.
97. Stark T, Di Martino S, Drago F, Wotjak CT, Micale V. Phytocannabinoids and schizophrenia: focus on adolescence as a critical window of enhanced vulnerability and opportunity for treatment. *Pharmacol Res* 2021;**174**, 105938.
98. Briscoe J, Therond PP. The mechanisms of Hedgehog signalling and its roles in development and disease. *Nat Rev Mol Cell Biol* 2013;**14**:416–29.
99. Lo HF, Hong M, Szutorisz H, Hurd YL, Krauss RS. Delta-9-tetrahydrocannabinol inhibits Hedgehog-dependent patterning during development. *Development* 2021;**148**.
100. Weimar HV, Wright HR, Warrick CR, et al. Long-term effects of maternal *Cannabis* vapor exposure on emotional reactivity, social behavior, and behavioral flexibility in offspring. *Neuropharmacology* 2020;**179**, 108288.
101. Scheyer AF, Borsoi M, Pelissier-Alicot AL, Manzoni OJJ. Perinatal THC exposure via lactation induces lasting alterations to social behavior and prefrontal cortex function in rats at adulthood. *Neuropsychopharmacology* 2020;**45**:1826–33.
102. Scheyer AF, Borsoi M, Wager-Miller J, et al. Cannabinoid exposure via lactation in rats disrupts perinatal programming of the gamma-aminobutyric acid trajectory and select early-life behaviors. *Biol Psychiatry* 2020;**87**:666–77.
103. Scheyer AF, Borsoi M, Pelissier-Alicot AL, Manzoni OJJ. Maternal exposure to the cannabinoid agonist WIN 55,12,2 during lactation induces lasting behavioral and synaptic alterations in the rat adult offspring of both sexes. *eNeuro* 2020;**7**.
104. Dalterio SL, deRooij DG. Maternal cannabinoid exposure. Effects on spermatogenesis in male offspring. *Int J Androl* 1986;**9**:250–8.
105. Murphy SK, Itchon-Ramos N, Visco Z, et al. Cannabinoid exposure and altered DNA methylation in rat and human sperm. *Epigenetics* 2018;**13**:1208–21.
106. Szutorisz H, DiNieri JA, Sweet E, et al. Parental THC exposure leads to compulsive heroin-seeking and altered striatal synaptic plasticity in the subsequent generation. *Neuropsychopharmacology* 2014;**39**:1315–23.
107. Nashed MG, Hardy DB, Laviolette SR. Prenatal cannabinoid exposure: emerging evidence of physiological and neuropsychiatric abnormalities. *Front Psych* 2020;**11**, 624275.
108. Smith A, Kaufman F, Sandy MS, Cardenas A. *Cannabis* exposure during critical windows of development: epigenetic and molecular pathways implicated in neuropsychiatric disease. *Curr Environ Health Rep* 2020;**7**:325–42.
109. Szutorisz H, Egervari G, Sperry J, Carter JM, Hurd YL. Cross-generational THC exposure alters the developmental sensitivity of ventral and dorsal striatal gene expression in male and female offspring. *Neurotoxicol Teratol* 2016;**58**:107–14.

Chapter 6

Correlates and consequences of cannabinoid exposure on adolescent brain remodeling: Focus on glial cells and epigenetics

Zamberletti Erica[a], Manenti Cristina[a], Gabaglio Marina[a], Rubino Tiziana[a], and Parolaro Daniela[b]
[a]Deptartment of Biotechnology and Life Sciences and Neuroscience Center, University of Insubria, Busto Arsizio, Italy, [b]Zardi-Gori Foundation, Milan, Italy

Introduction

As discussed in other chapters of this book (10 and 11), prolonged exposure to cannabinoids during adolescence has been associated with long-term detrimental effects on emotional and cognitive functions, possibly contributing to the development of psychiatric disorders. Current literature supports the hypothesis that the persistent behavioral deficits observed following adolescent cannabinoid exposure may be driven by long-lasting molecular changes induced by exogenous cannabinoid exposure in the developing brain. Although the great majority of studies have focused on cannabinoid-induced alterations in neuronal function and structure, recent evidence suggests that glial cells and epigenetic remodeling may represent additional neurobiological underpinnings of cannabis-induced dysfunctions. This chapter will describe available data in support of a contribution of glial cells and epigenetic mechanisms to the detrimental behavioral effects associated with adolescent cannabinoid exposure.

Glial cells

The endocannabinoid system in glial cells

EC system in astrocytes

Astrocytes represent a large population of brain cells that actively contributes to fundamental metabolic, structural and protective functions in the central nervous system. These cells are able to detect synaptic signals coming from neurons and readily respond to these stimuli by modulating neuronal activity.[1]

Although at expression levels much lower than those present in neurons, recent studies demonstrated the presence of CB1 receptors in astrocytes both at the plasma membrane level and on mitochondria in several brain regions including the hippocampus, the neocortex, and the caudate putamen.[2] CB1 receptors located on astrocyte plasma membrane are mainly coupled to Gαq/11 proteins that activate phospholipase C, leading to the production of inositol 1,4,5-trisphosphate (IP3) and intracellular calcium release, and their activation has been associated with an increase in soma size and processes of astrocytes.[3] In contrast, activation of mitochondrial CB1 receptors coupled to Gαi/o proteins, leads to inhibition of adenylyl cyclase and reduction of cellular respiration and ATP production.[4]

Both in vitro and ex vivo studies revealed that astrocytes are also capable of producing and releasing endocannabinoids,[2,5,6] and recent evidence in animal models supports the idea that astrocyte-produced endocannabinoids can affect neuronal transmission and behavior.[5,7] However, further studies are still needed to clarify the functional relevance of endocannabinoid production by astrocytes and its impact on neuronal functioning.

EC system on microglia

Microglia are the resident immune cells of the brain that continuously survey the parenchyma and rapidly respond to changes in the brain.[8] In response to pro-inflammatory cytokines and other signaling molecules, they can transition to an active state by altering glial-specific intermediate filament proteins in the cytoplasm, resulting in diminished cellular processes and soma hypertrophy.[9] Many studies have revealed that microglia cells express both cannabinoid CB1 and CB2 receptors and possess all the machinery to produce endocannabinoids. Both CB1 and CB2 receptors are found at constitutively low levels in resting microglia whereas upregulation of CB2 receptor expression has been found following activation of microglia by inflammatory insult.[6] This effect is thought to help limit the inflammatory process, as activation of CB2 receptors by cannabinoids modulates microglial immune function by increasing microgliosis, migration, phagocytosis, and by reducing their production of pro-inflammatory mediators such as tumor necrosis factor (TNF)-α and free radicals.[10] Microglia cells also express the endocannabinoid synthetic and degrading enzymes and were shown to exceed neuron and astrocyte production of endocannabinoids in vitro,[11,12] suggesting that this cell type might be the main cellular source of endocannabinoids in neuroinflammatory conditions. Increasing the production of endocannabinoids is thought to amplify the anti-inflammatory and protective microglial phenotype mostly through activation of CB2 receptors.[13] A central role of CB2 receptors in microglia activation processes has been confirmed by a recent study demonstrating that CB2 receptor signaling can modulate microglial polarization.[14] Overall, available evidence suggests that targeting the endocannabinoid system on microglia cells could be a useful strategy to maintain or enhance the protective microglial phenotype and to relieve neuroinflammatory states.

EC system on oligodendrocytes

In the central nervous system, oligodendrocytes assemble myelin, and provide metabolic support to myelinated axons.[15,16] The oligodendrocyte developmental program begins with the specification of oligodendrocyte precursor cells (OPC), which emerge and expand in the late prenatal and post-natal periods. OPC differentiate to become the mature myelinating oligodendrocyte cells, which produce axonal myelin.[17] Progression through the mature oligodendrocyte lineage is tightly regulated by a multitude of intrinsic and extrinsic cues, which also play roles during adult lifespan and under pathological conditions.[18]

The expression of both CB1 and CB2 cannabinoid receptors on OPC and mature oligodendrocytes has been demonstrated both at the gene level and at the protein level,[19,20] suggesting that cannabinoid receptors could regulate cell function at multiple stages of oligodendrocyte lineage progression. Changes in CB1 expression levels have been reported in human multiple sclerosis samples,[21] possibly suggesting that the endocannabinoid system could also play a role in cells of the oligodendrocyte lineage in pathological conditions.

Additionally, it has been shown that both OPC and oligodendrocytes express the enzymes responsible for endocannabinoid metabolism. In particular, high levels of DAGL have been found in OPC whereas mature oligodendrocytes express higher level of MAGL, suggesting that OPC might produce a greater amount of 2-AG than mature oligodendrocytes.[22] In contrast, levels of anandamide are low and do not differ between the cell stages of oligodendrocyte lineage progression.[22] At the functional level, 2-AG has been highlighted as an important modulator of oligodendrogenesis and oligodendrocyte functions in different conditions.[20] Elevation of 2-AG levels by pharmacological inhibition of MAGL induces cell proliferation in cultured OPC, and promotes/accelerates oligodendrocyte maturation.[23–25] A role for 2-AG has also been highlighted in animal models of demyelination, where it attenuates myelin degeneration and inflammation and prevents mitochondrial dysfunction.[26]

Thus, current evidence demonstrates that both OPC and oligodendrocytes express components of the endocannabinoid system, and suggests that endocannabinoids might drive oligodendrocyte maturation and activity.

Consequences of adolescent THC exposure on glial cells

To date the majority of clinical and pre-clinical research investigating the molecular and cellular consequences of adolescent THC exposure on the brain has mainly focused on alterations in neuronal function and morphology.[27] However, the observation that also glial cells express cannabinoid receptors clearly indicates that chronic THC treatment would target not only neurons but also glial cells, possibly suggesting that the behavioral outcomes associated with THC could arise as a consequence of its actions on both neurons and glia. The following paragraphs will discuss the available studies that have investigated the long-term consequences of adolescent THC exposure on glia cell populations.

Effects of adolescent THC exposure on astrocytes

Adolescent THC treatment increases the expression of the astrocyte marker glial fibrillary acidic protein (GFAP) in the hilus of the dentate gyrus of both male and female rats,[28] suggesting a possible effect of THC on the astrocyte cell population. This has been confirmed by a subsequent study demonstrating the presence of astrocyte reactivity in the hippocampus of male rats after adolescent THC treatment.[29] Increased GFAP levels were associated with increased expression of some pro-inflammatory mediators, including TNF-α and iNOS, and reduced expression of the anti-inflammatory cytokine IL-10 in the same brain region.[29] In contrast, no changes in astrocyte reactivity have been observed in female rats after adolescent THC treatment in the pre-frontal cortex (PFC),[30] possibly suggesting that astrocytes within the hippocampus could display a greater vulnerability to THC. Interestingly, astrocyte CB1 receptors seem to mediate cognitive effects of THC and stimulation of astrocyte CB1 receptors by THC has been associated with enhanced inflammatory signaling, glutamate release, abnormal neuronal activities.[31]

Further supporting a role of CB1 receptors on astrocyte in mediating the effects of THC is the observation that the expression of a mutation of disrupted in schizophrenia 1 (DN-DISC1) gene, associated with an increased risk of developing psychiatric disorders selectively in astrocytes, worsened the effects of adolescent THC exposure on recognition memory evaluated at adulthood.[32] At the cellular level, this effect was associated with the activation of the pro-inflammatory NF-KB-COX-2 pathway in astrocytes and dysfunction of GABAergic neurons in the hippocampus.[32] Blocking the NF-KB-COX-2 pathway prevented the development of THC-induced cognitive deficit in this model,[32] supporting a causal role between astrocyte genetic risk factors and activation of hippocampal NF-KB-COX-2 pathway and cognitive effects of adolescent THC exposure.

Changes in astrocyte reactivity in the hippocampus following chronic cannabis use during adolescence have recently been confirmed in humans.[33] Compared with non-users, adolescent-onset regular cannabis users had significantly lower levels of the glial marker myoinositol in a brain voxel that included parts of the left hippocampus.[33] Moreover, myoinositol levels in cannabis users were positively correlated with glutamate levels in the hippocampus, an association that was not found in non-users,[33] suggesting that glia dysfunction could contribute to lower levels of glutamate availability and hippocampal-mediated cognitive impairments associated with cannabis use.

Overall, available data from both animal models and humans provide evidence for THC-induced altered astrocyte reactivity in the hippocampus, and seem to suggest that disruption of CB1 receptor signaling on astrocyte following THC intake could underlie hippocampal dysfunction and memory impairment.

Effects of adolescent THC exposure on microglia

Altered microglia morphology, increased expression of the pro-inflammatory markers, TNF-α, iNOS and COX-2, reduction of the anti-inflammatory cytokine, IL-10, and up-regulation of CB2 receptors on microglial cells have been described in the PFC of adult female rats chronically exposed to THC during adolescence.[30] Similar to what has been observed for astrocytes, also THC-induced microglia activation seems to be region-specific. While astrocyte alterations have been mainly observed in the hippocampus, microglia activation seems to preferentially involve cortical regions, since no alterations were detected in the nucleus accumbens (NAc), hippocampus, and amygdala.[30] The administration of an inhibitor of glia activation, Ibudilast, during THC treatment prevented short-term memory impairment present in adult rats, and the increases in TNF-α, iNOS, COX-2 levels as well as the up-regulation of CB2 receptors on microglial cells.[30] These data suggest that adolescent THC treatment can induce lasting alterations in microglia reactivity in the PFC of female rats, which might contribute to long-term cognitive impairments associated with the treatment. In contrast, microglia reactivity was not observed when the same treatment protocol was applied to male rats,[29] suggesting a sex-dimorphism in the effects of chronic THC administration during adolescence on microglia alterations. Changes in microglial reactivity and enhanced expression of specific pro-inflammatory genes, including IL-1β, have been also demonstrated after chronic THC administration in adulthood.[34] These alterations were observed in the cerebellum of male mice and were causally associated with CB1 receptor downregulation.[34] In adult animals, no changes were observed in the hippocampus, striatum or PFC,[34] suggesting that the susceptibility of microglia cells to THC might differ across different brain regions depending on the time of exposure.

Altogether, it appears that adolescent THC treatment can affect microglia reactivity in the long-term in a sex- and region-dependent manner. The few available studies seem to suggest that THC-induced microglia alterations could contribute to the development of cognitive deficits associated with the treatment both in adolescent and in adult animals.

Effects of adolescent THC exposure on oligodendrocytes

Despite the observation that myelination and white matter development play dominant roles during adolescence,[35] no study has yet investigated possible effects of adolescent THC exposure on oligodendrocytes and myelin formation. However, the possibility that THC might impact myelin formation in the adolescent brain is suggested also in light of a recent study that evaluated the effect of acute

THC on oligodendrocyte development during post-natal myelination of the central nervous system.[19] Acute THC was shown to induce oligodendrocyte development in the post-natal white matter by promoting OPC cell cycle exit and differentiation in vivo, and favored oligodendrocyte maturation and myelination.[19]

As the process of myelination continues to occur well into adolescence,[36] the investigation of possible long-term effect of adolescent THC exposure on white matter formation should represent a major goal for future studies.

Conclusions

Altogether, the (few) available data on the effect of chronic THC exposure on astrocytes and microglia suggest important roles played by these cells in response to chronic stimulation of CB1 receptors, supporting the hypothesis that both cell populations might contribute to the behavioral alterations associated with the treatment, especially on measures of cognitive defects. Current literature suggests that adolescent THC treatment affects astrocyte and microglia reactivity in a sex- and region-dependent manner. There is instead a complete lack of information on possible effects of adolescent THC exposure on oligodendrocytes and myelin development, which might also have potential relationship to behavior, and learning alterations. Further studies are needed to thoroughly comprehend the role of glial cells in mediating the behavioral effects of adolescent THC exposure, and how these cells might contribute to the synaptic alterations that have been highlighted following chronic THC treatment during adolescence.

Epigenetics

The long-term nature of the effects triggered by adolescent exposure to cannabinoids depicted in other chapters of this book (10 and 11) suggests the involvement of alterations at the level of gene expression. Emerging evidence has demonstrated a crucial role for epigenetic mechanisms in driving lasting changes in gene expression. Accordingly, these mechanisms are the ones that allow experiences to modify brain circuits in order to guide future adaptive behavior. The term "epigenetics" was coined by Waddington in 1942 to explain phenotypic changes that occur with cellular differentiation throughout development.[37] We know that all our cells share the same DNA and thus the same genes; however, these genes are differentially expressed in both time and space. In other words, in an organism, different cell types may express different genes keeping other ones silent. This different expression is obtained by epigenetic mechanisms, that constrain expression by adapting regions of the genome to promote either gene silencing or gene activity.[38] These mechanisms entail (i) direct chemical modification of DNA (DNA methylation) that, when present in promoter regions and transcriptional regulatory sequences, has frequently been associated with gene silencing; (ii) modifications of proteins that are closely associated with DNA, the histone proteins, which can increase or decrease the accessibility of genomic regions and the binding of transcription factors to the DNA; (iii) and the production of non-coding RNAs that may regulate gene expression at the transcriptional and post-transcriptional levels.[39] Epigenetic mechanisms are fundamental for driving developmental processes, but they have also critical functions in differentiated cells, where they allow cells to appropriately fine-tune their physiologies in response to environmental signals. For example, in mature neurons, epigenetic mechanisms act as key modulators of synaptic plasticity and memory.[40]

Epigenetic alterations induced by adolescent THC exposure

Despite the growing knowledge gained on these mechanisms, few studies have dealt with this topic in relation to adolescent cannabis/THC exposure.

The first paper addressing this issue was published in 2012 by Hurd's group.[41] The authors were interested in the neurobiological underpinnings of the enhanced heroin self-administration observed in adult rats after adolescent exposure to THC. They discovered that increased expression of the proenkefalin opioid neuropeptide in the NAc was fundamental in sustaining the self-administration behavior present at adulthood. Furthermore, this increased expression was paralleled by a reduction in di- and tri-methylation of histone H3 lysine 9 (H3K9), histone markers associated with transcriptional repression, in the Penk promotor region, suggesting that this histone modification may have a role in the proenkefalin upregulation.

More recently, Prini and co-workers described alterations in histone modifications 2, 24, and 48 h after the end of a protocol resembling moderate to heavy adolescent cannabis exposure in different brain areas of female rats, namely the PFC, hippocampus, NAc, and amygdala.[42,43] In order to investigate the actual vulnerability of the adolescent brain to such treatment, adult female rats were submitted to the same protocol and their response evaluated. In the adolescent brain, immediately after the end of the treatment (2–24 h), a general repressive effect was described, mediated mainly by an increase in di- and tri-methylation of H3K9. In the PFC, this wave of transcriptional repression mostly impacted the expression of genes closely associated with synaptic plasticity, and was mediated by the histone methyltransferase SUV39H1. Remarkably, pharmacological inhibition of SUV39H1, and thus blockade of methylation of H3K9 during adolescent THC treatment, prevented the development of long-lasting THC-induced cognitive deficits.[43] In the adolescent brain, this primary repressive effect was followed by alterations in other histone modifications producing a homeostatic response that promoted gene expression, to counterbalance the initial transcriptional repression. This was observed 48 h after the end of treatment and was likely driven by the absence of THC. In contrast, adult THC exposure induced milder changes, restricted to just one histone modification within 24 h after the cessation of the treatment and without homeostatic changes. Furthermore, in the adult brain these histone modifications often had an opposite outcome at transcriptional level when compared to the ones described in the adolescent brain. However, in the amygdala, a more complex picture was present at adulthood, not during adolescence. It is known that the amygdala is activated during exposure to aversive stimuli, functioning as a "behavioral brake."[44] This different response between the adult and adolescent amygdala could represent the biological basis of the adolescent propensity for risk-taking and novelty-seeking behaviors. On the other hand, a greater responsiveness of the adolescent NAc was observed when compared to the adult one. This higher NAc/lower amygdala responsiveness of the adolescent brain could fit well with the hypothesis of an imbalance in the tension between reward and threat systems in favor of reward in this developmental window.[45]

In male rats, even chronic low THC doses, administered starting from the very early adolescence, were able to disrupt morphological and transcriptional trajectory of PFC pyramidal neurons.[46] Indeed, in THC-treated animals, a profound reorganization of the transcriptome mainly related to histone modification, chromatin remodeling, and synaptic plasticity gene networks was described in layer III pyramidal neurons.

Reprogramming of key brain regions to make them respond differently to future insults has also been shown after synthetic cannabinoid exposure during adolescence.[47] Indeed, WIN 55,212-2 pre-exposure in male rats resulted in cross-sensitization to cocaine which was paralleled by an increased global histone hyperacetylation and decreased levels of HDAC6 in the PFC. Remarkably, this cocaine effect was not observed when the cannabinoid exposure was performed at adulthood.

The presence of alterations in DNA methylation or non-coding RNAs after adolescent cannabinoid exposure is even less investigated than histone modifications.

The only papers dealing with non-coding RNAs after adolescent cannabinoid exposure were published by Hollins and colleagues.[48,49] In the 2014 paper, using a miRNA microarray, they discovered that adolescent exposure to the cannabinoid agonist HU210 in male rats produced a differential expression of seven miRNAs in the enthorinal cortex, an effect that was magnified by the co-occurrence of a pre-natal environmental risk factor such as maternal immune activation (MIA). Intriguingly, a high proportion of these differentially expressed miRNA were located in the long arm of chromosome 6 (6q32), with a potential role in the regulation of pathways involved in synaptic remodeling, learning and memory. These results were substantiated in the more recent paper,[49] where gene expression was also investigated as well as its relationship with the differentially expressed miRNAs, suggesting the occurrence of a genomic remodeling with implications relevant for schizophrenia and other neuropsychiatric disorders.

As far as DNA methylation is concerned, Tomas-Roig and colleagues[50] demonstrated that adolescent administration of the synthetic cannabinoid agonist WIN 55,212-2 in male mice induced an increase in DNA methylation in an intragenic region of Rgs7 gene at adulthood, paralleled by a small but significant decrease in the mRNA levels of Rgs7. Remarkably, the protein coded by this gene is involved in hippocampus-dependent learning and memory, a cognitive domain impaired by the cannabinoid exposure.[50]

Recently, methylation studies have been carried out also in human adolescents. Indeed, Clark and colleagues[51] investigated the association between problematic adolescent cannabis use and methylation differences in blood, using a genome-wide methylation approach. When whole blood was assessed, results indicated an increased DNA methylation of the gene encoding the protein calmin, involved in neuronal development and function, and in the gene encoding a regulatory protease of SUMO proteins, involved in neuronal function as well. However, different blood cell types may have different methylation profiles, thus the same approach was also applied to study the response in the different blood cell sub-populations. In granulocytes, it was found increased methylation of Slit2-Robo1 signaling, likely impacting axon guidance and inflammatory responses, whereas in B cells hypermethylation was observed in regions involved in DNA repair, with potential downstream effects on immune and neurological functioning. As the authors suggest, further studies are needed to validate the relevance of these findings, nonetheless the data they obtained support the idea that cannabis-associated methylation can impact neurological development and inflammation response.

Conclusions

As a whole, these data suggest that the adolescent encounter with THC/cannabis, if precocious and intense enough, is able to modify the neuronal circuit responsiveness to further experiences, through a mechanism involving epigenetics. The resulting altered gene expression is indeed responsible for persistent changes in neural circuits. The complex interplay between circuit activity and neuronal gene regulation is vital to neuronal plasticity essential to proper learning and memory processes. Environmental risk factors, such as adolescent cannabis exposure, may disrupt the correct occurring of this interplay, thus increasing the vulnerability to develop debilitating psychiatric conditions.

References

1. Verkhratsky A, Nedergaard M. Physiology of astroglia. *Physiol Rev* 2018;**98**(1): 239–389. https://doi.org/10.1152/physrev.00042.2016.
2. Covelo A, Eraso-Pichot A, Fernández-Moncada I, Serrat R, Marsicano G. CB1R-dependent regulation of astrocyte physiology and astrocyte-neuron interactions. *Neuropharmacology* 2021;**195**. https://doi.org/10.1016/j.neuropharm.2021.108678, 108678.
3. Navarrete M, Araque A. Endocannabinoids mediate neuron-astrocyte communication. *Neuron* 2008;**57**(6):883–93. https://doi.org/10.1016/j.neuron.2008.01.029.

4. Jimenez-Blasco D, Busquets-Garcia A, Hebert-Chatelain E, Serrat R, Vicente-Gutierrez C, Ioannidou C, et al. Glucose metabolism links astroglial mitochondria to cannabinoid effects. *Nature* 2020;**583**(7817):603–8. https://doi.org/10.1038/s41586-020-2470-y.
5. Schuele LL, Glasmacher S, Gertsch J, Roggan MD, Transfeld JL, Bindila L, et al. Diacylglycerol lipase alpha in astrocytes is involved in maternal care and affective behaviors. *Glia* 2021;**69**(2):377–91. https://doi.org/10.1002/glia.23903.
6. Stella N. Cannabinoid and cannabinoid-like receptors in microglia, astrocytes, and astrocytomas. *Glia* 2010;**58**(9):1017–30. https://doi.org/10.1002/glia.20983.
7. Smith NA, Bekar LK, Nedergaard M. Astrocytic endocannabinoids mediate hippocampal transient heterosynaptic depression. *Neurochem Res* 2020;**45**(1):100–8. https://doi.org/10.1007/s11064-019-02834-0.
8. Duffy SS, Hayes JP, Fiore NT, Moalem-Taylor G. The cannabinoid system and microglia in health and disease. *Neuropharmacology* 2021;**190**. https://doi.org/10.1016/j.neuropharm.2021.108555.
9. Sierra A, Gottfried-Blackmore AC, Mcewen BS, Bulloch K. Microglia derived from aging mice exhibit an altered inflammatory profile. *Glia* 2007;**55**(4):412–24. https://doi.org/10.1002/glia.20468.
10. Cabral GA, Griffin-Thomas LT. Emerging role of the cannabinoid receptor CB2 in immune regulation: therapeutic prospects for neuroinflammation. *Expert Rev Mol Med* 2009;**11**. https://doi.org/10.1017/S1462399409000957.
11. Araujo DJ, Tjoa K, Saijo K. The endocannabinoid system as a window into microglial biology and its relationship to autism. *Front Cell Neurosci* 2019;**13**. https://doi.org/10.3389/fncel.2019.00424.
12. Walter L, Franklin A, Witting A, Wade C, Xie Y, Kunos G, et al. Nonpsychotropic cannabinoid receptors regulate microglial cell migration. *J Neurosci* 2003;**23**(4):1398–405. https://doi.org/10.1523/jneurosci.23-04-01398.2003.
13. Cabral GA, Rogers TJ, Lichtman AH. Turning over a new leaf: cannabinoid and endocannabinoid modulation of immune function. *J Neuroimmune Pharmacol* 2015;**10**(2):193–203. https://doi.org/10.1007/s11481-015-9615-z.
14. Tanaka M, Sackett S, Zhang Y. Endocannabinoid modulation of microglial phenotypes in neuropathology. *Front Neurol* 2020;**11**. https://doi.org/10.3389/fneur.2020.00087.
15. Lee Y, Morrison BM, Li Y, Lengacher S, Farah MH, Hoffman PN, et al. Oligodendroglia metabolically support axons and contribute to neurodegeneration. *Nature* 2012;**487**(7408):443–8. https://doi.org/10.1038/nature11314.
16. Nave KA, Werner HB. Myelination of the nervous system: mechanisms and functions. *Annu Rev Cell Dev Biol* 2014;**30**:503–33. Annual Reviews Inc https://doi.org/10.1146/annurev-cellbio-100913-013101.
17. Monje M. Myelin plasticity and nervous system function. *Annu Rev Neurosci* 2018;**41**:61–76. https://doi.org/10.1146/annurev-neuro-080317-061853.
18. Elbaz B, Popko B. Molecular control of oligodendrocyte development. *Trends Neurosci* 2019;**42**(4):263–77. https://doi.org/10.1016/j.tins.2019.01.002.
19. Huerga-Gómez A, Aguado T, Sánchez-de la Torre A, Bernal-Chico A, Matute C, Mato S, et al. Δ9-tetrahydrocannabinol promotes oligodendrocyte development and CNS myelination in vivo. *Glia* 2021;**69**(3):532–45. https://doi.org/10.1002/glia.23911.
20. Ilyasov AA, Milligan CE, Pharr EP, Howlett AC. The endocannabinoid system and oligodendrocytes in health and disease. *Front Neurosci* 2018. https://doi.org/10.3389/fnins.2018.00733.
21. Benito C, Romero JP, Tolón RM, Clemente D, Docagne F, Hillard CJ, et al. Cannabinoid CB1 and CB2 receptors and fatty acid amide hydrolase are specific markers of plaque cell subtypes in human multiple sclerosis. *J Neurosci* 2007;**27**(9):2396–402. https://doi.org/10.1523/JNEUROSCI.4814-06.2007.

22. Gomez O, Arevalo-Martin A, Garcia-Ovejero D, Ortega-Gutierrez S, Cisneros JA, Almazan G, et al. The constitutive production of the endocannabinoid 2-arachidonoylglycerol participates in oligodendrocyte differentiation. *Glia* 2010;**58**(16):1913–27. https://doi.org/10.1002/glia.21061.
23. Alpár A, Tortoriello G, Calvigioni D, Niphakis MJ, Milenkovic I, Bakker J, et al. Endocannabinoids modulate cortical development by configuring Slit2/Robo1 signalling. *Nat Commun* 2014;**5**. https://doi.org/10.1038/ncomms5421.
24. Arévalo-Martín A, García-Ovejero D, Rubio-Araiz A, Gómez O, Molina-Holgado F, Molina-Holgado E. Cannabinoids modulate Olig2 and polysialylated neural cell adhesion molecule expression in the subventricular zone of post-natal rats through cannabinoid receptor 1 and cannabinoid receptor 2. *Eur J Neurosci* 2007;**26**(6):1548–59. https://doi.org/10.1111/j.1460-9568.2007.05782.x.
25. Molina-Holgado E, Vela JM, Arévalo-Martín A, Almazán G, Molina-Holgado F, Borrell J, et al. Cannabinoids promote oligodendrocyte progenitor survival: involvement of cannabinoid receptors and phosphatidylinositol-3 kinase/Akt signaling. *J Neurosci* 2002;**22**(22):9742–53. https://doi.org/10.1523/jneurosci.22-22-09742.2002.
26. Bernal-Chico A, Canedo M, Manterola A, Victoria Sánchez-Gómez M, Pérez-Samartín A, Rodríguez-Puertas R, et al. Blockade of monoacylglycerol lipase inhibits oligodendrocyte excitotoxicity and prevents demyelination in vivo. *Glia* 2015;**63**(1):163–76. https://doi.org/10.1002/glia.22742.
27. Dhein S. Different effects of cannabis abuse on adolescent and adult brain. *Pharmacology* 2020;**105**(11−12):609–17. https://doi.org/10.1159/000509377.
28. Lopez-Rodriguez AB, Llorente-Berzal A, Garcia-Segura LM, Viveros MP. Sex-dependent long-term effects of adolescent exposure to THC and/or MDMA on neuroinflammation and serotoninergic and cannabinoid systems in rats. *Br J Pharmacol* 2014;**171**(6):1435–47. https://doi.org/10.1111/bph.12519.
29. Zamberletti E, Gabaglio M, Grilli M, Prini P, Catanese A, Pittaluga A, et al. Long-term hippocampal glutamate synapse and astrocyte dysfunctions underlying the altered phenotype induced by adolescent THC treatment in male rats. *Pharmacol Res* 2016;**111**:459–70. https://doi.org/10.1016/j.phrs.2016.07.008.
30. Zamberletti E, Gabaglio M, Prini P, Rubino T, Parolaro D. Cortical neuroinflammation contributes to long-term cognitive dysfunctions following adolescent delta-9-tetrahydrocannabinol treatment in female rats. *Eur Neuropsychopharmacol* 2015;**25**(12):2404–15. https://doi.org/10.1016/j.euroneuro.2015.09.021.
31. Chen R, Zhang J, Fan N, Teng Z, Wu Y, Yang H, et al. Δ^9-THC-caused synaptic and memory impairments are mediated through COX-2 signaling. *Cell* 2013;1154–65. https://doi.org/10.1016/j.cell.2013.10.042.
32. Jouroukhin Y, Zhu X, Shevelkin AV, Hasegawa Y, Abazyan B, Saito A, et al. Adolescent Δ^9-tetrahydrocannabinol exposure and astrocyte-specific genetic vulnerability converge on nuclear factor-κB-cyclooxygenase-2 signaling to impair memory in adulthood. *Biol Psychiatry* 2019;**85**(11):891–903. https://doi.org/10.1016/j.biopsych.2018.07.024.
33. Blest-Hopley G, O'Neill A, Wilson R, Giampietro V, Lythgoe D, Egerton A, et al. Adolescent-onset heavy cannabis use associated with significantly reduced glial but not neuronal markers and glutamate levels in the hippocampus. *Addict Biol* 2020;**25**(6). https://doi.org/10.1111/adb.12827.
34. Cutando L, Busquets-Garcia A, Puighermanal E, Gomis-González M, Delgado-García JM, Gruart A, et al. Microglial activation underlies cerebellar deficits produced by repeated cannabis exposure. *J Clin Investig* 2013;**123**(7):2816–31. https://doi.org/10.1172/JCI67569.
35. Corrigan NM, Yarnykh VL, Hippe DS, Owen JP, Huber E, Zhao TC, et al. Myelin development in cerebral gray and white matter during adolescence and late childhood. *Neuroimage* 2021;**227**. https://doi.org/10.1016/j.neuroimage.2020.117678.

36. Williamson JM, Lyons DA. Myelin dynamics throughout life: an ever-changing landscape? *Front Cell Neurosci* 2018;**12**. https://doi.org/10.3389/fncel.2018.00424.
37. Mews P, Calipari ES, Day J, Lobo MK, Bredy T, Abel T. From circuits to chromatin: the emerging role of epigenetics in mental health. *J Neurosci* 2021;**41**(5):873–82. Society for Neuroscience https://doi.org/10.1523/JNEUROSCI.1649-20.2020.
38. Jaenisch R, Bird A. Epigenetic regulation of gene expression: how the genome integrates intrinsic and environmental signals. *Nat Genet* 2003;**33**(3S):245–54. https://doi.org/10.1038/ng1089.
39. Stewart AF, Fulton SL, Maze I. Epigenetics of drug addiction. *Cold Spring Harb Perspect Med* 2021;**11**(7). https://doi.org/10.1101/cshperspect.a040253, a040253.
40. Campbell RR, Wood MA. How the epigenome integrates information and reshapes the synapse. *Nat Rev Neurosci* 2019;**20**(3):133–47. https://doi.org/10.1038/s41583-019-0121-9.
41. Tomasiewicz HC, Jacobs MM, Wilkinson MB, Wilson SP, Nestler EJ, Hurd YL. Proenkephalin mediates the enduring effects of adolescent cannabis exposure associated with adult opiate vulnerability. *Biol Psychiatr* 2012;**72**(10):803–10. https://doi.org/10.1016/j.biopsych.2012.04.026.
42. Prini P, Penna F, Sciuccati E, Alberio T, Rubino T. Chronic Δ^9-THC exposure differently affects histone modifications in the adolescent and adult rat brain. *Int J Mol Sci* 2017;**18**(10). https://doi.org/10.3390/ijms18102094.
43. Prini P, Rusconi F, Zamberletti E, Gabaglio M, Penna F, Fasano M, et al. Adolescent THC exposure in female rats leads to cognitive deficits through a mechanism involving chromatin modifications in the prefrontal cortex. *J Psychiatr Neurosci* 2018;87–101. https://doi.org/10.1503/jpn.170082.
44. Zald DH. The human amygdala and the emotional evaluation of sensory stimuli. *Brain Res Rev* 2003;**41**(1):88–123. https://doi.org/10.1016/S0165-0173(02)00248-5.
45. Ernst M, Pine DS, Hardin M. Triadic model of the neurobiology of motivated behavior in adolescence. *Psychol Med* 2006;**36**(3):299–312. https://doi.org/10.1017/S0033291705005891.
46. Miller ML, Chadwick B, Dickstein DL, Purushothaman I, Egervari G, Rahman T, et al. Adolescent exposure to Δ^9-tetrahydrocannabinol alters the transcriptional trajectory and dendritic architecture of prefrontal pyramidal neurons. *Mol Psychiatr* 2019;**24**(4):588–600. https://doi.org/10.1038/s41380-018-0243-x.
47. Scherma M, Qvist JS, Asok A, Huang SSC, Masia P, Deidda M, et al. Cannabinoid exposure in rat adolescence reprograms the initial behavioral, molecular, and epigenetic response to cocaine. *Proc Natl Acad Sci U S A* 2020;**117**(18):9991–10002. https://doi.org/10.1073/pnas.1920866117.
48. Hollins SL, Zavitsanou K, Walker FR, Cairns MJ. Alteration of imprinted Dlk1-Dio3 miRNA cluster expression in the entorhinal cortex induced by maternal immune activation and adolescent cannabinoid exposure. *Transl Psychiatr* 2014;**4**(9). https://doi.org/10.1038/tp.2014.99.
49. Hollins SL, Zavitsanou K, Walker FR, Cairns MJ. Alteration of transcriptional networks in the entorhinal cortex after maternal immune activation and adolescent cannabinoid exposure. *Brain Behav Immun* 2016;**56**:187–96. https://doi.org/10.1016/j.bbi.2016.02.021.
50. Tomas-Roig J, Benito E, Agis-Balboa RC, Piscitelli F, Hoyer-Fender S, Di Marzo V, et al. Chronic exposure to cannabinoids during adolescence causes long-lasting behavioral deficits in adult mice. *Addict Biol* 2017;**22**(6):1778–89. https://doi.org/10.1111/adb.12446.
51. Clark SL, Chan R, Zhao M, Xie LY, Copeland WE, Aberg KA, et al. Methylomic investigation of problematic adolescent cannabis use and its negative mental health consequences. *J Am Acad Child Adolesc Psychiatry* 2021. https://doi.org/10.1016/j.jaac.2021.02.008.

Chapter 7

Effects of prenatal THC exposure on the mesolimbic dopamine system: Unveiling an endophenotype of sensory information processing deficits

Roberto Frau[a,b] and Miriam Melis[a]
[a]*Department of Biomedical Sciences, Division of Neuroscience and Clinical Pharmacology, University of Cagliari, Cagliari, Italy,* [b]*"Guy Everett" Laboratory, University of Cagliari, Cagliari, Italy*

Introduction

Mental health of children and adolescents has been one of the areas of major concern of the World Health Organization (WHO) as mental disorders during the pediatric age may have lifelong consequences and reduce the safety and productivity of societies.[1] Nowadays, the overall state of mental health in children and adolescents has never been so alarming. The latest reports from the WHO and the US Centers for Disease Control and Prevention (CDC) give us a clear picture of the constant rise of mental disorders among people aged 3–19 years.[2,3] Current estimates of mental illness in pediatric populations account for 10% of children and adolescents with half of these conditions starting by the age of 14 years, though they remain mostly undetected and untreated.[3] Although it is difficult to identify the exact age of onset of any given mental condition, statistics indicate the emergence of emotional (e.g., depression, anxiety) and behavioral disorders (e.g., attention-deficit/hyperactivity disorder—ADHD, conduct disorders) during adolescence and childhood, respectively. Conversely, disorders including psychotic symptoms typically manifest during late adolescence or at young adulthood, albeit early forms of psychoses may rarely occur during childhood.

To tackle this issue, management of mental health is moving toward preventative and early intervention strategies. There is now a broad consensus on the effectiveness of early detection interventions to downsize the burden of mental illness and improve treatment efficacy. Early intervention during the "prodrome" phase (i.e., the stage characterized by sub-threshold symptoms not sufficient for a categorical diagnosis of mental disorder) is regarded as the most effective healthcare strategy to prevent the onset of any given mental disorder. As such, children and adolescents become the target population for primary prevention of mental disorders. Yet, three out of four children experiencing a mental health problem do not receive the help needed with tremendous life-altering repercussions, including unemployment and criminal propensity. In addition, 10% of individuals diagnosed with psychiatric conditions throughout childhood and adolescence, if untreated, are more resistant to treatment and exhibit a disease course that is more difficult to manage. The identification of transitory though clinically relevant phenomena in vulnerable individuals, along with their pathophysiological trajectories, is therefore highly relevant for the development of preventative and disease-modifying therapeutic strategies.

Beyond the well-established involvement of genetic components in the etiology of neuropsychiatric disorders, other determinants may contribute to the rise of mental illness in children and adolescents. In particular, during discrete sensitive developmental windows, the interaction between inherited predispositions and the environment sensitizes brain circuits, as conceptualized in the theory of developmental origins of health and disease (DOHaD). DOHaD theory postulates that exposure to certain environmental insults during specific sensitive windows of development elicits negative outcomes on individual mental health later in life.[4,5] Although originally focused on the impact of maternal malnutrition on the prevalence of cardiovascular diseases in adult offspring,[6,7] DOHaD theory has extended toward mental health through the investigations of the impact of early life adversities, including the exposure to stress, xenobiotics (e.g., environmental chemicals and prescription, legal, and illegal drugs) and pathogenic microorganisms and viruses on mental health outcomes later in life. In this perspective, the developing brain exposed to insults responds with a series of adaptive responses altering its developmental trajectory, which may lead to subtle pre-clinical (endo)phenotypes that are susceptible to subsequent environmental challenges. These latter challenges will ultimately allow the transition to the onset of the condition. Accordingly, the "two-hit" hypothesis of schizophrenia and complex diseases[8] posits that a perinatal genetic or environmental insult acts as a "first hit" by interfering with neurodevelopment, and increases vulnerability to a "second hit" occurring later in life. Neither insult per se would be sufficient to induce the disease. However, the "first hit" by altering developmental trajectories primes the brain for the second hit (e.g., environmental challenges), which would be otherwise irrelevant.

A large body of animal and human research provides significant evidence that pre-natal cannabis exposure (PCE) is likely one of the environmental factors that better recapitulates both this new psychopathological conceptualization of DOHaD[9] and the "two-hit" hypothesis of mental illness. In fact, PCE elicits a multifaceted sequalae of molecular, epigenetic, neurophysiological, and brain circuit alterations that positions the offspring at a high risk of developing mental disorders. Remarkably, the majority of these changes are subtle, though they make the offspring more susceptible to the effects of subsequent adverse experiences during development (childhood and adolescence). In particular, PCE progeny manifest psychotic-like experiences emerging especially during childhood and preadolescence.[10–12] In this framework, dopamine as a key neurotransmitter involved in motivation, emotional regulation, reward, and cognition, has long been tied to a variety of neuropsychiatric disorders characterized by childhood and adolescent onset, such as ADHD, substance use disorders, and psychosis.[13–16]

In this chapter, we present evidence from animal studies for how PCE affects, in a sex-dependent manner, mesolimbic dopamine function, and its related behavioral domains, which trigger adverse mental outcomes upon subsequent environmental challenges. Not only can these studies help provide mechanistic insights into the heightened risk of mental health problems observed in children who were exposed to cannabis during gestation, but they also may help with identifying molecular targets that serve as entry points for novel pharmacological and/or behavioral interventions early in life. The focus on early life is of particular relevance as the majority of mental disorders tied to PCE manifest from infancy throughout adolescence, thereby affecting life-course of mental well-being.

Because of the expanding legalization and liberalization policies regarding cannabis use as well as the underestimation and misconceptions about its negative effects on fetal neurodevelopment, the current estimates of PCE-induced adverse mental outcomes will inevitably rise in the next years. As with cigarettes and alcohol, therefore, it is paramount that public policies strive to disseminate potential risks of maternal cannabis use on mental health of the progeny. If no action is taken, the rates of offspring with no genetic predisposition, but "primed" with PCE, may increase as well as those cohorts at a high-risk of developing severe mental disorders associated with aberrant mesolimbic dopamine system function.

Pre-natal cannabis exposure as an environmental risk factor for neuropsychiatric disorders

The majority of neuropsychiatric disorders result from a complex interplay between multiple genetic components and environmental factors.[17] Indeed, while considerable progress has been made in understanding genetic contributions to most psychiatric disorders, genetics alone cannot fully explain the complex etiology as well as the different pathological phenotypes associated with these illnesses. Thus, non-genetic risk factors provide a potential explanation of the etiology and clinical manifestation of neuropsychiatric disorders. Multiple lines of pre-clinical and clinical evidence have indeed shown that different environmental factors, such as infectious agents, chemical and psychosocial risk factors, contribute to heightened risk for developing major neuropsychiatric disorders, especially those with neurodevelopmental origin. Given that brain development is a dynamic process finely driven by genetic programming, it is per se under a constant environmental influence/pressure. While adverse environmental events lead to a greater liability for psychopathology across the lifespan, the interplay between genetic and environmental factors occurs early in life.

Ample clinical literature suggests PCE is one of the environmental factors affecting cognitive and neurobehavioral domains and eliciting discrete neuropsychiatric disorders in offspring. In fact, pioneering data from three prospective longitudinal human cohorts of PCE and offspring outcomes—The Ottawa Pre-natal Prospective Study (OPPS),[18–21] The Maternal Health Practices and Child Development Study (MHPCD),[22–24] and The Generation R Study (GenR)[10,25–27]—reveal its neurodevelopmental sequelae. Children and adolescents with PCE exhibit a series of deficits in attentional and cognitive performance, including sustained attention, visuospatial working memory, and verbal reasoning. Children and adolescents with PCE also show psychomotor agitation, impulsivity, greater sensitivity to drugs of abuse, and aggressive behaviors.[28–30] In addition, the more recent cross-sectional study Adolescent Brain Cognitive Development (ABCD) reports an association between PCE and adverse neurodevelopmental outcomes during childhood including psychotic-like experiences, depression and anxiety, impulsivity, social problems as well as sleep disturbance.[11,12,31,32] Collectively, these epidemiological findings indicate that in utero cannabis exposure impacts the developing brain, thereby leading to susceptibility to a broad spectrum of psychiatric symptoms later on in life.

The relevance of PCE as a risk factor for vulnerability to neuropsychiatric disorders may lie in the strategy and use of intermediate phenotypes in psychiatric research, i.e., endophenotypes. Endophenotypes are quantitative neurobehavioral and neurophysiological traits of heritable vulnerability,[33] such as electroencephalographic signatures, neurocognitive performance deficits, and impaired facial emotion recognition,[34–36] which can be detected during early phases of neurodevelopmental alterations. Unveiling endophenotypes is pivotal as it allows definition of the etiopathogenesis of neuropsychiatric disorders and the design of more effective therapeutic interventions to prevent the transition to the disorder. Accordingly, as PCE is a modifiable predictor for diverse psychiatric conditions, public health interventions regarding the awareness of the harm associated with maternal cannabis use should be implemented to prevent the potential onset of psychiatric symptoms, particularly following exposure to environmental risk factors. This cannot always occur, unfortunately, especially given the escalation of risk-taking behaviors, curiosity, and desire for experimentation that emerge during the transition from childhood to adolescence. In this regard, it is important to point out that early exposure to drugs, from caffeine to cannabis, tobacco and alcohol, may concomitantly alter (e.g., via accelerating or delaying the onset of processes, by stalling, synchronizing or uncoupling the progression of brain maturation) neurodevelopmental trajectories.[37] Additionally, one cannot avoid the exposure to stress, such as that one resulting from relationships with peers, from school needs, from safety problems or from family tensions.

Diverse animal models recapitulate the DOHaD theory by examining the interaction of multiple environmental risk factors, without any genetic endowment. For example, pre-natal or post-natal stress manipulations are able to produce long-term neurodevelopmental changes that resemble (endo)phenotypic alterations reported in humans.[38,39] As such, in line with DOHaD theory, cannabis may perturb endocannabinoid (ECB) signaling pathways during two sensitive developmental windows: the first one in the fetal life, thus leading to a derangement from typical development toward psychopathological trajectories; the second one may be in (pre)adolescence, during which cannabis acts on diverse neuronal circuits finely modulated by ECB system, previously altered by PCE. Whether PCE triggers homeostatic adaptations of ECB signaling affecting either ECB production/degradation or cannabinoid receptor type 1 (CB1) expression/function within the mesolimbic dopamine pathway is still under investigation (Ref. 40 and next paragraph). Of note, PCE might induce different activation states of the ECB system (i.e., prior, tonic, or persistent)[41] that might also be dissimilar depending on the specific time points of neurodevelopment (Box 1). These interconnected functional activation states of ECB signaling on the dopamine system can either trigger metaplasticity or represent a means for metaplastic control of the ECB system, and may possibly influence any subsequent plasticity induction (Box 1).[41] Within this framework, the study of the neurobiological underpinnings of PCE in pre-clinical settings has the advantage of avoiding many social and environmental factors usually influencing clinical investigation results. In this context, it is worth noting that both human behavioral repertoire typical of developing ages (e.g., novelty and sensation seeking, impulsivity, risk-taking behaviors) and key stages of neurodevelopment are conserved across mammals. In rodents, it is therefore possible to reveal and examine an endophenotype, characterized by a normal behavioral repertoire, which is unmasked when flagged by environmental challenges (e.g., THC, stress). As such, the emerging picture is that PCE is one of those environmental risk factors for the onset of neuropsychiatric disorders of developmental origin.[30]

Several excellent reviews about PCE sequelae in preclinical models are available nowadays.[44–49] In the next paragraphs, we describe how PCE negatively affects dopamine signaling in pre-adolescent rats, whose dysfunctions result in an endophenotype of sensory information deficits, a *neutral cross-disorder trait* for neuropsychiatric disorders with early onset and shared heritability, such as ADHD, bipolar disorder, major depressive disorder, and schizophrenia.[50]

BOX 1 Endocannabinoids and regulation of synaptic plasticity
- **Synaptic plasticity** is the extraordinary capability of the brain to change its structure and function in an activity-dependent manner and in response to an ever-changing environment. This ability is critical in normal development, learning and memory, repairing mechanisms and is impaired during disease states. This fascinating property of the brain allow us to modify feelings, thoughts, and behavior on the bases of precedent experiences. Different forms of synaptic plasticity exist spanning discrete temporal domains (long- vs short-term plasticity) serving diverse functions and ascribed to various mechanisms.[42] Virtually all synapses in the brain simultaneously express discrete forms of synaptic plasticity.
- **Metaplasticity** is defined as the ability to generate synaptic plasticity in response to activity-dependent changes in neuronal states, synapses, and networks.[43] Metaplasticity requires a change in neuronal function upon a "priming" (pre-synaptic or post-synaptic activity). The priming stimulus produces an effect that must persist after the removal of the priming stimulus, and that influences the ability of a subsequent stimulus to induce synaptic plasticity.
- **Interconnected functional activation states of endocannabinoid signaling** have been proposed to contribute to metaplastic control of synaptic and behavioral functions in healthy and disease states.[41] The three functional states affecting endocannabinoid-mediated synaptic plasticity and dynamically affect neuronal circuits and behavior are ascribed to prior, tonic, and persistent engagement of endocannabinoid signaling. *Prior* engagement of endocannabinoid signaling (e.g., increased release) has been shown to affect short-term forms of synaptic plasticity thereby triggering metaplasticity. *Tonic* endocannabinoid signaling (e.g., decreased efficacy of degrading enzymes) regulates basal neurotransmission, sets different thresholds for gating subsequent synaptic plasticity, and ultimately altering postsynaptic cell excitability. *Persistent* engagement of endocannabinoid signaling (e.g., CB1 receptor desensitization or downregulation) triggers homeostatic neuroadaptations may enable metaplastic mechanisms.

Pre-natal cannabis exposure impacts on the offspring mesolimbic dopamine system

Preclinical and clinical studies have clearly demonstrated that the consumption of cannabis during pregnancy and lactation is harmful for the fetus and its development (but see Ref. 51). THC, the major psychoactive ingredient of cannabis derivatives, easily crosses the placenta, enters the fetal bloodstream and interferes with the ECB system,[52] which plays a key role in driving and shaping neuronal development.[53,54] Indeed, molecular components of ECB signaling, such as anandamide and 2-arachidonoylglycerol, their targeting receptors as well as the enzymatic machinery for synthesis, transport and catabolism, can already be detected at the earliest stages of neurodevelopment (Refs. 55,56; and chapter TBD in this book). In particular, during the pre-natal temporal window, the expression of CB1 receptors follows a tropism toward mesocorticolimbic structures, where they influence the release and function of key neurotransmitters involved in the etiopathogenesis of neuropsychiatric disorders, including dopamine.[57,58] Dopamine is abundant during neurodevelopment and participates in a wide range of processes, especially those related to the differentiation and maturation of forebrain structures, such as the (pre)frontal cortex (PFC) and striatum.[59,60] Within the frontal and striatal neurocircuitries, dopamine signaling is key for the proper regulation of emotional, motivational, and cognitive processing. Dopamine dysregulation during neurodevelopment, secondary to complex circuit abnormalities, may be one of the starting points in search for endophenotypes of individual differences in behavioral domains leading to early onset of neuropsychiatric disorders in life (i.e., differential susceptibility[61–63]).

Josè Ramos and colleagues have pioneered research in this specific field and showed that perinatal cannabinoid administration biases diverse dopaminergic pathways in rodents.[64–66] Their first studies, on the impact of hashish crude extract on neuronal dopamine function, demonstrated sex-dependent effects on the activity of hypothalamic dopaminergic systems.[67] Remarkably, these adverse outcomes disappeared when the progeny reached adulthood though they could be overt upon a challenge with THC.[68,69] Following up on these studies, the same research group and others extended the susceptibility of hypothalamic dopamine system to other drugs but also to other dopaminergic pathways, including nigrostriatal and mesolimbic ones.[69,70] In line with these studies and with the available epidemiological data,[10–12] in an animal model of PCE the effects are observed at pre-adolescence exclusively in the male progeny: PCE disrupts sensorimotor gating (Box 2) and alters dopamine neuronal activity of the ventral tegmental area (VTA), which displays a biased phenotype.[40,71,72] Sensorimotor gating is a function involved in processing incoming sensory information and can be measured across species via the pre-pulse inhibition (PPI) of the startle reflex[73] (Box 2).

This function, regulated by dopamine within a complex circuit,[74–76] represents an important feature of healthy individuals that can inhibit behavioral responses to incoming sensory stimuli. The observation that, in preadolescent rat cohorts, only PCE male subjects display deficits in PPI following a "second hit," such as an acute stress or a single exposure to THC, is relevant as PCE male progeny only exhibit a hyperdopaminergia along with hypersensitive mesolimbic dopamine transmission. Remarkably, females do not show any of these alterations.[40,72]

Male hyperdopaminergia is exemplified by VTA dopamine cells spontaneously firing at higher frequencies, displaying depolarized resting membrane potentials, lower voltage membrane thresholds, and an increased spike fidelity.[40] Indeed, PCE induces a complex circuit rearrangement resulting in an increased balance between excitatory and inhibitory inputs (i.e., excitatory to

BOX 2 Sensorimotor gating function and pre-pulse Inhibition of the Startle Reflex (PPI)

- **Sensorimotor gating** is a neurological process aimed at filtering irrelevant incoming sensory stimuli. It provides the primary (pre)attentional mechanism—"the front door"—that the central nervous system enacts to screen sensory information on the basis of their salience. It prevents higher-order functions from being overloaded through the inhibition or the suppression of the response to incoming sensory inputs that are irrelevant or redundant, thereby allowing the proper evaluation of salient sensory stimuli. As such, deficits of this function result in information overload, misinterpretation of incoming sensory inputs, increased emotional reactivity, cognitive fragmentation and disorganized thinking.
- **Pre-pulse Inhibition of the startle reflex (PPI)** is a cross-species neurophysiological measure of sensorimotor gating function. It can be easily assessed across species, including human and non-human primates, rodents, zebrafish, and invertebrates. In this paradigm, a weaker prestimulus inhibits the reaction of an organism to a subsequent reflex eliciting stimulus, namely the startle reflex. This paradigm is based upon the responses to two stimuli: a pulse and a pre-pulse. The pulse is a sudden, loud (pulse or click) stimulus (e.g., visual, acoustic, tactile or olfactory) eliciting the startle reflex that is measured, in rodents, as a whole-body muscle contraction or, in humans and nonhuman primates, as eye-blink responses. Routinely, the pulse is immediately preceded (30–500ms) by a lower-intensity-non-startling stimulus (pre-pulse), which would alert the individual to inhibit the subsequent startle response. As an operational measure of sensorimotor gating function, PPI is thought to reflect the ability of an individual to screen out distracting stimuli, thus allowing the preservation of selective attention and management of stimuli from an everchanging environment.
- **Neuronal circuits** contributing to PPI involve different areas of the brain stem, such as the inferior and superior colliculus, the pedunculopontine tegmental/laterodorsal tegmental complex and caudal pontine reticular nucleus. These brain regions are under a top-down influence of cortico-limbic structures, including medial PFC, ventral hippocampus, nucleus accumbens shell/core, basolateral amygdala, and ventral pallidum (cortico-striato-pallido-pontine network).
- **The Dopamine system** plays a prominent role in the regulation of PPI: direct and indirect dopaminergic agonists disrupt PPI in both humans and rodents by engaging dopamine D2 receptors in the nucleus accumbens.
- **PPI deficits**, originally identified in patients with schizophrenia, have been observed across categorically distinct neuropsychiatric disorders, including obsessive compulsive disorder, attention-deficit/hyperactivity disorder, Tourette's syndrome, and post-traumatic stress-disorders. Although PPI is not a diagnostic means, evidence suggests that PPI deficits may represent a relevant subclinical/vulnerability trait within the spectrum of neuropsychiatric manifestation, namely an endophenotype. Thus, dysregulation of PPI serves as an important **cross-disease trait** of those disorders characterized by abnormal sensory information processing.

Effects of prenatal THC exposure on the mesolimbic DA system **Chapter | 7** 115

inhibitory ratio) on VTA dopamine neurons. The maturation of glutamatergic ionotropic receptors (i.e., AMPA and NMDA) on dopamine neurons also shows a delay, which confers heightened postsynaptic responsiveness to incoming inputs, despite the observed marked reduction in the afferent input density (∼45%). Collectively, these changes may contribute to the shift in the threshold for subsequent synaptic plasticity induction, as excitatory "immature" synapses on dopamine neurons respond to stimuli usually inducing a form of long-term depression by expressing a long-term potentiation.[40] To further heighten dopamine cell susceptibility, PCE also dampens synaptic inhibition onto dopamine neurons without affecting their afferent input density, but instead by re-modeling presynaptic nanoarchitecture of the active zone at release sites (Fig. 1).

The observed molecular crowding at vesicle release sites[77] limiting GABA diffusion into the synaptic cleft helps promote dopamine cell excitability. In addition, nanoscale super-resolution analyses of inhibitory synapses also reveal that PCE shifts the ratio between pre-synaptic CB1 receptors and their molecular effectors (i.e., voltage gated calcium channels), a phenomenon that might contribute to the reduced probability of GABA release, and ultimately to the susceptible phenotype. This sex-dependent effect also occurs in vivo and it is mirrored by larger acute THC-induced enhancement of dopamine levels in the shell of the nucleus accumbens,[40,71,72] i.e., the target region of dopamine cells located in the lateral portion of the posterior VTA.[78] Notably, these larger

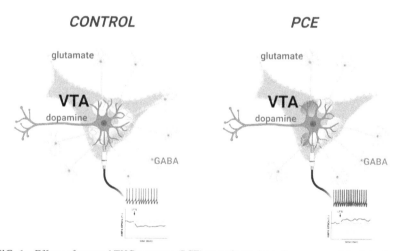

FIG. 1 Effects of prenatal THC exposure (PCE) on male rat dopamine neurons of the lateral portion of the ventral tegmental area (VTA) at preadolescence. In PCE animals, the hyperdopaminergic endophenotype results from both a disinhibition and an upscaling at excitatory synapses. Susceptible dopamine neurons respond to low frequency stimulation (LFS) of excitatory synapses with a long-term potentiation whereas control and naïve animals a long-term depression. Number of input afferents and color shading depict changes (decrease/increase) in afferent input density and synaptic input efficacy, respectively.

extracellular dopamine levels measured in the nucleus accumbens shell correlated with THC-induced deficits of sensorimotor gating functions, which required an enhanced mesolimbic dopamine signaling.[40]

The PCE male endophenotype may result from alterations in mesolimbic circuit function during computation that are secondary to modifications in the sensitivity of the cell to subtle changes in firing patterns of its inputs.[40] Of note, the dynamic accuracy of input information is guaranteed by ECBs through the regulation of the initial probability of neurotransmitter release and by mediating diverse forms of synaptic plasticity and homo-synaptic and hetero-synaptic metaplasticity.[41,79] At this stage, as the plethora of ECB actions spans from restricted synapse-specific control to wide (i.e., hetero-synaptic) regulation of long-lasting changes in the neural circuits or networks used in information processing,[41] we cannot rule out that other adaptive responses involving dysregulated ECB signaling might take place at discrete afferent inputs impinging on dopamine cells following PCE. Indeed, in males exposed to PCE, dopamine neurons that regulate the selection of relevant environmental stimuli respond to these stimuli once they are flagged by developmental challenges (e.g., acute stress, cannabis use), and their behavioral readout are also abnormal.[40,71,72] Of note, these sex-specific and disease-relevant phenotypes, such as locomotor hyperactivity (i.e., a sign of psychomotor agitation), proneness to risk-taking behavior (i.e., an index of impulsivity) and deficits in information processing (i.e., reduced PPI) disappear at adulthood.[40] Indeed, THC challenge fails to elicit PPI deficits and increase locomotion in young male adult PCE offspring[40] (Frau and Melis, unpublished data). These observations further highlight the importance of revealing risk factors for sensitive developmental windows that often precede the prodrome and that are frequently revealed by environmental challenges (e.g., stress, traumas, experimentation with recreational drugs), ultimately and potentially leading to a "distress disorder" accompanied by misattribution of salience to irrelevant stimuli.[80] As mentioned above, these maladaptive changes underpinning a susceptible (endo)phenotype are absent in female preadolescent individuals.

Sex differences in the effects of pre-natal cannabis exposure on mesolimbic dopamine system function

In contrast to the paucity of clinical data, sexual dimorphism in response to prenatal cannabinoid exposure is well recognized in experimental animals,[45,48,49] with males being mostly the affected offspring.

Male progeny are more susceptible than females to neurobehavioral dysregulation concerning cognitive and emotional domains induced by PCE via sex-specific alterations of dopamine, GABA and glutamate systems in mesolimbic and mesocortical pathways (Refs. 28,40,81–90; and also see for an exhaustive review[49]). In vivo and ex vivo electrophysiological recordings along with neurochemical and behavioral assays indicate that PCE does not affect mesolimbic dopamine system function and its related behavioral readouts in female offspring in the prepubertal period.[40,72] In particular, unlike their male counterparts, female PCE offspring do not display any spontaneous or THC-induced behavioral disease-relevant phenotypes: they exhibit natural locomotor activity and proclivity to stay near the perimeter of the open field arena, even upon a THC challenge.[72] Similarly, THC does not elicit any deficits in sensorimotor gating function in females.[40] PCE females are also less responsive to acute THC when accumbal dopamine levels are measured by microdialysis. This resilient phenotype exhibited by PCE females is also mirrored by electrophysiological features of VTA dopamine neurons displaying similar profiles to their control counterparts.[72] At pre-adolescence, PCE females also manifest a risk-averse phenotype along with normal social behavior and adaptive coping strategies to an acute stress, further suggesting that the sex variable "female" may confer protective factor(s) to PCE deleterious effects.[72]

Similar sex-specific outcomes of pre-natal cannabinoid exposure are also described at different stages of development (i.e., infancy and adulthood) and are associated with alterations in other brain region functions implicated in the regulation of mesolimbic dopamine pathway, such as the PFC and hippocampus.[81,84,85,87–89] In particular, in adulthood, PCE disrupts social interactions in males but not females, along with male-specific alterations of neuronal excitability and synaptic plasticity in the PFC (Refs. 81,83; and chapter TBD in this book); fetal exposure to cannabinoids elicits male-specific deficits in social communication and locomotor activity in infant offspring.[85] PCE males display disruptions of ECB long-term depression and heightened excitability of PFC pyramidal neurons. Similarly, PCE exerts a sex-dependent interference with the development of cholecystokinin-containing interneurons in the hippocampus, an effect solely observed in male progeny that depends upon CB1 receptor activation.[84,91] This sex-specific bias in GABAergic neurotransmission in the basket cell of the male hippocampus is also accompanied by impairments in spatial cognition and disruption in social behavior (Ref. 84; and chapter TBD in this book). These studies strongly suggest sex dimorphisms to detrimental effects of cannabinoid exposure in the womb and during lactation.

The scenario emerging from these pre-clinical observations suggests that male sex may be a risk factor for discrete neuropsychiatric disorders produced by PCE. As such, female sex may endow the progeny with intrinsic/innate biological resilience factors allowing them to adapt in the face of adverse environmental insults, such as stress or THC. Alternatively, PCE female progeny may (i) display adverse neurobehavioral outcomes that have yet to be discovered, (ii) differ from those observed in males, and/or (iii) eventually exhibit other disease-relevant phenotypes. These conceivable explanations warrant further investigations to uncover potential detrimental effects of PCE on other pathological outcomes in females that may include rearrangement of other brain circuits in a different sensitive developmental window, such as adolescence (see also chapter(s) TBD).

Clinical investigations on gender differences in the effects of PCE are mostly overlooked. Of note, only one of the abovementioned longitudinal studies (i.e., The OPPS study) includes gender as a confounding factor, reporting a number of long-lasting gender-dimorphisms in neurobehavioral outcomes, including heightened tremors, increased startle responsiveness and deficits in executive function (e.g., attention, cognitive flexibility, problem-solving, impulse control) (Refs. [19,92]; for a review see Traccis et al.[93]). It is, therefore, paramount to also include sex as a biological in human studies[94] when analyzing the risks and harms associated with cannabis use during pregnancy. Similarly, it is critical to implement preclinical investigations aimed at deciphering the mechanisms contributing to this sex-dependent effect of intrauterine THC exposure. Interestingly, growing evidence suggests that these might be secondary to perturbations of placental functions.[95] Of note, adverse intrauterine conditions or environmental factors have been shown to affect placental orchestration of fetal brain programming in a sex-dependent manner.[96] Environmental challenges can affect placental organogenesis and function, including efficiency of the transplacental barrier, nutrient and oxygen exchange, endocrine regulation, and gene expression.[97–99] As drugs of abuse target and impair the proper functioning of the placenta,[100] and molecular components of the ECB system are abundantly expressed in this organ (Ref. 101; Chapter TBD in this book), it is plausible that sex-specific differences in the outcomes of in utero exposure to THC, and more generally to (phyto)cannabinoids, might begin during fetal programming under fine control of the placenta.

Measuring vulnerability to mental illness induced by pre-natal cannabis exposure: A path toward an early clinical staging

Despite the significant advances made in the understanding of neurobiological substrates and brain circuit function and connectivity, the majority of neuropsychiatric disorders lack appropriate interventions capable of affecting their prognosis and preventing disease progression. One of the main reasons for this therapeutic failure lies on the scarce attention toward the development of psychopathology, as a phenomenological continuum of severity of symptoms, at the expenses of traditional symptom-based practices. The clinical staging model in psychiatry, proposed by McGorry et al.,[102] promotes an early intervention by integrating biological, social and environmental factors and an accurate distinction between earlier stage of pathophysiological trajectory—with its milder yet detectable clinical phenomena—from those that commonly identify illness progression and extension.[103] In this context, the identification of risk factors and at-risk individuals at an early stage of mental illness represent an ideal opportunity for defining the genesis of mental disorders and for designing preventative and targeted therapeutic strategies. As such, the observation that disorders of sensory information processing are observed in neurodevelopmental disorders (e.g., ADHD, autism),[104,105] other psychiatric disorders,[106,107] and also in the general pediatric population,[108,109] especially in males[110,111] is highly relevant. Sensory information processing is proposed as an endophenotype of different mental disorders or a *neutral cross-disorder trait* that should be examined across (1) a temporal and phenomenological continuum of increasing severity, and (2) persistence of symptoms between the healthy and clinical populations[80,107,112] (Fig. 2).

Within this framework, PCE may be regarded as one of those environmental determinants to be adopted into clinical staging for neuropsychiatric disorders, as children with PCE display more proneness to psychotic-like experiences as well as adverse neurodevelopmental outcomes.[10-12,31,32] Accordingly, PCE bestows an "at-risk" state, which can be termed, according to McGorry et al.,[102] as stage 0, exemplified by a derangement of dopamine signaling in brain structures deeply involved in the development of neuropsychiatric conditions.[81-83,91,113,114] Of note, the individual appears asymptomatic until psychopathological traits emerge upon environmental challenges, which act as a "second hit" by triggering aberrant behaviors at preadolescence (stage 1) (Fig. 3). Further progression of the disease (stage 2) and the persistence of symptoms (stage 3) may occur, with the emergence of other psychopathological symptoms (i.e., comorbidity) later in life and/or enduring neurobehavioral abnormalities in the exposed offspring. Mechanistic insights into these different stages can be inferred from preclinical investigations.[40,48,71,72,81-83,87-89,91,113–115]

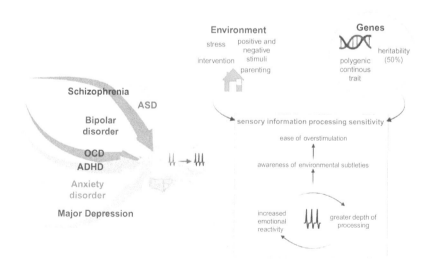

FIG. 2 Substantial inter-individual differences in sensory information processing exist across the population secondary to inherited predispositions and exposure to positive and/or negative environment. *ASD*, autism spectrum disorder; *OCD*, obsessive-compulsive disorder; *ADHD*, attention deficit hyperactivity disorder.

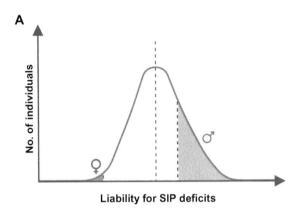
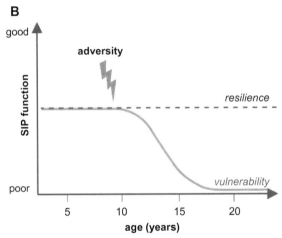

FIG. 3 Pre-natal cannabis exposure induces susceptibility to deficits in sensory information processing. (A) Sex-differential biology in the liability for deficits of sensory information processing (SIP). *Shaded area* represents diagnosed individuals. (B) Differential susceptibility theory applied to the sex-differential biological sensitivity to the context.

As such, not only is PCE a modifiable risk factor, but it can also be therapeutically targeted at different levels of stage transition of the disease. For instance, by increasing the awareness of the exposed progeny of their at-risk condition (stage 0), individuals have the chance to avoid and/or prevent the dangers of potential second hits (i.e., the transition to stage 1). Alternatively, individuals with PCE may benefit from an early therapeutic strategy aimed at the prevention of disease onset and/or its progression, and by lowering the risk of harm associated with potential exposure to second hits. As research in this field advances, the list of available medications increases. For example, the neurosteroid pregnenolone—a drug approved by the US Food and Drug Administration (FDA)—has been proposed as a therapeutic aid to protect the brain from acute cannabis intoxication.[116,117] In naïve rodents, pregnenolone, as a negative allosteric signal specific-inhibitor of CB1 receptors,[116] acutely antagonizes the addiction-related effects of THC by dampening THC-induced increases in VTA firing activity as well as extracellular dopamine levels in the nucleus accumbens.[116]

Additionally, pregnenolone acutely blocks THC-induced psychotic-like states, psychomotor agitation, reduction of social interaction and the mismatch between perception and reality.[118] Furthermore, a sub-chronic administration of pregnenolone in PCE offspring during preadolescence restores mesolimbic dopamine circuit function and rescues disease-relevant behavioral phenotypes,[40] thus suggesting that pregnenolone may act as a disease-modifying drug since it prevents the transition from stage 0 to 1. Similarly, the fatty acid amide (FAAH) inhibitor URB-597 and the positive allosteric modulator of metabotropic glutamate receptor type 5 (mGluR5), 3-cyano-N-(1,3-diphenyl-1H-pyrazol-5-yl)benzamide (CDPPB) have been identified as novel therapeutic strategies for counteracting synaptic and behavioral deficits induced by PCE.[81,85] Enhanced circulating anandamide levels, by means of preventing its degradation (i.e., FAAH inhibition), rescued the deficits in endocannabinoid mediated-LTD and social interaction of male rats pre-natally exposed to cannabinoids.[81] mGluR5 regulates the activity of neuronal networks, including the intrinsic properties of pyramidal cells in pre-frontal cortex (PFC), one of the target dopaminergic regions that is impaired in neuropsychiatric disorders. Indeed, PCE sex-specifically affects mGluR5 signaling leading to a heightened excitability of PFC pyramidal neurons, along with profound deficits in social interaction, social communication, and locomotion at different developmental periods.[81,85] Hence, pharmacological enhancement of either mGluR5 or anandamide signaling rectified PCE-induced synaptic defects and rescued behavioral deficits only in PCE male progeny,[81,85] consistently with the view that regulating the endocannabinoidome may act as disease-modifying therapy for endophenotypes at risk of developing neuropsychiatric conditions.

Concluding remarks

As outlined in this chapter, PCE is a modifiable risk factor inducing an endophenotype (i.e., disorders of sensory information processing) susceptible to subsequent challenges (e.g., cannabis use, negative environment) during sensitive neurodevelopmental windows (e.g., preadolescence), especially in male offspring. To date, many questions remain unanswered. First, although dopamine regulation is key for proper PPI, the precise identity of afferent inhibitory inputs accounting for the mesolimbic hyperdopaminergia tied to deficits in sensorimotor gating functions has never been investigated in health and disease states. Second, whether or not there is a sex bias in the regulation of inhibitory inputs contributing to dopamine modulation of sensory information processing and PPI is currently unknown. Third, whether sex causes categorically distinct changes related to mesocortical vs mesolimbic dopamine pathways has never been assessed in the healthy and PCE population. Additionally, since dysregulation of gene expression influences the timing of neurodevelopmental processes and confers risk for mental illness, which shows differential prevalence between sexes, it is paramount to determine sex differences in molecular signature profiles of dopamine neurons projecting to distinct target regions. Differential post-natal dynamics may represent susceptibility and/or resiliency factors that could be exploited as potential biomarkers, or that could uncover novel candidates for targeted therapeutics. Equally important, it is imperative to understand whether different states of engagement of ECB signaling (i.e., prior, tonic, and persistent) represent adaptive or maladaptive changes, in order to gain knowledge about pathophysiological mechanisms related to PCE and to identify and exploit new therapeutic targets. Future investigations are warranted to design evidence-based mechanistic therapeutic intervention points for mental illness associated with PCE and implicated in aberrant assignment of salience to environmental stimuli, which may contribute to life-course vulnerability to mental illness.

References

1. WHO. World Health Organization. *Caring for children and adolescents with mental disorders: Setting WHO directions.* WHO Library Cataloguing-in-Publication Data; 2002.
2. CDC. *Centers for Disease Control and Prevention. Data and statistics on children's mental health*; 2020.
3. WHO. World Health Organization. *Adolescent mental health*; 2020.
4. Mandy M, Nyirenda M. Developmental origins of health and disease: the relevance to developing nations. *Int Health* 2018;**10**:66–70.
5. Wadhwa PD, Buss C, Entringer S, Swanson JM. Developmental origins of health and disease: brief history of the approach and current focus on epigenetic mechanisms. *Semin Reprod Med* 2009;**27**:358–68.
6. Barker DJ, Osmond C. Infant mortality, childhood nutrition, and ischaemic heart disease in England and Wales. *Lancet* 1986;**1**:1077–81.
7. Barker DJ, Gluckman PD, Godfrey KM, Harding JE, Owens JA, Robinson JS. Fetal nutrition and cardiovascular disease in adult life. *Lancet* 1993;**341**:938–41.
8. Bayer TA, Falkai P, Maier W. Genetic and non-genetic vulnerability factors in schizophrenia: the basis of the "two hit hypothesis". *J Psychiatr Res* 1999;**33**:543–8.
9. Fleiss B, Rivkees SA, Gressens P. Early origins of neuropsychiatric disorders. *Pediatr Res* 2019;**85**:113–4.
10. Bolhuis K, Kushner SA, Yalniz S, Hillegers MHJ, Jaddoe VWV, Tiemeier H, et al. Maternal and paternal cannabis use during pregnancy and the risk of psychotic-like experiences in the offspring. *Schizophr Res* 2018;**202**:322–7.
11. Fine JD, Moreau AL, Karcher NR, Agrawal A, Rogers CE, Barch DM, et al. Association of prenatal cannabis exposure with psychosis proneness among children in the adolescent brain cognitive development (ABCD) study. *JAMA Psychiatry* 2019;**76**:762–4.
12. Paul SE, Hatoum AS, Fine JD, Johnson EC, Hansen I, Karcher NR, et al. Associations between prenatal cannabis exposure and childhood outcomes: results from the ABCD study. *JAMA Psychiatry* 2021;**78**:64–76.

13. Arias-Carrión O, Stamelou M, Murillo-Rodríguez E, Menéndez-González M, Pöppel E. Dopaminergic reward system: a short integrative review. *Int Arch Med* 2010;**3**:24.
14. Blum K, Chen AL, Braverman ER, et al. Attention-deficit-hyperactivity disorder and reward deficiency syndrome. *Neuropsychiatr Dis Treat* 2008;**4**:893–918.
15. Kesby JP, Eyles DW, McGrath JJ, Scott JG. Dopamine, psychosis and schizophrenia: the widening gap between basic and clinical neuroscience. *Transl Psychiatry* 2018;**8**:30.
16. Wu CS, Jew CP, Lu HC. Lasting impacts of prenatal cannabis exposure and the role of endogenous cannabinoids in the developing brain. *Future Neurol* 2011;**6**:459–80.
17. Assary E, Vincent JP, Keers R, Pluess M. Gene-environment interaction and psychiatric disorders: review and future directions. *Semin Cell Dev Biol* 2018;**77**:133–43.
18. Fried PA. The Ottawa Prenatal Prospective Study (OPPS): methodological issues and findings—it's easy to throw the baby out with the bath water. *Life Sci* 1995;**56**:2159–68.
19. Fried PA, Smith AM. A literature review of the consequences of prenatal marihuana exposure. An emerging theme of a deficiency in aspects of executive function. *Neurotoxicol Teratol* 2001;**23**:1–11.
20. Fried PA, Watkinson B, Gray R. Differential effects on cognitive functioning in 13- to 16-year-olds prenatally exposed to cigarettes and marihuana. *Neurotoxicol Teratol* 2003;**25**:427–36.
21. Smith AM, Fried PA, Hogan MJ, Cameron I. Effects of prenatal marijuana on visuospatial working memory: an fMRI study in young adults. *Neurotoxicol Teratol* 2006;**28**:286–95.
22. Day NL, Richardson GA. Prenatal marijuana use: epidemiology, methodologic issues, and infant outcome. *Clin Perinatol* 1991;**18**:77–91.
23. Goldschmidt L, Richardson GA, Larkby C, Day NL. Early marijuana initiation: the link between prenatal marijuana exposure, early childhood behavior, and negative adult roles. *Neurotoxicol Teratol* 2016;**58**:40–5.
24. Richardson GA, Ryan C, Willford J, Day NL, Goldschmidt L. Prenatal alcohol and marijuana exposure: effects on neuropsychological outcomes at 10 years. *Neurotoxicol Teratol* 2002;**24**:309–20.
25. El Marroun H, Tiemeier H, Steegers EA, et al. Intrauterine cannabis exposure affects fetal growth trajectories: the generation R study. *J Am Acad Child Adolesc Psychiatry* 2009;**48**:1173–81.
26. El Marroun H, Hudziak JJ, Tiemeier H, et al. Intrauterine cannabis exposure leads to more aggressive behavior and attention problems in 18-month-old girls. *Drug Alcohol Depend* 2011;**118**:470–4.
27. El Marroun H, Bolhuis K, Franken IHA, et al. Preconception and prenatal cannabis use and the risk of behavioural and emotional problems in the offspring; a multi-informant prospective longitudinal study. *Int J Epidemiol* 2019;**48**:287–96.
28. Calvigioni D, Hurd YL, Harkany T, Keimpema E. Neuronal substrates and functional consequences of prenatal cannabis exposure. *Eur Child Adolesc Psychiatry* 2014;**23**:931–41.
29. McLemore GL, Richardson KA. Data from three prospective longitudinal human cohorts of prenatal marijuana exposure and offspring outcomes from the fetal period through young adulthood. *Data Brief* 2016;**9**:753–7.
30. Richardson KA, Hester AK, McLemore GL. Prenatal cannabis exposure—the "first hit" to the endocannabinoid system. *Neurotoxicol Teratol* 2016;**58**:5–14.
31. Corsi DJ, Donelle J, Sucha E, Hawken S, Hsu H, El-Chaâr D, et al. Maternal cannabis use in pregnancy and child neurodevelopmental outcomes. *Nat Med* 2020;**26**:1536–40.
32. Roffman JL, Sipahi ED, Dowling KF, Hughes DE, Hopkinson CE, Lee H, et al. Association of adverse prenatal exposure burden with child psychopathology in the adolescent brain cognitive development (ABCD) study. *PLoS ONE* 2021;**16**, e0250235.

33. Gottesman II, Gould TD. The endophenotype concept in psychiatry: etymology and strategic intentions. *Am J Psychiatry* 2003;**160**:636–45.
34. Braff DL, Light GA. The use of neurophysiological endophenotypes to understand the genetic basis of schizophrenia. *Dialogues Clin Neurosci* 2005;**7**:125–35.
35. Harper J, Liu M, Malone SM, McGue M, Iacono WG, Vrieze SI. Using multivariate endophenotypes to identify psychophysiological mechanisms associated with polygenic scores for substance use, schizophrenia, and education attainment. *Psychol Med* 2021;1–11.
36. Turetsky BI, Dress EM, Braff DL, et al. The utility of P300 as a schizophrenia endophenotype and predictive biomarker: clinical and socio-demographic modulators in COGS-2. *Schizophr Res* 2015;**163**:53–62.
37. Hensch TK, Bilimoria PM. Re-opening windows: manipulating critical periods for brain development. *Cerebrum* 2012;**2012**:11.
38. Van den Bergh BRH, van den Heuvel MI, Lahti M, Braeken M, de Rooij SR, Entringer S, et al. Prenatal developmental origins of behavior and mental health: the influence of maternal stress in pregnancy. *Neurosci Biobehav Rev* 2020;**117**:26–64.
39. Viveros MP, Llorente R, Suarez J, Llorente-Berzal A, López-Gallardo M, de Fonseca FR. The endocannabinoid system in critical neurodevelopmental periods: sex differences and neuropsychiatric implications. *J Psychopharmacol* 2012;**26**:164–76.
40. Frau R, Miczàn V, Traccis F, Aroni S, Pongor CI, Saba P, et al. Prenatal THC exposure produces a hyperdopaminergic phenotype rescued by pregnenolone. *Nat Neurosci* 2019;**22**:1975–85.
41. Melis M, Greco B, Tonini R. Interplay between synaptic endocannabinoid signaling and metaplasticity in neuronal circuit function and dysfunction. *Eur J Neurosci* 2014;**39**:1189–201.
42. Citri A, Malenka RC. Synaptic plasticity: multiple forms, functions, and mechanisms. *Neuropsychopharmacology* 2008;**33**:18–41.
43. Abraham WC, Bear MF. Metaplasticity: the plasticity of synaptic plasticity. *Trends Neurosci* 1996;**19**:126–30.
44. Alpar A, Di Marzo V, Harkany T. At the tip of an iceberg: prenatal marijuana and its possible relation to neuropsychiatric outcome in the offspring. *Biol Psychiatry* 2016;**79**:e33–45.
45. Hurd YL, Manzoni OJ, Pletnikov MV, Lee FS, Bhattacharyya S, Melis M. Cannabis and the developing brain: insights into its long-lasting effects. *J Neurosci* 2019;**39**:8250–8.
46. Melis M, Frau R, Kalivas PW, Spencer S, Chioma V, Zamberletti E, et al. New vistas on cannabis use disorder. *Neuropharmacology* 2017;**124**:62–72.
47. Pinky PD, Bloemer J, Smith WD, Moore T, Hong H, Suppiramaniam V, et al. Prenatal cannabinoid exposure and altered neurotransmission. *Neuropharmacology* 2019;**149**:181–94.
48. Scheyer AF, Melis M, Trezza V, Manzoni OJJ. Consequences of perinatal cannabis exposure. *Trends Neurosci* 2019;**42**:871–84.
49. Tirado Munoz J, Belen Lopez-Rodriguez A, Fonseca F, Farre M, Torrens M, Viveros MP. Effects of cannabis exposure in the prenatal and adolescent periods: preclinical and clinical studies in both sexes. *Front Neuroendocrinol* 2020;**57**, 100841.
50. Brainstorm Consortium, Anttila V, Bulik-Sullivan B, et al. Analysis of shared heritability in common disorders of the brain. *Science* 2018;**360**, eaap8757.
51. Torres CA, Medina-Kirchner C, O'Malley KY, Hart CL. Totality of the evidence suggests prenatal cannabis exposure does not lead to cognitive impairments: a systematic and critical review. *Front Psychol* 2020;**11**:816.
52. Jaques SC, Kingsbury A, Henshcke P, et al. Cannabis, the pregnant woman and her child: weeding out the myths. *J Perinatol* 2014;**34**:417–24.

53. Harkany T, Keimpema E, Barabás K, Mulder J. Endocannabinoid functions controlling neuronal specification during brain development. *Mol Cell Endocrinol* 2008;**286**:S84–90.
54. Lubman DI, Cheetham A, Yücel M. Cannabis and adolescent brain development. *Pharmacol Ther* 2015;**148**:1–16.
55. Basavarajappa BS, Nixon RA, Arancio O. Endocannabinoid system: emerging role from neurodevelopment to neurodegeneration. *Mini-Rev Med Chem* 2009;**9**:448–62.
56. Harkany T, Guzmán M, Galve-Roperh I, Berghuis P, Devi LA, Mackie K. The emerging functions of endocannabinoid signaling during CNS development. *Trends Pharmacol Sci* 2007;**28**:83–92.
57. Brisch R, Saniotis A, Wolf R, et al. The role of dopamine in schizophrenia from a neurobiological and evolutionary perspective: old fashioned, but still in vogue. *Front Psychiatry* 2014;**5**:110.
58. Howes OD, Kapur S. The dopamine hypothesis of schizophrenia: version III—the final common pathway. *Schizophr Bull* 2009;**35**:549–62.
59. Money KM, Stanwood GD. Developmental origins of brain disorders: roles for dopamine. *Front Cell Neurosci* 2013;**7**:260.
60. Spencer GE, Klumperman J, Syed NI. Neurotransmitters and neurodevelopment. Role of dopamine in neurite outgrowth, target selection and specific synapse formation. *Perspect Dev Neurobiol* 1998;**5**:451–67.
61. Homberg JR, Jagiellowicz J. A neural model of vulnerability and resilience to stress-related disorders linked to differential susceptibility. *Mol Psychiatry* 2021. https://doi.org/10.1038/s41380-021-01047-8.
62. Jolicoeur-Martineau A, Belsky J, Szekely E, Widaman KF, Pluess M, Greenwood C, et al. Distinguishing differential susceptibility, diathesis-stress, and vantage sensitivity: beyond the single gene and environment model. *Dev Psychopathol* 2020;**32**:73–83.
63. Simons RL, Beach SR, Barr AB. Differential susceptibility to context: a promising model of the interplay of genes and the social environment. *Adv Group Process* 2012;**29**. https://doi.org/10.1108/S0882-6145.
64. Bonnin A, de Miguel R, Rodríguez-Manzaneque JC, Fernández-Ruiz JJ, Santos A, Ramos JA. Changes in tyrosine hydroxylase gene expression in mesencephalic catecholaminergic neurons of immature and adult male rats perinatally exposed to cannabinoids. *Brain Res Dev Brain Res* 1994;**81**:147–50.
65. Bonnin A, de Miguel R, Hernández ML, Ramos JA, Fernández-Ruiz JJ. The prenatal exposure to delta 9-tetrahydrocannabinol affects the gene expression and the activity of tyrosine hydroxylase during early brain development. *Life Sci* 1995;**56**:2177–84.
66. Bonnin A, de Miguel R, Castro JG, Ramos JA, Fernandez-Ruiz JJ. Effects of perinatal exposure to delta 9-tetrahydrocannabinol on the fetal and early postnatal development of tyrosine hydroxylase-containing neurons in rat brain. *J Mol Neurosci* 1996;**7**:291–308.
67. Rodriguez de Fonseca F, Cebeira M, Fernandez-Ruiz JJ, Navarro M, Ramos JA. Effects of pre- and perinatal exposure to hashish extracts on the ontogeny of brain dopaminergic neurons. *Neuroscience* 1991;**43**:713–23.
68. Fernandez-Ruiz JJ, Bonnin A, Cebeira M, Ramos JA. Ontogenic and adult changes in the activity of hypothalamic and extra hypothalamic dopamine neurons after perinatal cannabinoid exposure. In: Palomo T, Archer T, editors. *Strategies for studying brain disorders*. vol. I. England: Farrand Press; 1994. p. 357–90.
69. García L, de Miguel R, Ramos JA, Fernàndez-Ruiz JJ. Perinatal delta 9-tetrahydrocannabinol exposure in rats modifies the responsiveness of midbrain dopaminergic neurons in adulthood to a variety of challenges with dopaminergic drugs. *Drug Alcohol Depend* 1996;**42**:155–66.

70. García-Gil L, De Miguel R, Muñoz RM, et al. Perinatal delta(9)-tetrahydrocannabinol exposure alters the responsiveness of hypothalamic dopaminergic neurons to dopamine-acting drugs in adult rats. *Neurotoxicol Teratol* 1997;**19**:477–87.
71. Sagheddu C, Traccis F, Serra V, Congiu M, Frau R, Cheer JF, et al. Mesolimbic dopamine dysregulation as a signature of information processing deficits imposed by prenatal THC exposure. *Prog Neuro-Psychopharmacol Biol Psychiatry* 2021;**105**, 110128.
72. Traccis F, Serra V, Sagheddu C, Congiu M, Saba P, Giua G, et al. Prenatal THC does not affect female mesolimbic dopaminergic system in preadolescent rats. *Int J Mol Sci* 2021;**22**:1666.
73. Swerdlow NR, Braff DL, Geyer MA. Cross-species studies of sensorimotor gating of the startle reflex. *Ann N Y Acad Sci* 1999;**877**:202–16.
74. Davies PL, Chang WP, Gavin WJ. Maturation of sensory gating performance in children with and without sensory processing disorders. *Int J Psychophysiol* 2009;**72**:187–97.
75. Swerdlow NR, Light GA. Sensorimotor gating deficits in schizophrenia: advancing our understanding of the phenotype, its neural circuitry and genetic substrates. *Schizophr Res* 2018;**198**: 1–5.
76. Vargas JP, Diaz E, Portavella M, Lopez JC. Animal models of maladaptive traits: disorders in sensorimotor gating and attentional quantifiable responses as possible endophenotypes. *Front Psychol* 2016;**7**:206.
77. Glebov OO, Jackson RE, Winterflood CM, Owen DM, Barker EA, Doherty P, et al. Nanoscale structural plasticity of the active zone matrix modulates presynaptic function. *Cell Rep* 2017;**18**:2715–28.
78. Lammel S, Ion DI, Roeper J, Malenka RC. Projection-specific modulation of dopamine neuron synapses by aversive and rewarding stimuli. *Neuron* 2011;**70**:855–62.
79. Abraham WC, Richter-Levin G. From synaptic metaplasticity to behavioral metaplasticity. *Neurobiol Learn Mem* 2018;**154**:1–4.
80. Bolhuis K, Koopman-Verhoeff ME, Blanken LME, Cibrev D, Jaddoe VWV, Verhulst FC, et al. Psychotic-like experiences in pre-adolescence: what precedes the antecedent symptoms of severe mental illness? *Acta Psychiatr Scand* 2018;**138**:15–25.
81. Bara A, Manduca A, Bernabeu A, et al. Sex-dependent effects of in utero cannabinoid exposure on cortical function. *elife* 2018;**7**. https://doi.org/10.7554/eLife.36234, e36234.
82. DiNieri JA, Wang X, Szutorisz H, et al. Maternal cannabis use alters ventral striatal dopamine D2 gene regulation in the offspring. *Biol Psychiatry* 2011;**70**:763–9.
83. de Salas-Quiroga A, Díaz-Alonso J, García-Rincón D, et al. Prenatal exposure to cannabinoids evokes long-lasting functional alterations by targeting CB1 receptors on developing cortical neurons. *Proc Natl Acad Sci U S A* 2015;**112**:13693–8.
84. de Salas-Quiroga A, García-Rincón D, Gómez-Domínguez D, et al. Long-term hippocampal interneuronopathy drives sex-dimorphic spatial memory impairment induced by prenatal THC exposure. *Neuropsychopharmacology* 2020;**45**:877–86.
85. Manduca A, Servadio M, Melancia F, Schiavi S, Manzoni OJ, Trezza V. Sex-specific behavioural deficits induced at early life by prenatal exposure to the cannabinoid receptor agonist WIN55, 212-2 depend on mGlu5 receptor signalling. *Br J Pharmacol* 2020;**177**: 449–63.
86. Saez TM, Aronne MP, Caltana L, Brusco AH. Prenatal exposure to the CB1 and CB2 cannabinoid receptor agonist WIN 55,212-2 alters migration of early-born glutamatergic neurons and GABAergic interneurons in the rat cerebral cortex. *J Neurochem* 2014;**129**:637–48.

87. Scheyer AF, Borsoi M, Wager-Miller J, et al. Cannabinoid exposure via lactation in rats disrupts perinatal programming of the gamma-aminobutyric acid trajectory and select early-life behaviors. *Biol Psychiatry* 2020;**87**:666–77.
88. Scheyer AF, Borsoi M, Pelissier-Alicot AL, Manzoni OJJ. Perinatal THC exposure via lactation induces lasting alterations to social behavior and prefrontal cortex function in rats at adulthood. *Neuropsychopharmacology* 2020;**45**:1826–33.
89. Scheyer AF, Borsoi M, Pelissier-Alicot AL, Manzoni OJJ. Maternal exposure to the cannabinoid agonist WIN 55,12,2 during lactation induces lasting behavioral and synaptic alterations in the rat adult offspring of both sexes. *eNeuro* 2020;**7**, ENEURO.0144-20.2020.
90. Spano MS, Ellgren M, Wang X, Hurd YL. Prenatal cannabis exposure increases heroin seeking with allostatic changes in limbic enkephalin systems in adulthood. *Biol Psychiatry* 2007;**61**:554–63.
91. Vargish GA, Pelkey KA, Yuan X, et al. Persistent inhibitory circuit defects and disrupted social behaviour following in utero exogenous cannabinoid exposure. *Mol Psychiatry* 2017;**22**:56–67.
92. Fried PA, Makin JE. Neonatal behavioural correlates of prenatal exposure to marihuana, cigarettes and alcohol in a low risk population. *Neurotoxicol Teratol* 1987;**9**(1):1–7.
93. Traccis F, Frau R, Melis M. Gender differences in the outcome of offspring prenatally exposed to drugs of abuse. *Front Behav Neurosci* 2020;**14**(72). https://doi.org/10.3389/fnbeh.2020.00072.
94. Clayton JA. Applying the new SABV (sex as a biological variable) policy to research and clinical care. *Physiol Behav* 2018;**187**:2–5.
95. Bronson SL, Bale TL. The placenta as a mediator of stress effects on neurodevelopmental reprogramming. *Neuropsychopharmacology* 2016;**41**:207–18.
96. Walsh K, McCormack CA, Webster R, et al. Maternal prenatal stress phenotypes associate with fetal neurodevelopment and birth outcomes. *Proc Natl Acad Sci U S A* 2019;**116**:23996–4005.
97. Fowden AL, Sferruzzi-Perri AN, Coan PM, Constancia M, Burton GJ. Placental efficiency and adaptation: endocrine regulation. *J Physiol* 2009;**587**:3459–72.
98. Jansson T, Powell TL. Role of the placenta in fetal programming: underlying mechanisms and potential interventional approaches. *Clin Sci (Lond)* 2007;**113**:1–13.
99. Watson ED, Cross JC. Development of structures and transport functions in the mouse placenta. *Physiology (Bethesda)* 2005;**20**:180–93.
100. Ganapathy V. Drugs of abuse and human placenta. *Life Sci* 2011;**88**:926–30.
101. Park B, Gibbons HM, Mitchell MD, Glass M. Identification of the CB1 cannabinoid receptor and fatty acid amide hydrolase (FAAH) in the human placenta. *Placenta* 2003;**24**:990–5.
102. McGorry PD, Hickie IB, Yung AR, Pantelis C, Jackson HJ. Clinical staging of psychiatric disorders: a heuristic framework for choosing earlier, safer and more effective interventions. *Aust N Z J Psychiatry* 2006;**40**:616–22.
103. McGorry P, Keshavan M, Goldstone S, et al. Biomarkers and clinical staging in psychiatry. *World Psychiatry* 2014;**13**:211–23.
104. Giakoumaki SG, Roussos P, Rogdaki M, Karli C, Bitsios P, Frangou S. Evidence of disrupted prepulse inhibition in unaffected siblings of bipolar disorder patients. *Biol Psychiatry* 2007;**62**:1418–22.
105. Molholm S, Murphy JW, Bates J, Ridgway EM, Foxe JJ. Multisensory audiovisual processing in children with a sensory processing disorder (I): behavioral and electrophysiological indices under speeded response conditions. *Front Integr Neurosci* 2020;**14**:4.

106. Acevedo B, Aron E, Pospos S, Jessen D. The functional highly sensitive brain: a review of the brain circuits underlying sensory processing sensitivity and seemingly related disorders. *Philos Trans R Soc Lond Ser B Biol Sci* 2018;**373**, 20170161.
107. Greven CU, Lionetti F, Booth C, Aron EN, Fox E, Schendan HE, et al. Sensory processing sensitivity in the context of environmental sensitivity: a critical review and development of research agenda. *Neurosci Biobehav Rev* 2019;**98**:287–305.
108. Aron EN, Aron A. Sensory-processing sensitivity and its relation to introversion and emotionality. *J Pers Soc Psychol* 1997;**73**:345–68.
109. Aron EN, Aron A, Jagiellowicz J. Sensory processing sensitivity: a review in the light of the evolution of biological responsivity. *Personal Soc Psychol Rev* 2012;**16**:262–82.
110. Hollis C, Rapoport J. Child and adolescent schizophrenia. In: *Schizophrenia*. 3rd ed; 2008. p. 22.
111. Jussila K, Junttila M, Kielinen M, Ebeling H, Joskitt L, Moilanen I, et al. Sensory abnormality and quantitative autism traits in children with and without autism spectrum disorder in an epidemiological population. *J Autism Dev Disord* 2020;**50**:180–8.
112. Linscott RJ, van Os J. An updated and conservative systematic review and meta-analysis of epidemiological evidence on psychotic experiences in children and adults: on the pathway from proneness to persistence to dimensional expression across mental disorders. *Psychol Med* 2013;**43**:1133–49.
113. Hurd YL, Michaelides M, Miller ML, Jutras-Aswad D. Trajectory of adolescent cannabis use on addiction vulnerability. *Neuropharmacology* 2014;**76**:416–24.
114. Szutorisz H, DiNieri JA, Sweet E, et al. Parental THC exposure leads to compulsive heroin-seeking and altered striatal synaptic plasticity in the subsequent generation. *Neuropsychopharmacology* 2014;**39**:1315–23.
115. Campolongo P, Trezza V, Ratano P, Palmery M, Cuomo V. Developmental consequences of perinatal cannabis exposure: behavioral and neuroendocrine effects in adult rodents. *Psychopharmacology* 2011;**214**:5–15.
116. Vallée M, et al. Pregnenolone can protect the brain from cannabis intoxication. *Science* 2014;**343**:94–8.
117. Vallée M. Neurosteroids and potential therapeutics: focus on pregnenolone. *J Steroid Biochem Mol Biol* 2016;**160**:78–87.
118. Busquets-Garcia A, Soria-Gómez E, Redon B, et al. Pregnenolone blocks cannabinoid-induced acute psychotic-like states in mice. *Mol Psychiatry* 2017;**22**:1594–603.

Chapter 8

Perinatal cannabis exposure and long-term consequences on synaptic programming

Gabriele Giua[a,b], Olivier JJ. Manzoni[b], and Andrew Scheyer[a,b]
[a]*Cannalab, Cannabinoids Neuroscience Research International Associated Laboratory, INSERM-Aix-Marseille University, Marseille, France*, [b]*Mediterranean Neurobiology Institute, INSERM U1249, Aix-Marseille University, Marseille, France*

Estimates of cannabis consumption during pregnancy are largely self-reported or determined via plasma analysis at mid-gestational physiological exams and at the time of birth. Self-reported cannabis consumption during pregnancy determined via the 2007–2012 National Surveys on Drug Use and Health was found to be 3.9% among women aged 18–44 years.[1] Elsewhere, a retrospective analysis of live births in Ontario, Canada, between 2007 and 2012, reported that cannabis use among mothers was significantly lower (0.6%).[2] Age may play a significant role in this variability, as Young-Wolff and colleagues, in an analysis of California health system screenings, found that between 2009 and 2016 the rate of cannabis consumption during the gestational period increased from 4.2% to 7.1%, with the highest increases found among women under 18 years of age (from 12.5% to 21.8%) and those aged 18–24 (from 9.8% to 19%).[3] However, using meconium analysis, the rate of cannabinoid exposure among 400 assayed infants in Glasgow was found to be even higher, at 13.25%, suggesting significant variability in rates of cannabis use during pregnancy based on regional factors, as well.[4]

Developmental cannabis exposure remains a significant concern during early post-natal development. Cannabinoids, including tetrahydrocannabinol (THC) and cannabidiol (CBD), are also excreted in human milk and thusly transferred to the nursing infant.[5,6] Crucially, THC, CBD, and associated metabolites remain in significant quantities in breast milk well after their acute half-lives. In chronic cannabis users, THC was found in 63% of breast milk samples up to 6 days after the last consumption, with CBD and the active THC metabolites 11-OH-THC further detected in 10% of these samples. Unlike the metabolism of cannabinoids in plasma, concentrations in breast milk appear to follow a linear degradation timeline.[7] The estimated dose delivered to the developing infant is 2.5% that of the peak maternal plasma concentration. Assuming an inhaled consumption of 0.1 g of cannabis containing 23.18 mg of THC, infant exposure for an exclusively breasted infant would be 8 μg/kg/day.[8] Thus during the crucial postnatal developmental period, maternal cannabis use results in significant infant cannabinoid exposure.

Early development is defined by critical windows of tightly regulated processes responsible for laying the foundational of functional neurocircuitry.[9,10] Among the primary regulatory actors in the developmental process, from early ontogenesis through neurodevelopmental maturation, is the endocannabinoid (eCB) system.[11,12] Neuronal expression of cannabinoid receptors is particularly temporally regulated. Restricted expression of CB1R in cholinergic neurons is required for a brief pre-natal developmental period, for instance, but not in post-natal life.[13] White matter expression of CB1R in the developing rat brain almost universally (save for the anterior commissure) peaks at PND1 and decreases progressively until disappearing at adulthood.[14] Analysis of human brain tissue has yielded similar findings, illustrating a both temporal and spatial selectivity of cannabinoid receptor expression over the course of both pre- and post-natal development.[15] In part, this may explain the peculiar nature of perinatal cannabinoid exposure (PCE) outcomes, which appear to impact select aspects of neuronal function rather than induce global insult.

Parental germline cannabis exposure could affect progeny through transgenerational epigenetic alterations

Endocannabinoid system and epigenetic re-programming cohabitation in re-productive tissues

There is now ample evidence that the eCB system (ECS) participates in mammalian reproductive tissue homeostasis and influences fertilization, embryo implantation, and gestational processes.[16–20] Epigenetic re-programming occurs extensively during gametogenesis and embryogenesis,[21,22] periods during which drugs of abuse can indeed lead to cross-generational epigenome modifications.[23,24] Multiple epigenetic mechanisms such as histone post-translational modifications, small non-coding RNAs, and DNA methylation lead to chromatin alteration and changes in gene expression.[25] Epigenetic processes such as DNA methylation have been implicated in synaptic plasticity and drug addiction, suggesting a meso-cortical-limbic vulnerability in epigenetic environmental.[26–29] In the instance of transgenerational epigenetic inheritance, these modifications are not completely erased from the germline and can be transmitted to subsequent generations.[30]

Direct effects of cannabinoids exposure on parent's germline

The ECS participates in male gametogenesis.[20] In mammals, eCBs have been detected in prostate epithelial cells,[31] sperm cells,[32] and testis,[33] while CB1R receptors have been identified in germ cells, testis, prostate, vase deferens, and Leydig cells,[33–36] CB2R in spermatozoa, Sertoli cells, and prostate epithelium,[37–40] and TRPV1 in sperm cells.[41] Accordingly, the activity of CB1R influences male fertility, while its activation has been shown to decrease sperm cell motility and acrosome reaction operation.[36,42] CB2R activity impacts sperm cell vitality.[40] More recently, TRPV1 channels have also been implicated in regulating sperm quality and functions.[41,43] Further, higher CB1R and CB2R mRNA levels were found in mature, rather than immature spermatozoa, and the largest levels were measured in fertile men.[44] Notably, large supraphysiological concentrations of the AEA analog met-AEA decreased human sperm motility in a CB1R-dependent manner.[45] Thus interference with the normal physiological functions of the ECS can potentially affect mammalian male fertility. In agreement with this idea, aberrant DNA methylation of testicular tissues has been associated with male infertility[46–48] and eCB receptor agonists alter sperm DNA methylation patterns in both rodents and humans models employing either THC[49] or CB2R-specific agonists.[50]

The ECS influences female gametogenesis, too.[18] An interplay between hormonal levels and ECS components seems to take place during the menstrual cycle in female reproductive tissues. Both progesterone and estrogen regulate FAAH expression levels and consequently AEA.[51] Concentrations of AEA, as well as CB1R and CB2R, fluctuate throughout the menstrual cycle.[52–54] During oocyte meiosis, CB1R may participate in oocyte maturation by reaching the cell surface[55] and, consequently, THC incubation of mice oocytes accelerates CB1R migration to the cell surface, further demonstrating cannabinoid-mediated regulation of oocyte homeostasis.[56] Recently, Castel and collaborators showed that in-utero exposure to THC or a synthetic cannabinoid agonist decreased the number on non-growing ovarian follicles (the "ovarian reserve") by about 40% in young adult females and that, conversely, in-utero exposure to a CB1R antagonist increased the ovarian reserve.[57] These data further illustrate the detrimental long-term effects of prenatal cannabis exposure on the reproductive process.

Pre-gestational cannabinoid exposure and transgenerational effects

THC's transgenerational epigenetic influences on striatal molecular/functional features are particularly well illustrated by a series of pioneering studies from the Hurd laboratory. Therein, male Long Evans rats born from parents who were treated throughout adolescence with THC showed an increase in heroin self-administration, dysregulated long-term plasticity in the dorsal striatum, and alterations to glutamate receptor subunits at both the mRNA and protein levels.[58] Further, male and female offspring presented altered DNA methylation in genes implicated in development, synaptic plasticity, and neurotransmission.[59] In the F1 generation, mRNA expression patterns of genes encoding sub-units of CB1R, NMDAR, and AMPAR were altered in a sex- and developmental stage-specific manner.[60] Significant transgenerational effects of exogenous cannabinoid exposure were also observed in the offspring of female rats treated with a CB1R/CB2R agonist (WIN 55,212-2) during adolescence and subsequently mated with drug-naïve partners. Their progeny exhibited heightened sensitivity to morphine during conditioned place preference and locomotor sensitization testing.[61,62] Elsewhere, transgenerational epigenetic effects of cannabinoids on DNA methylation pattern have been found in the pre-frontal cortex (PFC) and correlated with an altered stress response.[63] Thus exposure to exogenous cannabinoids outside of the reproductive window has the potential to induce transgenerational consequences via multiple mechanisms, highlighting the sensitivity of the reproductive system to manipulation of the ECS.

Cannabinoids exposure and peri-embryo implantation influences

Role of the endocannabinoid system in early embryo development and implantation

The ECS synchronizes the embryo peri implantation phase from oviductal transport to reception in the uterine wall.[16–19] Implantation processes and AEA levels are to be strongly correlated: downregulation of AEA increases the uterine predisposition to reception[64–67] and the blastocyst propensity for implantation,[68,69] with FAAH proving to be an important player in this reproductive process.[51,70,71]

CB1R and CB2R were detected in both human[72] and rodents uteri[73–75] as well as in pre-implantation embryo.[64,76]

The embryonic CB1R was shown to play a critical role in proper pre-implantation development, while the maternal CB1R exhibited crucial participation in controlling oviductal transport. Indeed, embryo exposure to high eCB levels delays development via CB1R activation.[68] In accordance, CB1R +/− but not CB1R KO embryos develop properly in CB1R KO dams.[77] Conversely, in WT dams treated with CB1R antagonists, embryos are retained in the oviduct, and CB1R −/− dams similarly exhibit a high retention rate of transferred WT embryo.[77]

Maternal exocannabinoid exposure and repercussions on embryo implantation and development

Despite the converging evidence of a role for the ECS in embryo implantation, the effects of exocannabinoids are still not completely understood. The observations that although THC injections during the pre-implantation period had no effect on embryo implantation,[78] systemic infusion of a synthetic agonist interfered with zona hatching and implantation,[65] may be explained by rapid cytochrome P450 metabolism of THC.[79] Indeed, THC can induce tuberoinfundibular dopaminergic neurons responses when cytochrome P450-linked monooxygenase enzymes are inhibited.[80] Accordingly, THC combined with metyrapone and clotrimazole inhibited implantation in 92% of tested mice in a CB1R-dependent manner.[81] Thus as with earlier stages of fertilization, the ECS heavily participates in embryo implantation and exposure to exogenous cannabinoids has the potential to substantially alter this process.

Perinatal cannabis exposure and long-term consequences on synaptic programming

Role of the ECS in perinatal neurodevelopment

Beyond its intricate role in fertility, fertilization, embryonic implantation, and throughout the gestational period, the ECS is a critical component of CNS developmental homeostasis during perinatal life.[82–84] Both CB1 and CB2 receptors have been detected in the post-natal mammalian brain.[84,85] On the other hand, while CB1R expression is widely demonstrated in early gestation CNS, the presence of CB2R at these stages is still controversial.[73,86] Further, a role for CB2R in neuronal progenitor cell proliferation and axons guidance has been reported[86–90]; however, its role in CNS early development remains marginal and poorly understood. In contrast, evidence of CB1R involvement throughout perinatal CNS development has been confirmed across species.[73,91–95] Indeed, CB1R activity influences neuronal progenitor proliferation, axonal growth, and fasciculation.[13,94,96–98] Expression of CB1R in the developing brain shows spatial and temporal specificity, in support of its participation in the developmental program underlying synaptic differentiation and maturation. For example, CB1R expression peaks during synaptic development of GABAergic interneurons and cortical glutamatergic cells,[94,96] which are both particularly impacted following PCE (as described later). Accordingly, the existence of temporal windows where exocannabinoid exposure exerts significant influence on both structural and functional neurodevelopment has been hypothesized.[92]

During CNS development, spatial and temporal patterns of CB1R expression are tightly correlated with eCB (e.g., 2-AG) levels. Indeed, the 2-AG synthesizing enzymes DAGLα/β are sequentially expressed in pre-synaptic and post-synaptic compartments during gestational development, influencing axonal growth and post-synaptic specialization.[99] This autocrine mechanism of early synaptic programming differs from the retrograde paracrine mechanism[100] that regulates synaptic plasticity in the adolescent and adult brain.[101] In parallel, the 2-AG degrading enzyme monoacylglycerol lipase (MGL) is excluded from the motile tip of the neurite outgrowth by a proteasomal degradation mechanism, allowing for fine regulation of neuronal protrusion through the area-specific action of 2-AG.[102] Despite the fundamental role of AEA in blastocyst, embryo implantation, and pregnancy maintenance, its role in perinatal development is comparatively minimal; 2-AG concentrations are approximately 1000 times higher than those of AEA in the perinatal brain, supporting the central role of 2-AG in early synaptic programming.[14,103]

Protracted effects of PCE on synaptic programming and the neurodevelopment trajectory

The global re-evaluation of cannabis acceptance, together with the perception that cannabis consumption is of little harmful consequence, has driven a rise in the number of women consuming cannabis during pregnancy for purported antiemetic, anxiolytic, and analgesic effects.[104–107] However, it is known that maternal cannabis consumption results in significant cannabinoid concentrations in both breast milk and offspring plasma,[7,8,108] and evidence continues to mount that PCE exerts significant influence on neurodevelopment.[109–111] Pre-clinical evidence exhibiting sex-specific neurodevelopmental perturbations following maternal cannabis consumption are indisputable.[110,112,113]

Here, we will summarize the impact of PCE on offspring neurodevelopment, focusing on deviations from the normal connectivity trajectory and altered synaptic plasticity throughout the progeny life span. We will mainly consider preclinical studies in which cannabinoid doses are comparable with moderate human consumption and do not induce signs of toxicity or malformations. The time window examined under the name of PCE includes studies in which exposure occurs during the gestational period, breastfeeding, or both.

The developmental trajectory of cortical glutamatergic synapses is one of the main targets through which PCE impacts neurodevelopment. The ECS is intimately involved in the development and wiring of glutamatergic synapses.[96,114,115] Subsequently, glutamate receptors play a significant role in neurodevelopment and adult neurogenesis.[116] As a consequence, exposure to exocannabinoids during embryonic life causes synaptic dysregulation.[117,118] Furthermore, glutamate interacts with eCB signaling in the mediation of various forms of synaptic plasticity,[101] the building block of learning and memory, behaviors which are consequently altered by PCE in both human epidemiological studies and pre-clinical animal models.[119,120]

Studying the maturational sequence of eCB synaptic plasticity at deep layer PFC pyramidal neurons, Bernabeu et al. discovered period- and sex-specific differences in development and in the receptors underlying endocannabinoid-mediated long-term depression (eCB-LTD) in the rat PFC.[121] These differences were specific to eCB plasticity and not apparent in the accumbens, favoring the idea of a specific role of the ECS in the sex-specific maturational trajectories of the PFC and providing novel substrates to the cellular and behavioral sex differences in the effects of cannabinoid exposure, including PCE (see later).

PCE decreases cortical extracellular glutamate levels and disrupts neuronal growth patterns and learning.[122] In the hippocampus, impaired glutamate release is accompanied by deficits in LTP and memory retention.[123] Cortical glutamate levels are reduced following PCE due the over-expression of GLT1 and EAAC1 glutamate transporters.[124] In the PFC, eCB-LTD is selectively disrupted in male (but not female) offspring exposed to cannabinoids during gestation. Reinforcing the notion of inter-twined roles for the ECS and glutamatergic development and function, positive allosteric modulation of mGlu5 receptors and pharmacological enhancement of circulating AEA levels have both been shown to restore the aforementioned plasticity deficits in male PCE offspring.[125] This line of evidence suggests a correlation between low cortical glutamate levels, hypo-stimulation of post-synaptic receptors, and impaired retrograde release of eCB mediators, which are essential for eCB-LTD induction and maintenance. Of note, these data identified a sexually differential participation of ECS components in plasticity: female plasticity is primarily mediated by TRPV1R, rather than CB1R, in both naïve and PCE offspring. Sex differences in PCE have also been linked to a sex-specific distribution of cannabinoid receptors, wherein sex hormones directly influence CB1R and TRPV1 function and distribution.[126,127]

In addition to gestational exposure, PCE (THC or the synthetic CB1/2 agonist WIN 55,212-2) has been shown to induce persistent behavioral and synaptic deficits in rat PFC when occurring exclusively during the breastfeeding period. Specifically, sociobehavioral alterations correlated with abolished eCB-LTD in PFC during adulthood. Intriguingly, the reported plasticity deficit was selective for that mediated by eCBs because the co-existing mGlu2/3-mediated form of LTD remained intact. Moreover, in adult rats exposed to THC via lactation, there was impairment of LTP in the PFC. Both LTD and LTP PFC deficits in PCE offspring were restored by increasing 2-AG levels.[128,129]

Early life exposure to cannabinoids also impedes GABAergic synapse function and development. GABAergic inhibitory control largely occurs through interneurons acting at the level of micro-circuits orchestrating glutamatergic synaptic communication.[130–132] The ECS is highly involved in GABAergic interneuron development and branching.[133,134] In the CA1 region of the hippocampus, GABAergic interneuron homeostasis is impacted in a sex-differentiated manner by prenatal THC exposure: pre-adolescent male (but not female) mice exposed in utero to THC showed CB1R downregulation and CCK+ basket cells' interneuropathy.[135,136] Further, perinatal exposure to THC reduced basal and K^+-evoked GABA outflow/uptake in the hippocampus of adult male rats.[137] GABA is the primary adult inhibitory neurotransmitter, but in immature brains, it acts as an excitatory agent due to high intracellular chloride owing to low levels of the potassium-chloride cotransporter 2 (KCC2). Subsequently, a developmental increase in the ratio of KCC2/NKCC1 (sodium-potassium-chloride transporter) membrane transporters mediates the transition to an inhibitory role for GABA. This transition is a fundamental to harmonious neurodevelopment.[138] Indeed, a wide range of neurodevelopmental and psychiatric disorders are associated with delays in this transition,[139,140] the timing of which is perturbed by maternal exposure to stress during pregnancy, as well as neonatal maternal separation.[141,142] Consistent with these observations, a marked delay in GABA polarity transition in the PFC has been noted in rats exposed to cannabinoids via lactation. This developmental delay was found to be mediated by CB1R and associated with the retarded upregulation of KCC2.[143]

The ECS also interacts with the dopaminergic system.[144] Indeed, dopaminergic transmission is directly impacted by cannabinoids[145] and cannabis use has been associated with dopamine-related disorders.[146-150] PCE offspring displays sex-specific alterations to the expression and function of crucial dopaminergic system components, such as tyrosine hydroxylase.[151-154] In utero exposure to THC in rats has also been observed to induce dopaminergic neuron hyperexcitability in male juvenile offspring.[155] As in the case of glutamatergic alterations, this same protocol was without effect in female offspring.[156] Considering the involvement of mesocorticostriatal circuit in reward and addiction,[157-159] cannabinoid induced alterations of the mesolimbic dopaminergic pathway[155] may precipitate or exacerbate substance use disorders in offspring. In fact, in the wake of growing evidence which reported interactions between cannabinoids and opioid system,[160-166] PCE progeny showed sex-specific alterations in reinforcement and reward to opioids.[167-169]

Conclusions

The ECS has been identified as an important player in the physiological processes subsequent to and underlying pregnancy outcomes, as well as in pre- and post-natal neuronal programming. This raises numerous questions in the face of the increasing global acceptance of cannabis consumption. In particular, the potential impact of an interaction between exocannabinoids and the ECS during early development requires careful evaluation. In this chapter, we review state-of-the-art research addressing these concerns, starting from the pre-fecundation phases, and continuing through early postnatal development.

Early exposure of the parental germ line to exocannabinoids shows transgenerational effects on the neuronal programming of progeny. These data suggest an important revision of the scientific thought wherein pre-fecundation parental experiences are considered of importance for the development of offspring.

At the post-fecundation phase of embryo peri-implantation, current evidence shows an intriguing bilateral role for the ECS. In fact, acting in maternal tissues as well as in embryonic ones, the ECS functions to synchronize embryonic implantation. These data therefore highlight the powerful influence that exocannabinoids could have in the initiation and maintenance of pregnancy from its earliest stages.

During perinatal temporal window where early neurodevelopment not only receives a crucial boost but also heightens vulnerability, biochemical insults may be deleterious for development with prolonged effects spanning throughout life. Interestingly, the ECS appears to be an assiduous participant in the complex range of physiological processes, which subtend CNS formation. In effect, perinatal exposure to exocannabinoids appears to dysregulate synaptic formation and modify the trajectory of its programming. Consequently, an altered behavioral phenotype emerges in exposed individuals during early life in both pre-clinical and clinical studies.

Throughout these developmental stages, important sex-specific differences have borne out from these findings, triggering a growing conviction of the importance of a bilateral sex-dependent research approach. Overall, reported data bring to the light a clear potential for the interaction between exocannabinoids and the ECS in influencing perinatal outcomes, particularly long-lasting CNS wiring.

Author contributions

Gabriele Giua: Conceptualization; Writing: review and editing.
Olivier JJ Manzoni: Conceptualization; Supervision; Funding acquisition; Writing: original draft, review, and editing; Project administration.
Andrew F. Scheyer: Conceptualization; Supervision; Writing: original draft, review, and editing.

Funding and disclosures

This work was supported by the Institut National de la Santé et de la Recherche Médicale (INSERM); the INSERM-NIH exchange program (to AFS); Agence Nationrale de la Recherche (2CUREXFra; ANR-18-CE12-0002-01 to G.G. and O.J.J.M.); Fondation pour la Recherche Médicale (Equipe FRM 2015 to O.J.J.M.); and the NIH (R01DA043982 and R01DA046196 to OJJM).

Declarations of interest

The authors declare no competing interests.

Acknowledgments

The authors are grateful to the Chavis-Manzoni team members for helpful discussions.

References

1. Ko JY, Farr SL, Tong VT, Creanga AA, Callaghan WM. Prevalence and patterns of marijuana use among pregnant and nonpregnant women of reproductive age. *Am J Obstet Gynecol* 2015;**213**(2):201.e1–201.e10. https://doi.org/10.1016/j.ajog.2015.03.021.
2. Corsi DJ, Donelle J, Sucha E, et al. Maternal cannabis use in pregnancy and child neurodevelopmental outcomes. *Nat Med* 2020;**26**(10):1536–40. https://doi.org/10.1038/s41591-020-1002-5.
3. Young-Wolff KC, Tucker LY, Alexeeff S, et al. Trends in self-reported and biochemically tested Marijuana use among pregnant females in California from 2009-2016. *J Am Med Assoc* 2017;**318**(24):2490–1. https://doi.org/10.1001/jama.2017.17225.
4. Williamson S, Jackson L, Skeoch C, Azzim G, Anderson R. Determination of the prevalence of drug misuse by meconium analysis. *Arch Dis Child Fetal Neonatal Ed* 2006;**91**(4). https://doi.org/10.1136/adc.2005.078642.
5. Perez-Reyes M, Wall ME. Presence of delta9-tetrahydrocannabinol in human milk. *N Engl J Med* 1982;**307**(13):819–20. https://doi.org/10.1056/NEJM198209233071311.
6. Garry A, Rigourd V, Amirouche A, Fauroux V, Aubry S, Serreau R. Cannabis and breastfeeding. *J Toxicol* 2009;**2009**:1–5. https://doi.org/10.1155/2009/596149.
7. Bertrand KA, Hanan NJ, Honerkamp-Smith G, Best BM, Chambers CD. Marijuana use by breastfeeding mothers and cannabinoid concentrations in breast milk. *Pediatrics* 2018;**142**(3). https://doi.org/10.1542/peds.2018-1076.
8. Baker T, Datta P, Rewers-Felkins K, Thompson H, Kallem RR, Hale TW. Transfer of inhaled cannabis into human breast milk. *Obstet Gynecol* 2018;**131**(5):783–8. https://doi.org/10.1097/AOG.0000000000002575.
9. Hensch TK. Critical period regulation. *Annu Rev Neurosci* 2004;**27**:549–79. https://doi.org/10.1146/annurev.neuro.27.070203.144327.
10. Hensch TK. Critical period plasticity in local cortical circuits. *Nat Rev Neurosci* 2005;**6**(11):877–88. https://doi.org/10.1038/nrn1787.
11. Taylor AH, Ang C, Bell SC, Konje JC. The role of the endocannabinoid system in gametogenesis, implantation and early pregnancy. *Hum Reprod Update* 2007;**13**(5):501–13. https://doi.org/10.1093/humupd/dmm018.
12. Fride E. Multiple roles for the endocannabinoid system during the earliest stages of life: pre- and oostnatal development. *J Neuroendocrinol* 2008;**20**:75–81. https://doi.org/10.1111/j.1365-2826.2008.01670.x.
13. Watson S, Chambers D, Hobbs C, Doherty P, Graham A. The endocannabinoid receptor, CB1, is required for normal axonal growth and fasciculation. *Mol Cell Neurosci* 2008;**38**(1):89–97. https://doi.org/10.1016/j.mcn.2008.02.001.
14. Berrendero F, Sepe N, Ramos JA, Di Marzo V, Fernández-Ruiz JJ. Analysis of cannabinoid receptor binding and mRNA expression and endogenous cannabinoid contents in the developing rat brain during late gestation and early postnatal period. *Synapse* 1999;**33**(3):181–91. https://doi.org/10.1002/(SICI)1098-2396(19990901)33:3<181::AID-SYN3>3.0.CO;2-R.
15. Biegon A, Kerman IA. Autoradiographic study of pre- and postnatal distribution of cannabinoid receptors in human brain. *NeuroImage* 2001;**14**(6):1463–8. https://doi.org/10.1006/nimg.2001.0939.
16. Correa F, Wolfson ML, Valchi P, Aisemberg J, Franchi AM. Endocannabinoid system and pregnancy. *Reproduction* 2016;**152**(6):R191–200. https://doi.org/10.1530/REP-16-0167.
17. Costa MA. The endocannabinoid system: a novel player in human placentation. *Reprod Toxicol* 2016;**61**:58–67. https://doi.org/10.1016/j.reprotox.2016.03.002.
18. Cecconi S, Rapino C, Di Nisio V, Rossi G, Maccarrone M. The (endo)cannabinoid signaling in female reproduction: what are the latest advances? *Prog Lipid Res* 2020;**77**. https://doi.org/10.1016/j.plipres.2019.101019.

19. Maia J, Fonseca BM, Teixeira N, Correia-Da-Silva G. The fundamental role of the endocannabinoid system in endometrium and placenta: implications in pathophysiological aspects of uterine and pregnancy disorders. *Hum Reprod Update* 2020;**26**(4):586–602. https://doi.org/10.1093/humupd/dmaa005.
20. Maccarrone M, Rapino C, Francavilla F, Barbonetti A. Cannabinoid signalling and effects of cannabis on the male reproductive system. *Nat Rev Urol* 2021;**18**(1):19–32. https://doi.org/10.1038/s41585-020-00391-8.
21. Wang Y, Liu Q, Tang F, Yan L, Qiao J. Epigenetic regulation and risk factors during the development of human gametes and early embryos. *Annu Rev Genomics Hum Genet* 2019;**20**:21–40. https://doi.org/10.1146/annurev-genom-083118-015143.
22. Xu R, Li C, Liu X, Gao S. Insights into epigenetic patterns in mammalian early embryos. *Protein Cell* 2021;**12**(1):7–28. https://doi.org/10.1007/s13238-020-00757-z.
23. Szutorisz H, Hurd YL. High times for cannabis: epigenetic imprint and its legacy on brain and behavior. *Neurosci Biobehav Rev* 2018;**85**:93–101. https://doi.org/10.1016/j.neubiorev.2017.05.011.
24. Goldberg LR, Gould TJ. Multigenerational and transgenerational effects of paternal exposure to drugs of abuse on behavioral and neural function. *Eur J Neurosci* 2019;**50**(3):2453–66. https://doi.org/10.1111/ejn.14060.
25. Skvortsova K, Iovino N, Bogdanović O. Functions and mechanisms of epigenetic inheritance in animals. *Nat Rev Mol Cell Biol* 2018;**19**(12):774–90. https://doi.org/10.1038/s41580-018-0074-2.
26. Kennedy AJ, Sweatt JD. Drugging the methylome: DNA methylation and memory. *Crit Rev Biochem Mol Biol* 2016;**51**(3):185–94. https://doi.org/10.3109/10409238.2016.1150958.
27. Lax E, Szyf M. The role of DNA methylation in drug addiction: implications for diagnostic and therapeutics. In: *Progress in molecular biology and translational science*. vol. 157. Elsevier B.V.; 2018. p. 93–104. https://doi.org/10.1016/bs.pmbts.2018.01.003.
28. Walker DM, Nestler EJ. Neuroepigenetics and addiction. In: *Handbook of clinical neurology*. vol. 148. Elsevier B.V.; 2018. p. 747–65. https://doi.org/10.1016/B978-0-444-64076-5.00048-X.
29. Browne CJ, Godino A, Salery M, Nestler EJ. Epigenetic mechanisms of opioid addiction. *Biol Psychiatry* 2020;**87**(1):22–33. https://doi.org/10.1016/j.biopsych.2019.06.027.
30. Liberman N, Wang SY, Greer EL. Transgenerational epigenetic inheritance: from phenomena to molecular mechanisms. *Curr Opin Neurobiol* 2019;**59**:189–206. https://doi.org/10.1016/j.conb.2019.09.012.
31. Ruiz-Llorente L, Ortega-Gutiérrez S, Viso A, et al. Characterization of an anandamide degradation system in prostate epithelial PC-3 cells: synthesis of new transporter inhibitors as tools for this study. *Br J Pharmacol* 2004;**141**(3):457–67. https://doi.org/10.1038/sj.bjp.0705628.
32. Lewis SEM, Rapino C, Di Tommaso M, et al. Differences in the endocannabinoid system of sperm from fertile and infertile men. *PLoS ONE* 2012;**7**(10). https://doi.org/10.1371/journal.pone.0047704.
33. Nielsen JE, Rolland AD, De Meyts ER, et al. Author correction: characterisation and localisation of the endocannabinoid system components in the adult human testis [Sci Rep, (2019), 9, 1, (12866), 10.1038/s41598-019-49177-y]. *Sci Rep* 2020;**10**(1). https://doi.org/10.1038/s41598-020-58153-w.
34. Pertwee RG, Joe-Adigwe G, Hawksworth GM. Further evidence for the presence of cannabinoid CB1 receptors in mouse vas deferens. *Eur J Pharmacol* 1996;**296**(2):169–72. https://doi.org/10.1016/0014-2999(95)00790-3.
35. Gye MC, Kang HH, Kang HJ. Expression of cannabinoid receptor 1 in mouse testes. *Arch Androl* 2005;**51**(3):247–55. https://doi.org/10.1080/014850190898845.
36. Rossato M, Popa FI, Ferigo M, Clari G, Foresta C. Human sperm express cannabinoid receptor Cb1, the activation of which inhibits motility, acrosome reaction, and mitochondrial function. *J Clin Endocrinol Metab* 2005;**90**(2):984–91. https://doi.org/10.1210/jc.2004-1287.

37. Brown SM, Wager-Miller J, Mackie K. Cloning and molecular characterization of the rat CB2 cannabinoid receptor. *Biochim Biophys Acta Gene Struct Expr* 2002;**1576**(3):255–64. https://doi.org/10.1016/S0167-4781(02)00341-X.
38. Maccarrone M, Cecconi S, Rossi G, Battista N, Pauselli R, Finazzi-Agrò A. Anandamide activity and degradation are regulated by early postnatal aging and follicle-stimulating hormone in mouse Sertoli cells. *Endocrinology* 2003;**144**(1):20–8. https://doi.org/10.1210/en.2002-220544.
39. Sarfaraz S, Afaq F, Adhami VM, Mukhtar H. Cannabinoid receptor as a novel target for the treatment of prostate cancer. *Cancer Res* 2005;**65**(5):1635–41. https://doi.org/10.1158/0008-5472.CAN-04-3410.
40. Agirregoitia E, Carracedo A, Subirán N, et al. The CB2 cannabinoid receptor regulates human sperm cell motility. *Fertil Steril* 2010;**93**(5):1378–87. https://doi.org/10.1016/j.fertnstert.2009.01.153.
41. Francavilla F, Battista N, Barbonetti A, et al. Characterization of the endocannabinoid system in human spermatozoa and involvement of transient receptor potential vanilloid 1 receptor in their fertilizing ability. *Endocrinology* 2009;**150**(10):4692–700. https://doi.org/10.1210/en.2009-0057.
42. Aquila S, Guido C, Santoro A, et al. Human sperm anatomy: ultrastructural localization of the cannabinoid1 receptor and a potential role of anandamide in sperm survival and acrosome reaction. *Anat Rec* 2010;**293**(2):298–309. https://doi.org/10.1002/ar.21042.
43. Martínez-León E, Osycka-Salut C, Signorelli J, et al. Fibronectin modulates the endocannabinoid system through the cAMP/PKA pathway during human sperm capacitation. *Mol Reprod Dev* 2019;**86**(2):224–38. https://doi.org/10.1002/mrd.23097.
44. Hazem NM, Zalata A, Alghobary M, Comhaire F, Elabbasy LM. Evaluation of cannabinoid receptors type 1 and type 2 mRNA expression in mature versus immature spermatozoa from fertile and infertile males. *Andrologia* 2020;**52**(4). https://doi.org/10.1111/and.13532.
45. Amoako AA, Marczylo TH, Marczylo EL, et al. Anandamide modulates human sperm motility: implications for men with asthenozoospermia and oligoasthenoteratozoospermia. *Hum Reprod* 2013;**28**(8):2058–66. https://doi.org/10.1093/humrep/det232.
46. Rajender S, Avery K, Agarwal A. Epigenetics, spermatogenesis and male infertility. *Mutat Res Rev Mutat Res* 2011;**727**(3):62–71. https://doi.org/10.1016/j.mrrev.2011.04.002.
47. Kropp J, Carrillo JA, Namous H, et al. Male fertility status is associated with DNA methylation signatures in sperm and transcriptomic profiles of bovine preimplantation embryos. *BMC Genomics* 2017;**18**(1). https://doi.org/10.1186/s12864-017-3673-y.
48. Santi D, De Vincentis S, Magnani E, Spaggiari G. Impairment of sperm DNA methylation in male infertility: a meta-analytic study. *Andrology* 2017;**5**(4):695–703. https://doi.org/10.1111/andr.12379.
49. Murphy SK, Itchon-Ramos N, Visco Z, et al. Cannabinoid exposure and altered DNA methylation in rat and human sperm. *Epigenetics* 2018;**13**(12):1208–21. https://doi.org/10.1080/15592294.2018.1554521.
50. Innocenzi E, De Domenico E, Ciccarone F, et al. Paternal activation of CB2 cannabinoid receptor impairs placental and embryonic growth via an epigenetic mechanism. *Sci Rep* 2019;**9**(1). https://doi.org/10.1038/s41598-019-53579-3.
51. Maccarrone M, De Felici M, Bari M, Klinger F, Siracusa G, Finazzi-Agrò A. Down-regulation of anandamide hydrolase in mouse uterus by sex hormones. *Eur J Biochem* 2000;**267**(10):2991–7. https://doi.org/10.1046/j.1432-1033.2000.01316.x.
52. El-Talatini MR, Taylor AH, Konje JC. The relationship between plasma levels of the endocannabinoid, anandamide, sex steroids, and gonadotrophins during the menstrual cycle. *Fertil Steril* 2010;**93**(6):1989–96. https://doi.org/10.1016/j.fertnstert.2008.12.033.

53. El-Talatini MR, Taylor AH, Elson JC, Brown L, Davidson AC, Konje JC. Localisation and function of the endocannabinoid system in the human ovary. *PLoS ONE* 2009;**4**(2). https://doi.org/10.1371/journal.pone.0004579.
54. El-Talatini MR, Taylor AH, Konje JC. Fluctuation in anandamide levels from ovulation to early pregnancy in in-vitro fertilization-embryo transfer women, and its hormonal regulation. *Hum Reprod* 2009;**24**(8):1989–98. https://doi.org/10.1093/humrep/dep065.
55. López-Cardona AP, Pérez-Cerezales S, Fernández-González R, et al. CB1 cannabinoid receptor drives oocyte maturation and embryo development via PI3K/Akt and MAPK pathways. *FASEB J* 2017;**31**(8):3372–82. https://doi.org/10.1096/fj.201601382RR.
56. Totorikaguena L, Olabarrieta E, López-Cardona AP, Agirregoitia N, Agirregoitia E. Tetrahydrocannabinol modulates in vitro maturation of oocytes and improves the blastocyst rates after in vitro fertilization. *Cell Physiol Biochem* 2019;**53**(3):439–52. https://doi.org/10.33594/000000149.
57. Castel P, Barbier M, Poumerol E, et al. Prenatal cannabinoid exposure alters the ovarian reserve in adult offspring of rats. *Arch Toxicol* 2020;**94**(12):4131–41. https://doi.org/10.1007/s00204-020-02877-1.
58. Szutorisz H, DiNieri JA, Sweet E, et al. Parental THC exposure leads to compulsive heroin-seeking and altered striatal synaptic plasticity in the subsequent generation. *Neuropsychopharmacology* 2014;**39**(6):1315–23. https://doi.org/10.1038/npp.2013.352.
59. Watson CT, Szutorisz H, Garg P, et al. Genome-wide DNA methylation profiling reveals epigenetic changes in the rat nucleus accumbens associated with cross-generational effects of adolescent THC exposure. *Neuropsychopharmacology* 2015;**40**(13):2993–3005. https://doi.org/10.1038/npp.2015.155.
60. Szutorisz H, Egervári G, Sperry J, Carter JM, Hurd YL. Cross-generational THC exposure alters the developmental sensitivity of ventral and dorsal striatal gene expression in male and female offspring. *Neurotoxicol Teratol* 2016;**58**:107–14. https://doi.org/10.1016/j.ntt.2016.05.005.
61. Byrnes JJ, Johnson NL, Schenk ME, Byrnes EM. Cannabinoid exposure in adolescent female rats induces transgenerational effects on morphine conditioned place preference in male offspring. *J Psychopharmacol* 2012;**26**(10):1348–54. https://doi.org/10.1177/0269881112443745.
62. Vassoler FM, Johnson NL, Byrnes EM. Female adolescent exposure to cannabinoids causes transgenerational effects on morphine sensitization in female offspring in the absence of in utero exposure. *J Psychopharmacol* 2013;**27**(11):1015–22. https://doi.org/10.1177/0269881113503504.
63. Ibn Lahmar Andaloussi Z, Taghzouti K, Abboussi O. Behavioural and epigenetic effects of paternal exposure to cannabinoids during adolescence on offspring vulnerability to stress. *Int J Dev Neurosci* 2019;**72**:48–54. https://doi.org/10.1016/j.ijdevneu.2018.11.007.
64. Paria BC, Das SK, Dey SK. The preimplantation mouse embryo is a target for cannabinoid ligand-receptor signaling. *Proc Natl Acad Sci U S A* 1995;**92**(21):9460–4. https://doi.org/10.1073/pnas.92.21.9460.
65. Schmid PC, Paria BC, Krebsbach RJ, Schmid HHO, Dey SK. Changes in anandamide levels in mouse uterus are associated with uterine receptivity for embryo implantation. *Proc Natl Acad Sci U S A* 1997;**94**(8):4188–92. https://doi.org/10.1073/pnas.94.8.4188.
66. Paria BC, Song H, Wang X, et al. Dysregulated cannabinoid signaling disrupts uterine receptivity for embryo implantation. *J Biol Chem* 2001;**276**(23):20523–8. https://doi.org/10.1074/jbc.M100679200.

67. Guo Y, Wang H, Okamoto Y, et al. N-acylphosphatidylethanolamine-hydrolyzing phospholipase D is an important determinant of uterine anandamide levels during implantation. *J Biol Chem* 2005;**280**(25):23429–32. https://doi.org/10.1074/jbc.C500168200.
68. Wang H, Matsumoto H, Guo Y, Paria BC, Roberts RL, Dey SK. Differential G protein-coupled cannabinoid receptor signaling by anandamide directs blastocyst activation for implantation. *Proc Natl Acad Sci U S A* 2003;**100**(25):14914–9. https://doi.org/10.1073/pnas.2436379100.
69. Liu WM, Duan EK, Cao YJ. Effects of anandamide on embryo implantation in the mouse. *Life Sci* 2002;**71**(14):1623–32. https://doi.org/10.1016/S0024-3205(02)01928-8.
70. Maccarrone M, Bisogno T, Valensise H, et al. Low fatty acid amide hydrolase and high anandamide levels are associated with failure to achieve an ongoing pregnancy after IVF and embryo transfer. *Mol Hum Reprod* 2002;**8**(2):188–95. https://doi.org/10.1093/molehr/8.2.188.
71. Wang H, Xie H, Guo Y, et al. Fatty acid amide hydrolase deficiency limits early pregnancy events. *J Clin Invest* 2006;**116**(8):2122–31. https://doi.org/10.1172/JCI28621.
72. Taylor AH, Abbas MS, Habiba MA, Konje JC. Histomorphometric evaluation of cannabinoid receptor and anandamide modulating enzyme expression in the human endometrium through the menstrual cycle. *Histochem Cell Biol* 2010;**133**(5):557–65. https://doi.org/10.1007/s00418-010-0695-9.
73. Buckley NE, Hansson S, Harta G, Mezey É. Expression of the CB1 and CB2 receptor messenger RNAs during embryonic development in the rat. *Neuroscience* 1997;**82**(4):1131–49. https://doi.org/10.1016/S0306-4522(97)00348-5.
74. Fonseca BM, Correia-da-Silva G, Taylor AH, Konje JC, Bell SC, Teixeira NA. Spatio-temporal expression patterns of anandamide-binding receptors in rat implantation sites: evidence for a role of the endocannabinoid system during the period of placental development. *Reprod Biol Endocrinol* 2009;**7**. https://doi.org/10.1186/1477-7827-7-121.
75. Li Y, Bian F, Sun X, Dey SK. Mice missing cnr1 and Cnr2 show implantation defects. *Endocrinology* 2019;**160**(4):938–46. https://doi.org/10.1210/en.2019-00024.
76. Yang ZM, Paria BC, Dey SK. Activation of brain-type cannabinoid receptors interferes with preimplantation mouse embryo development. *Biol Reprod* 1996;**55**(4):756–61. https://doi.org/10.1095/biolreprod55.4.756.
77. Wang H, Guo Y, Wang D, et al. Aberrant cannabinoid signaling impairs oviductal transport of embryos. *Nat Med* 2004;**10**(10):1074–80. https://doi.org/10.1038/nm1104.
78. Paria BC, Kapur S, Dey SK. Effects of 9-ene-tetrahydrocannabinol on uterine estrogenicity in the mouse. *J Steroid Biochem Mol Biol* 1992;**42**(7):713–9. https://doi.org/10.1016/0960-0760(92)90112-V.
79. Lucas CJ, Galettis P, Schneider J. The pharmacokinetics and the pharmacodynamics of cannabinoids. *Br J Clin Pharmacol* 2018;**84**(11):2477–82. https://doi.org/10.1111/bcp.13710.
80. Bonnin A, De Miguel R, Javier Fernández-Ruiz J, Maribel C, Ramos JA. Possible role of the cytochrome P450-linked monooxygenase system in preventing δ9-tetrahydrocannabinol-induced stimulation of tuberoinfundibular dopaminergic activity in female rats. *Biochem Pharmacol* 1994;**48**(7):1387–92. https://doi.org/10.1016/0006-2952(94)90561-4.
81. Paria BC, Ma W, Andrenyak DM, et al. Effects of cannabinoids on preimplantation mouse embryo development and implantation are mediated by brain-type cannabinoid receptors. *Biol Reprod* 1998;**58**(6):1490–5. https://doi.org/10.1095/biolreprod58.6.1490.
82. Harkany T, Guzmán M, Galve-Roperh I, Berghuis P, Devi LA, Mackie K. The emerging functions of endocannabinoid signaling during CNS development. *Trends Pharmacol Sci* 2007;**28**(2):83–92. https://doi.org/10.1016/j.tips.2006.12.004.

83. Harkany T, Keimpema E, Barabás K, Mulder J. Endocannabinoid functions controlling neuronal specification during brain development. *Mol Cell Endocrinol* 2008;**286**(1-2 SUPPL. 1): S84–90. https://doi.org/10.1016/j.mce.2008.02.011.
84. Lu HC, Mackie K. Review of the endocannabinoid system. *Biol Psychiatry Cogn Neurosci Neuroimaging* 2020. https://doi.org/10.1016/j.bpsc.2020.07.016.
85. Svíženská I, Dubový P, Šulcová A. Cannabinoid receptors 1 and 2 (CB1 and CB2), their distribution, ligands and functional involvement in nervous system structures—a short review. *Pharmacol Biochem Behav* 2008;**90**(4):501–11. https://doi.org/10.1016/j.pbb.2008.05.010.
86. Palazuelos J, Aguado T, Egia A, et al. Non-psychoactive CB 2 cannabinoid agonists stimulate neural progenitor proliferation. *FASEB J* 2006;**20**(13):2405–7. https://doi.org/10.1096/fj.06-6164fje.
87. Molina-Holgado F, Rubio-Araiz A, García-Ovejero D, et al. CB2 cannabinoid receptors promote mouse neural stem cell proliferation. *Eur J Neurosci* 2007;**25**(3):629–34. https://doi.org/10.1111/j.1460-9568.2007.05322.x.
88. Goncalves MB, Suetterlin P, Yip P, et al. A diacylglycerol lipase-CB2 cannabinoid pathway regulates adult subventricular zone neurogenesis in an age-dependent manner. *Mol Cell Neurosci* 2008;**38**(4):526–36. https://doi.org/10.1016/j.mcn.2008.05.001.
89. Palazuelos J, Ortega Z, Díaz-Alonso J, Guzmán M, Galve-Roperh I. CB 2 cannabinoid receptors promote neural progenitor cell proliferation via mTORC1 signaling. *J Biol Chem* 2012;**287**(2):1198–209. https://doi.org/10.1074/jbc.M111.291294.
90. Duff G, Argaw A, Cecyre B, et al. Cannabinoid receptor CB2 modulates axon guidance. *PLoS ONE* 2013;**8**(8). https://doi.org/10.1371/journal.pone.0070849.
91. McLaughlin CR, Abood ME. Developmental expression of cannabinoid receptor mRNA. *Dev Brain Res* 1993;**76**(1):75–8. https://doi.org/10.1016/0165-3806(93)90124-S.
92. Wang X, Dow-Edwards D, Keller E, Hurd YL. Preferential limbic expression of the cannabinoid receptor mRNA in the human fetal brain. *Neuroscience* 2003;**118**(3):681–94. https://doi.org/10.1016/S0306-4522(03)00020-4.
93. Begbie J, Doherty P, Graham A. Cannabinoid receptor, CB1, expression follows neuronal differentiation in the early chick embryo. *J Anat* 2004;**205**(3):213–8. https://doi.org/10.1111/j.0021-8782.2004.00325.x.
94. Mulder J, Aguado T, Keimpema E, et al. Endocannabinoid signaling controls pyramidal cell specification and long-range axon patterning. *Proc Natl Acad Sci U S A* 2008;**105**(25): 8760–5. https://doi.org/10.1073/pnas.0803545105.
95. Psychoyos D, Vinod KY, Cao J, et al. Cannabinoid receptor 1 signaling in embryo neurodevelopment. *Birth Defects Res B Dev Reprod Toxicol* 2012;**95**(2):137–50. https://doi.org/10.1002/bdrb.20348.
96. Berghuis P, Rajnicek AM, Morozov YM, et al. Hardwiring the brain: endocannabinoids shape neuronal connectivity. *Science* 2007;**316**(5828):1212–6. https://doi.org/10.1126/science.1137406.
97. Compagnucci C, Di Siena S, Bustamante MB, et al. Type-1 (CB1) cannabinoid receptor promotes neuronal differentiation and maturation of neural stem cells. *PLoS ONE* 2013;**8**(1). https://doi.org/10.1371/journal.pone.0054271.
98. Saez TMM, Aronne MP, Caltana L, Brusco AH. Prenatal exposure to the CB1 and CB2 cannabinoid receptor agonist WIN 55,212-2 alters migration of early-born glutamatergic neurons and GABAergic interneurons in the rat cerebral cortex. *J Neurochem* 2014;**129**(4): 637–48. https://doi.org/10.1111/jnc.12634.
99. Bisogno T, Howell F, Williams G, et al. Cloning of the first sn1-DAG lipases points to the spatial and temporal regulation of endocannabinoid signaling in the brain. *J Cell Biol* 2003;**163**(3):463–8. https://doi.org/10.1083/jcb.200305129.

100. Jung KM, Sepers M, Henstridge CM, et al. Uncoupling of the endocannabinoid signalling complex in a mouse model of fragile X syndrome. *Nat Commun* 2012;**3**:1080. https://doi.org/10.1038/ncomms2045.
101. Araque A, Castillo PE, Manzoni OJ, Tonini R. Synaptic functions of endocannabinoid signaling in health and disease. *Neuropharmacology* 2017;**124**:13–24. https://doi.org/10.1016/j.neuropharm.2017.06.017.
102. Keimpema E, Barabas K, Morozov YM, et al. Differential subcellular recruitment of monoacylglycerol lipase generates spatial specificity of 2-arachidonoyl glycerol signaling during axonal pathfinding. *J Neurosci* 2010;**30**(42):13992–4007. https://doi.org/10.1523/JNEUROSCI.2126-10.2010.
103. Wu CS, Jew CP, Lu HC. Lasting impacts of prenatal cannabis exposure and the role of endogenous cannabinoids in the developing brain. *Future Neurol* 2011;**6**(4):459–80. https://doi.org/10.2217/fnl.11.27.
104. Young-Wolff KC, Sarovar V, Tucker LY, et al. Trends in marijuana use among pregnant women with and without nausea and vomiting in pregnancy, 2009–2016. *Drug Alcohol Depend* 2019;**196**:66–70. https://doi.org/10.1016/j.drugalcdep.2018.12.009.
105. Young-Wolff KC, Sarovar V, Tucker LY, et al. Self-reported daily, weekly, and monthly cannabis use among women before and during pregnancy. *JAMA Netw Open* 2019;**2**(7). https://doi.org/10.1001/jamanetworkopen.2019.6471.
106. Dickson B, Mansfield C, Guiahi M, et al. Recommendations from cannabis dispensaries about first-trimester cannabis use. In: *Obstetrics and gynecology*. vol. 131. Lippincott Williams and Wilkins; 2018. p. 1031–8. https://doi.org/10.1097/AOG.0000000000002619.
107. Agrawal A, Rogers CE, Lessov-Schlaggar CN, Carter EB, Lenze SN, Grucza RA. Alcohol, cigarette, and cannabis use between 2002 and 2016 in pregnant women from a nationally representative sample. *JAMA Pediatr* 2019;**173**(1):95–6. https://doi.org/10.1001/jamapediatrics.2018.3096.
108. Bailey JR, Cunny HC, Paule MG, Slikker W. Fetal disposition of Δ9-tetrahydrocannabinol (THC) during late pregnancy in the rhesus monkey. *Toxicol Appl Pharmacol* 1987;**90**(2):315–21. https://doi.org/10.1016/0041-008X(87)90338-3.
109. Grant KS, Petroff R, Isoherranen N, Stella N, Burbacher TM. Cannabis use during pregnancy: pharmacokinetics and effects on child development. *Pharmacol Ther* 2018;**182**:133–51. https://doi.org/10.1016/j.pharmthera.2017.08.014.
110. Scheyer AF, Melis M, Trezza V, Manzoni OJJ. Consequences of perinatal cannabis exposure. *Trends Neurosci* 2019;**42**(12):871–84. https://doi.org/10.1016/j.tins.2019.08.010.
111. Nashed MG, Hardy DB, Laviolette SR. Prenatal cannabinoid exposure: emerging evidence of physiological and neuropsychiatric abnormalities. *Front Psychiatry* 2021;11. https://doi.org/10.3389/fpsyt.2020.624275.
112. Hurd YL, Manzoni OJ, Pletnikov MV, Lee FS, Bhattacharyya S, Melis M. Cannabis and the developing brain: insights into its long-lasting effects. *J Neurosci* 2019;**39**(42):8250–8. https://doi.org/10.1523/JNEUROSCI.1165-19.2019.
113. Tirado-Muñoz J, Lopez-Rodriguez AB, Fonseca F, Farré M, Torrens M, Viveros MP. Effects of cannabis exposure in the prenatal and adolescent periods: preclinical and clinical studies in both sexes. *Front Neuroendocrinol* 2020;57. https://doi.org/10.1016/j.yfrne.2020.100841.
114. Dow-Edwards D, Silva L. Endocannabinoids in brain plasticity: cortical maturation, HPA axis function and behavior. *Brain Res* 2017;**1654**(Pt B):157–64. https://doi.org/10.1016/j.brainres.2016.08.037.
115. Scheyer AF, Martin HGS, Manzoni OJ. The endocannabinoid system in prefrontal synaptopathies. In: *Endocannabinoids and lipid mediators in brain functions*. Springer International Publishing; 2017. p. 171–210. https://doi.org/10.1007/978-3-319-57371-7_7.

116. Jansson LC, Åkerman KE. The role of glutamate and its receptors in the proliferation, migration, differentiation and survival of neural progenitor cells. *J Neural Transm* 2014;**121**(8): 819–36. https://doi.org/10.1007/s00702-014-1174-6.
117. Tortoriello G, Morris CV, Alpar A, et al. Miswiring the brain: Δ9-tetrahydrocannabinol disrupts cortical development by inducing an SCG10/stathmin-2 degradation pathway. *EMBO J* 2014;**33**(7):668–85. https://doi.org/10.1002/embj.201386035.
118. De Salas-Quiroga A, Díaz-Alonso J, García-Rincón D, et al. Prenatal exposure to cannabinoids evokes long-lasting functional alterations by targeting CB1 receptors on developing cortical neurons. *Proc Natl Acad Sci U S A* 2015;**112**(44):13693–8. https://doi.org/10.1073/pnas.1514962112.
119. Trezza V, Cuomo V, Vanderschuren LJMJ. Cannabis and the developing brain: insights from behavior. *Eur J Pharmacol* 2008;**585**(2–3):441–52. https://doi.org/10.1016/j.ejphar.2008.01.058.
120. Trezza V, Campolongo P, Manduca A, et al. Altering endocannabinoid neurotransmission at critical developmental ages: impact on rodent emotionality and cognitive performance. *Front Behav Neurosci* 2012;**6**. https://doi.org/10.3389/fnbeh.2012.00002.
121. Bernabeu A, Bara A, Manduca A, et al. Sex-specific maturational trajectory of endocannabinoid plasticity in the rat prefrontal cortex. *bioRxiv* 2020.
122. Antonelli T, Tomasini MC, Tattoli M, et al. Prenatal exposure to the CB1 receptor agonist WIN 55,212-2 causes learning disruption associated with impaired cortical NMDA receptor function and emotional reactivity changes in rat offspring. *Cereb Cortex* 2005;**15**(12):2013–20. https://doi.org/10.1093/cercor/bhi076.
123. Mereu G, Fà M, Ferraro L, et al. Prenatal exposure to a cannabinoid agonist produces memory deficits linked to dysfunction in hippocampal long-term potentiation and glutamate release. *Proc Natl Acad Sci U S A* 2003;**100**(8):4915–20. https://doi.org/10.1073/pnas.0537849100.
124. Castaldo P, Magi S, Gaetani S, et al. Prenatal exposure to the cannabinoid receptor agonist WIN 55,212-2 increases glutamate uptake through overexpression of GLT1 and EAAC1 glutamate transporter subtypes in rat frontal cerebral cortex. *Neuropharmacology* 2007;**53**(3): 369–78. https://doi.org/10.1016/j.neuropharm.2007.05.019.
125. Bara A, Manduca A, Bernabeu A, et al. Sex-dependent effects of in utero cannabinoid exposure on cortical function. *elife* 2018;**7**. https://doi.org/10.7554/eLife.36234.
126. Castelli M, Fadda P, Casu A, et al. Male and female rats differ in brain cannabinoid CB1 receptor density and function and in behavioural traits predisposing to drug addiction: effect of ovarian hormones. *Curr Pharm Des* 2014;**20**(13):2100–13. https://doi.org/10.2174/13816128113199990430.
127. Artero-Morales M, González-Rodríguez S, Ferrer-Montiel A. TRP channels as potential targets for sex-related differences in migraine pain. *Front Mol Biosci* 2018;**5**. https://doi.org/10.3389/fmolb.2018.00073.
128. Scheyer AF, Borsoi M, Pelissier-Alicot AL, Manzoni OJJ. Maternal exposure to the cannabinoid agonist WIN 55,12,2 during lactation induces lasting behavioral and synaptic alterations in the rat adult offspring of both sexes. *eNeuro* 2020;**7**(5):1–11. https://doi.org/10.1523/ENEURO.0144-20.2020.
129. Scheyer AF, Borsoi M, Pelissier-Alicot A-L, Manzoni OJJ. Perinatal THC exposure via lactation induces lasting alterations to social behavior and prefrontal cortex function in rats at adulthood. *Neuropsychopharmacology* 2020;**45**(11):1826–33. https://doi.org/10.1038/s41386-020-0716-x.
130. Tremblay R, Lee S, Rudy B. GABAergic interneurons in the neocortex: from cellular properties to circuits. *Neuron* 2016;**91**(2):260–92. https://doi.org/10.1016/j.neuron.2016.06.033.
131. Pelkey KA, Chittajallu R, Craig MT, Tricoire L, Wester JC, McBain CJ. Hippocampal gabaergic inhibitory interneurons. *Physiol Rev* 2017;**97**(4):1619–747. https://doi.org/10.1152/physrev.00007.2017.

132. Lim L, Mi D, Llorca A, Marín O. Development and functional diversification of cortical interneurons. *Neuron* 2018;**100**(2):294–313. https://doi.org/10.1016/j.neuron.2018.10.009.
133. Berghuis P, Dobszay MB, Wang X, et al. Endocannabinoids regulate interneuron migration and morphogenesis by transactivating the TrkB receptor. *Proc Natl Acad Sci U S A* 2005;**102**(52):19115–20. https://doi.org/10.1073/pnas.0509494102.
134. Younts TJ, Castillo PE. Endogenous cannabinoid signaling at inhibitory interneurons. *Curr Opin Neurobiol* 2014;**26**:42–50. https://doi.org/10.1016/j.conb.2013.12.006.
135. de Salas-Quiroga A, García-Rincón D, Gómez-Domínguez D, et al. Long-term hippocampal interneuronopathy drives sex-dimorphic spatial memory impairment induced by prenatal THC exposure. *Neuropsychopharmacology* 2020;**45**(5):877–86. https://doi.org/10.1038/s41386-020-0621-3.
136. Vargish GA, Pelkey KA, Yuan X, et al. Persistent inhibitory circuit defects and disrupted social behaviour following in utero exogenous cannabinoid exposure. *Mol Psychiatry* 2017;**22**(1): 56–67. https://doi.org/10.1038/mp.2016.17.
137. Beggiato S, Borelli AC, Tomasini MC, et al. Long-lasting alterations of hippocampal GABAergic neurotransmission in adult rats following perinatal Δ9-THC exposure. *Neurobiol Learn Mem* 2017;**139**:135–43. https://doi.org/10.1016/j.nlm.2016.12.023.
138. Kaila K, Price TJ, Payne JA, Puskarjov M, Voipio J. Cation-chloride cotransporters in neuronal development, plasticity and disease. *Nat Rev Neurosci* 2014;**15**(10):637–54. https://doi.org/10.1038/nrn3819.
139. Hyde TM, Lipska BK, Ali T, et al. Expression of GABA signaling molecules KCC2, NKCC1, and GAD1 in cortical development and schizophrenia. *J Neurosci* 2011;**31**(30):11088–95. https://doi.org/10.1523/JNEUROSCI.1234-11.2011.
140. He Q, Nomura T, Xu J, Contractor A. The developmental switch in GABA polarity is delayed in fragile X mice. *J Neurosci* 2014;**34**(2):446–50. https://doi.org/10.1523/JNEUROSCI.4447-13.2014.
141. Veerawatananan B, Surakul P, Chutabhakdikul N. Maternal restraint stress delays maturation of cation-chloride cotransporters and GABAA receptor subunits in the hippocampus of rat pups at puberty. *Neurobiol Stress* 2016;**3**:1–7. https://doi.org/10.1016/j.ynstr.2015.12.001.
142. Furukawa M, Tsukahara T, Tomita K, et al. Neonatal maternal separation delays the GABA excitatory-to-inhibitory functional switch by inhibiting KCC2 expression. *Biochem Biophys Res Commun* 2017;**493**(3):1243–9. https://doi.org/10.1016/j.bbrc.2017.09.143.
143. Scheyer AF, Borsoi M, Wager-Miller J, et al. Cannabinoid exposure via lactation in rats disrupts perinatal programming of the gamma-aminobutyric acid trajectory and select early-life behaviors. *Biol Psychiatry* 2020;**87**(7):666–77. https://doi.org/10.1016/j.biopsych.2019.08.023.
144. Covey DP, Mateo Y, Sulzer D, Cheer JF, Lovinger DM. Endocannabinoid modulation of dopamine neurotransmission. *Neuropharmacology* 2017;**124**:52–61. https://doi.org/10.1016/j.neuropharm.2017.04.033.
145. Bloomfield MAP, Ashok AH, Volkow ND, Howes OD. The effects of δ9-tetrahydrocannabinol on the dopamine system. *Nature* 2016;**539**(7629):369–77. https://doi.org/10.1038/nature20153.
146. Murray RM, Englund A, Abi-Dargham A, et al. Cannabis-associated psychosis: neural substrate and clinical impact. *Neuropharmacology* 2017;**124**:89–104. https://doi.org/10.1016/j.neuropharm.2017.06.018.
147. Gage SH, Hickman M, Zammit S. Association between cannabis and psychosis: epidemiologic evidence. *Biol Psychiatry* 2016;**79**(7):549–56. https://doi.org/10.1016/j.biopsych.2015.08.001.
148. Gage SH, Jones HJ, Burgess S, et al. Assessing causality in associations between cannabis use and schizophrenia risk: a two-sample Mendelian randomization study. *Psychol Med* 2017;**47**(5):971–80. https://doi.org/10.1017/S0033291716003172.

149. Jutras-Aswad D, DiNieri JA, Harkany T, Hurd YL. Neurobiological consequences of maternal cannabis on human fetal development and its neuropsychiatric outcome. *Eur Arch Psychiatry Clin Neurosci* 2009;**259**(7):395–412. https://doi.org/10.1007/s00406-009-0027-z.
150. Fine JD, Moreau AL, Karcher NR, et al. Association of prenatal cannabis exposure with psychosis proneness among children in the adolescent brain cognitive development (ABCD) study. *JAMA Psychiatry* 2019;**76**(7):762–4. https://doi.org/10.1001/jamapsychiatry.2019.0076.
151. Bonnin A, de Miguel R, Rodríguez-Manzaneque JC, Fernández-Ruiz JJ, Santos A, Ramos JA. Changes in tyrosine hydroxylase gene expression in mesencephalic catecholaminergic neurons of immature and adult male rats perinatally exposed to cannabinoids. *Dev Brain Res* 1994;**81**(1):147–50. https://doi.org/10.1016/0165-3806(94)90079-5.
152. Bonnin A, De Miguel R, Castro JG, Ramos JA, Fernandez-Ruiz JJ. Effects of perinatal exposure to Δ9-tetrahydrocannabinol on the fetal and early postnatal development of tyrosine hydroxylase-containing neurons in rat brain. *J Mol Neurosci* 1996;**7**(4):291–308. https://doi.org/10.1007/BF02737066.
153. de Fonseca FR, Cebeira M, Fernández-Ruiz JJ, Navarro M, Ramos JA. Effects of pre- and perinatal exposure to hashish extracts on the ontogeny of brain dopaminergic neurons. *Neuroscience* 1991;**43**(2–3):713–23. https://doi.org/10.1016/0306-4522(91)90329-M.
154. García-Gil L, Ramos JA, Rubino T, Parolaro D, Fernández-Ruiz JJ. Perinatal Δ9-tetrahydrocannabinol exposure did not alter dopamine transporter and tyrosine hydroxylase mRNA levels in midbrain dopaminergic neurons of adult male and female rats. *Neurotoxicol Teratol* 1998;**20**(5):549–53. https://doi.org/10.1016/S0892-0362(98)00012-9.
155. Frau R, Miczán V, Traccis F, et al. Prenatal THC exposure produces a hyperdopaminergic phenotype rescued by pregnenolone. *Nat Neurosci* 2019;**22**(12):1975–85. https://doi.org/10.1038/s41593-019-0512-2.
156. Traccis F, Serra V, Sagheddu C, et al. Prenatal THC does not affect female mesolimbic dopaminergic system in preadolescent rats. *Int J Mol Sci* 2021;**22**(4):1–18. https://doi.org/10.3390/ijms22041666.
157. Volkow ND, Morales M. The brain on drugs: from reward to addiction. *Cell* 2015;**162**(4):712–25. https://doi.org/10.1016/j.cell.2015.07.046.
158. Koob GF, Volkow ND. Neurobiology of addiction: a neurocircuitry analysis. *Lancet Psychiatry* 2016;**3**(8):760–73. https://doi.org/10.1016/S2215-0366(16)00104-8.
159. Cooper S, Robison AJ, Mazei-Robison MS. Reward circuitry in addiction. *Neurotherapeutics* 2017;**14**(3):687–97. https://doi.org/10.1007/s13311-017-0525-z.
160. Viganò D, Rubino T, Parolaro D. Molecular and cellular basis of cannabinoid and opioid interactions. In: *Pharmacology biochemistry and behavior*. vol. 81. Elsevier Inc.; 2005. p. 360–8. https://doi.org/10.1016/j.pbb.2005.01.021.
161. Fattore L, Deiana S, Spano SM, et al. Endocannabinoid system and opioid addiction: behavioural aspects. In: *Pharmacology biochemistry and behavior*. vol. 81. Elsevier Inc.; 2005. p. 343–59. https://doi.org/10.1016/j.pbb.2005.01.031.
162. Solinas M, Goldberg SR. Motivational effects of cannabinoids and opioids on food reinforcement depend on simultaneous activation of cannabinoid and opioid systems. *Neuropsychopharmacology* 2005;**30**(11):2035–45. https://doi.org/10.1038/sj.npp.1300720.
163. Spano M, Fadda P, Fratta W, Fattore L. Cannabinoid-opioid interactions in drug discrimination and self-administration: effect of maternal, postnatal, adolescent and adult exposure to the drugs. *Curr Drug Targets* 2010;**11**(4):450–61. https://doi.org/10.2174/138945010790980295.
164. Abrams DI, Couey P, Shade SB, Kelly ME, Benowitz NL. Cannabinoid-opioid interaction in chronic pain. *Clin Pharmacol Ther* 2011;**90**(6):844–51. https://doi.org/10.1038/clpt.2011.188.

165. Befort K. Interactions of the opioid and cannabinoid systems in reward: insights from knockout studies. *Front Pharmacol* 2015;**6**. https://doi.org/10.3389/fphar.2015.00006.
166. Manduca A, Lassalle O, Sepers M, et al. Interacting cannabinoid and opioid receptors in the nucleus accumbens core control adolescent social play. *Front Behav Neurosci* 2016;**10**. https://doi.org/10.3389/fnbeh.2016.00211.
167. Vela G, Martín S, García-Gil L, et al. Maternal exposure to δ9-tetrahydrocannabinol facilitates morphine self-administration behavior and changes regional binding to central μ opioid receptors in adult offspring female rats. *Brain Res* 1998;**807**(1–2):101–9. https://doi.org/10.1016/S0006-8993(98)00766-5.
168. Singh ME, McGregor IS, Mallet PE. Perinatal exposure to Δ9-tetrahydrocannabinol alters heroin-induced place conditioning and Fos-immunoreactivity. *Neuropsychopharmacology* 2006;**31**(1):58–69. https://doi.org/10.1038/sj.npp.1300770.
169. Spano MS, Ellgren M, Wang X, Hurd YL. Prenatal cannabis exposure increases heroin seeking with allostatic changes in limbic enkephalin systems in adulthood. *Biol Psychiatry* 2007;**61**(4):554–63. https://doi.org/10.1016/j.biopsych.2006.03.073.

Chapter 9

Molecular and cellular principles of endocannabinoid signaling and their sensitivity to cannabis in the developing brain

Erik Keimpema[a] and Tibor Harkany[a,b]
[a]*Department of Molecular Neurosciences, Center for Brain Research, Medical University of Vienna, Vienna, Austria,* [b]*Department of Neuroscience, Biomedicum 7D, Karolinska Institutet, Solna, Sweden*

Introduction

The cultivation of hemp (*Cannabis* spp.) for seeds in food preparation, as well as its fiber for the production of garments, rope and fishing nets, dates back to at least 8000 BC.[1] Cannabis was heavily traded along the silk road with its first hints of religious and medicinal use discovered as early as 3000 BC in both ancient China and Egypt.[2–4] Its seeds, pollen and vegetative parts were used as an incense for ritualistic purposes, and a wide variety of medical conditions, both externally as topical treatments for wounds and infections, and internally for analgesia and inflammation.[3] While feral cannabis has low levels of its main psychoactive compound Δ^9-tetrahydrocannabinol (THC),[5] later archaeological dig sites revealed an increase of THC levels in combustible preparations,[2] suggestive of selective cultivation to enhance it psychoactive and medicinal effects. Thus, cannabis can be considered as one of the first cultivated plants in human history, not only for its use in nutrition and tools, but also for its mind expanding and healing effects.

In our current political climate, cannabis has become central in the dilemma if plant-derived recreational drugs can have therapeutic potential and, consequently, how to address their regulation and safe distribution. Reminiscent of the current opiate crisis in the United States where potent synthetic opiates are overshadowing weaker naturally occurring morphine,[6] cannabis nowadays contains substantially more THC through selective cultivation (from 4% in 1995 up to 35% in some strains[7]). In addition, promoting vaporizing compounds through e-cigarettes as safe smoking alternatives, as well as the cheap provision of technology for easy plant-derived lipid extraction, has accelerated the consumption of cannabis oil preparations[8,9] with THC concentrations reaching up to 95%. While the toxicity of high doses of THC to humans is being recognized[9] and THC-related diseases such as cannabinoid hyperemesis syndrome (HES) are starting to emerge,[10] it is unclear how copious amounts of THC (>300 µg/kg)[11] affect the developing brain. THC has a long half-life up to 24 h,[12] with some of its metabolites detectable for up to 6 days,[12] and is stored in fatty tissues (which can be released during lipolysis[13]) leaving both the central and peripheral nervous systems under its influence for prolonged periods of time, especially when large amounts are consumed.

Legal cannabis (<0.2% THC) has been recently cultivated to gain high levels of cannabidiol (CBD), traditionally regarded as a non-psychoactive cannabinoid (although recreational users anecdotally report relaxing and pleasurable effects[14,15]), as it gained a prominent role in the treatment of multiple diseases, including epilepsy, anxiety, and inflammation.[16] However, the effects of this "safe and healthy" cannabinoid are hardly understood, as it interacts with a wide variety of signaling systems including many G protein-coupled receptors, serotonin and opioid signaling,[17,18] ion channels,[19] fatty acid and glucose metabolism[20] and liver enzymes.[21] Complications with CBD consumption could further occur through its effective half-life of 10–17 h[21,22] and low bioavailability (~6%), which necessitates the intake of copious daily amounts of oral CBD (even as much as 1.5 g). Although bioavailability can be improved fourfold when taken with a high fat food source[22] or as an oil preparation (as CBD is also a lipophilic compound), these intake regimens affect CBD bioavailability as food contains fluctuating levels of fat and oil. Thus, its promiscuous signaling effects, the molecular characteristics and complex manner to provide steady-state plasma concentrations through high oral doses can result in an unpredictable and likely stochastic engagement of many signaling systems. While this characteristic can be of minimal significance in debilitating diseases such as epilepsy when a polypharmacology approach is desired to control neuronal activation and increase survival,[18] with benefits clearly outweighing the risks, it is questionable to provide high doses of CBD in the case of a more minor illnesses, which would necessitate specific pathway engagement. The advent of vaporizing lipophilic compounds in lower doses for rapid uptake through inhalation could be a step in the right direction because it vastly reduces the amount necessary to reach therapeutic effect size, most likely making this the preferred route of administration for future clinical treatment where patient stratification and molecular targeting objectives matter (although presenting its own set of complications such as lung damage due to other preparation constituents or hot vapor inhalation).[16,23]

A recent study on pregnant women revealed an increased opinion that regular cannabis use during pregnancy is safe,[24] increasingly exposing unborn children to both THC and CBD. This is not only important for the developing fetus and newborns, as THC can also be transferred through maternal milk,[25,26] but also for young adolescent cannabis consumers[27,28] and bystanders (both mothers and children) passively exposed to potent cannabis vapors.[29,30] The prolonged maturation of the human brain,[31] in combination with other disciplines such as sociology, led to the scientific proposal to reclassify the timing of adolescence until the age of 24 years.[32] Thus, the human brain is vulnerable for a long period of time to recreational drugs such as cannabis, during a critical developmental period mostly known for excessive risk taking (e.g., drug experimentation) and impulsivity.[33] Therefore, the aim of this chapter is to summarize current knowledge on the innate neuromodulator system that exerts effects similar to cannabis itself, the "endocannabinoid system," in the central nervous system, and how phytocannabinoids (plant-derived, THC, and CBD) can modify this delicate signaling entity.

Cannabis use during pregnancy

Research on the effects of cannabis on the pre-natal human brain was spurred by Fried in the Canadian Ottawa Prenatal Prospective Study (OPPS),[34–37] studying a cohort of pre-natally exposed children up to the age of 22 years. Regular use of cannabis throughout pregnancy was correlated with a lack of visual habituation, the inability to self-quiet and increased amounts of tremors and startles in newborns, indicating profound effects of cannabis on the developing central nervous system.[34] However, no significant changes to muscular tone, general activity, or alertness were observed. A follow-up study revealed a surprising positive association with the children's attitudes and interests at 1 year of age, which was obscured at 2 years of age due to a strong relationship of cognitive function by environmental influences.[37] Subsequently, a negative association with executive functions requiring impulse control and visual analysis in 9–12 year-old children was observed, which was maintained until 16 years of age, but no effect on global intelligence or the ability to read and interpret text.[35] Finally, an fMRI study on the remaining participants aged up to 22 years showed that increased pre-natal exposure to cannabis increased circuit activity in the bilateral prefrontal cortex and right premotor cortex, as well as attenuated neuronal activity in the left cerebellum during response inhibition,[38] suggestive of long-lasting alterations to cortical networks.

The American Maternal Health Practices and Child Development Study (MHPCD) started in1982 as the second longitudinal study following a cohort of children up to the age of 22 years. Between the ages of 3 and 6 years, children exposed to cannabis in utero scored lower on the Stanford-Binet Intelligence Scale, a strong predictor for inattention and impulsivity,[39,40] with males more affected than females. Throughout early adolescence, an increase in impulsivity, inattention, hyperactivity, delinquency, and increased use of cannabis was observed.[41,42] Furthermore, pre-natal cannabis exposure negatively affected reading and spelling scores and was correlated with underachievement.

Although modest, long-term psychotic symptoms were dose-dependently associated with cannabis exposure at age 22.[43]

The third longitudinal study (Generation R) started in 2006 in the Netherlands and followed a cohort of children up the age of 13 years.[44] Pre-natal cannabis use was associated with growth reductions, smaller head circumference and lower birth weight,[45] while at 18 months of age girls had a significant increase of risk for aggressive behavior and lack of attention.[46] While this study is still ongoing,[47] the commonalities of all three studies include the reduction in attention and impulse control, suggestive of improper circuit formation in the prefrontal cortex,[35] a structure critical for mediating inhibitory cognitive responses.[48] Since the maturing prefrontal cortex in children shows a similarly increased firing pattern as in adolescents exposed to preterm cannabis,[38] and children generally have lower attention and impulse control,[49] it is postulated that cannabis can hinder the proper development of prefrontal cortical networks leading to diminished cognition[38] (*see below*). Although the above data indicate a strong effect of cannabis on the developing human nervous system, they are severely limited in control groups for polydrug consumption, socioeconomic factors, and the rising concentrations of THC in cannabis preparations.[50] Therefore, animal research is pivotal in understanding the molecular underpinnings to precisely dissect the effects of pre-natal cannabis (THC) exposure to developing neuronal circuits in the fetal brain.

The endocannabinoid system

Endocannabinoid-sensing receptors

Even though the isolation and structural characterization of THC as the main psychoactive compound in cannabis was concluded in 1965,[5] the first cannabinoid receptor (CB1R) was not discovered until the early 90s.[51,52] The CB1R is considered to be one of the most abundant G protein-coupled receptors in the brain[53] and was historically thought to be restricted to the central nervous system. In contrast, the second cannabinoid receptor (CB2R) was mostly assigned to the periphery and involved in immune responses.[54,55] However, that idea has recently been invalidated as CB1R signaling can also suppress immune responses[56,57] and CB2Rs can modulate midbrain dopaminergic neuron activity.[58] A second layer of complexity is added through the presence of prospective cannabinoid receptors, including GPR55 involved in immunomodulation[59] and anxiety-like behavior,[60,61] GPR119 affecting food intake[62,63] and the "abnormal" CBD receptor GPR18 driving microglial migration.[64] Furthermore, activity of the transient receptor potential sub-family vanilloid type 1 ion channel (TRPV1), mostly known to respond to noxious stimuli and induce pain, can be modulated by mixed endovanilloids/endocannabinoids, resulting in the desensitization of TRPV1 and thus, inhibiting pain and inflammation.[65] Overall, the interaction of different cannabinoid receptors on a wide variety of both neuronal and non-neuronal cell types can create intricate state-changes to control brain function throughout life (Fig. 1).

Enzymatic production and degradation of endocannabinoids: Subcellular partitioning for intercellular communication

The expression of the above cannabinoid receptors predicts the existence of endogenous cannabinoids. Endocannabinoids are short-acting arachidonic acid-derived lipids,[66,67] synthesized on demand and degraded within minutes,[68,69] as compared to the relative long terminal half-life of 22h for THC.[70,71] Anandamide (AEA), the first discovered endocannabinoid named after the Sanskrit word for "bliss,"[66] is synthesized through cleavage of N-arachidonoyl phosphatidylethanolamine (NAPE) by either phospholipase C (NAPE-PLC)[72] or phospholipase D (NAPE-PLD) in a calcium-dependent manner.[73] In addition, AEA can be synthesized by the deacylation of NAPE through α,β-hydrolase 4 (ABHD4), and subsequent cleavage of the glycerophosphate moiety.[74] Its degradation occurs through fatty acid amide hydrolase (FAAH1/2) resulting in the formation of free arachidonic acid and ethanolamine. The second main endocannabinoid, 2-arachidonoyl glycerol (2-AG)[67] is synthesized by sn-1-diacylglycerol lipases α and β (DAGLα/β)[75] and chiefly degraded by monoacylglycerol lipase (MAGL),[76,77] as well as ABHD6 and ABHD12.[78] Although the main inactivation of AEA and 2-AG is thought to occur through hydrolysis, recent data show that they are also subject to the main oxidative metabolic pathways leading to the production of eicosanoids, signaling molecules derived from arachidonic acid, and poly-unsaturated fatty acids (PUFA).[79] Included in this oxidative pathway are lipoxygenases (LOX) and cyclooxygenases (COX) for both AEA and 2-AG, and cytochrome P450 for AEA.[79] As 2-AG levels are approximately 1000-fold higher in concentration in the brain as compared to AEA, it is considered to be the main signaling entity with AEA seen as a fine-tuner instead.[80]

In addition, atypical cannabinoid-like substances, such as N-palmitoylethanolamine (PEA), N-oleoylethanolamine (OEA), and N-stearoylethanolamine (SEA), do not directly induce cannabinoid receptor signaling but can allosterically modulate the effects of endocannabinoids, commonly known as the "entourage" effect.[81,82] For instance, PEA can downregulate FAAH,[83] while increasing 2-AG levels and potentiating 2-AG's action at TRPV1.[84] As (a)typical endocannabinoids are derived from ω-6 PUFAs (*for an in-depth review see* Ref. 85), dietary alterations to fatty acid composition during pre-natal development could affect endocannabinoid levels in the offspring. Indeed, pre-natal exposure to high levels of ω-6 PUFA desensitizes CB1Rs, leading to reduced cell adhesion molecule expression (most likely resulting in disturbances to axonal pathfinding and synaptogenesis, *see below*), as well as anxiety and depression-like behavior in affected offspring.[86] In turn, caloric restriction during pregnancy leads to underweight offspring, and sexually dimorphic changes to AEA, 2-AG and PEA levels in the hippocampus,

154 Cannabis and the developing brain

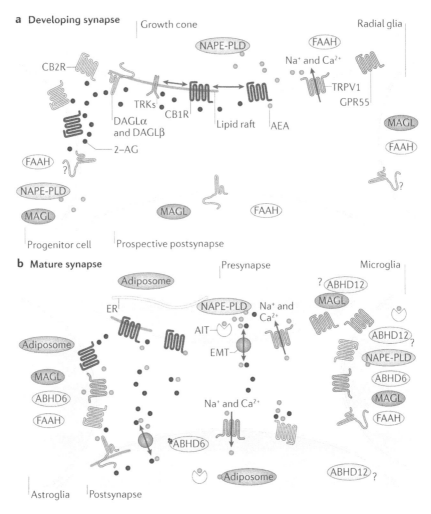

FIG. 1 Molecular architecture of the endocannabinoid system during synaptogenesis and at mature synapses. Neuronal and glial components of developing and mature synapses are shown. It is important to note that the molecular architecture shown here is typical. There are likely to be differences in neurotransmitter system-specific and developmentally regulated enzyme and/or receptor expression and function at different types of synapse and at different stages of development. Most prominently, monoacylglycerol lipase (MAGL) is excluded from motile growth cones until synaptogenesis commences. Strike-through symbols indicate catalytic enzyme activity throughout. (A) In developing synapse, AEA *(green circles)* and 2-AG *(brown circles)* orchestrate signaling by binding to their target receptors: CB1R, CB2R, GPR55, and TRPV1. Their availability is determined by biosynthesis enzymes (NAPE-PLD and DAGLα/β) and degrading enzymes (FAAH and MAGL). Note that information on alternative enzymes (e.g., ABHD6 and ABHD12) is at present not available. Unlike other endocannabinoid-binding receptors, CB1Rs are preferentially recruited to and signal within cholesterol-enriched membrane microdomains termed lipid rafts.

(Continued)

hypothalamus, and olfactory bulb.[87] Thus, endocannabinoid levels are highly dependent on available fatty acids, the distribution of both synthesis and degradation enzymes, as well as their allosteric modulation by atypical cannabinoids, throughout the nervous system.

In the adult brain, endocannabinoids are mainly released in a non-vesicular fashion "on demand"[68] from post-synaptic sites and act as retrograde messengers to inhibit presynaptic neurotransmitter release.[88] Stimulation of CB1Rs subsequently inhibits N-type voltage-gated calcium channels[89] and activates inward rectifying potassium channels,[90] blocking synaptic vesicle fusion, and neurotransmitter release. As the degrading enzymes are expressed in the presynaptic terminal, endocannabinoids in the synaptic cleft can be swiftly catabolized[91] allowing for rapid and precise temporal control of neurotransmitter signaling. THC cannot be catabolized by the endocannabinoid degradation machinery and is instead oxidized in the liver by the cytochrome P450 complex to the primary active 11-hydroxy-Δ^9-tetrahydrocannabinol (11-OH-THC) and the primary inactive metabolite, 11-nor-9-carboxy-Δ^9-tetrahydrocannabinol (THC-COOH).[92] Since 11-OH-THC has a circulating half-life of 19–24 h (and THC-COOH up to 6 days)[12] it can disrupt the fast and localized nature of physiological endocannabinoid signaling for long periods of time. This is further complicated as THC is a partial agonist for both CB1Rs and CB2Rs, with its effect size dependent on receptor availability, cell types and the presence of endocannabinoids, other phytocannabinoids and lipids. Given that CBD can inhibit cytochrome P450, its presence in recreational cannabis preparations can potentially prolonging the effects of THC.[93] Thus, the introduction of not only THC, but also other terpene phytocannabinoids[81] present in cannabis preparations, can cause a long-lasting impact on physiological endocannabinoid signaling and intercellular communication. However, the distribution of the molecular machinery producing endocannabinoids, as well as the circulating levels, differs vastly in fetal development as compared to the adult, ultimately predicting distinct outcomes.

FIG. 1, CONT'D (B) At the mature synapse, the availability of AEA and 2-AG is determined by ABHD6 and ABHD12 in addition to FAAH and MAGL, and also by transmembrane (EMT) and intracellular (AIT) transport mechanisms (e.g., fatty acid binding proteins, heat shock protein, and FAAH-like AEA transporter), and storage organelles (adiposomes or lipid droplets). There is compelling evidence that key receptor and enzyme components of the endocannabinoid system partition distinctly, both intracellularly and amongst pre- and post-synaptic neurons, microglia, and astrocytes. CB2Rs are expressed mainly upon brain injury. *ABHD6/12*, α/β-hydrolase domain-containing 6/12 hydrolases; *AIT*, AEA intracellular transporter; *CB1R/CB2R*, G protein-coupled type-1 and type-2 cannabinoid receptors; *DAGLα/β*, sn-1-diacylglycerol lipase α/β; *EMT*, putative endocannabinoid transmembrane transporter; *ER*, endoplasmic reticulum; *FAAH*, fatty-acid amide hydrolase; *GPR55*, G protein-coupled receptor 55; *MAGL*, monoacylglycerol lipase; *NAPE-PLD*, N-acylphosphatidylethanolamine-specific phospholipase D; *TRPV1*, transient receptor potential vanilloid 1 channel. *(Adapted from Maccarrone M, Guzmán M, Mackie K, Doherty P, Harkany T. Programming of neural cells by (endo)cannabinoids: from physiological rules to emerging therapies.* Nat Rev Neurosci *2014;15:786–801.)*

Pre-natal endocannabinoid signaling and the adverse developmental outcomes of its disruption

Embryo implantation and growth

Cannabinoid receptors are expressed as early as the fertilized oocyte,[94] and subsequently at the 2-cell and 1-cell blastocyst stage for CB1Rs and CB2Rs, respectively.[95] During uterine maturation, placental AEA levels are downregulated[96] for the precise differentiation of the fertilized oocyte toward the blastocyst,[94] with a subsequent decrease in blastocyst CB1Rs, which is considered essential for embryo implantation.[96] Increased signaling at CB1Rs through the exogenous application of specific CB1R/CB2R ligands or THC itself reduces uterine implantation and decreases embryonic viability and growth by autophagic activation and apoptosis,[95,97] indicating that cannabinoid receptors are critical from the earliest developmental time points. While endocannabinoid actions are important for many developing different organ systems, such GPR55 in bone formation[98] and CB1Rs for pancreas development,[99] we focus here on the developing central nervous system.

Neurulation and neurogenesis

After induction of the neural plate and the formation of the neural tube, neural progenitors divide and give rise to neurons and eventual protoglial lineages (note that microglia arise from the yolk sac[100] and migrate into the developing nervous system using, among others, endocannabinoid cues). Immature neurons sequentially migrate through the neuroepithelium to their final positions, project their axons toward their cellular targets, and commence synaptogenesis to wire their respective neurocircuitry. As endocannabinoid signaling is sparsely understood in the development of astrocytes, data below focuses on the control of neurogenesis and neuritogenesis/synaptogenesis. Cannabinoid receptors, as well as the endocannabinoid enzymatic machinery, are already present in the somites and neural tube during early fetal development in for example chick.[101] Alike, both 2-AG and AEA are detectable in the mouse at embryonic day (E) 10, with their concentrations rising toward birth.[101] CB1R mRNA increases in parallel with the endocannabinoid tone in total mouse central nervous system preparations,[101] as well as the rat neural tube, and peripheral nervous system from E11 onwards.[75,102] The presence of cannabinoid receptors in highly proliferating structures argues for a role in the control of the proliferation of tissue-resident progenitors[98,99,103] and/or the production of daughter cells destined toward differentiated cell fates. Indeed, CB1R signaling drives neural progenitor proliferation[104] and given their innate endocannabinoid programs can even be recruited in adult cytotoxic neurodegeneration (e.g., stroke) to induce neuroregeneration.[105] Genetic loss of CB1Rs results in diminished cortical (sub) ventricular zone pyramidal cell progenitor proliferation,[103] while ablation of

FAAH demonstrates the reverse.[104,106] Mechanistic insights arise from the exposure of human induced pluripotent stem cell (iPSC)-derived neurons in which THC reduces the cellular transcriptional response otherwise seen after a depolarization event (mimicking activity-dependent stimuli), and shows significant changes to synaptic, mitochondrial and glutamate signaling cascades.[107]

While research on CBD exposure to the embryonic nervous system is largely lacking, initial results demonstrate that low doses of perinatal CBD administration alters repetitive and hedonic behavior far into adulthood by downregulating AEA levels and impacting serotonergic signaling.[108] In addition, pre-natal CBD exposure increases anxiety and alters brain DNA methylation in affected offspring,[109] although large doses (20 mg/kg) were used for a prolonged period of time to reach these effects. Conversely, CBD treatment in adults ameliorates anxiety and increases hippocampal neurogenesis[110,111] and volume,[112] in line with the previously mentioned regenerative effect after cytotoxic stress.[105] If the expression of CB1Rs is maintained from neuronal progenitors into developing neurons (corresponding to mouse E8 onwards),[101] then it can be postulated that every neuron, both in the central and the peripheral nervous systems, shall go through an endocannabinoid sensitive stage during (early) cellular differentiation.[113]

Neurite outgrowth and target innervation

The physiological role of endocannabinoid signaling in neuronal differentiation is mainly described in pyramidal cells of the cerebral cortex (*but see also midbrain dopamine and cholinergic contingents below*[114,115]), which project their CB1R-containing axons through the corpus callosum toward the thalamus (corticothalamic tract), parallel to the extension of mostly CB1R-negative axons from the thalamus into the cerebral cortex (thalamocortical projection). CB1Rs accumulate in extending neurites and growth cones of pyramidal cells with their engagement regulating growth cone steering decisions and neurite extension.[106,114,116,117] The careful guidance of axons toward their future post-synaptic targets is controlled by the presence of DAGLα/β along the extending neurite and growth cone,[106,116,118] allowing for the presence of a local 2-AG microgradient-like build-up, through either by cell-autonomous 2-AG synthesis or neighboring processes being the source of this ligand.[119] MAGL is expressed in stabilized neurites poised for rapid 2-AG degradation, thus creating exclusion zones where cytoskeletal modifications[116,120] should no longer occur. Upon reaching a post-synaptic target, DAGLα/β are reduced in the presynaptic corticothalamic (or corticospinal) terminal and be replaced by MAGL,[114] which then acts as a "stop signal" for the cessation of endocannabinoid-induced growth cone motility. Thus, the nascent presynapse with the first synaptic vesicles[116] adopts an adult-like configuration for 2-AG-mediated retrograde signaling[118] at an early stage. Genetic models ablating CB1Rs demonstrate chaotic axonal pathfinding and disorganized axon fasciculation in the forebrain[106,121] and pyramidal decussation,[122] as both corticothalamic (CB1R-positive) and thalamocortical projections (CB1R-negative) depend on each other for proper formation (an interaction dubbed the cortico-thalamic handshake[123]). Other indices of DAGL dependence include the reorganization of cholinergic and glutamatergic synapses in the hippocampus,[119] precipitating the lack of retrograde synaptic suppression in the hippocampus, striatum, and cerebellum in DAGL $\alpha^{-/-}$ mice.[124] While the effects of 2-AG on the differentiation of neurons are well understood through available genetic tools allowing for the efficient manipulation of rate-limiting enzymes at any given time (*see also in Drosophila models*[125–127] *in which 2-linoleoyl glycerol signaling is an evolutionary alternative to 2-AG*), the role of AEA on axonal outgrowth and pathfinding through the classical cannabinoid receptors remains elusive to this date. This would be particularly important to discover, as both the AEA-sensitive GPR55 and TRPV1 channel regulate axonal growth and target innervation.[128,129] Furthermore, disruption of FAAH reduces axonal regeneration in *C. elegans*[130] and the specific inhibition of FAAH in embryonic zebra fish diminished primary motor neuron branching, leading to locomotor deficiencies.[131] Pre-natal exposure to URB597, a FAAH inhibitor resulted in depressive behavior and impaired working memory in the affected offspring.[132] Thus, even though AEA levels are low in the developing nervous system relative to 2-AG, AEA could be equally well poised to contribute to axonal pathfinding decisions.

Pre-natal exogenous cannabinoids perturb adolescent synapse refinement

The enzymatic machinery controlling spatially-restricted endocannabinoid availability is unable to process phytocannabinoids. Thus, the introduction of THC, with its long half-life,[12] can disrupt the temporal precision of CB1R engagement, normally protected from signaling by MAGL, altering neuronal migration and provoking erroneous axonal navigation.[118] Indeed, pre-natal exposure to THC, or synthetic CB1R ligands, drastically downregulate CB1Rs and thus phenocopies aberrant axonal placement as observed in CB1R null mutants,[106,133] leading to deregulated synaptogenesis,[133] diminished hippocampal long-term potentiation[134] and glutamate availability[134,135] in the offspring. Treatment with THC or the synthetic CB1R agonist WIN55,212-2,[136] also reduces the number of CB1R$^+$ interneurons in the hippocampus and their dendritic complexity. These structural deficits impact both interneuron-mediated feedback and feed-forward inhibition, which manifest as impaired social interactions in the offspring.[137] THC dose-dependently increases AEA through CB1Rs by mobilizing PLD,[138] indicating that THC also affects CB1R signaling indirectly.

Endocannabinoids in the reward circuitry

The endocannabinoid system is particularly important for the function of dopaminergic circuitry as it tightly controls dopamine release from midbrain dopaminergic cells.[139] Furthermore, dopaminergic neurons express cannabinoid receptors already in the 20th week of gestation in human fetal brain reward centers.[140] In rats, pre-natal THC significantly alters the excitatory-inhibitory balance of dopaminergic neurons and increased dopamine release after THC exposure.[115] Furthermore, in the human and rat nucleus accumbens,[141] dopamine D2 receptor mRNA was reduced, as well as D2R binding sites with an increased sensitivity to opiates in adulthood.[141] These data suggest that early pre-natal cannabis exposure can destabilize the brain to be more susceptible to dopamine-mediated rewarding behavior in later life. Indeed, continuous THC exposure from conception until birth in rats also led to a stronger response to low heroin doses as compared to the controls, and sought more heroin during mild stress and drug extinction,[142] reinforcing that exposure to THC can sensitize to later strong rewarding behavior. This effect has recently been described as the "double-hit hypothesis," in which pre-natal or even early adolescent exposure to an exogenous substance can both have a negative impact directly on the central nervous system and simultaneously sensitize brain circuitry to stimuli when occurring in later life. Recent clinical work on neuropsychiatric disorders, including schizophrenia[143] and attention deficit disorder[144] with both a strong underlying dopaminergic pathology, have found significant alterations to both cannabinoid receptors and enzymes levels in the adult.[143,144] Thus, a "first hit" during critical developmental windows could not only prime the reward system to drug-seeking behavior, but also contribute to the development of dopamine-mediated neuropsychiatric illnesses.

Post-natal endocannabinoid signaling and adolescent circuit disruption

The perinatal period

Similar to the embryonic brain, CB1Rs are found in the early post-natal brain, with increasing levels until adulthood in regions including the hippocampus, cerebral cortex, and cerebellum.[145] While AEA and NAPE concentrations are significantly lower in both the fetal and post-natal rat brain as compared to the adult, 2-AG levels are similar across developmental stages into adulthood, with a twofold peak during birth,[145] and thought to help stimulate the suckling response and milk intake in neonates.[146] This suggests that the newborn is particularly dependent on cannabinoid receptor signaling in the first days after birth. Indeed, exposure to the CB1R antagonist SR141716A on post-natal day 1 (P1) ablated the neonates suckling response, initiated by the somatosensory cortex to find the food source and the hypothalamus to regulate hunger sensations,[147] leading to growth delays[148] and even death.[146] A continuation of SR141716A throughout the first post-natal week led to severe growth retardation and eventual mortality.[146] Co-administration of THC completely reversed the growth retardation, while 2-AG supplementation only delayed mortality,[148] most likely attributed through the long half-life of THC as compared to 2-AG. Similarly, neonatal $CB1R^{-/-}$ mice do not show suckling until P2, indicating a compensatory endocannabinoid signaling system through other cannabinoid receptors to occur and initiate later suckling as CB1R-null mice display no premature mortality.[146]

Neonatal findings

As THC can be found in breast milk, not only through cannabis consumption after birth, but also through sustained availability when used before birth,[149] it can thus impact the neonatal brain for prolonged periods of time. Early neonatal THC exposure results in diminished social discrimination and altered exploratory behavior in adult offspring through an altered excitatory-inhibitory balance.[137] These behavioral changes were correlated to the lack of endocannabinoid-mediated long-term depression, an activity dependent reduction in neurotransmitter release, the loss of long-term potentiation and changes to intrinsic excitability of prefrontal cortical pyramidal cells.[150] Furthermore, early to late neonatal THC exposure disrupts the excitatory-inhibitory GABA switch through delayed expression of the potassium-chloride cotransporter 2 (KCC2). This switch occurs in the third pre-natal week in mice, and is necessary for the change of GABA's trophic and regulatory role in synaptogenesis,[151–153] to its function as an inhibitory neurotransmitter. In addition, CBD can influence serotonin 5-HT1A and transient receptor potential cation channel TRPV1 signaling,[154] both required for embryonic neurogenesis and brain circuit formation,[155,156] with both pre-natal and post-natal CBD exposure leading to increased anxiety and improved memory behavior in a sex dependent manner in adult mice.[109] A DNA methylation screening of the cerebral cortex and hippocampus revealed substantial differentially methylated loci associated with functional enrichment of neurogenesis and substance abuse,[109] again reinforcing that early exposure to phytocannabinoids can have long-lasting effects on both an epigenetic and physiological level.

Early adolescence toward adulthood

Childhood and adolescent behavior is strongly associated with excessive risk-taking (e.g., drug experimentation) and impulsivity.[33] Thus, they are particularly at risk to disrupt brain network formation through potent cannabis preparations. Exposure to the synthetic CB1R agonist WIN 55,212-2 in mice during early and mid-adolescence, but not late adolescence, resulted in a frequency-dependent disinhibition in the prefrontal cortex, similar as the juvenile state of disinhibition, a period reminiscent of early adolescence in humans. In the prefrontal cortex, a brain structure regulating impulse control,[48] GABAergic neurotransmission onto layer V pyramidal neurons was markedly decreased, and could be normalized by application of Indiplon, a GABA-Aα1 positive allosteric modulator.[157] This is in line with the above human and experimental data, as both pre- and post-natal exogenous cannabinoid exposure leads to disbalances in inhibitory neurotransmission and increased impulsivity.

Furthermore, exposure to THC in early adolescent mice, indiscriminately affected the number and position of CB1R+ excitatory pyramidal cells, and both CB1R+ and CB1R− inhibitory interneurons in the hippocampus.[158] Proteomic analysis revealed a particular alteration in mitochondrial proteins, which could be detected up to 4 months after the last exposure. The occlusion of mitochondrial respiration by THC could be compensated for by antioxidants, bypassing glycolysis, pH stabilization and modulation of CB1Rs downstream signaling cascade involving soluble adenylyl cyclase (sAC).[158] This is in line with findings in the adult, as CB1R signaling on mitochondria limits oxygen consumption both in vivo and in vitro, decreases respiration by affecting the mitochondrial electron transport chain and increases oxidative stress.[159–161] The depressing effects of mitochondrial CB1R is thought to results in the amnesic effect of cannabis, since ablation of mitochondria-specific CB1Rs, or modulating sAC, prevented the reduction of mitochondrial mobility, synaptic neurotransmission, and decreased memory formation.[162]

Thus, subcellular cannabinoid receptor signaling can be a strong contributor to its above described roles in proliferation, migration, differentiation, and synaptogenesis during fetal brain development.

Conclusion

The relaxing legislative restrictions, the commercialization of cultivation and production, as well as the positive change in public opinion on both medicinal and recreational cannabis consumption led to an ever increasing cannabis availability and use for the general population.[163] While the (ir-)reversible long-term effects of cannabis on the adult brain are increasingly understood, a growing caution emerges on cannabis exposure to the developing nervous system, similar to other drugs of abuse, including alcohol and nicotine.[164] During fetal development, critical time windows open to permit signaling cascades in the proper organization of both the early embryo, as well as later specific brain structures. Endocannabinoids are particularly important during these windows, as they regulate (i) embryo implantation, (ii) neuronal proliferation, (iii) migration, (iv) differentiation, and (v) synaptogenesis extending into late adulthood.[103] The widespread presence of the many types of cannabinoid receptors, their endogenous ligands and allosteric modulators during development outlines a complex signaling system tightly controlled to allow for specific spatiotemporal signals to drive intercellular communication for neurocircuit formation. As such, disruptions to this delicate signaling system, by either altering endocannabinoid levels[132,165] or exposure to phyto- and synthetic cannabinoids with long half-lives, can thus destabilize the local endocannabinoid tone leading to ectopic and prolonged signaling, or even receptor desensitization.

As both animal and human pre-natal cannabis exposure is associated with adverse long-term behavioral outcomes, especially related to impulse control,[38] we cautiously advice against cannabis consumption during both intrauterine and post-natal central nervous system development, at least for two reasons: First, the entourage effect revolves around the finding that phytocannabinoids can both signal directly through cannabinoid receptors as well as modulate endocannabinoid signaling resulting in a perplexing setting which is "more than the sum of its parts." Second, the double-hit hypothesis posits that sub-threshold alterations to signaling cascades and subsequent brain circuit formation can sensitize networks to exogenous triggers (e.g., drugs or stress), leading to behavioral responses otherwise well-controlled and tolerated. In addition to pre-natal exposure, cannabis consumption during adolescence can be equally disruptive, as this developmental stage is mostly known for increased impulsivity and risk-taking leading to drug experimentation. In sum, in light of the current data on both experimental work and human longitudinal studies, we strongly propose that cannabis use during critical developmental stages should be avoided.

Acknowledgments

This work was funded by FWF (P 34121-B; E.K.) and the European Research Council (SECRET-CELLS, ERC-2015-AdG-695136 and FOODFORLIFE, ERC-2020-AdG-101021016;T.H.).

Author contributions

E.K. wrote the manuscript with input from T.H.

Conflict of interest

There is no conflict of interest.

References

1. Long T, Wagner M, Demske D, Leipe C, Tarasov PE. Cannabis in Eurasia: origin of human use and Bronze Age trans-continental connections. *Veget Hist Archaeobot* 2017;**26**:245–58.
2. Ren M, et al. The origins of cannabis smoking: chemical residue evidence from the first millennium BCE in the Pamirs. *Sci Adv* 2019;**5**:eaaw1391.
3. Russo E. History of cannabis and its preparations in saga, science, and sobriquet. *Chem Biodivers* 2007;**4**:1614–48.
4. Russo EB, et al. Phytochemical and genetic analyses of ancient cannabis from Central Asia. *J Exp Bot* 2008;**59**:4171–82.

5. Mechoulam R, Gaoni Y. A total synthesis of DL-delta-1-tetrahydrocannabinol, the active constituent of hashish. *J Am Chem Soc* 1965;**87**:3273–5.
6. Chambers SA, DeSousa JM, Huseman ED, Townsend SD. The DARK side of total synthesis: strategies and tactics in psychoactive drug production. *ACS Chem Neurosci* 2018. https://doi.org/10.1021/acschemneuro.7b00528.
7. ElSohly MA, et al. Changes in cannabis potency over the last two decades (1995–2014) – analysis of current data in the United States. *Biol Psychiatry* 2016;**79**:613–9.
8. Zhang Z, Zheng X, Zeng DD, Leischow SJ. Tracking dabbing using search query surveillance: a case study in the United States. *J Med Internet Res* 2016;**18**.
9. Alzghari SK, Fung V, Rickner SS, Chacko L, Fleming SW. To dab or not to dab: rising concerns regarding the toxicity of cannabis concentrates. *Cureus* 2017;**9**, e1676.
10. Habboushe J, Rubin A, Liu H, Hoffman RS. The prevalence of cannabinoid hyperemesis syndrome among regular marijuana smokers in an urban public hospital. *Basic Clin Pharmacol Toxicol* 2018. https://doi.org/10.1111/bcpt.12962.
11. Ramaekers JG, et al. Cannabis and tolerance: acute drug impairment as a function of cannabis use history. *Sci Rep* 2016;**6**:26843.
12. Schwilke EW, et al. Δ9-Tetrahydrocannabinol (THC), 11-hydroxy-THC, and 11-nor-9-carboxy-THC plasma pharmacokinetics during and after continuous high-dose oral THC. *Clin Chem* 2009;**55**:2180–9.
13. Gunasekaran N, et al. Reintoxication: the release of fat-stored Δ9-tetrahydrocannabinol (THC) into blood is enhanced by food deprivation or ACTH exposure. *Br J Pharmacol* 2009;**158**:1330–7.
14. Golombek P, Müller M, Barthlott I, Sproll C, Lachenmeier DW. Conversion of cannabidiol (CBD) into psychotropic cannabinoids including tetrahydrocannabinol (THC): a controversy in the scientific literature. *Toxics* 2020;**8**:E41.
15. Spindle TR, et al. Pharmacodynamic effects of vaporized and oral cannabidiol (CBD) and vaporized CBD-dominant cannabis in infrequent cannabis users. *Drug Alcohol Depend* 2020;**211**, 107937.
16. Larsen C, Shahinas J. Dosage, efficacy and safety of cannabidiol administration in adults: a systematic review of human trials. *J Clin Med Res* 2020;**12**:129–41.
17. Kathmann M, Flau K, Redmer A, Tränkle C, Schlicker E. Cannabidiol is an allosteric modulator at mu- and delta-opioid receptors. *Naunyn Schmiedeberg's Arch Pharmacol* 2006;**372**:354–61.
18. Martínez-Aguirre C, et al. Cannabidiol acts at 5-HT1A receptors in the human brain: relevance for treating temporal lobe epilepsy. *Front Behav Neurosci* 2020;**14**:1–8.
19. Watkins AR. Cannabinoid interactions with ion channels and receptors. *Channels (Austin)* 2019;**13**:162–7.
20. O'Sullivan SE. An update on PPAR activation by cannabinoids. *Br J Pharmacol* 2016;**173**:1899–910.
21. Devinsky O, et al. Cannabidiol: pharmacology and potential therapeutic role in epilepsy and other neuropsychiatric disorders. *Epilepsia* 2014;**55**:791–802.
22. Taylor L, Gidal B, Blakey G, Tayo B, Morrison G. A phase I, randomized, double-blind, placebo-controlled, single ascending dose, multiple dose, and food effect trial of the safety, tolerability and pharmacokinetics of highly purified cannabidiol in healthy subjects. *CNS Drugs* 2018;**32**:1053–67.
23. Balmes JR. Vaping-induced acute lung injury: an epidemic that could have been prevented. *Am J Respir Crit Care Med* 2019;**200**:1342–4.
24. Jarlenski M, et al. Trends in perception of risk of regular marijuana use among US pregnant and nonpregnant reproductive-aged women. *Am J Obstet Gynecol* 2017;**217**:705–7.

25. Moss MJ, et al. Cannabis use and measurement of cannabinoids in plasma and breast milk of breastfeeding mothers. *Pediatr Res* 2021. https://doi.org/10.1038/s41390-020-01332-2.
26. Manduca A, Campolongo P, Trezza V. Cannabinoid modulation of mother-infant interaction: is it just about milk? *Rev Neurosci* 2012;**23**:707–22.
27. Rickner SS, Cao D, Kleinschmidt K, Fleming S. A little 'dab' will do ya' in: a case report of neuro-and cardiotoxicity following use of cannabis concentrates. *Clin Toxicol (Phila)* 2017;**55**:1011–3.
28. Calvigioni D, Hurd YL, Harkany T, Keimpema E. Neuronal substrates and functional consequences of prenatal cannabis exposure. *Eur Child Adolesc Psychiatry* 2014;**23**:931–41.
29. Cone EJ, et al. Non-smoker exposure to secondhand cannabis smoke. I. Urine screening and confirmation results. *J Anal Toxicol* 2015;**39**:1–12.
30. Wang X, et al. One minute of marijuana secondhand smoke exposure substantially impairs vascular endothelial function. *J Am Heart Assoc* 2016;**5**.
31. Petanjek Z, et al. Extraordinary neoteny of synaptic spines in the human prefrontal cortex. *PNAS* 2011;**108**:13281–6.
32. Sawyer SM, Azzopardi PS, Wickremarathne D, Patton GC. The age of adolescence. *Lancet Child Adolesc Health* 2018;**2**:223–8.
33. Romer D. Adolescent risk taking, impulsivity, and brain development: implications for prevention. *Dev Psychobiol* 2010;**52**:263–76.
34. Fried PA. Marihuana use by pregnant women: neurobehavioral effects in neonates. *Drug Alcohol Depend* 1980;**6**:415–24.
35. Fried PA, Watkinson B, Gray R. Differential effects on cognitive functioning in 9- to 12-year olds prenatally exposed to cigarettes and marihuana. *Neurotoxicol Teratol* 1998;**20**:293–306.
36. Fried PA. The Ottawa Prenatal Prospective Study (OPPS): methodological issues and findings—it's easy to throw the baby out with the bath water. *Life Sci* 1995;**56**:2159–68.
37. Fried PA, Watkinson B. 12- and 24-month neurobehavioural follow-up of children prenatally exposed to marihuana, cigarettes and alcohol. *Neurotoxicol Teratol* 1988;**10**:305–13.
38. Smith AM, Fried PA, Hogan MJ, Cameron I. Effects of prenatal marijuana on response inhibition: an fMRI study of young adults. *Neurotoxicol Teratol* 2004;**26**:533–42.
39. Day NL, et al. Effect of prenatal marijuana exposure on the cognitive development of offspring at age three. *Neurotoxicol Teratol* 1994;**16**:169–75.
40. Leech SL, Richardson GA, Goldschmidt L, Day NL. Prenatal substance exposure: effects on attention and impulsivity of 6-year-olds. *Neurotoxicol Teratol* 1999;**21**:109–18.
41. Goldschmidt L, Day NL, Richardson GA. Effects of prenatal marijuana exposure on child behavior problems at age 10. *Neurotoxicol Teratol* 2000;**22**:325–36.
42. Day NL, Goldschmidt L, Thomas CA. Prenatal marijuana exposure contributes to the prediction of marijuana use at age 14. *Addiction* 2006;**101**:1313–22.
43. Day NL, Goldschmidt L, Day R, Larkby C, Richardson GA. Prenatal marijuana exposure, age of marijuana initiation, and the development of psychotic symptoms in young adults. *Psychol Med* 2015;**45**:1779–87.
44. Jaddoe VWV, et al. The Generation R Study: design and cohort profile. *Eur J Epidemiol* 2006;**21**:475–84.
45. El Marroun H, et al. Intrauterine cannabis exposure affects fetal growth trajectories: the Generation R Study. *J Am Acad Child Adolesc Psychiatry* 2009;**48**:1173–81.
46. El Marroun H, et al. Intrauterine cannabis exposure leads to more aggressive behavior and attention problems in 18-month-old girls. *Drug Alcohol Depend* 2011;**118**:470–4.
47. Kooijman MN, et al. The Generation R Study: design and cohort update 2017. *Eur J Epidemiol* 2016;**31**:1243–64.

48. Kim S, Lee D. Prefrontal cortex and impulsive decision making. *Biol Psychiatry* 2011;**69**:1140–6.
49. Logue AW. *Self-control: waiting until tomorrow for what you want today.* vol. xvi. Prentice-Hall, Inc; 1995. p. 188.
50. Freeman TP, et al. Changes in delta-9-tetrahydrocannabinol (THC) and cannabidiol (CBD) concentrations in cannabis over time: systematic review and meta-analysis. *Addiction* 2021;**116**:1000–10.
51. Devane WA, et al. Isolation and structure of a brain constituent that binds to the cannabinoid receptor. *Science* 1992;**258**:1946–9.
52. Matsuda LA, Lolait SJ, Brownstein MJ, Young AC, Bonner TI. Structure of a cannabinoid receptor and functional expression of the cloned cDNA. *Nature* 1990;**346**:561–4.
53. Herkenham M. Cannabinoid receptor localization in brain: relationship to motor and reward systems. *Ann N Y Acad Sci* 1992;**654**:19–32.
54. Cabral GA, Raborn ES, Griffin L, Dennis J, Marciano-Cabral F. CB2 receptors in the brain: role in central immune function. *Br J Pharmacol* 2008;**153**:240–51.
55. Parlar A, et al. The exogenous administration of CB2 specific agonist, GW405833, inhibits inflammation by reducing cytokine production and oxidative stress. *Exp Ther Med* 2018;**16**:4900–8.
56. Kaplan BLF. The role of CB1 in immune modulation by cannabinoids. *Pharmacol Ther* 2013;**137**:365–74.
57. Almogi-Hazan O, Or R. Cannabis, the endocannabinoid system and immunity—the journey from the bedside to the bench and back. *Int J Mol Sci* 2020;**21**:1–17.
58. Zhang H-Y, et al. Cannabinoid CB2 receptors modulate midbrain dopamine neuronal activity and dopamine-related behavior in mice. *Proc Natl Acad Sci U S A* 2014;**111**: E5007–15.
59. Hill JD, et al. Activation of GPR55 induces neuroprotection of hippocampal neurogenesis and immune responses of neural stem cells following chronic, systemic inflammation. *Brain Behav Immun* 2019;**76**:165–81.
60. Shi Q, et al. The novel cannabinoid receptor GPR55 mediates anxiolytic-like effects in the medial orbital cortex of mice with acute stress. *Mol Brain* 2017;**10**:38.
61. Lauckner JE, et al. GPR55 is a cannabinoid receptor that increases intracellular calcium and inhibits M current. *Proc Natl Acad Sci U S A* 2008;**105**:2699–704.
62. Fredriksson R, Höglund PJ, Gloriam DEI, Lagerström MC, Schiöth HB. Seven evolutionarily conserved human rhodopsin G protein-coupled receptors lacking close relatives. *FEBS Lett* 2003;**554**:381–8.
63. Rodríguez de Fonseca F, et al. An anorexic lipid mediator regulated by feeding. *Nature* 2001;**414**:209–12.
64. McHugh D, et al. N-Arachidonoyl glycine, an abundant endogenous lipid, potently drives directed cellular migration through GPR18, the putative abnormal cannabidiol receptor. *BMC Neurosci* 2010;**11**:44.
65. Di Marzo V, De Petrocellis L. Why do cannabinoid receptors have more than one endogenous ligand? *Phil Trans R Soc B* 2012;**367**:3216–28.
66. Deutsch DG, Chin SA. Enzymatic synthesis and degradation of anandamide, a cannabinoid receptor agonist. *Biochem Pharmacol* 1993;**46**:791–6.
67. Stella N, Schweitzer P, Piomelli D. A second endogenous cannabinoid that modulates long-term potentiation. *Nature* 1997;**388**:773–8.
68. Lutz B. On-demand activation of the endocannabinoid system in the control of neuronal excitability and epileptiform seizures. *Biochem Pharmacol* 2004;**68**:1691–8.

69. Rouzer CA, Ghebreselasie K, Marnett LJ. Chemical stability of 2-arachidonylglycerol under biological conditions. *Chem Phys Lipids* 2002;**119**:69–82.
70. Lucas CJ, Galettis P, Schneider J. The pharmacokinetics and the pharmacodynamics of cannabinoids. *Br J Clin Pharmacol* 2018;**84**:2477–82.
71. Heuberger JAAC, et al. Population pharmacokinetic model of THC integrates oral, intravenous, and pulmonary dosing and characterizes short- and long-term pharmacokinetics. *Clin Pharmacokinet* 2015;**54**:209–19.
72. Liu J, et al. A biosynthetic pathway for anandamide. *PNAS* 2006;**103**:13345–50.
73. Di Marzo V, et al. Formation and inactivation of endogenous cannabinoid anandamide in central neurons. *Nature* 1994;**372**:686–91.
74. Liu J, et al. Multiple pathways involved in the biosynthesis of anandamide. *Neuropharmacology* 2008;**54**:1–7.
75. Bisogno T, et al. Cloning of the first sn1-DAG lipases points to the spatial and temporal regulation of endocannabinoid signaling in the brain. *J Cell Biol* 2003;**163**:463–8.
76. Karlsson M, Contreras JA, Hellman U, Tornqvist H, Holm C. cDNA cloning, tissue distribution, and identification of the catalytic triad of monoglyceride lipase: evolutionary relationship to esterases, lysophospholipases, and haloperoxidases*. *J Biol Chem* 1997;**272**:27218–23.
77. Dinh TP, et al. Brain monoglyceride lipase participating in endocannabinoid inactivation. *Proc Natl Acad Sci U S A* 2002;**99**:10819–24.
78. Navia-Paldanius D, Savinainen JR, Laitinen JT. Biochemical and pharmacological characterization of human α/β-hydrolase domain containing 6 (ABHD6) and 12 (ABHD12). *J Lipid Res* 2012;**53**:2413–24.
79. Rouzer CA, Marnett LJ. Endocannabinoid oxygenation by cyclooxygenases, lipoxygenases, and cytochromes P450: cross-talk between the eicosanoid and endocannabinoid signaling pathways. *Chem Rev* 2011;**111**:5899–921.
80. Buczynski MW, Parsons LH. Quantification of brain endocannabinoid levels: methods, interpretations and pitfalls. *Br J Pharmacol* 2010;**160**:423–42.
81. Russo EB. The case for the entourage effect and conventional breeding of clinical cannabis: No 'Strain,' No Gain. *Front Plant Sci* 2018;**9**:1969.
82. Ben-Shabat S, et al. An entourage effect: inactive endogenous fatty acid glycerol esters enhance 2-arachidonoyl-glycerol cannabinoid activity. *Eur J Pharmacol* 1998;**353**:23–31.
83. Di Marzo V, et al. Palmitoylethanolamide inhibits the expression of fatty acid amide hydrolase and enhances the anti-proliferative effect of anandamide in human breast cancer cells. *Biochem J* 2001;**358**:249–55.
84. Petrosino S, et al. The anti-inflammatory mediator palmitoylethanolamide enhances the levels of 2-arachidonoyl-glycerol and potentiates its actions at TRPV1 cation channels. *Br J Pharmacol* 2016;**173**:1154–62.
85. Bosch-Bouju C, Layé S. Dietary omega-6/omega-3 and endocannabinoids: implications for brain health and diseases. In: *Cannabinoids in health and disease*. IntechOpen; 2016. https://doi.org/10.5772/62498.
86. Cinquina V, et al. Life-long epigenetic programming of cortical architecture by maternal 'Western' diet during pregnancy. *Mol Psychiatry* 2020;**25**:22–36.
87. Ramírez-López MT, et al. A moderate diet restriction during pregnancy alters the levels of endocannabinoids and endocannabinoid-related lipids in the hypothalamus, hippocampus and olfactory bulb of rat offspring in a sex-specific manner. *PLoS One* 2017;**12**, e0174307.
88. Gulyas AI, et al. Segregation of two endocannabinoid-hydrolyzing enzymes into pre- and postsynaptic compartments in the rat hippocampus, cerebellum and amygdala. *Eur J Neurosci* 2004;**20**:441–58.

89. Mackie K, Hille B. Cannabinoids inhibit N-type calcium channels in neuroblastoma-glioma cells. *Proc Natl Acad Sci U S A* 1992;**89**:3825–9.
90. Guo J, Ikeda SR. Endocannabinoids modulate N-type calcium channels and G-protein-coupled inwardly rectifying potassium channels via CB1 cannabinoid receptors heterologously expressed in mammalian neurons. *Mol Pharmacol* 2004;**65**:665–74.
91. Tanimura A, et al. Synapse type-independent degradation of the endocannabinoid 2-arachidonoylglycerol after retrograde synaptic suppression. *Proc Natl Acad Sci U S A* 2012;**109**:12195–200.
92. Musshoff F, Madea B. Review of biologic matrices (urine, blood, hair) as indicators of recent or ongoing cannabis use. *Ther Drug Monit* 2006;**28**:155–63.
93. Yamaori S, Ebisawa J, Okushima Y, Yamamoto I, Watanabe K. Potent inhibition of human cytochrome P450 3A isoforms by cannabidiol: role of phenolic hydroxyl groups in the resorcinol moiety. *Life Sci* 2011;**88**:730–6.
94. López-Cardona AP, et al. CB1 cannabinoid receptor drives oocyte maturation and embryo development via PI3K/Akt and MAPK pathways. *FASEB J* 2017;**31**:3372–82.
95. Sun X, Dey SK. Aspects of endocannabinoid signaling in periimplantation biology. *Mol Cell Endocrinol* 2008;**286**:S3–11.
96. Schmid PC, Paria BC, Krebsbach RJ, Schmid HH, Dey SK. Changes in anandamide levels in mouse uterus are associated with uterine receptivity for embryo implantation. *Proc Natl Acad Sci U S A* 1997;**94**:4188–92.
97. Oh H-A, et al. Uncovering a role for endocannabinoid signaling in autophagy in preimplantation mouse embryos. *Mol Hum Reprod* 2013;**19**:93–101.
98. Whyte LS, et al. The putative cannabinoid receptor GPR55 affects osteoclast function in vitro and bone mass in vivo. *Proc Natl Acad Sci U S A* 2009;**106**:16511–6.
99. Malenczyk K, et al. Fetal endocannabinoids orchestrate the organization of pancreatic islet microarchitecture. *Proc Natl Acad Sci U S A* 2015;**112**:E6185–94.
100. Ginhoux F, Lim S, Hoeffel G, Low D, Huber T. Origin and differentiation of microglia. *Front Cell Neurosci* 2013;1–14.
101. Psychoyos D, et al. Cannabinoid receptor 1 signaling in embryo neurodevelopment. *Birth Defects Res B Dev Reprod Toxicol* 2012;**95**:137–50.
102. Buckley NE, Hansson S, Harta G, Mezey E. Expression of the CB1 and CB2 receptor messenger RNAs during embryonic development in the rat. *Neuroscience* 1998;**82**:1131–49.
103. Maccarrone M, Guzmán M, Mackie K, Doherty P, Harkany T. Programming of neural cells by (endo)cannabinoids: from physiological rules to emerging therapies. *Nat Rev Neurosci* 2014;**15**:786–801.
104. Aguado T, et al. The endocannabinoid system drives neural progenitor proliferation. *FASEB J* 2005;**19**:1704–6.
105. Aguado T, et al. The CB1 cannabinoid receptor mediates excitotoxicity-induced neural progenitor proliferation and neurogenesis. *J Biol Chem* 2007;**282**:23892–8.
106. Mulder J, et al. Endocannabinoid signaling controls pyramidal cell specification and long-range axon patterning. *Proc Natl Acad Sci U S A* 2008;**105**:8760–5.
107. Guennewig B, et al. THC exposure of human iPSC neurons impacts genes associated with neuropsychiatric disorders. *Transl Psychiatry* 2018;**8**:1–9.
108. Maciel IdS, et al. Perinatal CBD or THC exposure results in lasting resistance to fluoxetine in the forced swim test: reversal by fatty acid amide hydrolase inhibition. *Cannabis Cannabinoid Res* 2021. https://doi.org/10.1089/can.2021.0015.
109. Wanner NM, Colwell M, Drown C, Faulk C. Developmental cannabidiol exposure increases anxiety and modifies genome-wide brain DNA methylation in adult female mice. *Clin Epigenetics* 2021;**13**:4.

110. Luján MÁ, Valverde O. The pro-neurogenic effects of cannabidiol and its potential therapeutic implications in psychiatric disorders. *Front Behav Neurosci* 2020;**14**:1–11.
111. Fogaça MV, Campos AC, Coelho LD, Duman RS, Guimarães FS. The anxiolytic effects of cannabidiol in chronically stressed mice are mediated by the endocannabinoid system: role of neurogenesis and dendritic remodeling. *Neuropharmacology* 2018;**135**:22–33.
112. Beale C, et al. Prolonged cannabidiol treatment effects on hippocampal subfield volumes in current cannabis users. *Cannabis Cannabinoid Res* 2018;**3**:94–107.
113. Begbie J, Doherty P, Graham A. Cannabinoid receptor, CB1, expression follows neuronal differentiation in the early chick embryo. *J Anat* 2004;**205**:213–8.
114. Keimpema E, et al. Nerve growth factor scales endocannabinoid signaling by regulating monoacylglycerol lipase turnover in developing cholinergic neurons. *Proc Natl Acad Sci U S A* 2013;**110**:1935–40.
115. Frau R, et al. Prenatal THC exposure produces a hyperdopaminergic phenotype rescued by pregnenolone. *Nat Neurosci* 2019;**22**:1975–85.
116. Berghuis P, et al. Hardwiring the brain: endocannabinoids shape neuronal connectivity. *Science* 2007;**316**:1212–6.
117. Vitalis T, et al. The type 1 cannabinoid receptor is highly expressed in embryonic cortical projection neurons and negatively regulates neurite growth in vitro. *Eur J Neurosci* 2008;**28**: 1705–18.
118. Keimpema E, Mackie K, Harkany T. Molecular model of cannabis sensitivity in developing neuronal circuits. *Trends Pharmacol Sci* 2011;**32**:551–61.
119. Keimpema E, et al. Diacylglycerol lipase α manipulation reveals developmental roles for intercellular endocannabinoid signaling. *Sci Rep* 2013;**3**:2093.
120. Roland AB, et al. Cannabinoid-induced actomyosin contractility shapes neuronal morphology and growth. *elife* 2014;**3**, e03159.
121. Watson S, Chambers D, Hobbs C, Doherty P, Graham A. The endocannabinoid receptor, CB1, is required for normal axonal growth and fasciculation. *Mol Cell Neurosci* 2008;**38**:89–97.
122. Díaz-Alonso J, et al. The CB1 cannabinoid receptor drives corticospinal motor neuron differentiation through the Ctip2/Satb2 transcriptional regulation axis. *J Neurosci* 2012;**32**:16651–65.
123. Wu C-S, et al. Requirement of cannabinoid CB1 receptors in cortical pyramidal neurons for appropriate development of corticothalamic and thalamocortical projections. *Eur J Neurosci* 2010;**32**:693–706.
124. Tanimura A, et al. The endocannabinoid 2-arachidonoylglycerol produced by diacylglycerol lipase alpha mediates retrograde suppression of synaptic transmission. *Neuron* 2010;**65**: 320–7.
125. Tortoriello G, et al. Targeted lipidomics in Drosophila melanogaster identifies novel 2-monoacylglycerols and N-acyl amides. *PLoS One* 2013;**8**, e67865.
126. Murataeva N, et al. Where's my entourage? The curious case of 2-oleoylglycerol, 2-linolenoylglycerol, and 2-palmitoylglycerol. *Pharmacol Res* 2016;**110**:173–80.
127. Tortoriello G, et al. Genetic manipulation of sn-1-diacylglycerol lipase and CB1 cannabinoid receptor gain-of-function uncover neuronal 2-linoleoyl glycerol signaling in Drosophila melanogaster. *Cannabis Cannabinoid Res* 2021;**6**:119–36.
128. Balenga NAB, et al. GPR55 regulates cannabinoid 2 receptor-mediated responses in human neutrophils. *Cell Res* 2011;**21**:1452–69.
129. Cui K, Yuan X. TRP channels and axon pathfinding. In: Liedtke WB, Heller S, editors. *TRP ion channel function in sensory transduction and cellular signaling cascades*. CRC Press/Taylor & Francis; 2007.

130. Pastuhov SI, et al. Endocannabinoid-Goα signalling inhibits axon regeneration in Caenorhabditis elegans by antagonizing Gqα-PKC-JNK signalling. *Nat Commun* 2012;**3**:1136.
131. Sufian MS, Amin MR, Ali DW. Early suppression of the endocannabinoid degrading enzymes FAAH and MAGL alters locomotor development in zebrafish. *J Exp Biol* 2021. https://doi.org/10.1242/jeb.242635.
132. Wu C-S, et al. Long-term consequences of perinatal fatty acid amino hydrolase inhibition. *Br J Pharmacol* 2014;**171**:1420–34.
133. Tortoriello G, et al. Miswiring the brain: Δ9-tetrahydrocannabinol disrupts cortical development by inducing an SCG10/stathmin-2 degradation pathway. *EMBO J* 2014;**33**:668–85.
134. Mereu G, et al. Prenatal exposure to a cannabinoid agonist produces memory deficits linked to dysfunction in hippocampal long-term potentiation and glutamate release. *Proc Natl Acad Sci U S A* 2003;**100**:4915–20.
135. Castaldo P, et al. Prenatal exposure to the cannabinoid receptor agonist WIN 55,212-2 increases glutamate uptake through overexpression of GLT1 and EAAC1 glutamate transporter subtypes in rat frontal cerebral cortex. *Neuropharmacology* 2007;**53**:369–78.
136. Felder CC, et al. Comparison of the pharmacology and signal transduction of the human cannabinoid CB1 and CB2 receptors. *Mol Pharmacol* 1995;**48**:443–50.
137. Vargish GA, et al. Persistent inhibitory circuit defects and disrupted social behaviour following in utero exogenous cannabinoid exposure. *Mol Psychiatry* 2017;**22**:56–67.
138. Burstein S, Budrow J, Debatis M, Hunter SA, Subramanian A. Phospholipase participation in cannabinoid-induced release of free arachidonic acid. *Biochem Pharmacol* 1994;**48**:1253–64.
139. Melis M, Pistis P. Endocannabinoid signaling in midbrain dopamine neurons: more than physiology? *Curr Neuropharmacol* 2007;**5**:268–77.
140. Wang X, Dow-Edwards D, Keller E, Hurd YL. Preferential limbic expression of the cannabinoid receptor mRNA in the human fetal brain. *Neuroscience* 2003;**118**:681–94.
141. DiNieri JA, et al. Maternal cannabis use alters ventral striatal dopamine D2 gene regulation in the offspring. *Biol Psychiatry* 2011;**70**:763–9.
142. Spano MS, Ellgren M, Wang X, Hurd YL. Prenatal cannabis exposure increases heroin seeking with allostatic changes in limbic enkephalin systems in adulthood. *Biol Psychiatry* 2007;**61**:554–63.
143. Volk DW, Lewis DA. The role of endocannabinoid signaling in cortical inhibitory neuron dysfunction in schizophrenia. *Biol Psychiatry* 2016;**79**:595–603.
144. Dawson DA, Persad C. Targeting the endocannabinoid system in the treatment of ADHD. *Genet Mol Med* 2021;**3**(1):1–7.
145. Berrendero F, Sepe N, Ramos JA, Di Marzo V, Fernández-Ruiz JJ. Analysis of cannabinoid receptor binding and mRNA expression and endogenous cannabinoid contents in the developing rat brain during late gestation and early postnatal period. *Synapse* 1999;**33**:181–91.
146. Fride E, et al. Milk intake and survival in newborn cannabinoid CB1 receptor knockout mice: evidence for a 'CB3' receptor. *Eur J Pharmacol* 2003;**461**:27–34.
147. Toda T, Kawasaki H. The development of suckling behavior of neonatal mice is regulated by birth. *Mol Brain* 2014;**7**:8.
148. Fride E, et al. Critical role of the endogenous cannabinoid system in mouse pup suckling and growth. *Eur J Pharmacol* 2001;**419**:207–14.
149. Knopf A. THC persists in breast milk if marijuana is used in pregnancy. *Alcohol Drug Abuse Weekly* 2021;**33**:5–6.
150. Scheyer AF, et al. Cannabinoid exposure via lactation in rats disrupts perinatal programming of the GABA trajectory and select early-life behaviors. *Biol Psychiatry* 2020;**87**:666–77.

151. Cellot G, Cherubini E. Functional role of ambient GABA in refining neuronal circuits early in postnatal development. *Front Neural Circuits* 2013;**7**:136.
152. Ganguly K, Schinder AF, Wong ST, Poo M. GABA itself promotes the developmental switch of neuronal GABAergic responses from excitation to inhibition. *Cell* 2001;**105**:521–32.
153. Yuste R, Katz LC. Control of postsynaptic Ca^{2+} influx in developing neocortex by excitatory and inhibitory neurotransmitters. *Neuron* 1991;**6**:333–44.
154. Almeida DLd, Devi LA. Diversity of molecular targets and signaling pathways for CBD. *Pharmacol Res Perspect* 2020;**8**, e00682.
155. Ramírez-Barrantes R, et al. Perspectives of TRPV1 function on the neurogenesis and neural plasticity. *Neural Plasticity* 2016;**2016**, e1568145.
156. Lauder JM, Wallace JA, Krebs H. Roles for serotonin in neuroembryogenesis. *Adv Exp Med Biol* 1981;**133**:477–506.
157. Cass DK, et al. CB1 cannabinoid receptor stimulation during adolescence impairs the maturation of GABA function in the adult rat prefrontal cortex. *Mol Psychiatry* 2014;**19**:536–43.
158. Beiersdorf J, et al. *Adverse effects of Δ9-tetrahydrocannabinol on neuronal bioenergetics during postnatal development*; 2020. p. 1–20. https://doi.org/10.1172/jci.insight.135418. https://insight.jci.org/articles/view/135418/figure/2.
159. Wolff V, et al. Tetrahydrocannabinol induces brain mitochondrial respiratory chain dysfunction and increases oxidative stress: a potential mechanism involved in cannabis-related stroke. *Biomed Res Int* 2015;**2015**, 323706.
160. Fišar Z, Singh N, Hroudová J. Cannabinoid-induced changes in respiration of brain mitochondria. *Toxicol Lett* 2014;**231**:62–71.
161. Melser S, et al. Functional analysis of mitochondrial CB1 cannabinoid receptors (mtCB1) in the brain. *Meth Enzymol* 2017;**593**:143–74.
162. Hebert-Chatelain E, et al. A cannabinoid link between mitochondria and memory. *Nature* 2016;**539**:555–9.
163. Zuckermann AME, et al. Trends in youth cannabis use across cannabis legalization: data from the COMPASS prospective cohort study. *Prev Med Rep* 2021;**22**, 101351.
164. Polańska K, Jurewicz J, Hanke W. Smoking and alcohol drinking during pregnancy as the risk factors for poor child neurodevelopment – a review of epidemiological studies. *Int J Occup Med Environ Health* 2015;**28**:419–43.
165. Alpár A, et al. Endocannabinoids modulate cortical development by configuring Slit2/Robo1 signalling. *Nat Commun* 2014;**5**:4421.

Chapter 10

How adolescent cannabinoid exposure sets the stage for long-term emotional and cognitive dysregulation: Impacts on molecular and neuronal risk pathways

Steven R. Laviolette[a,b]

[a]Addiction Research Group, Department of Anatomy & Cell Biology, Schulich School of Medicine & Dentistry, University of Western Ontario, London, ON, Canada, [b]Department of Psychiatry, Schulich School of Medicine & Dentistry, University of Western Ontario, London, ON, Canada

Introduction

The mammalian brain is faced with a constant barrage of competing and complex sensory inputs that must be properly analyzed and understood in terms of their emotional and cognitive significance. Such high-level sensory processing fidelity require equally complex neurobiological mechanisms capable of accurately assigning emotional salience to this incoming information, along with specialized neurochemical signaling pathways that can register emotional sensory salience and allow for appropriate associative memory formation between environmental cues and their emotional meaning. The mature functioning of these mechanisms require precise and adaptive periods of brain development, particularly during the adolescent periods of brain maturation wherein critical functional connections are formed between higher order cortical control regions and sub-cortical emotional processing centers such as the pre-frontal cortex (PFC), amygdala and mesolimbic dopamine (DA) pathway.

Disturbances in emotional salience processing and associative memory formation represent a core endophenotype across various neuropsychiatric disorders, including addiction, mood and anxiety disorders, post-traumatic stress disorder (PTSD) and schizophrenia. For example, addictive behaviors including dependence, craving and relapse, are characterized by the formation of pathological associative memories between the reinforcing properties of a given drug and environmental cues associated with these effects. In disorders like schizophrenia, individuals ascribe aberrant emotional salience and/or are unable to filter out sensory cues in the environment that healthy individuals would ignore. These distortions in emotional processing can ultimately lead to psychotic ideation, paranoid belief structures and a break from normative reality. Decades of clinical and pre-clinical research has identified the limbic system, including structures like the PFC, amygdala, hippocampus, and mesolimbic DA pathway, as critical neural structures linked to these emotional processing pathologies.

Two of the most critical neurochemical signaling pathways for the processing of emotional salience are the brain's endocannabinoid (eCB) system, comprised primarily of CB1 and CB2 receptor sub-types and the DA pathway, which involves two broad classes of D1 (D1R, D5R) and D2-like receptor (D2R, D3R, D4R) families. Both systems are crucial regulators of emotional processing and memory formation across multiple brain regions, including important mesocorticolimbic structures such as the basolateral amygdala (BLA), PFC, ventral tegmental area (VTA), and nucleus accumbens (NAc).

Clinical and pre-clinical evidence has demonstrated that extrinsic drug compounds that act on DAergic signaling or the eCB system are used recreationally for their ability to alter perception and mood. For example, cannabis strongly alters emotional salience perception and distorts sensory experiences while inducing states of paranoia resembling psychotic symptoms in some individuals.[1] Similarly, compounds that potentiate DAergic activity can potently alter emotional salience attribution, induce states of euphoria and mania and like cannabis-related compounds, can evoke paranoia and psychotic-like episodes.[2] These shared functional characteristics demonstrate common neurobiological substrates upon which these compounds may act. Delta-9-tetrahydrocannabinol (THC), the primary psychoactive compound in cannabis is a pharmacologically promiscuous phytochemical. Nevertheless, its primary neuropharmacological target is believed to be the CB1 receptor (CB1R) where it acts as a partial agonist.[3] However, given that the CB1R system itself is ubiquitously distributed in the mammalian brain, a major challenge has been identifying the precise functional neural circuits by which cannabinoids can independently (or synergistically) modulate DAergic neuronal activity states, specifically, the A10 DA neurons in the VTA. Multiple pre-clinical studies have demonstrated that cannabinoids can potently regulate the transmission of DA signals directly in the mesolimbic pathway.[4-6] These effects are particularly relevant for the known clinical and pre-clinical links between cannabinoid exposure and schizophrenia-related disorders, given the importance of DA dysregulation as an underlying endophenotype in the etiology of schizophrenia-related illness.[7,8] Several important limbic and cortical regions have emerged through both clinical and pre-clinical investigations that have revealed how cannabis exposure (both acutely and during neurodevelopmental windows) can powerfully regulate DAergic signaling mechanisms.

How cannabinoids control emotional processing: Convergent impacts on mesocorticolimbic dopamine states

The ability of cannabinoids to distort emotional perception and salience has been reported anecdotally for millennia and represents its cardinal psychoactive effect and attraction for recreational consumption.[9] However, only recently have the underlying neurobiological mechanisms responsible for these effects been understood. Schizophrenia-related psychoses are associated with profound disturbances in emotional processing and associative memory formation[10,11] and clinical research has demonstrated that aberrant signaling through the brains CB1 receptor system is strongly correlated with schizophrenia-related psychopathology.[1,12] For example, several post-mortem studies have reported abnormal expression patterns of CB1Rs in frontal cortical regions of schizophrenia patients.[13,14] In addition, abnormal levels of eCB metabolites are reported in schizophrenia populations.[15] Such correlational evidence is limited by a lack of causal mechanisms and potential confounds with historical cannabis use and/or patient medication histories. To this end, considerable preclinical evidence has identified several neurobiological mechanisms and brain circuits by which signaling through CB1R substrates can strongly modulate and/or distort emotional perception and associative memory formation.

The mammalian PFC is a complex neural region comprising several distinct sub-regions critically involved in emotional processing and associative memory formation. The rodent homologue of the human dorsolateral PFC includes the pre-limbic, infralimbic, and anterior cingulate cortices. In rodents, there is compelling evidence that cannabinoid transmission within the PFC can strongly modulate the salience of emotionally significant information as well as associative memory formation. For example, Laviolette and Grace[16] first reported that systemic or direct activation of CB1R transmission in the rat PFC could strongly potentiate neuronal and behavioral encoding of fear-related associative memories, through functional interactions with the BLA. In this case, systemic administration of the synthetic CB1R agonist, WIN-55, 212-2, dramatically increased PFC associative neuronal firing and bursting activation in response to olfactory cues previously paired with sub-threshold footshock conditioning cues. These effects were limited to PFC neuronal sub-populations that responded to electrical stimulation of the BLA, demonstrating a functional link between the BLA > PFC circuit in the mediation of these effects. Similarly, direct, intra-PFC infusions of WIN-55, 212-2 potentiated the formation of associative fear-memories to sub-threshold conditioning events which were ignored by control rats. Furthermore, the effects of direct CB1R activation in the PFC on emotional salience processing were completely blocked by pharmacologically induced inactivation of the BLA. Subsequent work revealed that bi-directional CB1R transmission directly in the BLA could similarly strongly control emotional salience processing and associative fear memory formation through functional modulation of neuronal activity states in the PFC. Thus, direct pharmacological activation of BLA CB1R populations were shown to potentiate normally non-salient associative fear conditioning cues; effects that were associated with potentiation in the firing and bursting activity states of PFC neurons.[17] In contrast, pharmacological blockade of intra-BLA CB1 transmission blocked the formation of normally salient associative fear memories and similarly reduced spontaneous firing frequency and bursting states of pyramidal neuron populations in the PFC. Pre-clinical studies using reward-related behaviors have similarly found that CB1R activation can modulate associative memory formation and learning in the context of appetitive conditioning cues. For example, Brancato et al.[18] reported that systemic pharmacological stimulation of the cannabinoid system prior to reward-related associative conditioning was able to potentiate reward-related explicit memory in the presence of novelty, whereas post-conditioning activation increased approach behavior to novel stimuli. In Fig. 1, a schematic summary presents the currently known regulatory roles of CB1R transmission mechanisms and their bidirectional control of emotional salience and neuronal activity states within the BLA-PFC pathway.

Cannabinoid transmission in pre-frontal cortex regulates emotional processing through control of sub-cortical dopamine

In addition to cannabinoid regulation of emotional processing via amygdala-PFC interactions, pre-clinical studies have shown that CB1R transmission in the PFC can directly regulate sub-cortical DA states and simultaneously regulates both reward and aversion-related emotional memory processing.[20,21] Using intra-PFC microinfusions of WIN-55 combined with simultaneous in vivo neuronal recordings of DA neurons in the rat VTA, Draycott et al.[21] demonstrated a bi-phasic, dose-dependent regulation of CB1R activation states coupled with downstream regulation of spontaneous DA neuron firing frequency and increase neuronal shifts from tonic to bursting states. This study found that WIN-55 doses in a specific range caused strong increases in VTA DA neuron activity states and increased bursting and frequency activity corresponding to a distortion in emotional salience perception; footshock stimuli that would normally be ignored were now highly salient and produced strong associative fear memories. In contrast, higher WIN-55 concentrations which presumably overwhelmed the PFC CB1R population, strongly decreased spontaneous VTA DA activity, effectively shutting down sub-cortical DA activity and remarkably, completely blocked the formation of associative fear memories that would normally be highly salience. Such a bi-phasic regulation of cortical CB1R > sub-cortical DA states has critical implications for how the cannabinoid system (and exposure to high concentrations of CB1R activators like WIN-55 or THC) may account for cannabinoid-induced psychotomimetic effects.[22,23] Similar to activity patterns observed in PFC neurons,[16] DA neuron bursting states are important in various neuropsychiatric conditions as they are

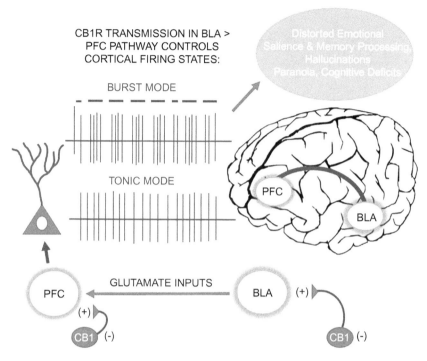

FIG. 1 Cannabinoid signaling in the BLA > PFC pathway modulates cortical neuronal activity states, emotional salience processing and associative memory formation: The BLA and PFC share functional, bi-direction connections. Cannabinoids can increase excitatory, glutamatergic outputs from either region by inhibiting local inhibitory neuronal bodies or terminals containing GABA. Pre-clinical studies have shown that direct (intra-BLA or intra-PFC) or systemic CB1R activation can potentiate normally non-salient associative fear memory cues[16,19] by potentiating pyramidal neuron frequency and bursting levels during associative learning tasks. These mechanisms may underlie the ability of cannabinoids to distort emotional processing and normal associative memory formation, leading to affective and cognitive phenotypes similar to schizophrenia, including anxiety, paranoia, hallucinations, and disruptions in normal cognitive function.

proposed to correlate with emotional stimulus salience and dysregulation of tonic versus bursting modes of activity may underlie emotional salience and associative learning disturbances in disorders like schizophrenia.[24,25] The modulatory role and translational implications of intra-PFC CB1R transmission on sub-cortical VTA DA and fear-related emotional salience processing is summarized in Fig. 2.

Beyond the regulation of the salience of aversive emotional states and associated memory formation, cannabinoid transmission in the PFC strongly modulates the processing of reward-related salience and associative memory formation. Using a combination of morphine conditioned place preference (CPP) procedures with direct, intra-PFC microinfusions of the CB1R agonist

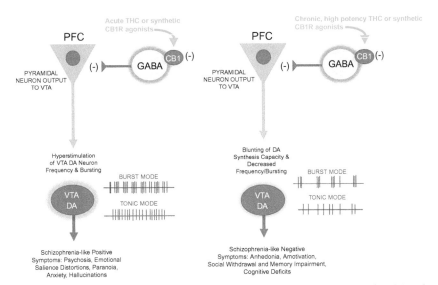

FIG. 2 Cannabinoid signaling through CB1Rs mediate biphasic effects on sub-cortical DA and emotional associative processing and memory. The PFC sends and receives functional inputs from the VTA and can modulate activity states of sub-cortical DA indirectly via glutamatergic inputs onto inhibitory VTA GABA neurons or via direct excitation of the A10 DA neurons. Pharmacological activation of PFC CB1Rs produce biphasic actions on DA neuron firing frequency and bursting states. Acute activation produces hyperactive VTA DA neuron activity and potentiates normally non-salient associative fear cues. In contrast, flooding the PFC with high dose CB1R agonists cause a shutting down of spontaneous VTA DA neuron firing and bursting states whilst simultaneously blocking the acquisition of normally salient conditioned fear memories.[21] These basic mechanisms may underlie the apparent bi-phasic effects of chronic cannabinoid exposure on the developing adolescent brain and associated disturbances in affective processing and associative memory formation.

WIN-55 or the antagonist, AM-251, Ahmad et al.[20] reported that CB1R transmission in the PFC could bi-directionally regulate the rewarding versus aversive effects of opioids via a mu versus kappa-opioid receptor-dependent gating mechanism in the VTA. Specifically, activation or blockade of CB1R transmission in the prelimbic cortical (PLC) division of the PFC was shown to bidirectionally gate the emotional valence of morphine. Thus, intra-PLC CB1R activation switched a normally rewarding morphine place preference into a place aversion. In contrast, pharmacological blockade of CB1R transmission potentiated the reward salience of normally sub-reward threshold morphine CPP effects. Interestingly, both of these effects were mediated through a DA-dependent mechanism since blockade of DAergic transmission prevented CB1R-mediated switching of morphine reward versus aversion behaviors. Finally, these cortical effects of PLC CB1R signaling were blocked by intra-VTA mu-opioid or kappa-receptor antagonists for reward versus aversion effects, respectively. Thus, these findings provided evidence for an equally important regulatory role of intra-PFC CB1R modulation of reward-related emotional memory processing.

Further evidence for acute control of cannabinoids over sub-cortical DAergic processing and emotional salience was reported by Fitoussi et al.[26] wherein the authors demonstrated that acute THC administration directly in the NAc, was able to potently potentiate the fear-related salience of associative footshock cues measured in an olfactory fear conditioning assay.[26] These intra-accumbens effects of THC were dependent upon functional modulation of the VTA DA system, as fear-potentiating doses of intra-NAc THC simultaneously caused dramatic increases in VTA DA neuronal firing frequency and bursting activity while simultaneously inhibiting the spontaneous activity of VTA GABAergic neurons. Indeed, blockade of intra-VTA GABA transmission with a mixed $GABA_A$, $GABA_B$ antagonist infusion was sufficient to completely block the effects of intra-accumbens THC on fear-related associative memory formation, demonstrating that cannabinoid signaling directly in the mesolimbic circuit can potently regulate emotional processing via functional regulation of VTA DA neuronal activity dynamics. Less is known in terms of how exposure to extrinsic cannabinoids like THC, may impact the PFC region in terms of schizophrenia-related outcomes. However, several clinical studies have reported abnormal expression patterns of CB1Rs in cortical and sub-cortical regions of schizophrenia patients,[13,14] suggesting that dysregulation of CB1R signaling within the frontal cortex may either represent an underlying cause or consequence of schizophrenia. In a recent study by Borgan et al.,[27] they observed reductions in CB1R binding levels across multiple brain regions including the anterior cingulate cortex (ACC), hippocampus, striatum, and thalamus, suggesting that abnormalities in CB1R expression patterns may be widespread beyond the frontal cortical areas. Regardless, it can be difficult to establish causality in these studies and disentangle the effects of medication and drug use history (particularly cannabis use) when retroactively assessing post-mortem samples. Hence, the use of well-controlled pre-clinical studies can effectively examine the roles of CBR signaling within these circuits and translate these mechanisms into behavioral outcomes reflective of symptom endophenotypes.

Thus, a wealth of pre-clinical and clinical evidence points to a critical modulatory role for acute cannabinoid receptor activation in causing a variety of neuropsychiatric symptoms, many of which resemble schizophrenia-related endophenotypes. Growing clinical and pre-clinical evidence has demonstrated that neurodevelopmental exposure to cannabinoid agonists, particularly during adolescent windows of brain development, can trigger multiple risk factors for the development of neuropsychiatric disorders. As will be discussed in the subsequent sections of this chapter, many of the long-term behavioral, neuronal, and molecular pathophysiological sequelae associated with neurodevelopmental cannabinoid exposure, closely resemble the mechanisms associated with acute cannabinoid modulation of emotional salience and memory formation and share similar brain circuits and pathological phenotypes associated with serious psychiatric pathologies, including schizophrenia, mood and anxiety and addictive disorders.

Translational rodent models of adolescent cannabinoid exposure

A substantial body of clinical evidence has found that adolescent exposure to cannabis is associated with a host of increased risk factors for various neuropsychiatric disorders, including schizophrenia-related psychoses and mood and anxiety-related disorders.[28–30] One of the earliest reports first reporting this link examined over 45,000 Swedish military conscripts and was able to directly correlate the frequency of adolescent cannabis consumption with increased risk of receiving a schizophrenia diagnosis in young adulthood.[28] However, given the inherently correlative nature of such studies and the reliance on potentially unreliable subject self-reporting, these studies have been questioned as to the actual causal implications of adolescent cannabis exposure on later neuropsychiatric risk. To this end, a growing body of pre-clinical studies, primarily using rodent models of adolescent brain development, have directly examined the causal and mechanistic relationships between adolescent exposure to cannabis and its primary psychoactive compound, THC, on the pathogenic effects of cannabinoid exposure during vulnerable periods of adolescent brain development.

Rodent models of adolescent cannabinoid exposure hold several advantages over retrospective clinical studies. First, the experimenter is in complete control over the dosing and temporal sequence of cannabinoid exposure in animal experimental cohorts. Second, rodent studies allow for combining cannabinoid exposure with potential interventional strategies either in tandem with, or after the period of vulnerable drug exposure. Third, experimenters can precisely control which specific chemical element of cannabis will serve as the drug target, such as using THC, CBD or combinations thereof during the drug exposure windows. This issue has important translational impacts considering that the relative ratio of THC:CBD in a given cannabis strain has been linked to the neuropsychiatric risks of exposure. For example, adolescent exposure to strains of cannabis such as sinsemilla (skunk), which contain exceptionally potent amounts of THC relative to lower CBD concentrations, are more likely to be associated with psychiatric risk for psychosis.[31,32] Nevertheless, several major disadvantages exist for rodent models of cannabinoid exposure as well. For example, rodents metabolize cannabinoids at different rates from human consumers and thus matching bioequivalent dosing for comparing human versus rodent cannabis exposure variables can be challenging. In addition, non-voluntary cannabinoid consumption procedures suffer from face validity when modeling chronic cannabis consumption patterns in human subjects. Nevertheless, many of these limitations have been mitigated with the development of cannabinoid vapor self-administration models and/or cannabis edible consumption procedures which can be self-administered in rodent strains.[33,34]

From a translational perspective, the rodent brain undergoes many of the same critical stages of brain development during adolescence as does the human brain, including critical periods of synaptic pruning and maturation and organization of cortical and sub-cortical circuitry. All of these critical neural changes taking place within the context of pubertal hormonal development. In rodents, adolescence is generally considered to start around post-natal day 28 (PND28) and is considered complete once the animal has reached full sexual maturity at approximately PND60. In addition, the period of adolescence can be further categorized into specific epochs such as early adolescence (starting around PND28), middle adolescence (starting around PND38), and finally, late adolescence (starting around PND49).[35] In pre-clinical adolescent cannabinoid exposure models, there have generally been three pharmacological agents used to mimic the actions of cannabis on the cannabinoid receptor type 1 (CB1R). These include purified isolates of THC[12,36,37] or synthetic cannabinoids such as WIN-55[38] or CP55,940 (CP).[39,40] THC acts as a partial agonist for CB1R, while WIN and CP represent more potent, full agonists at the CB1R.[41] These pharmacological tools may also represent a weakness for pre-clinical adolescent cannabinoid exposure models in that they do not represent the entire phytochemical spectrum of the cannabis plant in either smoked or edible formats. In addition, most studies rely on passive systemic injections rather than voluntary inhalation or edible administration routes, which are the most typical exposure formats for human consumption.

Nevertheless, the advantages offered by rodent adolescent cannabinoid exposure models in terms of precise control over independent variables and timing have resulted in a wide body of pre-clinical research. As will be reviewed herein, many of these studies have reported remarkable consistencies with the pathophysiology of various human neuropsychiatric conditions, including mood and anxiety disorders, addiction, and schizophrenia.

Effects of neurodevelopmental cannabinoid exposure on emotional processing circuits: Addiction implications

The mesocorticolimbic system serves as a central processor of emotional information in the mammalian brain and disruptions to the normal maturation of this system, particularly during vulnerable periods of adolescent neurodevelopment, is believed to set the stage for a wide variety of later neuropsychiatric risks, including anxiety and mood disturbances, addictive behaviors and schizophrenia. Given the central role of DA transmission in the processing of emotionally salient information and reward and aversion-related learning and memory,[42,43] many studies have examined how adolescent exposure to cannabinoids, in particular, THC, may cause long-lasting alterations in baseline activity states of the brains DA system, in the context of mood and anxiety-related behaviors, addiction and schizophrenia-like behavioral, molecular, and neuronal endophenotypes.

DA transmission is fundamentally involved in the addictive properties of most drugs of abuse. While the acute rewarding effects of many dependence-producing drugs, including opioids, can occur via DA-independent mechanisms,[44] continued and chronic exposure to many drug classes, including opioids, nicotine, and cannabinoids,[19,44–48] is known to sensitize the brains DA pathways, leading to pathological amplification of drug-related salience and increased vulnerability to relapse following withdrawal.[49,50] Given the reported ability of adolescent cannabinoid exposure to lead to long-lasting increases in spontaneous mesolimbic DA activity states and associated VTA DA neuron hyper-sensitization,[47,48] an interesting question is whether the ability of THC exposure to hyper-sensitize the brain's DA pathways may in turn increase vulnerability to subsequent drug dependence and drug seeking behaviors. Several lines of research have addressed this important question.

In terms of adolescent cannabis use, the gateway drug hypothesis has suggested that chronic cannabis use during adolescent brain development may increase the likelihood of using and developing dependence upon other, potentially "harder" drugs of abuse, such as opioids or amphetamines.[51] While there is little evidence to suggest that human cannabis exposure pre-disposes individuals to the use of other forms of dependence, several lines of pre-clinical rodent evidence has suggested that adolescent THC exposure may indeed pre-dispose the mesocorticolimbic system to increased vulnerability to drug reward effects. For example, Ellgren et al.[36] reported that chronic exposure to THC during adolescence led to heightened expression of levels of the pro-enkephalin peptide directly in the NAc. Interestingly, this alteration in the endogenous opioid signaling system corresponded to increased self-administration rates of opioids during adulthood. Similar effects were observed in adult rats following adolescent THC exposure measuring opioid reward conditioning in the CPP procedure and these effects were associated with hyperactive mesolimbic DAergic activity states in response to subsequent heroin exposure.[52] Beyond opioids, adolescent exposure to the synthetic CB1R agonist CP55,940, caused a selective potentiation in the rewarding effects of cocaine, only in female rats.[53] Thus, emerging evidence points to an important role for adolescent cannabinoid exposure in sensitizing drug reward related neural pathways to the addictive properties of various drug classes. Given the sensitizing effects of adolescent cannabinoid exposure on the mesolimbic DA and opioid receptor signaling pathways,[36,47,48] future studies should more closely examine the causal relationship between these linked phenomena. More importantly, longitudinal clinical studies are required to interrogate the potential relationships between adolescent cannabinoid use and whether such exposure is mechanistically linked to an increased vulnerability to other forms of drug dependence in later life.

Adolescent cannabinoid exposure alters dopaminergic, glutamatergic and GABAergic signaling and associated molecular pathways: Implication for schizophrenia vulnerability

Schizophrenia is a complex, uniquely human disorder that comprises a wide array of symptom profiles and is known to impact various neural circuits, including the mesocorticolimbic pathways. While the underlying neurochemical pathologies of schizophrenia involve multiple substrates, DA, glutamate (GLUT), and GABA signaling in particular have received the greatest amount of clinical and pre-clinical attention for their causal involvement in the pathophysiological sequelae involving both positive (i.e., hallucinations, paranoia, psychosis) and negative (i.e., cognitive impairments, anhedonia, social withdrawal) symptom clusters related to schizophrenia syndromes. Cannabinoids and in particular, THC, are capable of strongly modulating these neurochemical signaling pathways both acutely and following chronic exposure.[47,48,54,55] A large body of clinical and pre-clinical evidence has demonstrated that the adolescent brain is particularly susceptible to the effects of cannabinoid exposure for inducing long-term alterations in these signaling pathways, thereby setting the stage for increased risk of developing serious neuropsychiatric symptoms associated with schizophrenia in later life.[28,29,38,47,48] However, the relative importance of these cannabinoid-induced plasticity effects on DA, GLUT or GABAergic transmission in the developing human brain are not entirely understood, largely due to the intrinsic experimental limitations of measuring these long-term changes in clinical populations and problems with controlling specific cannabinoid exposure profiles using historical usage recall surveys. Preclinical rodent models have been able to provide a wealth of information regarding the neuronal and molecular mechanisms that may underlie the neurodevelopmental effects on these schizophrenia-related neurochemical pathological phenotypes.

Effects of adolescent cannabinoid exposure on long-term mesolimbic dopamine activity states

In terms of the effects of adolescent cannabinoid exposure on long-term alterations to the mesolimbic DA system, several pre-clinical studies have demonstrated that chronic exposure to THC isolates, can strongly sensitize spontaneous DA states when measured in later life. However, less is known regarding the precise neurobiological and underlying molecular mechanisms that may underlie the ability of cannabinoids to trigger these long-term adaptations in the DA system. For example, how does THC exposure ultimately lead to DAergic overdrive persisting into adulthood? Does THC exposure lead to a long-term loss of normal inhibitory mechanisms controlling VTA DA neurons, either via modulation of local inhibitory GABAergic neurons in the VTA or via removal of extrinsic inhibitory inputs to VTA DA neuronal populations? What associated molecular signaling alterations may co-occur with THC exposure during adolescent brain development leading to DAergic sensitization?

Using a 10-day adolescent escalating THC dose exposure protocol in male rats, Renard et al.[47] reported that neurodevelopmental THC exposure produced an enduring hyperactive VTA DA neuron state that persisted beyond early adulthood. This phenotype was reflected by dramatically increased spontaneous VTA DA neuron firing rates and bursting levels recorded in vivo with well-established extracellular recording procedures and neuronal classification techniques. This hyperactive DAergic phenotype occurred simultaneously with a variety of cognitive and affective behavioral abnormalities. First, THC exposed rats tested in early adulthood displayed elevated anxiety levels measured in a light-dark box anxiety test and anxiety-like thighmotaxic behaviors in the open field test. In addition, rats showed depressive-like phenotypes measured in the Porsolt Forced Swim Test (unpublished observations). THC exposed cohorts also displayed decreased social motivation and corresponding social memory deficits, consistent with schizophrenia-like endophenotypes.[56] In addition THC exposed cohorts displayed significant memory impairments demonstrated in the novel object recognition task, consistent with memory impairments present in schizophrenia[57,58] and significant deficits in sensory filtering measured in a paired-pulse inhibition (PPI) protocol, remarkably consistent with PPI deficits displayed in human schizophrenia patients.[59,60] Remarkably, the effects observed by Renard et al.[47] were entirely restricted to the effects of THC during adolescences (THC exposure during post-natal days 35–45) as cohorts of rats exposed to the identical THC regiment during early adulthood (post-natal days 65–75) displayed none of these behavioral abnormalities nor any evidence for hyperactive DAergic activity in the VTA.

Interestingly, a clinical study reported by Bloomfield et al.[55] reported that in chronic cannabis users experiencing psychosis, there was a significant reduction in DA synthesis capacity observed in the striatum. These findings suggested that chronic cannabis use may be associated with reduced striatal DA synthesis capacity and seemed inconsistent with the notion that cannabis-induced psychosis was related to the induction of a hyperactive mesolimbic DA state, as suggested by pre-clinical studies. However, it is possible that the observed decreased synthesis capacity was occurring independently of previously occurring midbrain DAergic activity and/or may have represented a compensatory down-regulation of DA function following an extended period of increased DAergic activity. In addition, this study did not focus on adolescent usage onset and thus did not take into consideration potential plastic adaptations induced by neurodevelopmental cannabinoid exposure. Alternatively, pre-clinical observations of hyperactive VTA DAergic cellular activity may represent a persistent 'adaptive' state whereby the system attempts to compensate for a loss of DA synthesis capacity by increasing spontaneous activity rates of mesolimbic DA neurons, specifically in the VTA. Interestingly, a pre-clinical study demonstrated that acute THC can directly act in the NAc (ventral striatum) to strongly potentiate DAergic neuron activity in the VTA via functional inputs from the NAc to the VTA, and inhibition of VTA GABA substrates.[26] Future studies are needed to identify the specific neuronal mechanisms within the mesolimbic pathway responsible for long-term dysregulation of DA activity states in order to bridge the mechanistic gaps between clinical and pre-clinical reports and determine precisely how dysregulation of VTA DA neuron activity states functionally relate to levels of DA receptor expression substrates and synthesis capacity in the striatum.

Effects of adolescent cannabinoid exposure on pre-frontal cortical regulation of sub-cortical dopamine activity: Impacts on molecular biomarkers for schizophrenia and other neuropsychiatric disorders

Findings demonstrating dysregulation of spontaneous activity states of DAergic neuronal populations following adolescent THC exposure represent highly informative phenotypes, particularly in the context of understanding increased vulnerability to addictive behaviors and psychiatric disorders like schizophrenia. Nevertheless, adolescent cannabinoid exposure and in particular, THC, will inevitably have effects beyond the DAergic (D2) and central CB1 receptor, impacting the downstream signaling pathways linked to the function of these receptors and their roles in intracellular homeostasis and plasticity. Importantly, cannabinoid exposure will impact multiple receptor sub-types beyond DA and CB1, including DA D1 and CB2 substrates. However, this chapter will focus selectively on pathways associated with the D2-like DA receptor and CB1, for the sake of brevity. While the entire scope of molecular signaling pathways linked to DAergic receptors and CB1 receptors is not fully understood, both receptors share common effects on downstream molecular pathways that have been linked to a variety of neuropsychiatric conditions, including addiction, schizophrenia, mood and anxiety disorders and cognitive impairments. These signaling pathways include the Wnt signaling pathway comprising glycogen-synthase kinase 3 (GSK-3) and beta-catenin; the protein kinase B (Akt) and mammalian target of rapamycin (mTOR) and Ribosomal protein S6 kinase beta-1 (S6K1), also known as p70S6 kinase (P70S6KA) pathways. A simplified summary of these main pathways and some of their established effects is presented in Fig. 3.

A comprehensive proteomic analysis of the effects of adolescent THC exposure on the expression patterns of these pathways, specifically in the rat PFC, was first reported by Renard et al.[47] In this study, tissue punches from the PFC were obtained at the conclusion of experiments performed in adulthood, meaning that the observed effects on protein expression patterns (analyzed with standard Western Blot procedures) were recorded over 30 days after the initial period of adolescent THC exposure. In addition, similar to the observed effects on DAergic activity states and behavioral endophenotypes, the effects of THC on multiple PFC signaling pathways (Fig. 3) were selective to adolescent exposure and either produced no effects or in some cases opposing effects during adulthood THC exposure.

First, examining the Wnt signaling pathway revealed significant reductions both GSK-3 and beta-catenin protein expression patterns in the PFC. Specifically, the levels of phosphorylated GSK-3 alpha and beta isoforms were dramatically reduced following THC exposure, without influencing levels of total GSK-3 proteins. This distinction is important since GSK-3 is considered a constitutively active molecule such that it is normally inactivated by

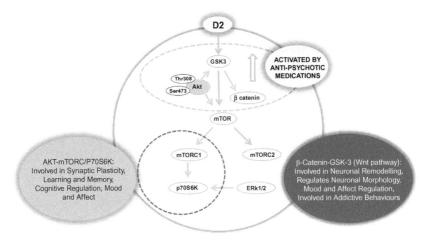

FIG. 3 Summary of the major molecular signaling pathways regulated by the DA D2R and impacted by adolescent cannabinoid exposure and neuropsychiatric phenotypes. The DA D2R regulates a variety of downstream molecular signaling pathways associated with various neurobiological and homeostatic mechanisms required for healthy brain function. Clinical and pre-clinical evidence had revealed several overlapping pathways shared with both the effects of neurodevelopmental cannabinoid exposure and molecular biomarkers linked to schizophrenia, addiction and mood and anxiety disorders. Most typical anti-psychotics produce their therapeutic effects by activation of the Wnt-GSK-3-β-catenin and Akt-mTOR signaling pathways in various neural regions, including the PFC suggesting that schizophrenia-related psychiatric phenotypes are associated with down-regulation of the same pathways affected by chronic THC exposure during adolescent brain development, which is a pre-disposing factor for developing schizophrenia in young adulthood.[28,32]

phosphorylation. Importantly, GSK-3 and Akt are functionally regulated via DA D2 receptor signaling and previous research has shown that administration of anti-psychotic D2 antagonists (e.g., raclopride) can increase phosphorylated Akt, total GSK-3, phosphorylated GSK-3 and β-catenin, whereas sub-chronic treatment with DA D2R agonists (e.g., quinpirole) produces the opposite effect.[61] This evidence provided a molecular correlate for a THC-induced hyperactive VTA DA drive to the PFC, as increased stimulation of PFC DA D2R receptors would be predicted to cause decreased GSK-3 phosphorylation. There is also confirmatory evidence from post-mortem frontal cortical tissue samples from schizophrenia patient cohorts, which have demonstrated that schizophrenia is associated with significant loss of GSK-3 protein levels.[62,63] While the interpretation of post-mortem brain tissue analyses, particularly from schizophrenia patient cohorts, is complicated by potential historical medication exposure confounds, the pre-clinical findings reported by Renard et al.[47] provided strong translational evidence for how chronic exposure to THC during adolescent neurodevelopment may trigger long-term adaptations in pre-frontal cortical GSK-3 signaling pathways that may in turn, set up the brain for increased vulnerability to schizophrenia-related molecular pathophysiology, particularly due to dysregulation of underlying DA D2 signaling mechanisms in the mesocorticolimbic circuitry. Future studies are required to examine these potential links between GSK-3-related biomarkers and underlying DAergic dysregulation in the human brain, particularly during the early, prodromal phases of schizophrenia etiology and how cannabis exposure may in turn modulate these developmental processes (Fig. 4).

Akt is a cytosolic molecule that remains inactive until a cell is stimulated and Akt translocates to the plasma membrane. Akt activation requires the phosphorylation of two major residues: threonine 308 (Thr308) in the activation loop and serine 473 (Ser473) in the C-terminal hydrophobic motif. Similar to the Wnt-GSK-3 signaling pathway, Akt is tightly regulated by DA transmission states and is a downstream target of DA receptor signaling that may be inhibited or dephosphorylated in response to pharmacological activation of specific DA receptors.[64] Interestingly, the functional modulation of Akt by DA receptor signaling is selective to the D2/D3 receptor sub-types and appears to be independent of D1/D4 signaling events.[64] Similarly, like GSK-3, Akt activation states are increased or decreased by pharmacological blockade or activation with select D2 receptor agonists/antagonists, or non-selective DA agonists like amphetamine.[61,64] Renard et al.[47] reported that adolescent THC exposure had no effect on expression levels of Akt Ser473 in the PFC. However, in stark contrast, adolescent THC exposure led to a near total abolition of Akt Thr308 levels in the PFC, to the point that loss of Akt Thr308 protein expression was so extreme that quantification of protein levels could not be performed within the linear exposure range of the expression film. Interestingly, reductions in Akt expression levels are observed in post-mortem schizophrenia tissue samples from the dorsolateral pre-frontal cortex (DLPFC),[65] suggesting that such alterations in this pathway may be a consequence of chronic DAergic signaling dysregulation, which is common to both schizophrenia and adolescent THC exposure. This dramatic and persistent Akt phenotype was consistent with the observed sub-cortical DA overdrive and again suggested that hyperactive DA signaling in the PFC following chronic adolescent THC exposure could lead to this schizophrenia-like phenotype. In addition, abnormalities in the genetic regulation of Akt Thr308 is associated with an increased vulnerability to the neuropsychiatric side effects of chronic cannabis exposure.[66] In this study, it was reported that the effect of lifetime cannabis use and associated risk of developing psychosis was significantly influenced by the genetic locus for Akt Thr308. Furthermore, subjects with this genotype and a corresponding lifetime history of cannabis use had a twofold increased risk of developing a psychotic disorder. In self-reported daily cannabis users this genotype was associated with a sevenfold increase in the likelihood of developing psychosis relative to non-carrier control subjects. Thus, modulation of Akt signaling represents a critical molecular biomarker that intersects both schizophrenia risk and the long-term effects of adolescent cannabinoid exposure during neurodevelopment. Allelic polymorphisms in Akt are also linked to cognitive

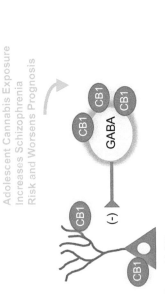

FIG. 4 Schematic summary highlighting the major neuronal and molecular adaptations observed in the frontal cortex and shared phenotypes between schizophrenia and adolescent THC exposure. A comparison of clinical and pre-clinical studies reveals several remarkable similarities between the pathophysiological correlates of schizophrenia and adolescent THC exposure in the frontal cortical regions. These similarities are reflected in shared dysregulation of GABAergic and glutamatergic balance in frontal cortical neuronal populations. In addition, multiple shared molecular abnormalities are observed when comparing post-mortem tissue samples from schizophrenia patient cohorts and the long-term changes observed in the frontal cortical regions of rats exposed to THC during adolescent neurodevelopment.

disturbances in PFC neuronal networks[67] consistent with the cognitive disturbances and loss of PFC Akt expression observed in animal models of adolescent THC exposure.[47] Finally, disturbances in normal Akt protein signaling has been shown to induce morphological abnormalities in the PFC and schizophrenia-like phenotypes in animal models,[68] consistent with effects reported in rat models of adolescent THC exposure. In addition, dysregulation of the Akt signaling pathway may relate to other schizophrenia-related neurobiological phenotypes, such as disturbances cellular neurodevelopmental differentiation or mitochondrial dysmetabolism.[69] Future studies in both clinical and pre-clinical approaches will be required to more fully elucidate the precise mechanisms underlying both the effects induced by genetic polymorphisms in the Akt signaling pathway and how these may modulate environmental exposure effects of cannabis and related impacts on DAergic transmission states.

The mTOR pathway is an important central regulator for neuronal activity states and regulation of synaptic inputs that is involved in the pathophysiology of numerous psychiatric disorders including schizophrenia, anxiety and mood disorders.[65,70,71] Renard et al.[47] reported significant reductions in mTOR phosphorylation states but no changes in total mTOR levels in the PFC of adolescent THC exposed rats. Interestingly, exposure to THC in adulthood induced the opposite pattern, with THC exposure leading to long-term upregulation of mTOR phosphorylation in the PFC. Similar to the previously described effects seen with GSK-3 and Akt signaling pathways in the PFC, clinical evidence has demonstrated that reductions in mTOR signaling levels is associated with serious neuropsychiatric disorders, including major depression and schizophrenia. For example, Inés et al.[70] analyzed DLPFC samples from schizophrenia patients and demonstrated that ribosomal protein S6, the major downstream effector of mTORC1, was significantly attenuated, suggesting upstream functional disturbances in the mTOR signaling pathway. Similarly, Chadha and Meador-Woodruff[65] reported a significant reduction in AKT and mTOR protein expression and/or phosphorylation states in DLPFC post-mortem samples from schizophrenia patients, along with attenuated protein expression levels of GβL, a sub-unit protein common to both mTORC1 and mTORC2 signaling complexes. Again, given that the rodent PFC represents a mammalian homologue of the human DLPFC, this demonstrates remarkable similarities between the effects of adolescent THC exposure in a pre-clinical rodent model, with observed phenotypes in the brains of schizophrenia patients.

Beyond schizophrenia, mTOR and the P70S6K signaling pathways have been implicated in the pathogenesis of major depression. For example, Jernigan et al.[71] reported significant reductions in levels of both mTOR and P70S6K measured in post-mortem DLPFC samples from patient cohorts diagnosed with major depression. This is relevant for adolescent cannabinoid exposure as several clinical studies have demonstrated associations between chronic cannabis exposures in adolescence with increased risk for major depression in later life. Indeed, Gobbi et al.[72] in a recent systematic review and meta-analysis comprised of 11 independent studies and 23,317 individuals, reported that adolescent cannabis exposure was correlated with significantly increased risks for major depression and suicidal behaviors later in life, independently of any pre-morbid conditions. While these studies are limited to correlative interpretations, combined with pre-clinical evidence demonstrating THC induced disturbances in frontal cortical mTOR-P70S6K signaling, this would suggest a causal mechanism linking pathophysiology of these pathways following chronic THC exposure. Future studies are required to clarify the precise mechanisms underlying these signaling relationships and in addition, to determine how disturbances in frontal cortical activity states may link with these molecular phenotypes and associated sub-cortical disturbances in DAergic activity states.

Effects of adolescent cannabinoid exposure on pre-frontal cortical GABAergic and glutamatergic functional balance: Implications for the pathophysiology of schizophrenia

Dysregulation of excitatory versus inhibitory control mechanisms in the PFC is one of the most consistently reported phenotype observed in schizophrenia. Generally, this presents as a loss of inhibitory GABAergic substrates reported in post-mortem analyses of cortical samples taken from schizophrenia patient samples. For example, many clinical studies have reported significant attenuation in the GABA synthesizing enzyme, glutamic acid decarboxylase-67 (GAD67), which in turn relates to the loss of GABAergic parvalbumin (PV) interneurons observed in frontal cortical tissue samples.[73,74] Disturbances in the expression levels of specific sub-units of the $GABA_A$ receptor are also reported in post-mortem analyses from schizophrenia samples. Specifically, these studies have reported abnormally elevated expression levels of the $GABA_A$ receptor α2 sub-unit and corresponding attenuation in the expression levels of γ2 and δ sub-units.[75,76] This dysregulation of cortical GABA function is believed to be associated with neuronal electrophysiological signatures of excitatory/inhibitory synchrony in the PFC region. Specifically, the power levels of γ-band oscillatory activity states (~40–80 Hz), typically measured with electroencephalogram (EEG) recordings, has been associated with both affective and cognitive disturbances observed in schizophrenia.[77] To this end, several studies have reported abnormally elevated g-band power in schizophrenia populations which may be associated with the presence of positive symptoms, such as hallucinations.[78–80] Interestingly, some evidence suggests that reductions in inhibitory GABA control in the PFC may be associated with dysregulation of sub-cortical mesolimbic DAergic activity states.[81,82] For example, genetic knockdown of the a3 sub-unit of the $GABA_A$ receptor led to a state of sub-cortical DAergic overdrive and associated schizophrenia-like phenotypes, remarkably similar to those observed following adolescent THC exposure.[47] As discussed previously, CB1R transmission in the PFC has been shown to acutely control the biphasic activity of sub-cortical VTA DA states and regulate DAergic control over emotional salience processing and associative memory formation (Fig. 2; Ref. 21). However, the precise mechanisms by which loss of GABAergic control in the PFC and associated sub-cortical dysregulation of VTA DAergic activity states may be linked to adolescent THC exposure are not well understood.

To address this important question, Renard et al.[48] performed a series of molecular, electrophysiological and behavioral studies in rats to determine how adolescent THC exposure may modulate this PFC-VTA pathway via regulation of PFC excitatory/inhibitory balance dynamics. Using a 10-day escalating THC dose during post-natal days 35–44 (described in Ref. 47) and beginning a series of behavioral, molecular, and electrophysiological analyses within the PFC-VTA circuitry starting at young adulthood (PND 75), they examined the effects of adolescent THC exposure on several molecular GABA markers in tissue samples from the PFC. GAD is an enzyme that catalyzes the decarboxylation of glutamate to GABA and CO_2. In the mammalian brain, GAD exists at two molecular weights, 67 kDA (GAD-67) and 65 kDA (GAD-65). GAD-67 produces GABA for neuronal processes not involved directly in neurotransmission, including regulation of synaptogenesis and various neuroprotective functions. In contrast, GAD-65 in required for the synthesis of GABA involved in neurotransmission and thus primarily localized in cell terminals and at synaptic locations. In schizophrenia, most studies find a selective loss in GAD-67, but not GAD-65.[83] This distinction suggests that GAD-67 is importantly involved in regulating the synaptogenesis of GABAergic synapses and that a loss in this function may lead to decreased inter-neuronal transmission of inhibitory signals. Such an effect would be consistent with the various $GABA_A$ receptor sub-units that are also found deficient in cortical regions from schizophrenia patient samples.[75,76]

Using Western Blot analyses of PFC tissue samples, Renard et al.[48] reported a selective loss of GAD-67 following adolescent THC exposure, a phenotype persisting into adulthood. In contrast, they observed no loss in GAD-65 nor in levels of parvalbumin, a neuronal marker for cortical interneurons, suggesting no loss in the total number of inhibitory cortical interneurons, but rather, a selective loss in GAD-67 that was remarkably similar to phenotypes observed in human schizophrenia samples. Importantly, several other biomarkers were indicative of a loss of inhibitory/excitatory cortical balance following adolescent THC exposure. First, in vivo recordings of PFC pyramidal neurons revealed a dramatically elevated level of spontaneous firing rates and in bursting neurons, rates of bursting events were strongly elevated. This effect was particularly interesting given that acute administration of CB1R agonists have been shown to strongly potentiate associative bursting levels directly in PFC pyramidal neurons,[16,19] suggesting that adolescent THC exposure may lead to similar neuronal cortical phenotypes associated with emotional processing dysregulation. In addition, Renard et al.[48] reported abnormally elevated power levels of the high γ-band wave (61–80 Hz) suggesting strong desynchronization of

GABAergic and glutamatergic activity and consistent with the spontaneously elevated levels of firing and bursting activity in single-unit recordings of PFC pyramidal neurons. Once again, this represented yet another consistent biomarker linking adolescent THC exposure to well established schizophrenia biomarkers in the frontal cortex.

In addition, THC exposed rats displayed behavioral abnormalities consistent with schizophrenia-related endophenotypes and previously reported findings using this same model of adolescent THC exposure.[47] THC exposed rats showed a hyperactive sub-cortical DA drive characterized by hyperactive VTA DA neuron firing rates and bursting activity. Behaviorally, THC exposed cohorts displayed significant memory impairments, social motivation and social cognition deficits and increased anxiety levels when examined in early adulthood. Given the previously described GABAergic abnormalities, Renard et al.[48] directly examined the potential role of $GABA_A$ receptor transmission in these observed effects by performing direct microinfusions of the selective $GABA_A$ receptor agonist, muscimol, into the PFC region. Remarkably, they found that intra-PFC stimulation of $GABA_A$ receptor transmission was able to completely reverse THC induced behavioral deficits when tested in adulthood. In addition, stimulation of $GABA_A$ transmission in the PFC reversed hyperactive VTA DA neuron firing states, demonstrating that the long-term neurodevelopmental effects of adolescent THC exposure could be reversed with a simple pharmacological manipulation of the PFC $GABA_A$ receptor system, all the way into adulthood.

Thus, there is a remarkable convergence between the observed dysregulation of excitatory/inhibitory functional balance in the frontal cortex in schizophrenia and the phenotypes observed in pre-clinical rodent models of adolescent THC exposure. Important questions remain. Notably, it will be critical to determine the mechanisms by which the simultaneous dysregulation of sub-cortical DAergic activity states may correspond, both causally and temporally with dysregulation of GABAergic and glutamatergic substrates in the frontal cortex. This is particularly important given the interconnectedness of these regions and the importance of regulatory control over these cortical-sub-cortical connections during critical periods of adolescent brain development. The pre-clinical demonstration that the long-term neuronal and behavioral disturbances induced by adolescent THC exposure may be reversible through pharmacological intervention in the PFC has important translational implications for the potential development of GABAergic pharmacotherapies that may prevent or reverse the long-term pathophysiological sequelae associated with adolescent cannabinoid exposure.

Conclusions and future directions

The findings discussed in this chapter collectively demonstrate a strong and growing convergence between well-established clinical biomarkers linked to long-term, adolescent cannabis exposure and their shared substrates and mechanisms with the pathophysiology of schizophrenia. Future research is required to more mechanistically link the basic, pre-clinical neurobiological mechanisms caused by adolescent THC exposure to shared pathways in the human brain. In addition, it will be critical to improve our understanding of the temporal sequence of events that occur in the human brain in terms of how specific pathologies may develop and at what point in the drug exposure timeline do these plastic changes become pathological, setting up the brain for future neuropsychiatric disorders. Finally, perhaps the most pressing question facing clinical and pre-clinical researchers in these shared domains is: *what is it about the adolescent brain that makes it so exquisitely sensitive to extrinsic cannabinoid exposure?* Indeed, pre-clinical research has thus far suggested that cannabinoid exposure in the mature adult brain poses little risk,[47] at least in terms of long-term vulnerability to mental health symptoms. Thus, what specific neuronal, molecular, and plastic events are taking place during adolescence that particularly pre-dispose the mammalian brain to the damaging effects of THC exposure? Equally important, how might these vulnerabilities differentially impact the male versus female brain during periods of adolescence? Answers to these important questions will lead to more effective public health policy and potentially to the identification of new pharmacotherapeutic targets aimed at preventing or reversing the pathophysiological effects of chronic cannabinoid exposure during windows of adolescent brain development. In addition, the continued identification of molecular and genetic biomarkers associated with vulnerability to the neurodevelopmental effects of cannabis exposure will provide tools to identify at risk individuals and allow for more informed choices for cannabis consumption.

References

1. Murray RM, Englund A, Abi-Dargham A, Lewis DA, Di Forti M, Davies C, et al. Cannabis-associated psychosis: neural substrate and clinical impact. *Neuropharmacology* 2017;**124**:89–104.
2. Bell DS. The experimental reproduction of amphetamine psychosis. *Arch Gen Psychiatry* 1973;**29**:35–40.
3. Bow EW, Rimoldi JM. The structure-function relationships of classical cannabinoids: CB1/CB2 modulation. *Perspectives Med Chem* 2016;**8**:17–39.
4. Cheer JF, Kendall DA, Mason R, Marsden CA. Differential cannabinoid-induced electrophysiological effects in rat ventral tegmentum. *Neuropharmacology* 2003;**44**:633–41.
5. Cheer JF, Wassum KM, Heien ML, Phillips PE, Wightman RM. Cannabinoids enhance subsecond dopamine release in the nucleus accumbens of awake rats. *J Neurosci* 2004;**24**:4393–400.
6. Diana M, Melis M, Gessa GL. Increase in meso-prefrontal dopaminergic activity after stimulation of CB1 receptors by cannabinoids. *Eur J Neurosci* 1998;**10**:2825–30.
7. Brisch R, Saniotis A, Wolf R, Bielau H, Bernstein HG, Steiner J, et al. The role of dopamine in schizophrenia from a neurobiological and evolutionary perspective: old fashioned, but still in vogue. *Front Psychiatry* 2014;**5**:47.

8. Kesby J, Eyles D, McGrath J, et al. Dopamine, psychosis and schizophrenia: the widening gap between basic and clinical neuroscience. *Transl Psychiatr* 2018;**8**:30.
9. Abel E.L. (1980) "Cannabis in the Ancient World". Marihuana: the first twelve thousand years. New York City: Plenum Publishers. ISBN 978-0-306-40496-2.
10. Laviolette SR. Dopamine modulation of emotional processing in cortical and subcortical neural circuits: evidence for a final common pathway in schizophrenia? *Schizophr Bull* 2007;**33**(4):971–81.
11. Oertel V, Kraft D, Alves G, Knöchel C, Ghinea D, Storchak H, et al. Associative memory impairments are associated with functional alterations within the memory network in schizophrenia patients and their unaffected first-degree relatives: an fMRI study. *Front Psychiatr* 2019;**10**:33.
12. Renard J, Rushlow WJ, Laviolette SR. What can rats tell us about adolescent cannabis exposure? Insights from preclinical research. *Can J Psychiatr* 2016;**61**:328–34.
13. Dalton VS, Long LE, Weickert CS, Zavitsanou K. Paranoid schizophrenia is characterized by increased CB1 receptor binding in the dorsolateral prefrontal cortex. *Neuropsychopharmacology* 2011;**36**:1620–30.
14. Newell KA, Deng C, Huang XF. Increased cannabinoid receptor density in the posterior cingulate cortex in schizophrenia. *Exp Brain Res* 2006;**172**:556–60.
15. Desfossés J, Stip E, Bentaleb LA, Potvin S. Endocannabinoids and schizophrenia. *Pharmaceuticals* 2010;**3**:3101–26.
16. Laviolette SR, Grace AA. Cannabinoids potentiate emotional learning plasticity in neurons of the medial prefrontal cortex through basolateral amygdala inputs. *J Neurosci* 2006;**26**:6458–68.
17. Tan H, Lauzon NM, Bishop SF, Chi N, Bechard M, Laviolette SR. Cannabinoid transmission in the basolateral amygdala modulates fear memory formation via functional inputs to the prelimbic cortex. *J Neurosci* 2011;**31**:5300–12.
18. Brancato A, Cavallaro A, Lavanco G, Plescia F, Cannizzaro C. Reward-related limbic memory and stimulation of the cannabinoid system: an upgrade in value attribution? *J Psychopharmacol* 2018;**32**:204–14.
19. Tan H, Bishop SF, Lauzon NM, Sun N, Laviolette SR. Chronic nicotine exposure switches the functional role of mesolimbic dopamine transmission in the processing of nicotine's rewarding and aversive effects. *Neuropharmacology* 2009;**56**:741–51.
20. Ahmad T, Lauzon NM, de Jaeger X, Laviolette SR. Cannabinoid transmission in the prelimbic cortex bidirectionally controls opiate reward and aversion signaling through dissociable kappa versus μ-opiate receptor dependent mechanisms. *J Neurosci* 2013;**33**:15642–51.
21. Draycott B, Loureiro M, Ahmad T, Tan H, Zunder J, Laviolette SR. Cannabinoid transmission in the prefrontal cortex bi-phasically controls emotional memory formation via functional interactions with the ventral tegmental area. *J Neurosci* 2014;**34**:13096–109.
22. Renard J, Rushlow WJ, Laviolette SR. Effects of adolescent THC exposure on the prefrontal GABAergic system: implications for schizophrenia-related psychopathology. *Front Psychiatr* 2018;**9**:281.
23. Renard J, Rushlow WJ, Laviolette SR. Effects of adolescent THC exposure on the prefrontal GABAergic system: implications for schizophrenia-related psychopathology. *Front Psychiatr* 2018;**9**:281.
24. Floresco SB, West AR, Ash B, Moore H, Grace AA. Afferent modulation of dopamine neuron firing differentially regulates tonic and phasic dopamine transmission. *Nat Neurosci* 2003;**6**:968–73.
25. Grace AA. Phasic versus tonic dopamine release and the modulation of dopamine system responsivity: a hypothesis for the etiology of schizophrenia. *Neuroscience* 1991;**41**:1–24.

26. Fitoussi A, Zunder J, Tan H, Laviolette SR. Delta-9-tetrahydrocannabinol potentiates fear memory salience through functional modulation of mesolimbic dopaminergic activity states. *Eur J Neurosci* 2018;**47**:1385–400.
27. Borgan F, Laurikainen H, Veronese M, et al. In vivo availability of cannabinoid 1 receptor levels in patients with first-episode psychosis. *JAMA Psychiatry* 2019;**76**:1074–84.
28. Andreasson S, Allebeck P, Engstrom A, Rydberg U. Cannabis and schizophrenia. A longitudinal study of Swedish conscripts. *Lancet* 1987;**2**:1483–6.
29. Arseneault L, Cannon M, Poulton R, Murray R, Caspi A, Moffitt TE. Cannabis use in adolescence and risk for adult psychosis: longitudinal prospective study. *BMJ* 2002;**325**:1212–3.
30. Moore TH, Zammit S, Lingford-Hughes A, Barnes TR, Jones PB, Burke M, et al. Cannabis use and risk of psychotic or affective mental health outcomes: a systematic review. *Lancet* 2007;**370**:319–28.
31. Di Forti M, Quattrone D, Freeman TP, Tripoli G, Gayer-Anderson C, Quigley H, et al. The contribution of cannabis use to variation in the incidence of psychotic disorder across Europe (EU-GEI): a multicentre case-control study. *Lancet Psychiatry* 2019;**6**:427–36.
32. Murray RM, Quigley H, Quattrone D, Englund A, Di Forti M. Traditional marijuana, high-potency cannabis and synthetic cannabinoids: increasing risk for psychosis. *World Psychiatr* 2016;**15**:195–204.
33. Freels TG, Baxter-Potter LN, Lugo JM, Glodosky NC, Wright HR, Baglot SL, et al. Vaporized cannabis extracts have reinforcing properties and support conditioned drug-seeking behavior in rats. *J Neurosci* 2020;**40**(9):1897–908.
34. Smoker MP, Mackie K, Lapish CC, Boehm SL. Self-administration of edible Δ^9-tetrahydrocannabinol and associated behavioral effects in mice. *Drug Alcohol Depend* 2019;**199**:106–15.
35. Spear LP. The adolescent brain and age-related behavioral manifestations. *Neurosci Biobehav Rev* 2000;**24**:417–63.
36. Ellgren M, Spano SM, Hurd YL. Adolescent cannabis exposure alters opiate intake and opioid limbic neuronal populations in adult rats. *Neuropsychopharmacology* 2007;**32**:607–15.
37. Rubino T, Viganò D, Realini N, et al. Chronic delta-9-tetrahydrocannabinol during adolescence provokes sex-dependent changes in the emotional profile in adult rats: behavioral and biochemical correlates. *Neuropsychopharmacology* 2008;**33**:2760–71.
38. Macías-Triana L, Romero-Cordero K, Tatum-Kuri A, Vera-Barrón A, Millán-Aldaco D, Arankowsky-Sandoval G, et al. Exposure to the cannabinoid agonist WIN 55, 212-2 in adolescent rats causes sleep alterations that persist until adulthood. *Eur J Pharmacol* 2020;**874**, 172911.
39. Biscaia M, Marin S, Fernandez B, et al. Chronic treatment with CP 55,940 during the peri-adolescent period differentially affects the behavioural responses of male and female rats in adulthood. *Psychopharmacology* 2003;**170**:301–8.
40. Renard J, Krebs MO, Le Pen G, Jay TM. Long-term consequences of adolescent cannabinoid exposure in adult psychopathology. *Front Neurosci* 2014;**8**:361.
41. Pertwee RG. Pharmacology of cannabinoid receptor ligands. *Curr Med Chem* 1999;**6**:635–64.
42. Lauzon NM, Laviolette SR. Dopamine D4-receptor modulation of cortical neuronal network activity and emotional processing: implications for neuropsychiatric disorders. *Behav Brain Res* 2010;**208**:12–22.
43. Wise RA, Robble MA. Dopamine and addiction. *Annu Rev Psychol* 2020;**71**:79–106.
44. Laviolette SR, Gallegos RA, Henriksen SJ, van der Kooy D. Opiate state controls bi-directional reward signaling via GABAA receptors in the ventral tegmental area. *Nat Neurosci* 2004;**7**:160–9.

45. Grieder TE, George O, Tan H, George SR, Le Foll B, Laviolette SR, et al. Phasic D1 and tonic D2 dopamine receptor signaling double dissociate the motivational effects of acute nicotine and chronic nicotine withdrawal. *Proc Natl Acad Sci U S A* 2012;**109**:3101–6.
46. Laviolette SR, van der Kooy D. GABA$_A$ receptors signal bidirectional reward transmission from the ventral tegmental area to the tegmental pedunculopontine nucleus as a function of opiate state. *Eur J Neurosci* 2004;**20**:2179–87.
47. Renard J, Rosen LG, Loureiro M, De Oliveira C, Schmid S, Rushlow WJ, et al. Adolescent cannabinoid exposure induces a persistent sub-cortical hyper-dopaminergic state and associated molecular adaptations in the prefrontal cortex. *Cereb Cortex* 2017;**27**(2):1297–310. https://doi.org/10.1093/cercor/bhv335.
48. Renard J, Szkudlarek HJ, Kramar CP, Jobson CEL, Moura K, Rushlow WJ, et al. Adolescent THC exposure causes enduring prefrontal cortical disruption of GABAergic inhibition and dysregulation of sub-cortical dopamine function. *Sci Rep* 2017;**7**:11420.
49. Robinson TE, Berridge KC. The neural basis of drug craving: an incentive-sensitization theory of addiction. *Brain Res Brain Res Rev* 1993;**18**:247–91.
50. Vanderschuren LJ, Pierce RC. Sensitization processes in drug addiction. *Curr Top Behav Neurosci* 2010;**3**:179–95.
51. Kandel D. Stages in adolescent involvement in drug use. *Science* 1975;**190**:912–4.
52. Cadoni C, Simola N, Espa E, Fenu S, Di Chiara G. Strain dependence of adolescent *Cannabis* influence on heroin reward and mesolimbic dopamine transmission in adult Lewis and Fischer 344 rats. *Addict Biol* 2015;**20**:132–42.
53. Higuera-Matas A, Soto-Montenegro ML, del Olmo N, et al. Augmented acquisition of cocaine self-administration and altered brain glucose metabolism in adult female but not male rats exposed to a cannabinoid agonist during adolescence. *Neuropsychopharmacology* 2008;**33**:806–13.
54. Bloomfield MA, Ashok AH, Volkow ND, Howes OD. The effects of Δ^9-tetrahydrocannabinol on the dopamine system. *Nature* 2016;**539**:369–77.
55. Bloomfield MA, Morgan CJ, Egerton A, Kapur S, Curran HV, Howes OD. Dopaminergic function in cannabis users and its relationship to cannabis-induced psychotic symptoms. *Biol Psychiatry* 2014;**75**:470–8.
56. Harvey PO, Lepage M. Neural correlates of recognition memory of social information in people with schizophrenia. *J Psychiatr Neurosci* 2014;**39**:97–109.
57. Guo JY, Ragland JD, Carter CS. Memory and cognition in schizophrenia. *Mol Psychiatr* 2019;**24**:633–42.
58. Tek C, Gold J, Blaxton T, Wilk C, McMahon RP, Buchanan RW. Visual perceptual and working memory impairments in schizophrenia. *Arch Gen Psychiatr* 2002;**59**:146–53.
59. Geyer MA, Krebs-Thomson K, Braff DL, Swerdlow NR. Pharmacological studies of prepulse inhibition models of sensorimotor gating deficits in schizophrenia: a decade in review. *Psychopharmacology (Berlin)* 2001;**156**:117–54.
60. Mena A, Ruiz-Salas JC, Puentes A, Dorado I, Ruiz-Veguilla M, De la Casa LG. Reduced prepulse inhibition as a biomarker of schizophrenia. *Front Behav Neurosci* 2016;**10**:202.
61. Sutton LP, Rushlow WJ. The dopamine D2 receptor regulates Akt and GSK-3 via Dvl-3. *Int J Neuropsychopharmacol* 2012;**15**:965–79.
62. Beasley C, Cotter D, Khan N, Pollard C, Sheppard P, Varndell I, et al. Glycogen synthase kinase-3-beta immunoreactivity is reduced in the prefrontal cortex in schizophrenia. *Neurosci Lett* 2001;**302**:117–20.
63. Kozlovsky N, Belmaker RH, Agam G. Low GSK-3 activity in frontal cortex of schizophrenic patients. *Schizophr Res* 2001;**52**:101–5.

64. Beaulieu JM, Tirotta E, Sotnikova TD, Masri B, Salahpour A, Gainetdinov RR, et al. Regulation of Akt signaling by D2 and D3 dopamine receptors in vivo. *J Neurosci* 2007;**27**:881–5.
65. Chadha R, Meador-Woodruff JH. Downregulated AKT-mTOR signaling pathway proteins in dorsolateral prefrontal cortex in schizophrenia. *Neuropsychopharmacology* 2020;**45**:1059–67.
66. Di Forti M, Iyegbe C, Sallis H, Kolliakou A, Falcone MA, Paparelli A, et al. Confirmation that the AKT1 (rs2494732) genotype influences the risk of psychosis in cannabis users. *Biol Psychiatry* 2012;**72**:811–6.
67. Pietilainen OP, Paunio T, Loukola A, Tuulio-Henriksson A, Kieseppa T, Thompson P, et al. Association of AKT1 with verbal learning, verbal memory, and regional cortical gray matter density in twins. *Am J Med Genet B Neuropsychiatr Genet* 2009;**150b**:683–92.
68. Lai WS, Xu B, Westphal KG, Paterlini M, Olivier B, Pavlidis P, et al. Akt1 deficiency affects neuronal morphology and predisposes to abnormalities in prefrontal cortex functioning. *Proc Natl Acad Sci U S A* 2006;**103**:16906–11.
69. Kwon B, Gamache T, Lee HK, Querfurth HW. Synergistic effects of β-amyloid and ceramide-induced insulin resistance on mitochondrial metabolism in neuronal cells. *Biochim Biophys Acta* 2015;**1852**:1810–23.
70. Inés I-L, Rebeca D-A, Benito M, Javier MJ, Callado LF, Leyre U. Ribosomal protein S6 hypofunction in postmortem human brain links mTORC1-dependent signaling and schizophrenia. *Frontiers Pharmacol* 2020;**11**:344.
71. Jernigan CS, Goswami DB, Austin MC, Iyo AH, Chandran A, Stockmeier CA, et al. The mTOR signaling pathway in the prefrontal cortex is compromised in major depressive disorder. *Prog Neuropsychopharmacol Biol Psychiatry* 2011;**35**:1774–9.
72. Gobbi G, Atkin T, Zytynski T, Wang S, Askari S, Boruff J, et al. Association of *Cannabis* use in adolescence and risk of depression, anxiety, and suicidality in young adulthood: a systematic review and meta-analysis. *JAMA Psychiatr* 2019;**76**:426–34.
73. Straub RE, Lipska BK, Egan MF, Goldberg TE, Callicott JH, Mayhew MB, et al. Weinberger DR Allelic variation in GAD1 (GAD67) is associated with schizophrenia and influences cortical function and gene expression. *Mol Psychiatr* 2007;**12**:854–69.
74. Woo TU, Whitehead RE, Melchitzky DS, Lewis DA. A subclass of prefrontal gamma-aminobutyric acid axon terminals are selectively altered in schizophrenia. *Proc Natl Acad Sci U S A* 1998;**95**:5341–6.
75. Huntsman MM, Tran BV, Potkin SG, Bunney Jr WE, Jones EG. Altered ratios of alternatively spliced long and short gamma2 subunit mRNAs of the gamma-amino butyrate type A receptor in prefrontal cortex of schizophrenics. *Proc Natl Acad Sci U S A* 1998;**95**:15066–71.
76. Vawter MP, Crook JM, Hyde TM, Kleinman JE, Weinberger DR, Becker KG, et al. Microarray analysis of gene expression in the prefrontal cortex in schizophrenia: a preliminary study. *Schizophr Res* 2002;**58**(1):11–20.
77. Aoki F, Fetz EE, Shupe L, Lettich E, Ojemann GA. Increased gamma-range activity in human sensorimotor cortex during performance of visuomotor tasks. *Clin Neurophysiol* 1999;**110**:524–37.
78. Gordon E, Williams L, Haig AR, Wright J, Meares RA. Symptom profile and 'gamma' processing in schizophrenia. *Cogn Neuropsychiatr* 2001;**6**:7–19.
79. Lee KH, Williams LM, Breakspear M, Gordon E. Synchronous gamma activity: a review and contribution to an integrative neuroscience model of schizophrenia. *Brain Res Brain Res Rev* 2003;**41**:57–78.
80. Sohal VS, Zhang F, Yizhar O, Deisseroth K. Parvalbumin neurons and gamma rhythms enhance cortical circuit performance. *Nature* 2009;**459**:698–702.

81. Nakazawa K, Zsiros V, Jiang Z, Nakao K, Kolata S, Zhang S, et al. GABAergic interneuron origin of schizophrenia pathophysiology. *Neuropharmacology* 2012;**62**:1574–83.
82. Yee BK, Keist R, von Boehmer L, Studer R, Benke D, Hagenbuch N, et al. A schizophrenia-related sensorimotor deficit links alpha 3-containing $GABA_A$ receptors to a dopamine hyperfunction. *Proc Natl Acad Sci U S A* 2005;**102**:17154–9.
83. Kimoto S, Bazmi HH, Lewis DA. Lower expression of glutamic acid decarboxylase 67 in the prefrontal cortex in schizophrenia: contribution of altered regulation by Zif268. *Am J Psychiatr* 2014;**171**:969–78.

Chapter 11

Molecular mechanisms underlying cannabis-induced risk of psychosis

Paula Unzueta-Larrinaga[a], Luis F. Callado[a,b], and Leyre Urigüen[a,b]
[a]Department of Pharmacology, School of Medicine, University of the Basque Country UPV-EHU and BioCruces Bizkaia Health Research Institute, Barakaldo, Bizkaia, Spain, [b]Biomedical Research Networking Center for Mental Health Network (CIBERSAM), Madrid, Spain

Introduction

Schizophrenia is a chronic, disabling, and early-onset disease with a lifetime prevalence up to 1% of the population worldwide. The disease confers substantial mortality and morbidity, and life expectancy of patients with schizophrenia is reduced by 10 to 15 years with no cure available yet. It is estimated to be the seventh most costly illness due to its early onset, frequency of hospitalizations, the need for psychosocial support and lost productivity.[1] Several hypotheses regarding etiopathogenesis of schizophrenia have been suggested but none has been consistently confirmed. Although symptomatic onset of schizophrenia occurs normally during late adolescence and early adulthood, the disorder arises from genetic[2] and/or environmental factors (i.e., maternal infection during pregnancy[3]) that are present prior to disease onset. In this context, the injuries to the brain during the neurodevelopment could be considered a key etiopathogenical mechanism that would lead to the morphofunctional dysfunctions observed in schizophrenia.[4] In this context, in the etiopathogenesis of schizophrenia, adouble-hit phenomenon has been suggested: a pre-natal priming event (i.e., maternal infection during pregnancy) that would induce vulnerability, followed by a second hit in peripuberty (i.e., drug abuse, stress…).

Cannabis consumption in early adolescence, a period of increased vulnerability to its effects, increases the risk of developing schizophrenia in vulnerable subjects.[5,6] Moreover, both cannabis extracts and Δ^9-tetrahydrocannabinol (THC) can evoke transient psychotic states in healthy subjects[7] and worsen symptoms in schizophrenia patients.[8] However, the neuronal mechanisms underlying vulnerability to develop schizophrenia after cannabis abuse during adolescence remain partially unknown.

In this context, there is a significant body of evidence supporting a role for abnormal serotonergic activity in the pathophysiology of schizophrenia.[9] The involvement of the serotonergic system in brain disorders is perhaps not surprising given its role in cortical development and the control of cognition, mood, and impulsivity. In this context, several findings suggest that serotonin 2A receptors (5-HT2AR) are involved in the molecular mechanisms underlying psychotic symptoms. Hallucinogenic and non-hallucinogenic serotonergic drugs activate the same population of 5-HT2AR in cortical pyramidal neurons, but differ in the receptor-dependent pattern of G-protein signaling and gene transcription induction that they elicit. In human and murine cortical neurons, non-hallucinogenic 5-HT2AR agonists induce the activation of Gαq/11 proteins while hallucinogenic 5-HT2AR agonists, such as ($\pm$)-DOI or LSD, also induce a Gαi/o-protein-dependent response. This signaling pattern has been proposed as a specific fingerprint of hallucinogenic pharmacological properties.[10]

In this context, it has been recently demonstrated that chronic administration of THC increases the pro-hallucinogenic signaling of 5-HT2AR in brain cortex of young mice.[11] Moreover, the Akt/mTOR/S6 pathway could mediate the THC effects to turn the signaling of 5-HT2AR from the canonical non-hallucinogenic Gαq/11-proteins activation to the hallucinogenic Gαi/o-protein-dependent response.

Schizophrenia

General aspects

Schizophrenia is a chronic and disabling mental disorder characterized by a decreased ability to perceive reality. The illness profoundly disrupts individual ability to think clearly, manage emotions, make decisions, and interact with other people. It is regarded as the most severe and disabling psychiatric disorder, affecting more than 20 million people worldwide.[12]

Lifetime prevalence of schizophrenia is estimated between 0.3% and 0.7% of the population worldwide[13] and it is associated with significant health, social, and economic concerns,[14] such as premature mortality, disproportionately high financial costs in terms of health care, loss of productivity, and social service needs. These, among other factors, make schizophrenia one of the top 20 leading causes of disease-related disability worldwide, being also the seventh most costly medical illness in our society.[15,16]

The average age of onset is in the late teens to the early 20s for men, and the late 20s to early 30s for women.[17,18]

The characteristic symptoms of schizophrenia fall into three dimensions:

Positive symptoms refer to hallucinations, delusions, disorganized speech, or disordered thinking. This cluster of symptoms is also referred as psychosis.

Negative symptoms refer to absent or diminished abilities, such as anhedonia (inability to experience pleasure from positive stimuli), alogia (decrease in verbal output or expressiveness), affective flattening (lack of facial and emotional expression), or avolition (reduction of self-initiated and purposeful acts, difficulties following through with commitments).

Cognitive symptoms include memory problems, difficulties with focus, attention and making decisions, and deficits in working memory.

Psychotic symptoms are usually episodic over time, and their emergence or worsening often requires temporary hospitalization. Meanwhile, negative and cognitive symptoms tend to be more stable over time, and contribute significantly to functional impairment. Moreover, psychotic symptoms of schizophrenia are treatment responsive, while current pharmacological treatment is practically ineffective on negative and cognitive symptoms.

The disorder usually has a gradual, insidious onset that takes place over about 5 years, beginning with the emergence of negative symptoms followed shortly by cognitive and social impairment. In this first stage of the disorder, called the prodromal period, some first signs can appear (social isolation, unusual thoughts and suspicions, change of friends, academic failure, sleep alterations, irritability, ...) (see Fig. 1). Yet these symptoms are common and non-specific, they are not diagnostic, although neither are they typical of the mentally healthy state of the individual. This period is then followed by the emergence of psychotic symptoms and first psychiatric contact.[19] This debut in positive symptoms is known as First Episode Psychosis (FEP).

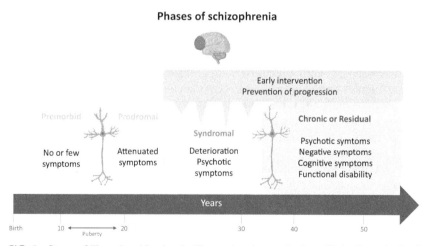

FIG. 1 Stages of illness in schizophrenia. The syndromic stage begins with the first episode of psychosis and continues through the progressive stage. After the onset of the first episode of psychosis, decline in functioning leads to the chronic effects of the disease. *(Adapted from Lieberman JA, First MB. Psychotic disorders. N Engl J Med 2018;379:270–80. https://doi.org/10.1056/NEJMra1801490.)*

Etiology and pathogenesis

What causes schizophrenia—etiology—and how it develops—pathogenesis—are questions that have occupied the minds of every psychiatric researcher for over a century. These crucial aspects have stimulated successive cycles of proof and disproof, and excited much controversy. However, a wealth of evidence has accumulated in the recent decades, leading to a virtual universal acceptance of schizophrenia as a complex neurodevelopmental disease.[20] One of the few features that are certainly accepted about schizophrenia is that it cannot be attributed solely to one factor. Instead, both genetic and environmental factors are known to take part in the onset and development of the illness.[21,22] Each of them has to combine necessarily with others in order to develop the disease, and individually accounts to a minimum extent of the disorder variance. Nevertheless, none of them is either necessary or sufficient in schizophrenia.

Current knowledge about schizophrenia has pointed to genetics as the undoubtedly main factor in the etiology of the disease. Recent studies estimate a proportion of variance explained by additive genetic effects around 80%.[23,24] This variance is explained by thousands of genetic loci with minimal prevalence differences from healthy people, most of which have not been discovered yet.

Despite this genetic association, the identification of specific molecular or structural variation has not been easy. Lack of clear *post-mortem* differences in the brains of subjects with schizophrenia was likely one of the most puzzling scenario back in time. Modification of diagnostic criteria along the years has also hampered the progress in the knowledge of the basis of the disease. Nowadays, however, the development of more sensitive techniques is demonstrating subtle but perceivable variances that offer more clues to the disorder pathophysiology.

Neuroimaging research in twins and first-degree relatives of patients has shown some heritable traits that underlie the illness. These include gray matter volume reductions in hippocampus, cortex (mainly frontal and, specifically, dorsolateral prefrontal), caudate nucleus or thalamus,[25,26] as well as ventricular enlargement,[27] and abnormalities in late brain maturation, with a disrupted trajectory of volume change with age.[28] A 18 year longitudinal study has also shown a progressive brain gray and white matter decrease, as well as cerebrospinal fluid increase in schizophrenia, being more severe during the early stages and correlating mainly with cognitive impairment.[29] Ventricles enlargement and gray matter reduction seen to continue over time, probably contributing to the pathogenesis of the chronic disease.[30]

Functional brain imaging studies point to a prefrontal and temporal defective connectivity.[31] Moreover, some *post-mortem* histological studies have reported cortical cytoarchitecture abnormalities, suggesting a defective neuronal migration during early developmental stage that could lead to an abnormal neuronal connectivity and circuitry postulated to underlie the illness.[32] Reduced spine densities,[33] smaller dendritic arbors, and reduced neuropil on the pyramidal cells of the prefrontal cortex (PFC)[20] also point to an aberrant pruning in schizophrenia.[34] Apart from brain morphological studies, neurophysiological studies have consistently disclosed alterations such as abnormal auditory P300, P50 amplitudes, or pre-pulse inhibition (PPI) deficits.[35]

According to the neurodevelopmental hypothesis (see Fig. 2), this defective neural circuitry is then vulnerable to dysfunction when unmasked by certain developmental processes, and the exposure to stressors or drugs as the individual moves through the age of risk.[20] Moreover, several neurotransmitter systems and circuits within the brain seem to be affected in patients with schizophrenia. This overall view offers only some clues about this complex disease. Knowing the molecular mechanisms underlying schizophrenia is critical for drug discovery, as well as for stratification of the patients and the improving of pharmacological treatment.

Genetics

The genetic component of schizophrenia was historically supposed due to its tendency to run in families, and schizophrenia has been shown to present higher heritability than other psychiatric diseases.[22] The studies in families, twin and adoption carried out in the 80s and 90s provided increasing proofs that genetics plays a major etiological role, and became the main foundation for the search of genetic risk factors.[36–40] However, carrying a heritable genetic vulnerability for schizophrenia is not sufficient for expressing the disease while environmental factors contribute to the multifactorial etiology of schizophrenia.[41,42]

The first large-scale common single-nucleotide polymorphisms (SNPs) association study in schizophrenia was published in 2006.[43] Since then, large genome-wide association studies (GWAS) of thousands of SNPs, demonstrated that schizophrenia is significantly associated with a substantial number of common variants of small effect size (also known as "common disease-common variant" model). Meanwhile, genome-wide copy number variations (CNV) studies have also demonstrated that some rare, highly penetrating CNVs— 22q11.2, 1q21.1 and 15q13.3 deletions, to name some—can play an important role in schizophrenia susceptibility.[44]

Later GWAS,[45,46] large-scale analyses of CNV,[47] and Next Generation Sequencing (NGS) studies[48,49] suggest the implication of a number of genes involved in several neural functions, such as synaptic plasticity, neurogenesis, and glutamatergic signaling.

The most recent GWAS in schizophrenia performed a meta-analysis of around 41,000 cases and 65,000 controls[50] and identified 145 risk loci, adding 50 new loci to the largest report. The study showed an association of brain-relevant functional gene sets involved in synaptic networks, neurogenesis and cortical development, glutamate ion channels or abnormal long-term potentiation. Some of the most replicated gene associations involve postsynaptic density (PSD) proteins, activity-regulated cytoskeleton-associated protein (Arc),

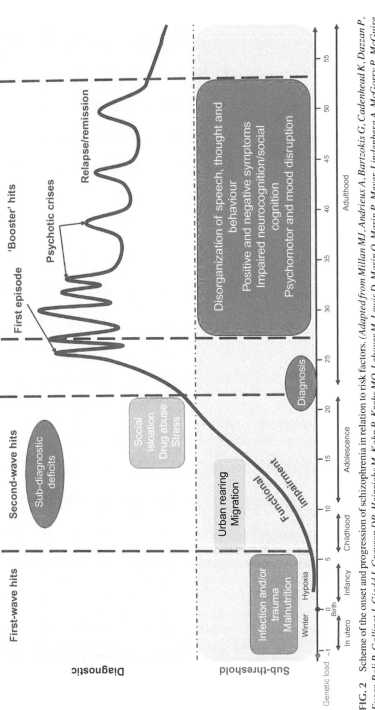

FIG. 2 Scheme of the onset and progression of schizophrenia in relation to risk factors. (Adapted from Millan MJ, Andrieux A, Bartzokis G, Cadenhead K, Dazzan P, Fusar-Poli P, Gallinat J, Giedd J, Grayson DR, Heinrichs M, Kahn R, Krebs MO, Leboyer M, Lewis D, Marin O, Marin P, Meyer-Lindenberg A, McGorry P, McGuire P, Owen MJ, Patterson P, Sawa A, Spedding M, Uhlhaas P, Vaccarino F, Wahlestedt C, Weinberger D. Altering the course of schizophrenia: progress and perspectives. Nat Rev Drug Discov 2016;**15**:485–515. https://doi.org/10.1038/nrd.2016.28.)

N-methyl-D-aspartate receptor (NMDAR), Fragile X mental retardation protein (FMRP) targets, voltage-gated calcium (Ca^{2+}) channels, and neural cell adhesion moleculNEURe 1 (NCAM1).

Neurotransmission systems

A major proportion of the studies in schizophrenia have been directed toward understanding the involvement of the different neurotransmitter systems in the pathology. Indeed, it was mainly through neuropharmacological observations using psychoactive drugs that upheld the formulation of hypotheses on the pathology of schizophrenia.[51] These hypotheses have implicated different neurotransmission systems, such as the dopaminergic, serotonergic, glutamatergic, or γ-aminobutyric acid (GABA)ergic systems in the pathology of schizophrenia (see Fig. 3).

Dopamine

The dopaminergic hypothesis has been the most enduring one to explain the etiology of schizophrenia. This should not be surprising, as since the discovery of chlorpromazine on the 50s, all the antipsychotic drugs available on the market target this system. This hypothesis is also the most thoroughly studied, and perhaps the one most commonly accepted.

Few years after the introduction of chlorpromazine, which is considered the first antipsychotic, dopamine (DA) receptor blockade was presumed to be the basis of the antipsychotic effects of this drug and the recently discovered at that time haloperidol.[52] The "antipsychotic" DA receptor,[53,54] now known as the DA D2 receptor (D2R), was confirmed to be the primary site of action for all antipsychotics, and their clinical potency was found to be highly correlated with their affinity for the receptor[55,56]. Furthermore, stimulant drugs, such as cocaine and amphetamine, which increase synaptic DA, can induce and aggravate psychotic symptoms.[57,58] These studies, among others, led to the classic DA hypothesis that maintained that schizophrenia was a result of excessive DA activity.

Whereas DA alterations are directly associated with the manifestations of the symptoms, the DA theory of schizophrenia has several flaws. The role of DA in the brain is complex, and DA alterations are probably the endpoint of a number of events involving other transmitters such as serotonin, GABA, and glutamate. Thus, investigation of the pathophysiology of schizophrenia has extended its field of inquiry beyond this system to include other neurotransmitters.

Serotonin

The serotonergic hypothesis of schizophrenia arose from early observational studies with lysergic acid diethylamide (LSD). After commercialized in 1949 for research purposes, it was evidenced that LSD and other related compounds

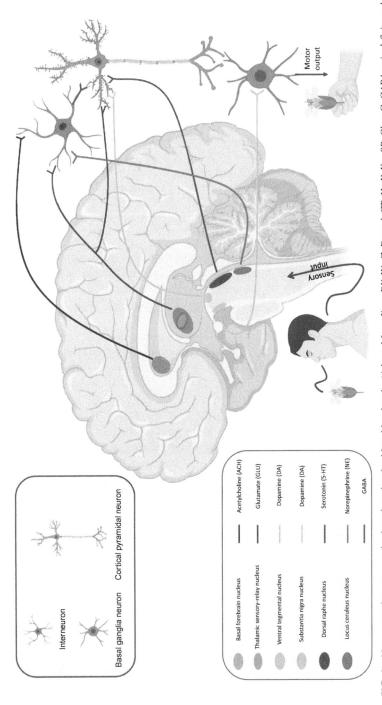

FIG. 3 Neurotransmitter systems that have been involved in schizophrenia. (Adapted from Jiang DY, Wu Z, Forsyth CT, Hu Y, Yee SP, Chen G. GABAergic deficits and schizophrenia-like behaviors in a mouse model carrying patient-derived neuroligin-2 R215H mutation. Mol Brain 2018;**11**:31. https://doi.org/10.1186/s13041-018-0375-6.)

produced mental disturbances resembling those occurring at the onset of schizophrenia. Moreover, these compounds had some chemical similarities to serotonin (5-HT),[59] thus raising what it is now known as the serotonergic hypothesis of schizophrenia. Some years later, in a study with several psychoactive agents, binding affinities for the so-called 5-HT2 receptor were found to correlate with human hallucinogenic potencies.[60] In the 90s and posterior years, it was demonstrated that LSD, psilocybin, 2,5-dimethoxy-4-iodoamphetamine (DOI), or mescaline agonism on serotonin 2A receptors (5-HT2AR) was responsible for their psychotic-like effects.[10,61–63]

In relation with the previous facts, atypical antipsychotic drugs were demonstrated to show higher affinity for 5-HT2 receptors than for D2 receptors.[64] Indeed, the atypical antipsychotic clozapine has been demonstrated to be the most effective in treatment-resistant patients with schizophrenia.[65,66]

Despite these quite consistent reports associating 5-HT system and psychosis, studies in human brains from patients with schizophrenia are quite inconclusive and the findings have not been consistently replicated.[67] Moreover, patients that have received antipsychotic medications for many years are known to confound observations, hindering even more the discerning between etiological alterations and the ones due to the course and the treatment of the disease.[68] Overall, a comprehensive model of 5-HT transmission in schizophrenia has not yet emerged and additional research is needed in order to establish its role in symptomatology and treatment opportunity.

As mentioned above, the discovery of the high affinity of some antipsychotic drugs for 5-HT2AR made this target one of the most widely studied in schizophrenia.

Glutamate

Like serotonergic theory, glutamate theory in schizophrenia also arose from causal observation of side effects. Phencyclidine (PCP) was legally manufactured and licitly used on humans as a short-acting analgesic and for general anesthesia. However, the regulation curtailed its use to animals because of post-operative side effects. First reported observation of its psychotomimetic effect, as well as of exacerbating thought disturbances in patients with schizophrenia, was in 1959.[69] Shortly after, together with ketamine, they were revealed to differ from other general anesthetics and, along with other related agents, they were classified as "dissociative anesthetics".[70]

Ketamine and PCP were classified as glutamate NMDAR non-competitive antagonists.[71,72] It was further described that at subanesthetic doses they mimic some positive, negative, and cognitive symptoms of schizophrenia in healthy people.[73] Thus, the glutamate hypothesis of schizophrenia postulated that NMDAR mediated dysfunction of neurotransmission might represent a primary deficit in the illness. Nowadays, it is well-established that administration of NMDAR antagonists can induce psychosis-like states resembling most of the symptoms seen in patients with schizophrenia.[74,75] Nevertheless, studies regarding NMDAR subunits alterations in schizophrenia have shown discrepancies. In this way, decreased mRNA and protein expression of the obligatory GRIN1 subunit of this receptor in the cortex and hippocampus is probably the most consistent finding.[76,77] However, the status of each NMDAR sub-unit in the brain of subjects with schizophrenia remains controversial.

An early report with patients with schizophrenia showed lower glutamate levels in cerebrospinal fluid, indicating that hypofunction of the glutamate neurotransmitter may contribute to the pathophysiology of this disease.[78] Glutamate dysfunction has been mainly associated with cognitive disorders.

Glutamate hypothesis is indeed closely related to DA hypothesis. Interestingly, specific GABAergic interneurons in the PFC undergo dramatic changes in NMDAR/AMPAR ratio during the adolescent period.[79] Blockade of NMDAR on these interneurons are thought to disinhibit glutamatergic neurons, increasing glutamate release and subsequently increasing DA neuron firing and release in projection targets such as the striatum and cortex.[80,81]

Although both ketamine and PCP are interesting pharmacological tools for the study of schizophrenia, it has been reported that both can bind also to D2 and 5-HT2 receptors. Thus, these models may be reproducing a non-selective neurochemical perturbation, rather than an exclusive glutamatergic alteration.[82]

GABA

Early studies on this neurotransmitter showed a reduction in thalamic and nucleus accumbens GABA content in *post-mortem* tissue of subjects with schizophrenia.[83,84] Later on, other studies suggested a reduction of GABA reuptake sites in left temporal lobe, hippocampus and amygdala,[85] and decreased GABA content and glutamate decarboxylase enzyme (GAD) activity—the enzyme that catalyzes GABA formation—in frontal and temporal areas, as well as in the putamen.[86,87]

Environmental factors

As stated before, environment unequivocally plays a role in the etiopathogenesis of schizophrenia. The general believe and more certainly proven fact is that some environmental factors can trigger the disease in people who already have a genetic predisposition, also known as "stress-vulnerability" model. This genetic vulnerability thus operates by making individuals selectively vulnerable for environmental risks (gene-environment interaction). An extensive review on the environmental influence on schizophrenia was carried out some years ago,[21] and a recent study has demonstrated the additive interaction between genetic risk for schizophrenia and several environmental exposures.[88] The environmental factors that have been associated most frequently with schizophrenia are the following:

Urbanicity

Urbanicity refers to the presence of conditions that are particular to urban areas or present more largely than in nonurban areas at a given time. Numerous studies have consistently reported an increased incidence of schizophrenia in urban areas across countries and cultures, especially when focusing on urban upbringing and showing a dose-response relationship, leading to an odds ratio (OR) of about 2,[89,90] and suggesting that the association may be causal.[91] Urbanicity seems to act synergistically with genetic liability[92] and, while studies addressing urbanicity lack of description consensus, it has been proposed that the social aspects of urbanicity may account for the major extent of the risk.[93]

Migration and minority group

Well-established evidence shows that some visible immigrant minority groups have a higher risk of developing psychotic disorders than the native-born individuals.[94] It has been proposed that the higher psychosis risk for certain migrant or ethnic minority groups is due to the chronic social adversity and discrimination.[95] This chronic social adversity results in a state of social "defeat" (chronic experience of an inferior position or social exclusion),[96] as it also applies to internal migrants and ethnic minorities without migration history, and not when living in a high own-group ethnic density area during puberty. These facts suggest that it is not the ethnic group per se that increases the risk, but rather the degree to which one stands out in relation to the wider social environment and find himself in a stressful outsider position.[97]

The association between both urbanicity and migration may indicate a common environmental influence of chronic social disadvantage and isolation[98] on schizophrenia etiopathogenesis, although further work is needed for the identification of specific mechanisms underlying the proxy risk factors.

Pre-natal stressful environmental exposures and obstetric complications

A wide variety of environmental exposures in the mother during the first trimester of pregnancy (from the death of a first degree relative to earthquakes) has been reported to increase the risk for adult psychotic outcomes.[99–101] Moreover, it has been also suggested that pre-natal and perinatal events—including maternal viral infections, pre-natal nutritional deficiency, rh incompatibility, or pregnancy and birth complications—increase the risk for psychosis. A meta-analysis of the prospective population-based studies[102] revealed that some obstetrics complications, when pooled, might account for a relatively small proportion of incidence of schizophrenia (effect sizes around 2). However, history of obstetric complications was found to be associated with increased risk of transition to schizophrenia (OR around 6) in "at risk" individuals,[103] pointing again to a gene-environment interaction. Studies addressing specifically pre-natal infections and schizophrenia have moved from ecologic designs based on epidemics in populations, to investigations based on reliable biomarkers in individual pregnancies, and have targeted specific infections—influenza, *Toxoplasma gondii*—as schizophrenia risk factors.[3]

Early childhood trauma and adversity

A meta-analysis including prospective cohort studies, cross-sectional, and case-control studies revealed that childhood adversities are associated with an increased risk for developing psychosis (OR around 2.8). When adversities were subdivided, emotional abuse was the one that increased most the risk (OR around 3.4).[104] It has been also reported that childhood abuse increases the risk for psychotic symptoms in adulthood in a dose-response fashion (increasing from 2 to 48 times more likely in people who had experienced child abuse of increasing severity).[105] An extensive review concluded that childhood abuse and neglect are related to some symptoms of schizophrenia, specifically hallucinations.[106] Moreover, childhood trauma has been related with less insight and poorer outcome in schizophrenia patients.[107] Interestingly, a study has shown a large shared effect of adversities including sexual, physical, emotional abuse, physical and emotional neglect, separation and institutionalization on the risk of psychosis, suggesting a common mechanism underlying traumatization.[108]

Cannabis use

Various lines of evidence point to associations between cannabis use and psychosis and suggest that early chronic exposure to cannabis is associated with a higher risk for psychotic outcomes, including schizophrenia in later life. This association is still a subject of study nowadays[109] and, as it supposes the main topic of this chapter, it will be thoroughly addressed in an independent section.

Comorbidity in schizophrenia

Comorbidity rates are very high in psychiatry, and most individuals who meet diagnostic criteria for one disorder meet diagnostic criteria for a second one, in some cases even for a third. Thus, when exclusion rules are bypassed, up to half of schizophrenia patients meet criteria for a co-occurring syndrome.

Given the evidence showing that symptoms are continuous rather than categorical and the overlap between psychiatric disorders, several empirical models of symptoms clustering tried to explain the comorbidity observed between closely related disorders. Clusters are usually categorized as internalizing (anxiety, depressive symptoms), externalizing (substance dependence, conduct disorder), psychotic/thought disorders (hallucinations, disorganized speech), pathological introversion… among others.[110–112] Recently, a general factor model was proposed, suggesting that there is one common liability to all forms of psychopathology and that co-variation transcends these dimensions.[113,114]

Irrespective of the theory, the truth is that there is an increased prevalence of a noteworthy amount of disorders among patients with schizophrenia compared with the general population, including anxiety disorders, depressive, and substance abuse disorders, among others. These co-morbidities occur at all phases of the course of illness and their presence is generally associated with increased severity of psychopathology and poorer outcomes, including more psychotic relapses, poor adherence to antipsychotic treatment, and greater use of health services. Moreover, they often require specific treatment and management approaches.[115]

Anxiety disorders are estimated to have a mean prevalence rate of 38% for at least one type.[116] These disorders are known to occur spontaneously, intermittently, in direct response to psychotic symptoms or even as a side effect of antipsychotic medications.[117]

Although schizophrenia is conceptualized as a non-affective psychotic syndrome, it is often associated with a variety of depressive states that are intrinsic to the illness and import a poorer outcome, including more psychotic relapses.[118] The lifetime prevalence have been determined to range from 7% to 75%, depending on the nature of the study, the diagnostic system and rating scales applied.[119] Although the heterogeneity is even bigger in depression symptoms than in anxiety, review of the studies addressing this issue indicate a modal frequency of around 25%.[120] It has been suggested that depression and schizophrenia, as well as bipolar disorder, may share some early-life risk factors, as well as some aspects of etiopathophysiology.[121,122] Moreover, evidence suggests that depression is linked to poorer outcomes in schizophrenia, such as worse quality of life, and suicide.[123]

Substance misuse is the most common co-occurring disorder in schizophrenia and worsens the course of illness. Data from the Epidemiologic Catchment Area (ECA) Study in United States in the 90s described that patients with schizophrenia are 4.6 times more likely to have a concomitant substance use disorder (excluding nicotine and caffeine) than the rest of the population. With a lifetime prevalence of 47%, odds for alcohol disorders are up to 3 times as high, and for other drug disorders they are 6 times as high.[124] The prevalence of smoking in schizophrenia is especially higher than among general population (88% versus 25%–30%), even among psychiatric patients in general (52%).[125] The specific comorbidity of cannabis use disorders in schizophrenia will be addressed in a separate section of this chapter. Despite the high co-occurring rates, patients with comorbid substance use disorders are often excluded from clinical trials, a fact that limits the generalization of results and ignores the potential effects of the intervention on substance use.

All of these co-morbidities are not only common, but also important determinants of the patient's degree of suffering, disability, and even survival. Understanding this relationship is complicated, although the available evidence in genetics is providing an increasingly clearer picture of the boundaries of the schizophrenia spectrum.[126,127]

Cannabis and schizophrenia

Historical overview

Back in the early 19th century, the French psychiatrist Jacques-Joseph Moreau provided the first systematic work on the effects of cannabis intoxication, on his monography "*Du hachisch et de l'aliénation mentale: études psychologiques, 1845.*" On that novel work, Moreau tried to model and understand mental illness using cannabis.[128] Since then, many scientists are still making efforts for elucidating the relationship between cannabis and psychosis.

Back in the 70s and 80s, several case reports started to suggest that cannabis could be a complicating factor in schizophrenia.[129–131] In addition, several studies reporting psychotic episodes triggered by cannabis consumption in non-psychiatric subjects started to appear on scientific literature.[132,133]

In 1987, it was described for the first time that cannabis could play a role in the onset of schizophrenia in a larger longitudinal study.[134] This 15-year follow up cohort-study of Swedish conscripts found that heavy use of marijuana increased the risk of schizophrenia later in life by up to six times in a dose-dependent manner. However, the majority of patients with schizophrenia never consumed cannabis, and only 3% of heavy users developed schizophrenia. These observations evidenced the already known fact that cannabis use is neither necessary nor sufficient to cause schizophrenia, and suggested that cannabis might play a role only in a subgroup of individuals with a preexisting vulnerability to schizophrenia. The Swedish conscript survey is, to date, the longest follow-up of psychotic patients with data on cannabis use prior to incidence of psychosis.

To date, there is rational evidence that in those patients with psychosis, continued use of cannabis is associated with more positive symptoms, more frequent and earlier relapses, and poorer outcome.[135–138]

Cannabis use as a risk factor for schizophrenia

Adolescence and cannabis use

The use of narcotic substances, and in particular the use of cannabis often begins during adolescence and is increasingly pervasive among adolescents today, even more common than cigarette smoking. However, why is cannabis so harmful during adolescence?

Scientific evidence shows that in adolescence the brain is still in the maturation process (neuronal maturation, myelination, synaptic pruning, dendritic plasticity, volumetric growth)[139] and that heavy cannabis consumption during adolescence appears to induce subtle changes in brain circuits such as thinning of cortices in temporal and frontal regions[140] and reduced volumes of orbitofrontal gyri.[141] Furthermore, early disease development and the increased psychopathology are also related to an early onset of consumption.[5]

The consumption of cannabis during adolescence or pre-adolescence is associated with an altered adolescent neurodevelopment, shifting the brain's developmental trajectory toward a disease-vulnerable state. Thus, the most disastrous consequence of cannabis in adolescence is the increased risk of developing psychotic events or schizophrenia later in life. An increasing evidence suggest a link between the use of cannabis and schizophrenia.[142] These epidemiological data suggest the existence of a relationship between cannabis use and the development of schizophrenia since the risk of developing the disease increases with the frequency of use.[143] Using cannabis has been also associated with an earlier age at the first psychotic event.[144] This psychotic potential of cannabis has been shown in healthy subjects with no prior history of exposure to cannabis. However, although cannabis abuse is associated with an increased risk of developing a psychotic disorder, most cannabis users do not develop such a disorder, suggesting that other factors are also involved in whether or not the disease develops.

This is because the endocannabinoid system (ECS) plays a crucial role in the neurodevelopmental processes occurring during this period, both cannabinoid receptors and endocannabinoids themselves appear at early stages of brain development, and appear to be involved in the synaptogenesis of neural circuits during development. During early phases of neuronal development, endocannabinoid signaling is integral for an array of processes including the proliferation and differentiation of progenitor cells, neuronal migration, axonal guidance, fasciculation, positioning of cortical interneurons, neurite outgrowth, and morphogenesis. In addition to being critically involved in neural development during pre-natal and early post-natal life, there is growing evidence that the ECS continues to undergo functional development during adolescence, and may play a role in regulating neurogenesis into adulthood.[145]

The maturation of the PFC is one of the most important processes during adolescence and it is possible that cannabis use in this critical period of development predominantly affects the consolidation processes of certain neurocircuits in this brain region.[146,147] It has been demonstrated the relationship between heavy use of cannabis in adolescence with a reduction in prefrontal volumes.[148] Thus, the PFC is a key region that not only regulates impulsivity, but is also the anatomical basis of the majority of higher cognitive processes. During this period, key processes in the remodeling and functional consolidation of the different brain regions take place. The consumption during this period of any exogenous toxic substance (such as cannabis, but also other drugs of abuse such as alcohol, amphetamines, or cocaine), can alter the normal maturation processes of the brain, producing alterations that may hinder the normal functioning of this key organ in the future which will increase the risk of psychopathology. In support of this hypothesis, there is evidence that alterations in the ECS are associated with a number of mental disorders, including schizophrenia and depression.[149] Other studies conducted with adolescent cannabis users have provided evidence that heavy or regular use is associated with a range of cognitive deficits, including impairments in attention,[150] learning and memory[151] and response perseveration.[152]

Association between cannabis and schizophrenia

Cannabis may represent one of the most potentially modifiable risk factors for the development of schizophrenia, so establishing the nature of this association and accurately estimating its magnitude still suppose the main aim of several human studies.[153,154] Despite the numerous studies in the literature, causality requires certain scientific evidence criteria, some of which remain undemonstrated. Non-causal explanations for associations may include reverse causation—where associations actually reflect psychosis increasing risk of cannabis use—confounding factors—where other variables that increase risk of both cannabis use and psychosis lead to false associations—and bias—where problems with measurement or sample selection lead to incorrect estimates.

It is important to note that causality cannot be proven by observational studies. The unique experimental design to demonstrate causality in humans is the randomized controlled trial. In this sense, experimental support for causality, in the sense of randomized allocation to exposure, has been demonstrated only for acute outcomes such as induction of transient delusions or hallucinations or cognitive impairment following experimental cannabis use.

Observational studies Since the first study from Andreasson and colleagues, numerous others have investigated the association between cannabis and schizophrenia. A subsequent follow-up study of the same cohort[155] still found a dose-related increase in risk of psychotic symptoms and schizophrenia with previous cannabis use, even when statistically controlling other potential confounding variables, such as psychiatric symptoms at baseline. Moreover, evidence from following cohort studies from New Zealand,[5,156,157] the United States,[158] the Netherlands,[143] Germany,[159] and the United Kingdom,[160] consistently reported an increased risk of psychosis in people using cannabis. The last study from the Swedish cohort, published in 2014,[161] evidenced a more severe course and poorer prognosis in cannabis users already diagnosed with schizophrenia.

Regarding the strength of the association, these cohort studies were reviewed in a careful meta-analysis that addressed studies quality,[142] which concluded that frequent cannabis consumption was associated with a twofold risk of a psychotic outcome. A more recent meta-analysis adding two cross-sectional studies,[162,163] and two case-control studies[6,164] confirmed a dose-response relationship, where the frequency of cannabis use correlated positively with the increase of the risk for psychosis.[165]

Some reports have pointed that among patients with a first-episode psychosis, current, and daily cannabis use is more prevalent, as well as lifetime use of high potent cannabis,[6,166,167] thus providing a preliminary evidence that regular cannabis use precedes psychotic outcomes. The most robust study carried out to date[168] is a large multi-centric case-control study that confirmed the contribution of frequent use of cannabis of high potency (more than 10% of THC) to the variation of the incidence of psychotic disorders. Moreover, data were fully adjusted for age, gender and ethnicity, level of education, employment status, tobacco use, stimulants, legal highs, and hallucinogens.

Despite all this scientific evidence regarding unspecific psychotic outcomes, whether cannabis use "causes" schizophrenia has been a matter for debate for decades. The fact that is consistent is that cannabis use may precipitate schizophrenia in people who are genetically vulnerable,[169] by means of a family history of schizophrenia[170] or because expressing subtle previous psychotic symptoms.[171] Importantly, a recent study has demonstrated that adolescents with a high genetic liability to schizophrenia that used cannabis most frequently had lower cortical thickness than those who never used cannabis.[172] In this context, it has been suggested an additive interaction between genetic risk state for schizophrenia and lifetime regular cannabis use, that indicates that the etiopathogenesis of schizophrenia involves genetic underpinnings that makes individuals more sensitive to the effects of some environmental exposures such as cannabis use.[88] There is also evidence that the genetic vulnerability for schizophrenia and for cannabis use partly overlap.[173–175]

Controlled studies in humans The first pharmacological observations with individual compounds present in cannabis date from the 40s,[176] although research into the pharmacology of cannabinoids increased markedly in the mid-1960s and early 1970s. Since the synthesis of THC, a huge number of studies have put efforts in elucidating the mechanisms underlying its psychoactive effect. Randomized controlled trials provided evidence that this compound induced a range of transient behavioral and cognitive effects in mentally healthy individuals similar to those seen in schizophrenia and other endogenous psychoses.[177] They have shown that both subjects with schizophrenia and first-degree relatives are more susceptible to its psychotropic effects than healthy controls.[178–180] Moreover, frequent users of cannabis show blunted responses to the psychotomimetic and cognitive impairing effects of THC but not to its euphoric effects.[181] Importantly, this group addressed important issues regarding THC psychotomimetic properties, such as the D2R implication in the effects,[182,183] and its impact on cortical processes such as γ-band neural oscillations, sensory gating, and working memory.[184–186]

Studies evaluating binocular depth inversion illusion (BDII) test, which is a measure of impaired visual processing that occurs in some psychotic states, found that cannabis resin, nabilone, and dronabinol (a synthetic form of THC) induce BDII similar to that observed in acute paranoid schizophrenia or schizophreniform psychotic patients.[187–189] Additionally, THC has also shown to disturb P300 waves amplitude, which has been related to cognitive impairment.[190]

This experimental evidence suggests that the THC present in cannabis plant may cause biological effects that resemble certain symptoms seen in subjects with schizophrenia. However, further work is still necessary to identify the factors that place individuals at high risk for chronic cannabis consumption, as well as the biological mechanisms underlying the association between this consumption and schizophrenia.

Mechanisms underlying cannabis and psychosis relationship

Apart from epidemiological studies, growing body of literature from pharmacological, genetic, and *post-mortem* approaches suggests that the consumption of exogenous cannabinoids may be involved in the pathophysiology of psychosis and/or schizophrenia.

Before the discovery of cannabinoid receptors, early studies carried out in humans with different THC analogs[191] provided the basis for correlating their psychotomimetic potency to a valid proxy in animal models, i.e., drug discrimination.[192] Subsequent studies led to hypothesize that the reinforcing and psychotropic effects of THC in humans are mediated through its agonist effect on CB1 receptor (CB1R), although it has modest affinity and low intrinsic activity over the receptor.[193–195] The question that remains unanswered is how their activation disrupts network dynamics and information processing to make psychosis more likely.

The CB1R is one of the most abundant G protein-coupled receptors (GPCR) in the central nervous system (CNS), and it is distributed pre-synaptically, with high density across many brain regions that are relevant to neural circuitry of psychosis and schizophrenia, such as frontal cortex, basal ganglia, anterior cingulate cortex and hippocampus.[196–198]

Across all the major brain structures where endocannabinoids signaling has been explored, both glutamatergic and GABAergic terminals are direct targets for exo- and endocannabinoids, and these compounds are known to modulate these excitatory and inhibitory inputs to dopaminergic neurons. Interestingly and as previously explained, striatal GABAergic projection neurons, cortical interneurons and glutamatergic pyramidal cells are the main cell types that have been recently associated with schizophrenia.[199]

There is suggestive evidence that the sensitization of the mesolimbic dopaminergic system may be one pathway by which the repeated use of cannabis may be related to the onset of psychotic symptoms. Thus, acute CB1R activation stimulates mesolimbic and mesoprefrontal dopaminergic transmission.[200–202] Dopaminergic activity is driven by excitatory and inhibitory inputs arising from numerous afferent structures and interneurons that are filtered by local endocannabinoid signaling, and CB1R on GABAergic terminals can facilitate this activity through suppression of inhibitory input on dopaminergic neurons.[203]

Negative effects of cannabinoids in working memory and other cognitive processes have been related to a disruption of neural synchronization in the PFC.[204] Activation of CB1Rs by cannabinoids in cortical interneurons inhibits GABA release, and may suppress the control that they exert over pyramidal neurons, thereby interfering with associative functions, disrupting normal gating mechanisms and resulting in poor integration of cortical inputs.[201] Cannabinoids, indeed, have also been shown to influence glutamatergic synaptic transmission and plasticity in the PFC.[205]

As said before, all of these effects are mainly driven by cannabinoid action upon CB1R. This receptor is known to downregulate after chronic activation,[206] and cannabis dependent subjects have shown decreased brain CB1R availability.[207] However, this downregulation is thought to be reversible and return to control values after around 3 weeks of cannabis abstinence in humans and rodents, although behavioral and molecular sensitization have also been reported.[208–210] In this regard, physiological neurodevelopmental processes especially of the frontal cortex and limbic system, that occur during adolescence,[211] are thought to be more sensitive to exogenous disturbances such as cannabis use.[212] Indeed, the ECS seems to have a crucial role in these neurodevelopmental processes,[213] as cannabis has particularly deleterious effects during adolescence.[146,214]

In summary, it has been shown that exposure to cannabinoids induces several alterations in the functioning of the brain in CB1R enriched regions and in neuromodulator systems relevant to cognition. Alterations in the functionality of the ECS, such as receptor downregulation, de-/sensitization and downstream effector changes accompanying the resultant regional neuroadaptations are supposed to underlie cognitive effects in relation with chronic cannabis use. Considering that cannabis regular use is highly prevalent in the population with psychosis, investigations of the factors influencing worse outcomes and early onset are needed in order to advance the search of new treatment options.

These studies suggest that a deregulation of the ECS can indeed interact with neurotransmitter systems that are already known to underlie schizophrenia.[215–217]

Serotonin 5-HT2A receptors and cannabinoids

Cannabinoid signaling has been involved in the regulation of some serotonergic functions[218] and, specifically, cannabinoids are able to modulate the signaling of the serotonin 2A receptor (5-HT2AR) in the brain.[11,219,220] Moreover, a dysregulation of serotonergic activity has been described in mice lacking CB1Rs.[221] Reciprocally, it seems that the activation of 5-HT2ARs induces the release of endocannabinoids in the brain.[222,223] Both 5-HT2AR and CB1R are expressed in brain structures involved in the regulation of emotions, learning, and memory; such as the amygdala, cerebral cortex, and hippocampus.[196,224,225] Interestingly, it has been recently suggested that cognitive impairment induced by THC may occur through heteromers between CB1R and 5-HT2AR.[226]

Moreover, 5-HT2ARs are GPCRs whose signaling appears to be involved in the molecular mechanisms underlying psychotic symptoms.[10,227,228] On the other hand, cannabis use may induce psychotic symptoms in vulnerable subjects.[138,189] In this context, accumulating data supports the hypothesis of a functional bidirectional interaction between cannabinoids and 5-HT2ARs in the brain, which could be of high relevance in the context of schizophrenia and other mental diseases.

5-HT2A receptors

The serotonin 5-HT2A receptor subtype is a member of the rhodopsin-like or class A superfamily of GPCR. This receptor is a member of the 14 5-HT receptors, classified in seven families based on their structural, operational and transductional features. All these receptors are differentially expressed throughout the organism, partly explaining the pharmacological complexity of 5-HT action.[229] So far, it has well-established fundamental role of serotonin and its receptors in mood control, food intake, or blood pressure regulation. Moreover, serotonin receptors have been related to many diseases, such as depression, anxiety, schizophrenia, obsessive-compulsive and panic disorders, migraine, or hypertension.[230]

In the human CNS, 5-HT2AR is abundantly expressed in several brain cortical areas, especially among layers III and V, and the hypothalamus. To a lesser extent, it is also expressed in the hippocampus and striatal structures.[225] A revealing study addressing specific location of these receptors in the PFC points predominantly to a postsynaptic location in the proximal apical dendrites of pyramidal neurons,[231] where they are suggested to play a crucial role on gating mechanisms implicated in latent inhibition processes and working memory.

Intracellular signaling

The effector that has been best characterized in 5-HT2AR signaling involves the stimulation of Gαq/11 heterotrimeric G-proteins. Upon activation, it promotes PLC-mediated catalysis of the hydrolysis of phosphatidylinositol-4,5-bisphosphate (PIP2) to IP3 and DAG, thus activating protein kinase C (PKC) and elevating cytosolic Ca^{2+}.[232,233] Although this signaling was the first to be established, 5-HT2AR is now known to activate several other signaling pathways that may involve other G proteins and alternative direct coupling to the receptor. For example, it mediates the release of arachidonic acid (AA) independently of PLC, and presumably through the activation of phospholipase A2 (PLA2).[234] This AA release is mediated by a complex mechanism that involves signaling through RhoA and p38 and p42/44 MAPK, as well as ERK signaling, all of them depending on Gα proteins different from Gαq/11, such as Gα12/13 and Gαi/o protein subtypes.[235]

5-HT2AR alterations in schizophrenia

Several reports have focused on the study of 5-HT2AR in schizophrenia, since the discovery of the fact that atypical antipsychotics block this receptor with high affinity.[236] Moreover, as indicated before, 5-HT2AR mediates the psychotic-like states exerted by some drugs, such as psilocybin, LSD, or DOI, in both humans and rodents.[63,227]

The disruption of 5-HT2AR functionality in psychosis is supported by genetic studies describing several gene polymorphisms and differential epigenetic methylation in subjects with schizophrenia.[237] Animal models of schizophrenia have shown also increased 5-HT2AR density and/or functionality.[238,239] Moreover, it has been demonstrated that the active conformation of the 5-HT2AR is upregulated in the PFC of antipsychotic-free schizophrenia subjects.[240] All these data demonstrate that upregulation and/or increased functionality of 5-HT2AR could predispose to psychosis or schizophrenia. Hallucinogenic and non-hallucinogenic serotonergic drugs activate the same population of 5-HT2AR in cortical pyramidal neurons, but differ in the receptor-dependent pattern of G-protein signaling and gene transcription induction that they elicit. As it has been previously explained, non-hallucinogenic 5-HT2AR agonists induce the activation of Gαq/11 proteins while hallucinogenic 5-HT2AR agonists, also induce a Gαi/o-protein-dependent response.[10]

Interaction between 5-HT2AR and cannabinoids

The serotonergic system shares a high level of overlap with the ECS in many physiological processes. In this context, both serotonergic and cannabinoid systems regulate mood, behavior, body temperature, feeding, or sleep.[241] Accumulating evidence suggests that there are functional interactions that occur between the cannabinoid system and the serotonergic system. ECS is already known to be a critical participant in neural homeostasis, and alterations of this system may affect not only GABA and glutamate, but also 5-HT release, either directly or indirectly. Early evidence for the interaction of endocannabinoid and 5-HT systems was suggested by behavioral studies showing a high level of functional overlap.[242–244] Thus, many studies have directly examined how endocannabinoid signaling modulates the function of the 5-HT system and vice versa, through different neurochemical and behavioral paradigms.

The first observations suggesting that cannabinoids were able to alter serotonergic transmission came in the late 1990s. The endocannabinoid oleamide has been reported to induce sleep, analgesia, hypothermia and hypolocomotion in rodent models.[245,246] Further studies have shown that oleamide interferes with anandamide hydrolysis and may act as a CB1R agonist.[247] Moreover, it can enhance 5-HT affinity for 5-HT2AR and modulate the signal transduction in rat brain.[248]

Exogenous, both natural and synthetic, cannabinoids such as WIN55,212-2, CP55,940, CBD, or rimonabant differently modulate 5-HT synthesis and release depending on the brain area, the ligand, and the dose.[221,249–251] It seems that, directly or indirectly, cannabinoids of different nature may control the excitability of dorsal raphe 5-HT neurons and modulate 5-HT release as well as the function and expression of various 5-HT receptors, including 5-HT1AR and 5-HT2AR.[11,248] Moreover, these serotonergic neurons can release endocannabinoids in an activity-dependent manner, which could mediate retrograde modulation of synaptic transmission.[222,223]

However, the 5-HT2AR sub-type has not been studied specifically in the major proportion of the reports, and its role on the different effects evoked by cannabinoids is uncertain. In this context, it has been recently demonstrated that chronic administration of THC increases the pro-hallucinogenic signaling of 5-HT2AR in brain cortex of young mice.[11] As it has been previously explained, hallucinogenic and non-hallucinogenic serotonergic drugs activate the same population of 5-HT2AR in cortical pyramidal neurons, but differ in the receptor-dependent pattern of G-protein signaling and gene transcription induction that they elicit.

This is the first demonstration that chronic THC leads to supersensitive coupling of 5-HT2AR to inhibitory G-proteins whereas $G\alpha q/11$-protein signaling pathway remains unaltered. Moreover, it seems that the Akt/mTOR/S6 pathway could mediate THC effects to turn the signaling of 5-HT2AR from the canonical non-hallucinogenic $G\alpha q/11$-proteins activation to the hallucinogenic $G\alpha i/o$-protein-dependent response. Thus, it seems that the Akt/mTOR intracellular signaling pathway is an important regulator of the 5-HT2AR pro-hallucinogenic signaling conversion after chronic THC administration.

The Akt/mTOR pathway

The Akt/mTOR/S6 molecular pathway plays an important role in protein synthesis, cell-cycle progression, cell growth, and proliferation.[252,253] Akt protein is an intermediator of different cellular pathways and acts in response to different extracellular stimuli. In neurons, the activation of Akt leads to the phosphorylation of different substrates that, therefore, regulate a wide variety of processes such as neuronal development, morphogenesis, dendritic development and synaptic plasticity. Deficiency in Akt has been linked to PFC alterations and schizophrenia-like phenotypes in rodent models.[254] Moreover, Akt phosphorylates and regulates the mammalian target of rapamycin (mTOR) signaling cascade, which has been proposed to be dysfunctional in schizophrenia.[255–257] One of the downstream effectors of mTOR is S6K, that regulates the activity of the ribosomal protein S6 (rpS6) by phosphorylating it at different sites. Interestingly, ablation of both phosphorylated (Ser473)-Akt and S6 kinase (S6K), is able to alter 5-HT2AR functionality and signaling.[258] In the brain, the phosphorylation of rpS6 is induced by different physiological and pharmacological events and has become a widely used marker of neuronal activity.[259] The exact role of these phosphorylation events remains unknown, although it has been proposed that the phosphorylation of rpS6 is involved in the translational machinery for protein synthesis and in the positive regulation of global protein translation.[260] Moreover, enhanced rpS6 phosphorylation has been associated with increased levels of proteins in some models of synaptic plasticity as well as in various mouse models of neurological and neurodevelopmental disorders which display an enhanced global protein synthesis.[261]

All these data suggest that the dysfunction of Akt/mTOR/S6 pathway may contribute to an aberrant dendritic reorganization and the loss of dendritic spines in the PFC, what finally would lead to dysfunctions in synaptic connectivity.

In the context of cannabis use, polymorphisms in the AKT1 gene are known to increase cannabis-induced psychotic responses.[262] At the same time, Akt/mTOR/S6 complex seem to be involved in the memory deficits[263] induced by cannabis.

In view of all these data, it is possible to suggest that dysfunctions in the signaling of Akt/mTOR/S6 pathway lead to a dendritic and synaptic disorganization that confers vulnerability to cannabis effects for the development of schizophrenia, in part by promoting a pro-hallucinogenic signaling of 5-HT2AR.

General aspects

Akt proteins, also known as protein kinase B, are one of the most versatile kinases in the human kinome. They are serine/threonine kinases that play a key role in several crucial aspects of cell physiology, such as cell cycle and survival, cell growth and proliferation, and glucose metabolism. Thus, its signaling has been classically related to prevention of apoptosis and cancer progression.[264–266] The Akt kinase family comprises three isoforms—Akt1 (PKBα), Akt2 (PKBβ), and Akt3 (PKBγ)—encoded by three different genes, *AKT1*, *AKT2* and *AKT3*, which are located in 14q32, 19q13.1-q13.2 and 1q43-q44 respectively,[267,268] and whose expression vary among different tissues. Akt1 is ubiquitously expressed, and its deficiency has been mechanistically related to alterations in PFC and working memory deficits.[254] Akt2 is predominantly expressed in brown fat, heart, and skeletal muscle,[269] correlating with its role in diabetes and glucose metabolism. Akt3 is most abundantly expressed in the brain, and it has an important role in attaining normal brain size.[270]

Signaling of Akt is one of the main outcomes of the activation of PI3K, and mediates a plethora of intracellular pathways that can be activated in response to different extracellular stimuli including growth factors, nutrients and insulin.[271] Akt can bind to several targets by virtue of an amino-terminal pleckstrin homology (pH) domain, such as phosphatidylinositol-3,4,5-trisphosphate (PIP3) produced by PI3K or phosphoinositide-dependent protein kinase 1 (PDK1) recruited to the plasma membrane by PIP3. Several stimuli have been demonstrated to activate PI3K, including α/γ subunits of heterotrimeric G proteins coupled to GPCR. This activation leads to Akt recruitment and activation.[272] Once recruited to the plasma membrane, Akt is phosphorylated at two sites (Thr308 residue by the PDK1 and Ser473 by the mechanistic target of rapamycin (mTOR) complex 2). Once activated, Akt phosphorylates substrates distributed throughout the cell to regulate multiple cellular functions[266] (see Fig. 4).

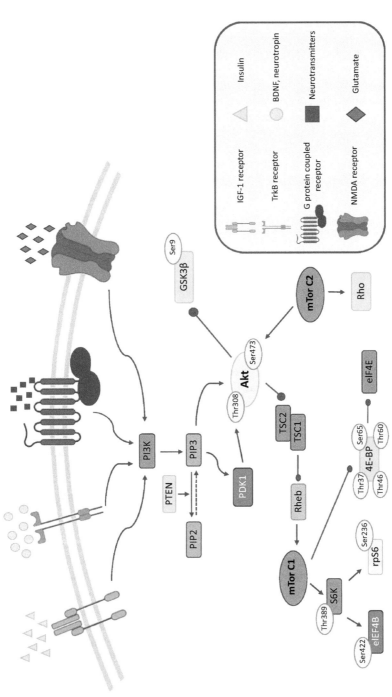

FIG. 4 Regulation of the mTOR signaling pathway in the brain. mTORC1 is activated by receptor signaling through the PI3K-Akt pathway. Through its downstream effectors the 4E-BPs and S6K1 and 2, mTORC1 controls neuronal protein synthesis. (*Adapted from Costa-Mattioli M, Monteggia LM. mTOR complexes in neurodevelopmental and neuropsychiatric disorders.* Nat Neurosci *2013;16:1537–1543. https://doi.org/10.1038/nn.3546.*)

In neurons, the activation of Akt is known to be a key outcome which regulates a wide variety of processes such as neural survival and architecture, axonal growth or synaptic strength control.[273,274] Once activated, Akt further stimulates mTOR complex 1 through inhibition of either tuberous sclerosis complex (TSC1/2), or proline-rich Akt substrate of 40 kDa (PRAS40), abolishing the tonic inhibition that they exert upon the complex (see Fig. 4). mTOR is a large (2549 amino acids, ~250 kDa), ubiquitously expressed multi-effector serine/threonine kinase that belongs to the PI3K-related kinase family and interacts with several proteins to form two distinct heteromeric complexes, mTOR complex 1 (mTORC1) and mTOR complex 2 (mTORC2). The mTOR pathway integrates various external signals and controls diverse cellular processes including translation, apoptosis, autophagy, energy metabolism and cell growth via the assembly of these multi-protein signaling complexes. Both mTOR complexes consist of numerous proteins that control mTOR signaling, dictate subcellular localization, and regulate substrate specificity; with mTORC1 having six and mTORC2 seven known protein components.[275] The mTOR-containing complexes have different sensitivities to rapamycin as well as upstream inputs and downstream outputs. While mTORC1 is acutely sensitive to rapamycin, chronic exposure to this compound also leads to mTORC2 disruption.[276]

The mTORC1 is the better characterized of the two mTOR complexes and its most remarkable feature is the number and diversity of upstream signals it senses. As previously mentioned, activation of mTORC1 is mediated mainly by Akt. By integrating inputs from major intracellular and extracellular cues, the best-characterized function of mTORC1 is the regulation of translation and protein synthesis, where it regulates two critical core components of the translation initiation machinery: p70 ribosomal protein S6 kinases (S6K) and the eukaryotic translation initiation factor 4E-binding proteins (4E-BPs). Once active, mTORC1 stimulates these two main downstream substrates, leading to the activation of ribosomal protein S6 (rpS6) and eukaryotic translation initiation factor 4E (eIF-4E), respectively, both involved in the translational machinery for protein synthesis.[277,278]

The mTOR signaling has emerged as a critical integrator of synaptic inputs that in turn affects many cellular processes, including autophagy, protein synthesis, transcription, actin dynamics, and neuronal morphology.[275] The increasing availability of mTOR inhibitors, such as rapamycin, has enabled to decipher most of what we know about mTOR function in brain disorders. So far, the mTOR pathway has been involved in several aspects of neural development and function including neuronal growth, maintenance and proliferation, dendrite and synapse formation, axonal elongation, and plasticity.[279–281] RpS6 phosphorylation is considered as a readout of mTORC1 activity, and it has been recently proposed that rpS6 could participate in the regulation of global translation in the CNS.[261]

During the last years, many studies have reported abnormalities in the expression and/or activity of its upstream and downstream components in several neurodevelopmental and neuropsychiatric disorders, such as autism spectrum disorders, drug addiction, intellectual disability, major depressive disorder, and schizophrenia.[277] In this context, mTORC1 pathway has been shown to be compromised in the PFC of patients with major depressive disorder.[282] In addition, it seems to mediate the therapeutic efficacy of the fast-acting antidepressant ketamine, enhancing the synthesis of excitatory synaptic proteins and the number of dendritic spines in the PFC.[283,284]

Akt/mTOR pathway alterations in schizophrenia

Diverse molecules that have been previously implicated in schizophrenia, such as glutamate, BDNF, or 5-HT, can lead to either over-activation or inhibition of Akt/mTOR signaling pathway. All of the knowledge that has been collected during the last years regarding the role of Akt/mTOR pathway in the CNS leds to hypothesize that alterations in mTOR pathway could have an etiopathological role in schizophrenia.[255,285,286] There are several genetic case-control and cohort studies demonstrating an association between certain SNPs—rs1130214, rs10149779, rs3730358 among others—in AKT1 gene and schizophrenia in several populations.[287–290] Other studies have proven that rs1130233 and rs1130214 AKT1 allelic variants are associated with cognitive impairments, as well as neuroanatomical and functional abnormalities in the PFC.[291] It has been also proposed that the dysfunction of mTORC1 pathway may contribute to an aberrant dendritic organization and a loss of dendritic spines that could lead to a connectivity dysfunction.[292]

Cannabis-induced 5-HT2AR pro-hallucinogenic signaling and Akt/mTOR pathway regulation

As it has been previously explained, chronic THC in young animals promotes a functional in vivo sensitization of 5-HT2AR responses to the hallucinogenic agonists. Sensorimotor gating alterations are recognized as an endophenotype of schizophrenia[293] and loss of normal pre-pulse inhibition (PPI) occurs in schizophrenia patients.[294]

Disruptions of PPI after acute (±)-DOI (an hallucinogenic 5-HT2AR agonist) administration are widely described.[295,296] In this context, it has been demonstrated a marked super-sensitivity to this disruption after chronic THC[11] administration. Moreover, the responses to (±)-DOI become increased together with an enhanced high-affinity conformation of the 5-HT2AR. Classically, the high-affinity conformation of G-protein receptors is assumed as the functionally active state and represents the receptor fraction coupled to G-proteins.[297] The chronic administration of THC in young mice selectively enhances the coupling toward inhibitory G-proteins, which is considered the pro-hallucinogenic signaling pathway of this receptor. Accordingly, an upregulation of cortical

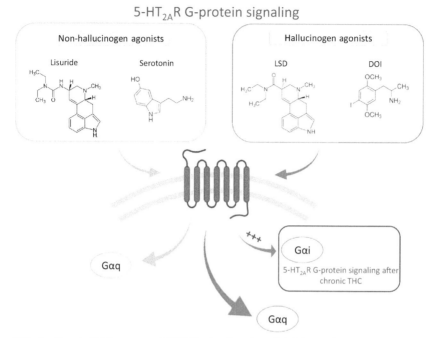

FIG. 5 Chronic THC promotes 5-HT2AR signaling through inhibitory G-proteins. Suggested model of the 5-HT2AR signaling modulation after chronic THC. *(Adapted from Ibarra-Lecue I, Mollinedo-Gajate I, Meana JJ, Callado LF, Diez-Alarcia R, Uriguen L. Chronic cannabis promotes pro-hallucinogenic signaling of 5-HT2A receptors through Akt/mTOR pathway.* Neuropsychopharmacology *2018;43(10):2028–2035. https://doi.org/10.1038/s41386-018-0076-y.)*

5-HT2AR mRNA and protein expression has been observed following a 7-day treatment with the synthetic cannabinoid CP55,940.[298]

These data provide the first evidence of a psychosis-like alteration of 5-HT2AR signaling after chronic THC administration[11] (Fig. 5). Thus, the specific alteration of agonist-induced signaling bias of 5-HT2AR observed after THC may explain the higher susceptibility to psychosis-like states. In fact, as previously explained, an increased density of the functional conformation of 5-HT2AR has been demonstrated in brain of schizophrenia patients.[240] The selective over-activation of inhibitory G-proteins by (±)-DOI after THC treatment provides a crucial aspect regarding how chronic THC exposure could increase vulnerability to psychosis. Remarkably, selective hypersensitive 5-HT2AR coupling to Gαi/o-proteins has also been demonstrated in postmortem frontal cortex of schizophrenia patients.[299] However, how THC increases the 5-HT2AR coupling to Gαi/o-proteins remains partially unknown yet. In this context, the Akt/mTOR signaling pathway seems to be involved in this regulation.

In the context of cannabis use, polymorphisms in the AKT1 gene are known to increase cannabis-induced psychotic responses. Different studies have demonstrated an association between variations in the Akt1 gene and the development of schizophrenia in cannabis users.[262,300,301] In this context, neuroimaging studies[302] have shown that healthy subjects carrying an allelic variant of the Akt1 gene present a greater psychotic response and greater striatal activation after the administration of THC. At the same time, Akt/mTOR/S6 complex seems to be involved in the memory deficits[263] induced by cannabis.

Results derived from the study carried out in mice[11] demonstrated that chronic treatment with THC leads to a hyperfunctional Akt and S6 status in brain cortex. Acute administration of THC is known to increase the phosphorylation of Akt and several downstream kinases in multiple brain areas, including cortex,[303] and the activation of this pathway has been involved in THC-induced memory impairment in animals.[304] Both short-term psychotomimetic effects of cannabis use in high-risk population and an increasing risk of being diagnosed with a psychotic disorder when having used cannabis is moderated by a polymorphism variant of AKT1 gene.[262,301] Moreover, AKT1 genotype seems to modulate the THC effect on striatal function[302] and acute psychotomimetic symptoms of cannabis use are predicted by that same variation in healthy young cannabis smokers.[305]

Apart from the THC effects on Akt kinase, sub-chronic THC is able to increase the phosphorylation of the hippocampal mTOR downstream target S6K, and produce cognitive deleterious effects in mice, whereas mTOR inhibition prevented both THC effects.[263] These results show an enduring hyperactive status of cortical Akt/mTOR pathway after chronic THC. Moreover, mTOR inhibition prevents chronic THC-induced modulation of 5-HT2AR signaling.[11] Interestingly, the mTOR downstream effector S6K2, whose target substrate is S6, has been shown to interact with 5-HT2AR and modulate its signaling by phosphorylation of specific serine residue, and seems to play a role in 5-HT2AR desensitization.[306,307] Moreover, ablation of the kinase elicits profound changes in patterns of agonist functional selectivity.[258] Neuronal ablation of phospho (Ser473)-Akt also lead to alterations in 5-HT2AR expression and functionality.[308]

Conclusions

The schizophrenia and cannabis interaction regarding both 5-HT2AR protein expression and Akt/mTOR signaling pathway status represents an interesting hypothesis that postulates the involvement of these two elements in the molecular mechanisms underlying chronic THC use during the adolescence and psychotic events that could lead to schizophrenia. Although further studies are needed to test all of the hypotheses previously mentioned in this chapter, data suggest, as a whole, that cannabis use and schizophrenia are pathologies whose underlying molecular mechanisms interact, and shed light to molecular pathways that may have a prognostic value in schizophrenia disease.

References

1. Ibrahim HM, Tamminga CA. Schizophrenia: treatment targets beyond monoamine systems. *Annu Rev Pharmacol Toxicol* 2011;**51**:189–209. https://doi.org/10.1146/annurev.pharmtox.010909.105851.
2. Owen MJ, Sawa A, Mortensen PB. Schizophrenia. *Lancet* 2016;**388**(10039):86–97. https://doi.org/10.1016/S0140-6736(15)01121-6.
3. Brown AS, Derkits EJ. Prenatal infection and schizophrenia: a review of epidemiologic and translational studies. *Am J Psychiatr* 2010;**167**:261–80. https://doi.org/10.1176/appi.ajp.2009.09030361.
4. Khandaker GM, Cousins L, Deakin J, Lennox BR, Yolken R, Jones PB. Inflammation and immunity in schizophrenia: implications for pathophysiology and treatment. *Lancet Psychiatry* 2015;**2**(3):258–70. https://doi.org/10.1016/S2215-0366(14)00122-9.
5. Arseneault L, Cannon M, Poulton R, Murray R, Caspi A, Moffitt TE. Cannabis use in adolescence and risk for adult psychosis: longitudinal prospective study. *BMJ* 2002;**325**(7374):1212–3. https://doi.org/10.1136/bmj.325.7374.1212.
6. Di Forti M, Morgan C, Dazzan P, et al. High-potency cannabis and the risk of psychosis. *Br J Psychiatry* 2009;**195**:488–91. https://doi.org/10.1192/bjp.bp.109.064220.
7. Hall W, Degenhardt L. Adverse health effects of non-medical cannabis use. *Lancet* 2009;**374**(9698):1383–91. https://doi.org/10.1016/S0140-6736(09)61037-0.
8. Foti DJ, Kotov R, Guey LT, Bromet EJ. Cannabis use and the course of schizophrenia: 10-year follow-up after first hospitalization. *Am J Psychiatry* 2010;**167**(8):987–93. https://doi.org/10.1176/appi.ajp.2010.09020189.
9. Celada P, Puig MV, Artigas F. Serotonin modulation of cortical neurons and networks. *Front Integr Neurosci* 2013;**7**:25. https://doi.org/10.3389/fnint.2013.00025.
10. Gonzalez-Maeso J, Yuen T, Ebersole BJ, et al. Transcriptome fingerprints distinguish hallucinogenic and nonhallucinogenic 5-hydroxytryptamine 2A receptor agonist effects in mouse somatosensory cortex. *J Neurosci* 2003;**23**:8836–43. https://doi.org/10.1523/JNEUROSCI.23-26-08836.2003.
11. Ibarra-Lecue I, Mollinedo-Gajate I, Meana JJ, Callado LF, Diez-Alarcia R, Uriguen L. Chronic cannabis promotes pro-hallucinogenic signaling of 5-HT2A receptors through Akt/mTOR pathway. *Neuropsychopharmacology* 2018;**43**(10):2028–35. https://doi.org/10.1038/s41386-018-0076-y.
12. Charlson FJ, Ferrari AJ, Santomauro DF, et al. Global epidemiology and burden of schizophrenia: findings from the global burden of disease study 2016. *Schizophr Bull* 2018;**44**:1195–203. https://doi.org/10.1093/schbul/sby058.
13. McGrath J, Saha S, Chant D, Welham J. Schizophrenia: a concise overview of incidence, prevalence, and mortality. *Epidemiol Rev* 2008;**30**:67–76. https://doi.org/10.1093/epirev/mxn001.
14. Whiteford HA, Degenhardt L, Rehm J, et al. Global burden of disease attributable to mental and substance use disorders: findings from the global burden of disease study 2010. *Lancet* 2013;**382**(9904):1575–86. https://doi.org/10.1016/S0140-6736(13)61611-6.
15. *Schizophrenia*; 2003.
16. James SL, Abate D, Abate KH, et al. Global, regional, and national incidence, prevalence, and years lived with disability for 354 diseases and injuries for 195 countries and territories, 1990-2017: a systematic analysis for the global burden of disease study 2017. *Lancet* 2018. https://doi.org/10.1016/S0140-6736(18)32279-7.
17. Üçok A, Gorwood P, Karadayi G. Employment and its relationship with functionality and quality of life in patients with schizophrenia: EGOFORS study. *Eur Psychiatry* 2012. https://doi.org/10.1016/j.eurpsy.2011.01.014.

18. Immonen J, et al. Age at onset and the outcomes of schizophrenia: a systematic review and meta-analysis. *Early Interv Psychiatry* 2017;**11**(6):453–60. https://doi.org/10.1111/eip.12412.
19. Häfner H, Löffler W, Maurer K, Hambrecht M, An Der Heiden W. Depression, negative symptoms, social stagnation and social decline in the early course of schizophrenia. *Acta Psychiatr Scand* 1999;**100**:105–18. https://doi.org/10.1111/j.1600-0447.1999.tb10831.x.
20. Selemon LD, Zecevic N. Schizophrenia: a tale of two critical periods for prefrontal cortical development. *Transl Psychiatry* 2015;**5**. https://doi.org/10.1038/tp.2015.115, e623.
21. van Os J, Kenis G, Rutten BPF. The environment and schizophrenia. *Nature* 2010;**468**:203–12. https://doi.org/10.1038/nature09563.
22. Sullivan PF, Daly MJ, O'Donovan M. Genetic architectures of psychiatric disorders: the emerging picture and its implications. *Nat Rev Genet* 2012;**13**:537–51. https://doi.org/10.1038/nrg3240.
23. Sullivan PF, Kendler KS, Neale MC. Schizophrenia as a complex trait. *Arch Gen Psychiatry* 2003;**60**:1187. https://doi.org/10.1001/archpsyc.60.12.1187.
24. Hilker R, Helenius D, Fagerlund B, et al. Heritability of schizophrenia and schizophrenia spectrum based on the Nationwide Danish twin register. *Biol Psychiatry* 2018;**83**:492–8. https://doi.org/10.1016/j.biopsych.2017.08.017.
25. Boos HBM, Aleman A, Cahn W, Hulshoff Pol H, Kahn RS. Brain volumes in relatives of patients with schizophrenia: a meta-analysis. *Arch Gen Psychiatry* 2007;**64**:297–304. https://doi.org/10.1001/archpsyc.64.3.297.
26. Goldman AL, Pezawas L, Mattay VS, et al. Heritability of brain morphology related to schizophrenia: a large-scale automated magnetic resonance imaging segmentation study. *Biol Psychiatry* 2008;**63**:475–83. https://doi.org/10.1016/j.biopsych.2007.06.006.
27. Kempton MJ, Stahl D, Williams SCR, DeLisi LE. Progressive lateral ventricular enlargement in schizophrenia: a meta-analysis of longitudinal MRI studies. *Schizophr Res* 2010;**120**:54–62. https://doi.org/10.1016/j.schres.2010.03.036.
28. van Haren NEM, Pol HEH, Schnack HG, et al. Progressive brain volume loss in schizophrenia over the course of the illness: evidence of maturational abnormalities in early adulthood. *Biol Psychiatry* 2008;**63**:106–13. https://doi.org/10.1016/j.biopsych.2007.01.004.
29. Andreasen NC, Nopoulos P, Magnotta V, Pierson R, Ziebell S, Ho BC. Progressive brain change in schizophrenia: a prospective longitudinal study of first-episode schizophrenia. *Biol Psychiatry* 2011;**70**:672–9. https://doi.org/10.1016/j.biopsych.2011.05.017.
30. Olabi B, Ellison-Wright I, McIntosh AM, Wood SJ, Bullmore E, Lawrie SM. Are there progressive brain changes in schizophrenia? A meta-analysis of structural magnetic resonance imaging studies. *Biol Psychiatry* 2011;**70**:88–96. https://doi.org/10.1016/j.biopsych.2011.01.032.
31. Mwansisya TE, Hu A, Li Y, et al. Task and resting-state fMRI studies in first-episode schizophrenia: a systematic review. *Schizophr Res* 2017;**189**:9–18. https://doi.org/10.1016/j.schres.2017.02.026.
32. Akbarian S, Viñuela A, Kim JJ, Potkin SG, Bunney WE, Jones EG. Distorted distribution of nicotinamide-adenine dinucleotide phosphate—diaphorase neurons in temporal lobe of schizophrenics implies anomalous cortical development. *Arch Gen Psychiatry* 1993;**50**:178–87. https://doi.org/10.1001/archpsyc.1993.01820150016002.
33. Glantz LA, Lewis DA. Decreased dendritic spine density on prefrontal cortical pyramidal neurons in schizophrenia. *Arch Gen Psychiatry* 2000;**57**:65–73. https://doi.org/10.1001/archpsyc.57.1.65.
34. McGlashan TH. Schizophrenia as a disorder of developmentally reduced synaptic connectivity. *Arch Gen Psychiatry* 2000;**57**:637–48. https://doi.org/10.1001/archpsyc.57.7.637.
35. Keshavan M, Tandon R, Boutros N, Nasrallah H. Schizophrenia, "just the facts": what we know in 2008Part 3: neurobiology. *Schizophr Res* 2008;**106**:89–107. https://doi.org/10.1016/j.schres.2008.07.020.

36. Frangos E, Athanassenas G, Tsitourides S, Katsanou N, Alexandrakou P. Prevalence of DSM III schizophrenia among the first-degree relatives of schizophrenic probands. *Acta Psychiatr Scand* 1985;**72**:382–6. https://doi.org/10.1111/j.1600-0447.1985.tb02625.x.
37. Kety SS. The significance of genetic factors in the etiology of schizophrenia: results from the national study of adoptees in Denmark. *J Psychiatr Res* 1987;**21**:423–9. https://doi.org/10.1016/0022-3956(87)90089-6.
38. Onstad S, Skre I, Torgersen S, Kringlen E. Twin concordance for DSM-III-R schizophrenia. *Acta Psychiatr Scand* 1991;**83**:395–401. https://doi.org/10.1111/j.1600-0447.1991.tb05563.x.
39. Tienari P. Interaction between genetic vulnerability and family environment: the Finnish adoptive family study of schizophrenia. *Acta Psychiatr Scand* 1991;**84**:460–5. https://doi.org/10.1111/j.1600-0447.1991.tb03178.x.
40. Kendler KS, Diehl SR. The genetics of schizophrenia: a current, genetic-epidemiologic perspective. *Schizophr Bull* 1993;**19**:261–85. https://doi.org/10.1093/schbul/19.2.261.
41. Gottesman II, Shields J. *Schizophrenia and genetics: a twin study vantage point*. New York: Academic Press, Inc.; 1972. p. 443.
42. Gottesman II. Confirming unexpressed genotypes for schizophrenia. *Arch Gen Psychiatry* 1989;**46**:867. https://doi.org/10.1001/archpsyc.1989.01810100009002.
43. Mah S, Nelson MR, DeLisi LE, et al. Identification of the semaphorin receptor PLXNA2 as a candidate for susceptibility to schizophrenia. *Mol Psychiatry* 2006;**11**:471–8. https://doi.org/10.1038/sj.mp.4001785.
44. Bassett AS, Scherer SW, Brzustowicz LM. Copy number variations in schizophrenia: critical review and new perspectives on concepts of genetics and disease. *Am J Psychiatr* 2010;**167**:899–914. https://doi.org/10.1176/appi.ajp.2009.09071016.
45. Schizophrenia Working Group of the Psychiatric Genomics Consortium. Biological insights from 108 schizophrenia-associated genetic loci. *Nature* 2014;**511**:421–7. https://doi.org/10.1038/nature13595.Biological.
46. Li Z, Chen J, Yu H, et al. Genome-wide association analysis identifies 30 new susceptibility loci for schizophrenia. *Nat Genet* 2017;**49**:1576–83. https://doi.org/10.1038/ng.3973.
47. Marshall CR, Howrigan DP, Merico D, et al. Contribution of copy number variants to schizophrenia from a genome-wide study of 41,321 subjects. *Nat Genet* 2017;**49**:27–35. https://doi.org/10.1038/ng.3725.
48. Fromer M, Pocklington AJ, Kavanagh DH, et al. De novo mutations in schizophrenia implicate synaptic networks. *Nature* 2014;**506**:179–84. https://doi.org/10.1038/nature12929.
49. Singh T, Kurki MI, Curtis D, et al. Rare loss-of-function variants in SETD1A are associated with schizophrenia and developmental disorders. *Nat Neurosci* 2016;**19**:571–7. https://doi.org/10.1038/nn.4267.
50. Pardiñas AF, Holmans P, Pocklington AJ, et al. Common schizophrenia alleles are enriched in mutation-intolerant genes and in regions under strong background selection. *Nat Genet* 2018;**50**:381–9. https://doi.org/10.1038/s41588-018-0059-2.
51. Steeds H, Carhart-Harris RL, Stone JM. Drug models of schizophrenia. *Ther Adv Psychopharmacol* 2015;**5**:43–58. https://doi.org/10.1177/2045125314557797.
52. Carlsson A, Lindqvist M. Effect of chlorpromazine or haloperidol on formation of 3-methoxytyramine and normetanephrine in mouse brain. *Acta Pharmacol Toxicol* 1963;**20**:140–4. https://doi.org/10.1111/j.1600-0773.1963.tb01730.x.
53. Seeman P, Lee T. Antipsychotic drugs: direct correlation between clinical potency and presynaptic action on dopamine neurons. *Science* 1975;**188**:1217–9. https://doi.org/10.1126/science.1145194.

54. Burt DR, Creese I, Snyder SH. Properties of [3H]haloperidol and [3H]dopamine binding associated with dopamine receptors in calf brain membranes. *Mol Pharmacol* 1976;**12**:800–12.
55. Creese I, Burt DR, Snyder SH. Dopamine receptor binding predicts clinical and pharmacological potencies of antischizophrenic drugs. *Science* 1976;**192**:481–3.
56. Seeman P, Lee T, Chau-Wong M, Wong K. Antipsychotic drug doses and neuroleptic/dopamine receptors. *Nature* 1976;**261**:717–9. https://doi.org/10.1038/261717a0.
57. Bell D. Comparison of amphetamine psychosis and schizophrenia. *Br J Psychiatry* 1965;**111**:701–7. https://doi.org/10.1192/bjp.111.477.701.
58. Angrist B, Peselov E, Rubinstein M, Wolkin A, Rotrosen J. Amphetamine response and relapse risk after depot neuroleptic discontinuation. *Psychopharmacology* 1985;**85**:277–83. https://doi.org/10.1007/BF00428187.
59. Woolley DW, Shaw E. A biochemical and pharmacological suggestion about certain mental disorders. *Proc Natl Acad Sci* 1954;**40**:228–31. https://doi.org/10.1073/pnas.40.4.228.
60. Glennon RA, Titeler M, McKenney JD. Evidence for 5-HT2 involvement in the mechanism of action of hallucinogenic agents. *Life Sci* 1984;**35**:2505–11. https://doi.org/10.1016/0024-3205(84)90436-3.
61. Marek GJ, Aghajanian GK. LSD and the phenethylamine hallucinogen DOI are potent partial agonists at 5-HT2A receptors on interneurons in rat piriform cortex. *J Pharmacol Exp Ther* 1996;**278**:1373–82.
62. Gonzalez-Maeso J, Sealfon SC. Psychedelics and schizophrenia. *Trends Neurosci* 2009;**32**:225–32. https://doi.org/10.1016/j.tins.2008.12.005.
63. Vollenweider FX, Vollenweider-Scherpenhuyzen MFI, Bäbler A, Vogel H, Hell D. Psilocybin induces schizophrenia-like psychosis in humans via a serotonin-2 agonist action. *Neuroreport* 1998;**9**:3897–902. https://doi.org/10.1097/00001756-199812010-00024.
64. Meltzer HY, Matsubara S, Lee JC. Classification of typical and atypical antipsychotic drugs on the basis of dopamine D-1, D-2 and serotonin2 pKi values. *J Pharmacol Exp Ther* 1989;**251**:238–46.
65. Kane J. Clozapine for the treatment-resistant schizophrenic. *Arch Gen Psychiatry* 1988;**45**:789. https://doi.org/10.1001/archpsyc.1988.01800330013001.
66. Davis JM, Chen N, Glick ID. A meta-analysis of the efficacy of second-generation antipsychotics. *Arch Gen Psychiatry* 2003;**60**:553–64. https://doi.org/10.1001/archpsyc.60.6.553.
67. Abi-Dargham A. Alterations of serotonin transmission in schizophrenia. *Int Rev Neurobiol* 2007;**78**:133–64.
68. Vita A, De Peri L, Deste G, Barlati S, Sacchetti E. The effect of antipsychotic treatment on cortical gray matter changes in schizophrenia: does the class matter? A meta-analysis and meta-regression of longitudinal magnetic resonance imaging studies. *Biol Psychiatry* 2015;**78**:403–12. https://doi.org/10.1016/j.biopsych.2015.02.008.
69. Luby ED. Study of a new schizophrenomimetic drug—sernyl. *Arch Neurol Psychiatr* 1959;**81**:363. https://doi.org/10.1001/archneurpsyc.1959.02340150095011.
70. Corssen G, Domino EF. Dissociative anesthesia: further pharmacologic studies and first clinical experience with the phencyclidine derivative CI-581. *Anesth Analg* 1966;**45**:29–40.
71. Anis NA, Berry SC, Burton NR, Lodge D. The dissociative anaesthetics, ketamine and phencyclidine, selectively reduce excitation of central mammalian neurones by N-methyl-aspartate. *Br J Pharmacol* 1983;**79**:565–75. https://doi.org/10.1111/j.1476-5381.1983.tb11031.x.
72. Thomson AM, West DC, Lodge D. An N-methylaspartate receptor-mediated synapse in rat cerebral cortex: a site of action of ketamine? *Nature* 1985;**313**:479–81. https://doi.org/10.1038/313479a0.

73. Krystal JH. Subanesthetic effects of the noncompetitive NMDA antagonist, ketamine, in humans. *Arch Gen Psychiatry* 1994;**51**:199. https://doi.org/10.1001/archpsyc.1994.03950030035004.
74. Newcomer J. Ketamine-induced NMDA receptor hypofunction as a model of memory impairment and psychosis. *Neuropsychopharmacology* 1999;**20**:106–18. https://doi.org/10.1016/S0893-133X(98)00067-0.
75. Stone JM, Erlandsson K, Arstad E, et al. Relationship between ketamine-induced psychotic symptoms and NMDA receptor occupancy—a [123I]CNS-1261 SPET study. *Psychopharmacology* 2008;**197**:401–8. https://doi.org/10.1007/s00213-007-1047-x.
76. Hu W, Macdonald ML, Elswick DE, Sweet RA. The glutamate hypothesis of schizophrenia: evidence from human brain tissue studies. *Ann N Y Acad Sci* 2015. https://doi.org/10.1111/nyas.12547.
77. Catts VS, Lai YL, Weickert CS, Weickert TW, Catts SV. A quantitative review of the post-mortem evidence for decreased cortical N-methyl-d-aspartate receptor expression levels in schizophrenia: how can we link molecular abnormalities to mismatch negativity deficits? *Biol Psychol* 2016;**116**:57–67. https://doi.org/10.1016/j.biopsycho.2015.10.013.
78. Kim JS, Kornhuber HH, Schmid-Burgk W, Holzmüller B. Low cerebrospinal fluid glutamate in schizophrenic patients and a new hypothesis on schizophrenia. *Neurosci Lett* 1980;**20**:379–82. https://doi.org/10.1016/0304-3940(80)90178-0.
79. Wang H-X, Gao W-J. Cell type-specific development of NMDA receptors in the interneurons of rat prefrontal cortex. *Neuropsychopharmacology* 2009;**34**:2028–40. https://doi.org/10.1038/npp.2009.20.
80. Moghaddam B, Javitt D. From revolution to evolution: the glutamate hypothesis of schizophrenia and its implication for treatment. *Neuropsychopharmacology* 2012;**37**:4–15. https://doi.org/10.1038/npp.2011.181.
81. Kokkinou M, Ashok AH, Howes OD. The effects of ketamine on dopaminergic function: meta-analysis and review of the implications for neuropsychiatric disorders. *Mol Psychiatry* 2018;**23**:59–69. https://doi.org/10.1038/mp.2017.190.
82. Kapur S, Seeman P. NMDA receptor antagonists ketamine and PCP have direct effects on the dopamine D2 and serotonin 5-HT2 receptors—implications for models of schizophrenia. *Mol Psychiatry* 2002;**7**:837–44. https://doi.org/10.1038/sj.mp.4001093.
83. Perry T, Buchanan J, Kish S, Hansen S. γ-Aminobutyric-acid deficiency in brain of schizphrenic patients. *Lancet* 1979;**313**:237–9. https://doi.org/10.1016/S0140-6736(79)90767-0.
84. Spokes EGS, Garrett NJ, Rossor MN, Iversen LL. Distribution of GABA in post-mortem brain tissue from control, psychotic and Huntington's chorea subjects. *J Neurol Sci* 1980. https://doi.org/10.1016/0022-510X(80)90103-3.
85. Simpson MDC, Slater P, Deakin JFW, Royston MC, Skan WJ. Reduced GABA uptake sites in the temporal lobe in schizophrenia. *Neurosci Lett* 1989;**107**:211–5. https://doi.org/10.1016/0304-3940(89)90819-7.
86. Simpson MDC, Slater P, Royston MC, Deakin JFW. Regionally selective deficits in uptake sites for glutamate and gamma-aminobutyric acid in the basal ganglia in schizophrenia. *Psychiatry Res* 1992;**42**:273–82. https://doi.org/10.1016/0165-1781(92)90119-N.
87. Sherman AD, Davidson AT, Baruah S, Hegwood TS, Waziri R. Evidence of glutamatergic deficiency in schizophrenia. *Neurosci Lett* 1991;**121**:77–80. https://doi.org/10.1016/0304-3940(91)90653-B.
88. Guloksuz S, Pries LK, Delespaul P, et al. Examining the independent and joint effects of molecular genetic liability and environmental exposures in schizophrenia: results from the EUGEI study. *World Psychiatry* 2019;**18**:173–82. https://doi.org/10.1002/wps.20629.

89. Pedersen CB. Evidence of a dose-response relationship between urbanicity during upbringing and schizophrenia risk. *Arch Gen Psychiatry* 2001;**58**:1039–46. https://doi.org/10.1001/archpsyc.58.11.1039.
90. Sundquist K, Frank G, Sundquist J. Urbanisation and incidence of psychosis and depression. *Br J Psychiatry* 2004;**184**:293–8. https://doi.org/10.1192/bjp.184.4.293.
91. Krabbendam L. Schizophrenia and urbanicity: a major environmental influence—conditional on genetic risk. *Schizophr Bull* 2005;**31**:795–9. https://doi.org/10.1093/schbul/sbi060.
92. van Os J, Pedersen CB, Mortensen PB. Confirmation of synergy between urbanicity and familial liability in the causation of psychosis. *Am J Psychiatr* 2004;**161**:2312–4. https://doi.org/10.1176/appi.ajp.161.12.2312.
93. March D, Hatch SL, Morgan C, et al. Psychosis and place. *Epidemiol Rev* 2008;**30**:84–100. https://doi.org/10.1093/epirev/mxn006.
94. Dykxhoorn J, Hollander AC, Lewis G, Magnusson C, Dalman C, Kirkbride JB. Risk of schizophrenia, schizoaffective, and bipolar disorders by migrant status, region of origin, and age-at-migration: a national cohort study of 1.8 million people. *Psychol Med* 2019;**49**:2354–63. https://doi.org/10.1017/S0033291718003227.
95. Morgan C, Charalambides M, Hutchinson G, Murray RM. Migration, ethnicity, and psychosis: toward a sociodevelopmental model. *Schizophr Bull* 2010;**36**:655–64. https://doi.org/10.1093/schbul/sbq051.
96. Selten J-P, Cantor-Graae E. Social defeat: risk factor for schizophrenia? *Br J Psychiatry* 2005;**187**:101–2. https://doi.org/10.1192/bjp.187.2.101.
97. van der Ven E, Selten J-P. Migrant and ethnic minority status as risk indicators for schizophrenia. *Curr Opin Psychiatry* 2018;**31**:231–6. https://doi.org/10.1097/YCO.0000000000000405.
98. McGrath J, Saha S, Welham J, El Saadi O, MacCauley C, Chant D. A systematic review of the incidence of schizophrenia: the distribution of rates and the influence of sex, urbanicity, migrant status and methodology. *BMC Med* 2004;**2**:13. https://doi.org/10.1186/1741-7015-2-13.
99. Khashan AS, Abel KM, McNamee R, et al. Higher risk of offspring schizophrenia following antenatal maternal exposure to severe adverse life events. *Arch Gen Psychiatry* 2008;**65**:146. https://doi.org/10.1001/archgenpsychiatry.2007.20.
100. Procopio M. Maternal exposure to death of a first degree relative during first trimester of pregnancy increases risk of schizophrenia in offspring. *Evid Based Ment Health* 2008;**11**:127. https://doi.org/10.1136/ebmh.11.4.127.
101. Guo C, He P, Song X, Zheng X. Long-term effects of prenatal exposure to earthquake on adult schizophrenia. *Br J Psychiatry* 2019;**15**:730–5. https://doi.org/10.1192/bjp.2019.114.
102. Cannon M, Jones PB, Murray RM. Obstetric complications and schizophrenia: historical and meta-analytic review. *Am J Psychiatr* 2002;**159**:1080–92. https://doi.org/10.1176/appi.ajp.159.7.1080.
103. Kotlicka-Antczak M, Pawełczyk A, Pawełczyk T, Strzelecki D, Żurner N, Karbownik MS. A history of obstetric complications is associated with the risk of progression from an at risk mental state to psychosis. *Schizophr Res* 2018;**197**:498–503. https://doi.org/10.1016/j.schres.2017.10.039.
104. Varese F, Smeets F, Drukker M, et al. Childhood adversities increase the risk of psychosis: a meta-analysis of patient-control, prospective- and cross-sectional cohort studies. *Schizophr Bull* 2012;**38**:661–71. https://doi.org/10.1093/schbul/sbs050.
105. Janssen I, Krabbendam L, Bak M, et al. Childhood abuse as a risk factor for psychotic experiences. *Acta Psychiatr Scand* 2004;**109**:38–45. https://doi.org/10.1046/j.0001-690X.2003.00217.x.

106. Read J, Os J, Morrison AP, Ross CA. Childhood trauma, psychosis and schizophrenia: a literature review with theoretical and clinical implications. *Acta Psychiatr Scand* 2005;**112**:330–50. https://doi.org/10.1111/j.1600-0447.2005.00634.x.
107. Pignon B, Lajnef M, Godin O, et al. Relationship between childhood trauma and level of insight in schizophrenia: a path-analysis in the national FACE-SZ dataset. *Schizophr Res* 2019;**208**:90–6. https://doi.org/10.1016/j.schres.2019.04.006.
108. Trauelsen AM, Bendall S, Jansen JE, et al. Childhood adversity specificity and dose-response effect in non-affective first-episode psychosis. *Schizophr Res* 2015;**165**:52–9. https://doi.org/10.1016/j.schres.2015.03.014.
109. Gage SH. Cannabis and psychosis: triangulating the evidence. *Lancet Psychiatry* 2019;**6**:364–5. https://doi.org/10.1016/S2215-0366(19)30086-0.
110. Cloninger CR. A systematic method for clinical description and classification of personality variants: a proposal. *Arch Gen Psychiatry* 1987;**44**:573–88. https://doi.org/10.1001/archpsyc.1987.01800180093014.
111. Markon KE. Modeling psychopathology structure: a symptom-level analysis of Axis I and II disorders. *Psychol Med* 2010;**40**:273–88. https://doi.org/10.1017/S0033291709990183.
112. Kotov R. New dimensions in the quantitative classification of mental illness. *Arch Gen Psychiatry* 2011;**68**:1003. https://doi.org/10.1001/archgenpsychiatry.2011.107.
113. Caspi A, Houts RM, Belsky H, Sandhya, et al. The p factor: one general psychopathology factor in the structure of psychiatric disorders? *Clin Psychol Sci* 2014;**2**:119–37. https://doi.org/10.1177/2167702613497473.
114. Caspi A, Moffitt TE. All for one and one for all: mental disorders in one dimension. *Am J Psychiatr* 2018;**175**:831–44. https://doi.org/10.1176/appi.ajp.2018.17121383.
115. Buonocore M, Bosia M, Bechi M, et al. Targeting anxiety to improve quality of life in patients with schizophrenia. *Eur Psychiatry* 2017;**45**:129–35. https://doi.org/10.1016/j.eurpsy.2017.06.014.
116. Achim AM, Maziade M, Raymond É, Olivier D, Mérette C, Roy MA. How prevalent are anxiety disorders in schizophrenia? A meta-analysis and critical review on a significant association. *Schizophr Bull* 2011;**37**:811–21. https://doi.org/10.1093/schbul/sbp148.
117. Aikawa S, Kobayashi H, Nemoto T, et al. Social anxiety and risk factors in patients with schizophrenia: relationship with duration of untreated psychosis. *Psychiatry Res* 2018;**263**:94–100. https://doi.org/10.1016/j.psychres.2018.02.038.
118. Conley RR, Ascher-Svanum H, Zhu B, Faries DE, Kinon BJ. The burden of depressive symptoms in the long-term treatment of patients with schizophrenia. *Schizophr Res* 2007;**90**:186–97. https://doi.org/10.1016/j.schres.2006.09.027.
119. Buckley PF, Miller BJ, Lehrer DS, Castle DJ. Psychiatric comorbidities and schizophrenia. *Schizophr Bull* 2009;**35**:383–402. https://doi.org/10.1093/schbul/sbn135.
120. Siris SG. Depression in schizophrenia: perspective in the era of "atypical" antipsychotic agents. *Am J Psychiatr* 2000;**157**:1379–89. https://doi.org/10.1176/appi.ajp.157.9.1379.
121. Smoller JW, Kendler K, Craddock N, et al. Identification of risk loci with shared effects on five major psychiatric disorders: a genome-wide analysis. *Lancet* 2013;**381**:1371–9. https://doi.org/10.1016/S0140-6736(12)62129-1.
122. Witt SH, Streit F, Jungkunz M, et al. Genome-wide association study of borderline personality disorder reveals genetic overlap with bipolar disorder, major depression and schizophrenia. *Transl Psychiatry* 2017;**7**. https://doi.org/10.1038/tp.2017.115, e1155.
123. Upthegrove R, Marwaha S, Birchwood M. Depression and schizophrenia: cause, consequence or trans-diagnostic issue? *Schizophr Bull* 2017;**43**:240–4. https://doi.org/10.1093/schbul/sbw097.

124. Regier DA. Comorbidity of mental disorders with alcohol and other drug abuse. *JAMA* 1990;**264**:2511. https://doi.org/10.1001/jama.1990.03450190043026.
125. Hughes JR, Hatsukami DK, Mitchell JE, Dahlgren LA. Prevalence of smoking among psychiatric outpatients. *Am J Psychiatr* 1986;**143**:993–7. https://doi.org/10.1176/ajp.143.8.993.
126. Cardno AG, Owen MJ. Genetic relationships between schizophrenia, bipolar disorder, and schizoaffective disorder. *Schizophr Bull* 2014;**40**:504–15. https://doi.org/10.1093/schbul/sbu016.
127. Maier RM, Visscher PM, Robinson MR, Wray NR. Embracing polygenicity: a review of methods and tools for psychiatric genetics research. *Psychol Med* 2018;**48**:1055–67. https://doi.org/10.1017/S0033291717002318.
128. Moreau J. *Du hachisch et de l'aliénation mentale: études psychologiques*; 1845. p. 431.
129. Breakey WR, Goodell H, Lorenz PC, Mchugh PR. Hallucinogenic drugs as precipitants of schizophrenia. *Psychol Med* 1974;**4**:255–61. https://doi.org/10.1017/S0033291700042938.
130. Treffert DA. Marijuana use in schizophrenia: a clear hazard. *Am J Psychiatr* 1978;**135**:1213–5. https://doi.org/10.1176/ajp.135.10.1213.
131. Knudsen P, Vilmar T. Cannabis and neuroleptic agents in schizophrenia. *Acta Psychiatr Scand* 1984;**69**:162–74. https://doi.org/10.1111/j.1600-0447.1984.tb02482.x.
132. PPålsson Å, Thulin SO, Tunving K. Cannabis psychoses in South Sweden. *Acta Psychiatr Scand* 1982;**66**:311–21. https://doi.org/10.1111/j.1600-0447.1982.tb00310.x.
133. Rottanburg D. Cannabis-associated psychosis with hypomanic features. *Lancet* 1982;**320**:1364–6. https://doi.org/10.1016/S0140-6736(82)91270-3.
134. Andréasson S, Allebeck P, Engström A, Rydberg U. Cannabis and schizophrenia. A longitudinal study of Swedish conscripts. *Lancet* 1987;**330**:1483–6. https://doi.org/10.1016/S0140-6736(87)92620-1.
135. Caspari D. Cannabis and schizophrenia: results of a follow-up study. *Eur Arch Psychiatry Clin Neurosci* 1999;**249**:45–9. https://doi.org/10.1007/s004060050064.
136. Grech A, Van Os J, Jones PB, Lewis SW, Murray RM. Cannabis use and outcome of recent onset psychosis. *Eur Psychiatry* 2005;**20**:349–53. https://doi.org/10.1016/j.eurpsy.2004.09.013.
137. Núñez C, Ochoa S, Huerta-Ramos E, et al. Cannabis use and cognitive function in first episode psychosis: differential effect of heavy use. *Psychopharmacology* 2016;**233**:809–21. https://doi.org/10.1007/s00213-015-4160-2.
138. Setién-Suero E, Neergaard K, Ortiz-García de la Foz V, et al. Stopping cannabis use benefits outcome in psychosis: findings from 10-year follow-up study in the PAFIP-cohort. *Acta Psychiatr Scand* 2019;**140**:349–59. https://doi.org/10.1111/acps.13081.
139. Malone DT, Hill MN, Rubino T. Adolescent cannabis use and psychosis: epidemiology and neurodevelopmental models. *Br J Pharmacol* 2010;**160**(3):511–22. https://doi.org/10.1111/j.1476-5381.2010.00721.x.
140. Jacobus J, Squeglia LM, Sorg SF, Nguyen-Louie TT, Tapert SF. Cortical thickness and neurocognition in adolescent marijuana and alcohol users following 28 days of monitored abstinence. *J Stud Alcohol Drugs* 2014;**75**(5):729–43. https://doi.org/10.15288/jsad.2014.75.729.
141. Filbey FM, Aslan S, Calhoun VD, et al. Long-term effects of marijuana use on the brain. *Proc Natl Acad Sci U S A* 2014;**111**(47):16913–8. https://doi.org/10.1073/pnas.1415297111.
142. Moore TH, Zammit S, Lingford-Hughes A, et al. Cannabis use and risk of psychotic or affective mental health outcomes: a systematic review. *Lancet* 2007;**370**:319–28. https://doi.org/10.1016/S0140-6736(07)61162-3.
143. van Os J. Cannabis use and psychosis: a longitudinal population-based study. *Am J Epidemiol* 2002;**156**:319–27. https://doi.org/10.1093/aje/kwf043.

144. Van Mastrigt S, Addington J, Addington D. Substance misuse at presentation to an early psychosis program. *Soc Psychiatry Psychiatr Epidemiol* 2004;**39**(1):69–72. https://doi.org/10.1007/s00127-004-0713-0.
145. Lubman DI, Cheetham A, Yucel M. Cannabis and adolescent brain development. *Pharmacol Ther* 2015;**148**:1–16. https://doi.org/10.1016/j.pharmthera.2014.11.009.
146. Rubino T, Parolaro D. The impact of exposure to cannabinoids in adolescence: insights from animal models. *Biol Psychiatry* 2016;**79**:578–85. https://doi.org/10.1016/j.biopsych.2015.07.024.
147. Renard J, Szkudlarek HJ, Kramar CP, et al. Adolescent THC exposure causes enduring prefrontal cortical disruption of gabaergic inhibition and dysregulation of sub-cortical dopamine function. *Sci Rep* 2017;**7**(1):11420. https://doi.org/10.1038/s41598-017-11645-8.
148. Churchwell JC, Lopez-Larson M, Yurgelun-Todd DA. Altered frontal cortical volume and decision making in adolescent cannabis users. *Front Psychol* 2010;**1**:225. https://doi.org/10.3389/fpsyg.2010.00225.
149. Ferretjans R, Moreira FA, Teixeira AL, Salgado JV. The endocannabinoid system and its role in schizophrenia: a systematic review of the literature. *Br J Psychiatry* 2012;**34**(suppl 2):S163–77. https://doi.org/10.1016/j.rbp.2012.07.003.
150. Harvey MA, Sellman JD, Porter RJ, Frampton CM. The relationship between non-acute adolescent cannabis use and cognition. *Drug Alcohol Rev* 2007;**26**(3):309–19. https://doi.org/10.1080/09595230701247772.
151. Medina KL, Nagel BJ, Park A, McQueeny T, Tapert SF. Depressive symptoms in adolescents: associations with white matter volume and marijuana use. *J Child Psychol Psychiatry* 2007;**48**(6):592–600. https://doi.org/10.1111/j.1469-7610.2007.01728.x.
152. Lane SD, Cherek DR, Tcheremissine OV, Steinberg JL, Sharon JL. Response perseveration and adaptation in heavy marijuana-smoking adolescents. *Addict Behav* 2007;**32**(5):977–90. https://doi.org/10.1016/j.addbeh.2006.07.007.
153. Hickman M, Vickerman P, MacLeod J, et al. If cannabis caused schizophrenia – how many cannabis users may need to be prevented in order to prevent one case of schizophrenia? England and Wales calculations. *Addiction* 2009. https://doi.org/10.1111/j.1360-0443.2009.02736.x.
154. Gage SH, Zammit S, Hickman M. Stronger evidence is needed before accepting that cannabis plays an important role in the aetiology of schizophrenia in the population. *F1000 Med Rep* 2013;**5**. https://doi.org/10.3410/M5-2.
155. Zammit S. Self reported cannabis use as a risk factor for schizophrenia in Swedish conscripts of 1969: historical cohort study. *BMJ* 2002;**325**:1199. https://doi.org/10.1136/bmj.325.7374.1199.
156. Fergusson DM, Horwood LJ, Ridder EM. Tests of causal linkages between cannabis use and psychotic symptoms. *Addiction* 2005;**100**(3):354–66. https://doi.org/10.1111/j.1360-0443.2005.01001.x.
157. Fergusson DM, Horwood LJ, Swain-Campbell NR. Cannabis dependence and psychotic symptoms in young people. *Psychol Med* 2003;**33**(1):15–21. https://doi.org/10.1017/s0033291702006402.
158. Tien AY, Anthony JC. Epidemiological analysis of alcohol and drug use as Risk factors for psychotic experiences. *J Nerv Ment Dis* 1990;**178**:473–80. https://doi.org/10.1097/00005053-199008000-00001.
159. Henquet C, Krabbendam L, Spauwen J, et al. Prospective cohort study of cannabis use, predisposition for psychosis, and psychotic symptoms in young people. *Br Med J* 2005;**330**:11–4. https://doi.org/10.1136/bmj.38267.664086.63.

160. Wiles NJ, Zammit S, Bebbington P, Singleton N, Meltzer H, Lewis G. Self-reported psychotic symptoms in the general population. *Br J Psychiatry* 2006;**188**:519–26. https://doi.org/10.1192/bjp.bp.105.012179.
161. Manrique-Garcia E, Zammit S, Dalman C, Hemmingsson T, Andreasson S, Allebeck P. Prognosis of schizophrenia in persons with and without a history of cannabis use. *Psychol Med* 2014;**44**:2513–21. https://doi.org/10.1017/S0033291714000191.
162. Miettunen J, Törmänen S, Murray GK, et al. Association of cannabis use with prodromal symptoms of psychosis in adolescence. *Br J Psychiatry* 2008;**192**:470–1. https://doi.org/10.1192/bjp.bp.107.045740.
163. McGrath J, Welham J, Scott J, et al. Association between cannabis use and psychosis-related outcomes using sibling pair analysis in a cohort of young adults. *Arch Gen Psychiatry* 2010;**67**:440. https://doi.org/10.1001/archgenpsychiatry.2010.6.
164. Di Forti M, Sallis H, Allegri F, et al. Daily use, especially of high-potency cannabis, drives the earlier onset of psychosis in cannabis users. *Schizophr Bull* 2014. https://doi.org/10.1093/schbul/sbt181.
165. Marconi A, Di Forti M, Lewis CM, Murray RM, Vassos E. Meta-analysis of the association between the level of cannabis use and risk of psychosis. *Schizophr Bull* 2016;**42**:1262–9. https://doi.org/10.1093/schbul/sbw003.
166. Larsen TK, Melle I, Auestad B, et al. Substance abuse in first-episode non-affective psychosis. *Schizophr Res* 2006;**88**:55–62. https://doi.org/10.1016/j.schres.2006.07.018.
167. Ferraro L, Murray RM, Di Forti M, et al. Premorbid adjustment and IQ in patients with first-episode psychosis: a multisite case-control study of their relationship with cannabis use. *Schizophr Res* 2019;**210**:81–8. https://doi.org/10.1016/j.schres.2019.06.004.
168. Di Forti M, Quattrone D, Freeman TP, et al. The contribution of cannabis use to variation in the incidence of psychotic disorder across Europe (EU-GEI): a multicentre case-control study. *Lancet Psychiatry* 2019;**6**:427–36. https://doi.org/10.1016/S2215-0366(19)30048-3.
169. Vaucher J, Keating BJ, Lasserre AM, et al. Cannabis use and risk of schizophrenia: a mendelian randomization study. *Mol Psychiatry* 2018;**23**:1287–92. https://doi.org/10.1038/mp.2016.252.
170. McGuire PK, Jones P, Harvey I, Williams M, McGuffin P, Murray RM. Morbid risk of schizophrenia for relatives of patients with cannabis-associated psychosis. *Schizophr Res* 1995;**15**:277–81. https://doi.org/10.1016/0920-9964(94)00053-B.
171. Verdoux H, Gindre C, Sorbara F, Tournier M, Swendsen JD. Effects of cannabis and psychosis vulnerability in daily life: an experience sampling test study. *Psychol Med* 2003;**33**:23–32. https://doi.org/10.1017/S0033291702006384.
172. French L, Gray C, Leonard G, et al. Early cannabis use, polygenic risk score for schizophrenia, and brain maturation in adolescence. *JAMA Psychiatry* 2015. https://doi.org/10.1001/jamapsychiatry.2015.1131.
173. Verweij KJH, Abdellaoui A, Nivard MG, et al. Short communication: genetic association between schizophrenia and cannabis use. *Drug Alcohol Depend* 2017;**171**:117–21. https://doi.org/10.1016/j.drugalcdep.2016.09.022.
174. Aas M, Melle I, Bettella F, et al. Psychotic patients who used cannabis frequently before illness onset have higher genetic predisposition to schizophrenia than those who did not. *Psychol Med* 2018;**48**:43–9. https://doi.org/10.1017/S0033291717001209.
175. Gurriarán X, Rodríguez-López J, Flórez G, et al. Relationships between substance abuse/dependence and psychiatric disorders based on polygenic scores. *Genes Brain Behav* 2019;**18**. https://doi.org/10.1111/gbb.12504, e12504.
176. Loewe S. Studies on the pharmacology and acute toxicity of compounds with marihuana activity. *J Pharmacol Exp Ther* 1946;**88**:154–61.

177. D'Souza DC, Perry E, MacDougall L, et al. The psychotomimetic effects of intravenous delta-9-tetrahydrocannabinol in healthy individuals: implications for psychosis. *Neuropsychopharmacology* 2004;**29**:1558–72. https://doi.org/10.1038/sj.npp.1300496.
178. D'Souza DC, Abi-Saab WM, Madonick S, et al. Delta-9-tetrahydrocannabinol effects in schizophrenia: implications for cognition, psychosis, and addiction. *Biol Psychiatry* 2005;**57**:594–608. https://doi.org/10.1016/j.biopsych.2004.12.006.
179. Veling W, Mackenbach JP, van Os J, Hoek HW. Cannabis use and genetic predisposition for schizophrenia: a case-control study. *Psychol Med* 2008;**38**:1251–6. https://doi.org/10.1017/S0033291708003474.
180. Kahn RS, Linszen DH, Van Os J, et al. Evidence that familial liability for psychosis is expressed as differential sensitivity to cannabis: an analysis of patient-sibling and sibling-control pairs. *Arch Gen Psychiatry* 2011;**68**:138–47. https://doi.org/10.1001/archgenpsychiatry.2010.132.
181. D'Souza DC, Ranganathan M, Braley G, et al. Blunted psychotomimetic and amnestic effects of Δ-9-tetrahydrocannabinol in frequent users of cannabis. *Neuropsychopharmacology* 2008;**33**:2505–16. https://doi.org/10.1038/sj.npp.1301643.
182. D'Souza DC, Braley G, Blaise R, et al. Effects of haloperidol on the behavioral, subjective, cognitive, motor, and neuroendocrine effects of Δ-9-tetrahydrocannabinol in humans. *Psychopharmacology* 2008;**198**:587–603. https://doi.org/10.1007/s00213-007-1042-2.
183. Gupta S, De Aquino JP, D'Souza DC, Ranganathan M. Effects of haloperidol on the delta-9-tetrahydrocannabinol response in humans: a responder analysis. *Psychopharmacology* 2019;**236**:2635–40. https://doi.org/10.1007/s00213-019-05235-x.
184. D'Souza DC, Fridberg DJ, Skosnik PD, et al. Dose-related modulation of event-related potentials to novel and target stimuli by intravenous δ 9-THC in humans. *Neuropsychopharmacology* 2012;**37**:1632–46. https://doi.org/10.1038/npp.2012.8.
185. Cortes-Briones J, Skosnik PD, Mathalon D, et al. Δ9-THC disrupts gamma (γ)-band neural oscillations in humans. *Neuropsychopharmacology* 2015;**40**:2124–34. https://doi.org/10.1038/npp.2015.53.
186. Skosnik PD, Hajós M, Cortes-Briones JA, et al. Cannabinoid receptor-mediated disruption of sensory gating and neural oscillations: a translational study in rats and humans. *Neuropharmacology* 2018;**135**:412–23. https://doi.org/10.1016/j.neuropharm.2018.03.036.
187. Emrich HM, Weber MM, Wendl A, Zihl J, Von Meyer L, Hanisch W. Reduced binocular depth inversion as an indicator of cannabis-induced censorship impairment. *Pharmacol Biochem Behav* 1991;**40**:689–90. https://doi.org/10.1016/0091-3057(91)90383-D.
188. Leweke FM, Schneider U, Radwan M, Schmidt E, Emrich HM. Different effects of nabilone and cannabidiol on binocular depth inversion in man. *Pharmacol Biochem Behav* 2000;**66**:175–81. https://doi.org/10.1016/S0091-3057(00)00201-X.
189. Koethe D, Gerth CW, Neatby MA, et al. Disturbances of visual information processing in early states of psychosis and experimental delta-9-tetrahydrocannabinol altered states of consciousness. *Schizophr Res* 2006;**88**:142–50. https://doi.org/10.1016/j.schres.2006.07.023.
190. Radhakrishnan R, Wilkinson ST, D'Souza DC. Gone to pot: a review of the association between cannabis and psychosis. *Front Psychiatry* 2014;**5**:54. https://doi.org/10.3389/fpsyt.2014.00054.
191. Hollister LE. Structure activity relationships in man of cannabis constituents, and homologs and metabolites of Δ9tetrahydrocannabinol. *Pharmacology* 1974;**11**:3–11. https://doi.org/10.1159/000136462.
192. Balster RL, Prescott WR. Δ9-tetrahydrocannabinol discrimination in rats as a model for cannabis intoxication. *Neurosci Biobehav Rev* 1992;**16**:55–62. https://doi.org/10.1016/S0149-7634(05)80051-X.

193. Wiley JL, Lowe JA, Balster RL, Martin BR. Antagonism of the discriminative stimulus effects of delta 9-tetrahydrocannabinol in rats and rhesus monkeys. *J Pharmacol Exp Ther* 1995;**275**:1–6.
194. Howlett AC. The cannabinoid receptors. *Prostaglandins Other Lipid Mediat* 2002;**68-69**: 619–31. https://doi.org/10.1016/S0090-6980(02)00060-6.
195. Spiller KJ, Bi G, He Y, Galaj E, Gardner EL, Xi ZX. Cannabinoid CB 1 and CB 2 receptor mechanisms underlie cannabis reward and aversion in rats. *Br J Pharmacol* 2019;**176**:1268–81. https://doi.org/10.1111/bph.14625.
196. Herkenham M, Lynn AB, Little MD, et al. Cannabinoid receptor localization in brain. *Proc Natl Acad Sci* 1990;**87**:1932–6. https://doi.org/10.1073/pnas.87.5.1932.
197. Glass M, Dragunow M, Faull RLM. Cannabinoid receptors in the human brain: a detailed anatomical and quantitative autoradiographic study in the fetal, neonatal and adult human brain. *Neuroscience* 1997;**77**:299–318. https://doi.org/10.1016/S0306-4522(96)00428-9.
198. Egertová M, Elphick MR. Localisation of cannabinoid receptors in the rat brain using antibodies to the intracellular C-terminal tail of CB1. *J Comp Neurol* 2000;**422**:159–71. https://doi.org/10.1002/(SICI)1096-9861(20000626)422:2<159::AID-CNE1>3.0.CO;2-1.
199. Skene NG, Bryois J, Bakken TE, et al. Genetic identification of brain cell types underlying schizophrenia. *Nat Genet* 2018;**50**:825–33. https://doi.org/10.1038/s41588-018-0129-5.
200. French ED, Dillon K, Wu X. Cannabinoids excite dopamine neurons in the ventral tegmentum and substantia nigra. *Neuroreport* 1997;**8**:649–52. https://doi.org/10.1097/00001756-199702100-00014.
201. Pistis M, Porcu G, Melis M, Diana M, Luigi Gessa G. Effects of cannabinoids on prefrontal neuronal responses to ventral tegmental area stimulation. *Eur J Neurosci* 2001;**14**: 96–102. https://doi.org/10.1046/j.0953-816x.2001.01612.x.
202. Voruganti LNP, Slomka P, Zabel P, Mattar A, Awad AG. Cannabis induced dopamine release: an in-vivo SPECT study. *Psychiatry Res Neuroimaging* 2001;**107**:173–7. https://doi.org/10.1016/S0925-4927(01)00104-4.
203. Covey DP, Mateo Y, Sulzer D, Cheer JF, Lovinger DM. Endocannabinoid modulation of dopamine neurotransmission. *Neuropharmacology* 2017;**124**:52–61. https://doi.org/10.1016/j.neuropharm.2017.04.033.
204. Gonzalez-Burgos G, Lewis DA. GABA neurons and the mechanisms of network oscillations: implications for understanding cortical dysfunction in schizophrenia. *Schizophr Bull* 2008;**34**:944–61. https://doi.org/10.1093/schbul/sbn070.
205. Auclair N, Otani S, Soubrie P, Crepel F. Cannabinoids modulate synaptic strength and plasticity at glutamatergic synapses of rat prefrontal cortex pyramidal neurons. *J Neurophysiol* 2000;**83**:3287–93. https://doi.org/10.1152/jn.2000.83.6.3287.
206. Sim-Selley LJ, Schechter NS, Rorrer WK, et al. Prolonged recovery rate of CB 1 receptor adaptation after cessation of long-term cannabinoid administration. *Mol Pharmacol* 2006;**70**: 986–96. https://doi.org/10.1124/mol.105.019612.
207. D'Souza DC, Cortes-Briones JA, Ranganathan M, et al. Rapid changes in cannabinoid 1 receptor availability in cannabis-dependent male subjects after abstinence from cannabis. *Biol Psychiatry* 2016;**1**:60–7. https://doi.org/10.1016/j.bpsc.2015.09.008.
208. Rubino T, Viganò D, Massi P, Parolaro D. The psychoactive ingredient of marijuana induces behavioural sensitization. *Eur J Neurosci* 2001;**14**:884–6. https://doi.org/10.1046/j.0953-816x.2001.01709.x.
209. Rubino T, Viganò D, Massi P, Parolaro D. Cellular mechanisms of Δ 9-tetrahydrocannabinol behavioural sensitization. *Eur J Neurosci* 2003;**17**:325–30. https://doi.org/10.1046/j.1460-9568.2003.02452.x.

210. Tournier BB, Tsartsalis S, Dimiziani A, Millet P, Ginovart N. Time-dependent effects of repeated THC treatment on dopamine D2/3 receptor-mediated signalling in midbrain and striatum. *Behav Brain Res* 2016;**311**:322–9. https://doi.org/10.1016/j.bbr.2016.05.045.
211. Giedd JN, Blumenthal J, Jeffries NO, et al. Brain development during childhood and adolescence: a longitudinal MRI study. *Nat Neurosci* 1999;**2**:861–3. https://doi.org/10.1038/13158.
212. Fuhrmann D, Knoll LJ, Blakemore SJ. Adolescence as a sensitive period of brain development. *Trends Cogn Sci* 2015;**19**:558–66. https://doi.org/10.1016/j.tics.2015.07.008.
213. Meyer HC, Lee FS, Gee DG. The role of the endocannabinoid system and genetic variation in adolescent brain development. *Neuropsychopharmacology* 2018;**43**:21–33. https://doi.org/10.1038/npp.2017.143.
214. Ashtari M, Avants B, Cyckowski L, et al. Medial temporal structures and memory functions in adolescents with heavy cannabis use. *J Psychiatr Res* 2011;**45**:1055–66. https://doi.org/10.1016/j.jpsychires.2011.01.004.
215. Fernandez-Espejo E, Viveros M-P, Núñez L, Ellenbroek BA, Rodriguez de Fonseca F. Role of cannabis and endocannabinoids in the genesis of schizophrenia. *Psychopharmacology* 2009;**206**:531–49. https://doi.org/10.1007/s00213-009-1612-6.
216. Volk DW, Lewis DA. The role of endocannabinoid signaling in cortical inhibitory neuron dysfunction in schizophrenia. *Biol Psychiatry* 2016;**79**:595–603. https://doi.org/10.1016/j.biopsych.2015.06.015.
217. Fakhoury M. Role of the endocannabinoid system in the pathophysiology of schizophrenia. *Mol Neurobiol* 2017;**54**:768–78. https://doi.org/10.1007/s12035-016-9697-5.
218. Haring M, Enk V, Aparisi Rey A, et al. Cannabinoid type-1 receptor signaling in central serotonergic neurons regulates anxiety-like behavior and sociability. *Front Behav Neurosci* 2015;**9**:235. https://doi.org/10.3389/fnbeh.2015.00235.
219. Franklin JM, Carrasco GA. Cannabinoid receptor agonists upregulate and enhance serotonin 2A (5-HT(2A)) receptor activity via ERK1/2 signaling. *Synapse* 2013;**67**(3):145–59. https://doi.org/10.1002/syn.21626.
220. Franklin JM, Carrasco GA. G-protein receptor kinase 5 regulates the cannabinoid receptor 2-induced up-regulation of serotonin 2A receptors. *J Biol Chem* 2013;**288**:15712–24. https://doi.org/10.1074/jbc.M113.454843.
221. Aso E, Renoir T, Mengod G, et al. Lack of CB 1 receptor activity impairs serotonergic negative feedback. *J Neurochem* 2009;**109**:935–44. https://doi.org/10.1111/j.1471-4159.2009.06025.x.
222. Best AR, Regehr WG. Serotonin evokes endocannabinoid release and retrogradely suppresses excitatory synapses. *J Neurosci* 2008;**28**:6508–15. https://doi.org/10.1523/JNEUROSCI.0678-08.2008.
223. Parrish JC, Nichols DE. Serotonin 5-HT 2A receptor activation induces 2-arachidonoylglycerol release through a phospholipase c-dependent mechanism. *J Neurochem* 2006;**99**:1164–75. https://doi.org/10.1111/j.1471-4159.2006.04173.x.
224. De Jesus ML, Salles J, Meana JJ, Callado LF. Characterization of CB1 cannabinoid receptor immunoreactivity in postmortem human brain homogenates. *Neuroscience* 2006;**140**(2):635–43. https://doi.org/10.1016/j.neuroscience.2006.02.024.
225. Pazos A, Probst A, Palacios JM. Serotonin receptors in the human brain—IV. Autoradiographic mapping of serotonin-2 receptors. *Neuroscience* 1987;**21**:123–39. https://doi.org/10.1016/0306-4522(87)90327-7.
226. Viñals X, Moreno E, Lanfumey L, et al. Cognitive impairment induced by Delta9-tetrahydrocannabinol occurs through heteromers between cannabinoid CB1 and serotonin 5-HT2A receptors. *PLoS Biol* 2015;**13**. https://doi.org/10.1371/journal.pbio.1002194, e1002194.

227. González-Maeso J, Weisstaub NV, Zhou M, et al. Hallucinogens recruit specific cortical 5-HT2A receptor-mediated Signaling pathways to affect behavior. *Neuron* 2007;**53**(3): 439–52. https://doi.org/10.1016/j.neuron.2007.01.008.
228. Geyer MA, Vollenweider FX. Serotonin research: contributions to understanding psychoses. *Trends Pharmacol Sci* 2008;**29**:445–53. https://doi.org/10.1016/j.tips.2008.06.006.
229. Hoyer D, Hannon JP, Martin GR. Molecular, pharmacological and functional diversity of 5-HT receptors. *Pharmacol Biochem Behav* 2002;**71**:533–54. https://doi.org/10.1016/S0091-3057(01)00746-8.
230. Pithadia A. 5-Hydroxytryptamine receptor subtypes and their modulators with therapeutic potentials. *J Clin Med Res* 2009;**1**:72–80. https://doi.org/10.4021/jocmr2009.05.1237.
231. Jakab RL, Goldman-Rakic PS. 5-Hydroxytryptamine2A serotonin receptors in the primate cerebral cortex: possible site of action of hallucinogenic and antipsychotic drugs in pyramidal cell apical dendrites. *Proc Natl Acad Sci U S A* 1998;**95**:735–40. https://doi.org/10.1073/pnas.95.2.735.
232. Conn PJ, Sanders-Bush E. Central serotonin receptors: effector systems, physiological roles and regulation. *Psychopharmacology* 1987;**92**:267–77. https://doi.org/10.1007/BF00210830.
233. Carr DB, Cooper DC, Ulrich SL, Spruston N, Surmeier DJ. Serotonin receptor activation inhibits sodium current and dendritic excitability in prefrontal cortex via a protein kinase C-dependent mechanism. *J Neurosci* 2002;**22**:6846–55. https://doi.org/10.1523/jneurosci.22-16-06846.2002.
234. Felder CC, Kanterman RY, Ma AL, Axelrod J. Serotonin stimulates phospholipase A2 and the release of arachidonic acid in hippocampal neurons by a type 2 serotonin receptor that is independent of inositolphospholipid hydrolysis. *Proc Natl Acad Sci U S A* 1990. https://doi.org/10.1073/pnas.87.6.2187.
235. Kurrasch-Orbaugh DM, Watts VJ, Barker EL, Nichols DE. Serotonin 5-hydroxytryptamine 2A receptor-coupled phospholipase C and phospholipase a 2 signaling pathways have different receptor reserves. *J Pharmacol Exp Ther* 2003;**304**:229–37. https://doi.org/10.1124/jpet.102.042184.
236. Meltzer HY, Bastani B, Ramirez L, Matsubara S. Clozapine: new research on efficacy and mechanism of action. *Eur Arch Psychiatry Neurol Sci* 1989;**238**(5–6):332–9. https://doi.org/10.1007/bf00449814.
237. Abdolmaleky HM, Yaqubi S, Papageorgis P, et al. Epigenetic dysregulation of HTR2A in the brain of patients with schizophrenia and bipolar disorder. *Schizophr Res* Jul 2011;**129**(2–3):183–90. https://doi.org/10.1016/j.schres.2011.04.007.
238. Malkova NV, Gallagher JJ, Yu CZ, Jacobs RE, Patterson PH. Manganese-enhanced magnetic resonance imaging reveals increased DOI-induced brain activity in a mouse model of schizophrenia. *Proc Natl Acad Sci U S A* 2014;**111**(24):E2492–500. https://doi.org/10.1073/pnas.1323287111.
239. Santini MA, Ratner C, Aznar S, Klein AB, Knudsen GM, Mikkelsen JD. Enhanced prefrontal serotonin 2A receptor signaling in the subchronic phencyclidine mouse model of schizophrenia. *J Neurosci Res* 2013;**91**(5):634–41. https://doi.org/10.1002/jnr.23198.
240. Muguruza C, Moreno JL, Umali A, Callado LF, Meana JJ, Gonzalez-Maeso J. Dysregulated 5-HT(2A) receptor binding in postmortem frontal cortex of schizophrenic subjects. *Eur Neuropsychopharmacol* 2013;**23**(8):852–64. https://doi.org/10.1016/j.euroneuro.2012.10.006.
241. Chaouloff F. Serotonin and stress. *Neuropsychopharmacology* 1999;**21**:28S–32S. https://doi.org/10.1016/S0893-133X(99)00008-1.

242. Arévalo C, de Miguel R, Hernández-Tristán R. Cannabinoid effects on anxiety-related behaviours and hypothalamic neurotransmitters. *Pharmacol Biochem Behav* 2001;**70**: 123–31. https://doi.org/10.1016/S0091-3057(01)00578-0.
243. Malone DT, Taylor DA. Modulation of Δ 9-tetrahydrocannabinol-induced hypothermia by fluoxetine in the rat. *Br J Pharmacol* 1998;**124**:1419–24. https://doi.org/10.1038/sj.bjp.0701980.
244. Marco EM, Pérez-Alvarez L, Borcel E, et al. Involvement of 5-HT1A receptors in behavioural effects of the cannabinoid receptor agonist CP 55,940 in male rats. *Behav Pharmacol* 2004;**15**:21–7. https://doi.org/10.1097/00008877-200402000-00003.
245. Huitrón-Reséndiz S, Gombart L, Cravatt BF, Henriksen SJ. Effect of oleamide on sleep and its relationship to blood pressure, body temperature, and locomotor activity in rats. *Exp Neurol* 2001;**172**:235–43. https://doi.org/10.1006/exnr.2001.7792.
246. Lichtman AH, Hawkins EG, Griffin G, Cravatt BF. Pharmacological activity of fatty acid amides is regulated, but not mediated, by fatty acid amide hydrolase in vivo. *J Pharmacol Exp Ther* 2002;**302**:73–9. https://doi.org/10.1124/jpet.302.1.73.
247. Leggett JD, Aspley S, Beckett SRG, D'Antona AM, Kendall DA, Kendall DA. Oleamide is a selective endogenous agonist of rat and human CB 1 cannabinoid receptors. *Br J Pharmacol* 2004;**141**:253–62. https://doi.org/10.1038/sj.bjp.0705607.
248. Cheer JF, Cadogan A-K, Marsden CA, Fone KCF, Kendall DA. Modification of 5-HT2 receptor mediated behaviour in the rat by oleamide and the role of cannabinoid receptors. *Neuropharmacology* 1999;**38**:533–41. https://doi.org/10.1016/S0028-3908(98)00208-1.
249. Laprairie RB, Bagher AM, Kelly MEM, Denovan-Wright EM. Cannabidiol is a negative allosteric modulator of the cannabinoid CB 1 receptor. *Br J Pharmacol* 2015;**172**: 4790–805. https://doi.org/10.1111/bph.13250.
250. Darmani NA, Janoyan JJ, Kumar N, Crim JL. Behaviorally active doses of the CB1 receptor antagonist SR 141716A increase brain serotonin and dopamine levels and turnover. *Pharmacol Biochem Behav* 2003;**75**:777–87. https://doi.org/10.1016/S0091-3057(03)00150-3.
251. Ortega JE, Gonzalez-Lira V, Horrillo I, Herrera-Marschitz M, Callado LF, Meana JJ. Additive effect of rimonabant and citalopram on extracellular serotonin levels monitored with in vivo microdialysis in rat brain. *Eur J Pharmacol* 2013;**709**:13–9. https://doi.org/10.1016/j.ejphar.2013.03.043.
252. Wullschleger S, Loewith R, Hall MN. TOR signaling in growth and metabolism. *Cell* 2006;**124**(3):471–84. https://doi.org/10.1016/j.cell.2006.01.016.
253. Li J, Kim SG, Blenis J. Rapamycin: one drug, many effects. *Cell Metab* 2014;**19**(3): 373–9. https://doi.org/10.1016/j.cmet.2014.01.001.
254. Lai WS, Xu B, Westphal KG, et al. Akt1 deficiency affects neuronal morphology and predisposes to abnormalities in prefrontal cortex functioning. *Proc Natl Acad Sci U S A* 2006;**103** (45):16906–11. https://doi.org/10.1073/pnas.0604994103.
255. Gururajan A, van den Buuse M. Is the mTOR-signalling cascade disrupted in schizophrenia? *J Neurochem* 2014;**129**(3):377–87. https://doi.org/10.1111/jnc.12622.
256. Dwyer JM, Maldonado-Aviles JG, Lepack AE, DiLeone RJ, Duman RS. Ribosomal protein S6 kinase 1 signaling in prefrontal cortex controls depressive behavior. *Proc Natl Acad Sci U S A* 2015;**112**(19):6188–93. https://doi.org/10.1073/pnas.1505289112.
257. Deli A, Schipany K, Rosner M, et al. Blocking mTORC1 activity by rapamycin leads to impairment of spatial memory retrieval but not acquisition in C57BL/6J mice. *Behav Brain Res* 2012;**229**(2):320–4. https://doi.org/10.1016/j.bbr.2012.01.017.
258. Strachan RT, Sciaky N, Cronan MR, Kroeze WK, Roth BL. Genetic deletion of p90 ribosomal S6 kinase 2 alters patterns of 5-Hydroxytryptamine 2A serotonin receptor functional selectivity. *Mol Pharmacol* 2010;**77**:327–38. https://doi.org/10.1124/mol.109.061440.

259. Knight ZA, Tan K, Birsoy K, et al. Molecular profiling of activated neurons by phosphorylated ribosome capture. *Cell* 2012;**151**(5):1126–37. https://doi.org/10.1016/j.cell.2012.10.039.
260. Ruvinsky I, Meyuhas O. Ribosomal protein S6 phosphorylation: from protein synthesis to cell size. *Trends Biochem Sci* 2006;**31**(6):342–8. https://doi.org/10.1016/j.tibs.2006.04.003.
261. Puighermanal E, Biever A, Pascoli V, et al. Ribosomal protein S6 phosphorylation is involved in novelty-induced locomotion, synaptic plasticity and mRNA translation. *Front Mol Neurosci* 2017;**10**:419. https://doi.org/10.3389/fnmol.2017.00419.
262. Di Forti M, Iyegbe C, Sallis H, et al. Confirmation that the AKT1 (rs2494732) genotype influences the risk of psychosis in cannabis users. *Biol Psychiatry* 2012;**72**(10):811–6. https://doi.org/10.1016/j.biopsych.2012.06.020.
263. Puighermanal E, Busquets-Garcia A, Gomis-Gonzalez M, Marsicano G, Maldonado R, Ozaita A. Dissociation of the pharmacological effects of THC by mTOR blockade. *Neuropsychopharmacology* 2013;**38**(7):1334–43. https://doi.org/10.1038/npp.2013.31.
264. Kennedy SG, Wagner AJ, Conzen SD, et al. The PI 3-kinase/Akt signaling pathway delivers an anti-apoptotic signal. *Genes Dev* 1997;**11**:701–13. https://doi.org/10.1101/gad.11.6.701.
265. Chang F, Lee JT, Navolanic PM, et al. Involvement of PI3K/Akt pathway in cell cycle progression, apoptosis, and neoplastic transformation: a target for cancer chemotherapy. *Leukemia* 2003;**17**:590–603. https://doi.org/10.1038/sj.leu.2402824.
266. Manning BD, Cantley LC. AKT/PKB signaling: navigating downstream. *Cell* 2007;**129**: 1261–74. https://doi.org/10.1016/j.cell.2007.06.009.
267. Coffer PJ, Jin J, Woodgett JR. Protein kinase B (c-Akt): a multifunctional mediator of phosphatidylinositol 3-kinase activation. *Biochem J* 1998;**335**:1–13. https://doi.org/10.1042/bj3350001.
268. Masure S, Haefner B, Wesselink J-J, et al. Molecular cloning, expression and characterization of the human serine/threonine kinase Akt-3. *Eur J Biochem* 1999;**265**:353–60. https://doi.org/10.1046/j.1432-1327.1999.00774.x.
269. Matheny RW, Geddis AV, Abdalla MN, et al. AKT2 is the predominant AKT isoform expressed in human skeletal muscle. *Phys Rep* 2018;**6**. https://doi.org/10.14814/phy2.13652, e13652.
270. Yang Z-Z, Tschopp O, Di-Poi N, et al. Dosage-dependent effects of Akt1/protein kinase B (PKB) and Akt3/PKB on thymus, skin, and cardiovascular and nervous system development in mice. *Mol Cell Biol* 2005;**25**:10407–18. https://doi.org/10.1128/MCB.25.23.10407-10418.2005.
271. Burgering BMT, Coffer PJ. Protein kinase B (c-Akt) in phosphatidylinositol-3-OH kinase signal transduction. *Nature* 1995;599–602.
272. Murga C, Laguinge L, Wetzker R, Cuadrado A, Gutkind JS. Activation of Akt/protein kinase B by G protein-coupled receptors. *J Biol Chem* 1998;**273**:19080–5. https://doi.org/10.1074/jbc.273.30.19080.
273. Huang H, Miao L, Yang L, et al. AKT-dependent and -independent pathways mediate PTEN deletion-induced CNS axon regeneration. *Cell Death Dis* 2019;**10**:203. https://doi.org/10.1038/s41419-018-1289-z.
274. Dudek H, Datta SR, Franke TF, et al. Regulation of neuronal survival by the serine-threonine protein kinase Akt. *Science* 1997;**275**:661–5. https://doi.org/10.1126/science.275.5300.661.
275. Laplante M, Sabatini DM. mTOR Signaling in growth control and disease. *Cell* 2012;**149**: 274–93. https://doi.org/10.1016/j.cell.2012.03.017.
276. Sarbassov DD, Ali SM, Sengupta S, et al. Prolonged rapamycin treatment inhibits mTORC2 assembly and Akt/PKB. *Mol Cell* 2006;**22**:159–68. https://doi.org/10.1016/j.molcel.2006.03.029.

277. Costa-Mattioli M, Monteggia LM. mTOR complexes in neurodevelopmental and neuropsychiatric disorders. *Nat Neurosci* 2013;**16**:1537–43. https://doi.org/10.1038/nn.3546.
278. Meyuhas O. Ribosomal protein S6 phosphorylation: four decades of research. *Int Rev Cell Mol Biol* 2015;**320**:41–73.
279. Hoeffer CA, Klann E. mTOR signaling: at the crossroads of plasticity, memory and disease. *Trends Neurosci* 2010;**33**:67–75. https://doi.org/10.1016/j.tins.2009.11.003.
280. Kitagishi Y, Kobayashi M, Kikuta K, Matsuda S. *Roles of PI3K/AKT/GSK3/mTOR pathway in cell signaling of mental illnesses*; 2012.
281. Takei N, Nawa H. mTOR signaling and its roles in normal and abnormal brain development. *Front Mol Neurosci* 2014;**7**:28. https://doi.org/10.3389/fnmol.2014.00028.
282. Jernigan CS, Goswami DB, Austin MC, et al. The mTOR signaling pathway in the prefrontal cortex is compromised in major depressive disorder. *Prog Neuro-Psychopharmacol Biol Psychiatry* 2011;**35**:1774–9. https://doi.org/10.1016/j.pnpbp.2011.05.010.
283. Li N, Lee B, Liu RJ, et al. mTOR-dependent synapse formation underlies the rapid antidepressant effects of NMDA antagonists. *Science* 2010;**329**(5994):959–64. https://doi.org/10.1126/science.1190287.
284. Abdallah CG, Sanacora G, Duman RS, Krystal JH. Ketamine and rapid-acting antidepressants: a window into a new neurobiology for mood disorder therapeutics. *Annu Rev Med* 2015;**66**:509–23. https://doi.org/10.1146/annurev-med-053013-062946.
285. English JA, Fan Y, Föcking M, et al. Reduced protein synthesis in schizophrenia patient-derived olfactory cells. *Transl Psychiatry* 2015;**5**. https://doi.org/10.1038/tp.2015.119, e663.
286. Pham X, Song G, Lao S, et al. The DPYSL2 gene connects mTOR and schizophrenia. *Transl Psychiatry* 2016;**6**. https://doi.org/10.1038/tp.2016.204, e933.
287. Emamian ES, Hall D, Birnbaum MJ, Karayiorgou M, Gogos JA. Convergent evidence for impaired AKT1-GSK3β signaling in schizophrenia. *Nat Genet* 2004;**36**:131–7. https://doi.org/10.1038/ng1296.
288. Schwab SG, Hoefgen B, Hanses C, et al. Further evidence for association of variants in the AKT1 gene with schizophrenia in a sample of European sib-pair families. *Biol Psychiatry* 2005;**58**:446–50. https://doi.org/10.1016/j.biopsych.2005.05.005.
289. Thiselton DL, Vladimirov VI, Kuo P-H, et al. AKT1 is associated with schizophrenia across multiple symptom dimensions in the Irish study of high density schizophrenia families. *Biol Psychiatry* 2008;**63**:449–57. https://doi.org/10.1016/j.biopsych.2007.06.005.
290. Mathur A, Law MH, Megson IL, Shaw DJ, Wei J. Genetic association of the AKT1 gene with schizophrenia in a British population. *Psychiatr Genet* 2010;**20**:118–22. https://doi.org/10.1097/YPG.0b013e32833a2234.
291. Tan H-Y, Nicodemus KK, Chen Q, et al. Genetic variation in AKT1 is linked to dopamine-associated prefrontal cortical structure and function in humans. *J Clin Investig* 2008;**118**:2200–8. https://doi.org/10.1172/JCI34725.
292. Dwyer JM, Duman RS. Activation of mammalian target of rapamycin and synaptogenesis: role in the actions of rapid-acting antidepressants. *Biol Psychiatry* 2013;**73**(12):1189–98. https://doi.org/10.1016/j.biopsych.2012.11.011.
293. Braff DL, Geyer MA. Sensorimotor gating and schizophrenia human and animal model studies. *Arch Gen Psychiatry* 1990;**47**:181–8.
294. Ludewig K, Geyer MA, Vollenweider FX. Deficits in prepulse inhibition and habituation in never-medicated, first-episode schizophrenia. *Biol Psychiatry* 2003;**54**:121–8. https://doi.org/10.1016/S0006-3223(02)01925-X.
295. de Oliveira RP, Nagaishi KY, Barbosa Silva RC. Atypical antipsychotic clozapine reversed deficit on prepulse inhibition of the acoustic startle reflex produced by microinjection of

DOI into the inferior colliculus in rats. *Behav Brain Res* 2017;**325**:72–8. https://doi.org/10.1016/j.bbr.2017.01.053.
296. Farid M, Martinez ZA, Geyer MA, Swerdlow NR. Regulation of sensorimotor gating of the startle reflex by serotonin 2A receptors: ontogeny and strain differences. *Neuropsychopharmacology* 2000;**23**:623–32. https://doi.org/10.1016/S0893-133X(00)00163-9.
297. De Lean A, Stadel JM, Lefkowitz RJ. A ternary complex model explains the agonist-specific binding properties of the adenylate cyclase-coupled beta-adrenergic receptor. *J Biol Chem* 1980;**255**(15):7108–17.
298. Franklin JM, Carrasco GA. Cannabinoid-induced enhanced interaction and protein levels of serotonin 5-HT2A and dopamine D2 receptors in rat prefrontal cortex. *J Psychopharmacol* 2012;**26**:1333–47. https://doi.org/10.1177/0269881112450786.
299. García-Bea A, Miranda-Azpiazu P, Muguruza C, et al. Serotonin 5-HT 2A receptor expression and functionality in postmortem frontal cortex of subjects with schizophrenia: selective biased agonism via G α i1 -proteins. *Eur Neuropsychopharmacol* 2019;**29**(12):1453–63. https://doi.org/10.1016/j.euroneuro.2019.10.013.
300. van Winkel R, Risk G. Outcome of P. Family-based analysis of genetic variation underlying psychosis-inducing effects of cannabis: sibling analysis and proband follow-up. *Arch Gen Psychiatry* 2011;**68**(2):148–57. https://doi.org/10.1001/archgenpsychiatry.2010.152.
301. van Winkel R, van Beveren NJM, Simons C. AKT1 moderation of cannabis-induced cognitive alterations in psychotic disorder. *Neuropsychopharmacology* 2011;**36**:2529–37. https://doi.org/10.1038/npp.2011.141.
302. Bhattacharyya S, Atakan Z, Martin-Santos R, et al. Preliminary report of biological basis of sensitivity to the effects of cannabis on psychosis: AKT1 and DAT1 genotype modulates the effects of δ-9-tetrahydrocannabinol on midbrain and striatal function. *Mol Psychiatry* 2012;**17**:1152–5. https://doi.org/10.1038/mp.2011.187.
303. Ozaita A, Puighermanal E, Maldonado R. Regulation of PI3K/Akt/GSK-3 pathway by cannabinoids in the brain. *J Neurochem* 2007;**102**:1105–14. https://doi.org/10.1111/j.1471-4159.2007.04642.x.
304. Puighermanal E, Marsicano G, Busquets-Garcia A, Lutz B, Maldonado R, Ozaita A. Cannabinoid modulation of hippocampal long-term memory is mediated by mTOR signaling. *Nat Neurosci* 2009;**12**:1152–8. https://doi.org/10.1038/nn.2369.
305. Morgan CJA, Freeman TP, Powell J, Curran HV. AKT1 genotype moderates the acute psychotomimetic effects of naturalistically smoked cannabis in young cannabis smokers. *Transl Psychiatry* 2016;**6**. https://doi.org/10.1038/tp.2015.219, e738.
306. Sheffler DJ, Kroeze WK, Garcia BG, et al. p90 ribosomal S6 kinase 2 exerts a tonic brake on G protein-coupled receptor signaling. *Proc Natl Acad Sci* 2006;**103**:4717–22. https://doi.org/10.1073/pnas.0600585103.
307. Strachan RT, Sheffler DJ, Willard B, Kinter M, Kiselar JG, Roth BL. Ribosomal S6 kinase 2 directly phosphorylates the 5-Hydroxytryptamine 2A (5-HT 2A) serotonin receptor, thereby modulating 5-HT 2A Signaling. *J Biol Chem* 2009;**284**:5557–73. https://doi.org/10.1074/jbc.M805705200.
308. Saunders C, Siuta M, Robertson SD, et al. Neuronal ablation of p-Akt at Ser473 leads to altered 5-HT1A/2A receptor function. *Neurochem Int* 2014;**73**:113–21. https://doi.org/10.1016/j.neuint.2013.09.015.

Chapter 12

Synthetic cannabinoids: State-of-the-art with a focus on fertility and development

A.-L. Pélissier-Alicot

Cannalab, Cannabinoids Neuroscience Research International Associated Laboratory, INSERM-Aix-Marseille University France/Indiana University, Bloomington, IN, United States; INMED, INSERM U1249, Marseille, France; Aix-Marseille University, Service de Medecine Legale, AP-HM, Marseille, France

The emergence of synthetic cannabinoids

Synthetic cannabinoids (SC), or synthetic cannabinoid receptor agonists, are defined as new psychoactive substances that mimic the effects of tetrahydrocannabinol (THC), the major psychoactive substance in cannabis.[1]

The first SC were synthesized in the 1970s by academic laboratories when researchers began to explore the signaling pathways of the endocannabinoid system.[2] The pharmaceutical industry quickly focused on these molecules to create new drugs that retain the biological activity of natural cannabinoids, especially antinociceptive properties, but lack psychoactive side effects.[2] The first selective cannabinoid receptor agonist, cyclohexylphenol (CP 55,940), was synthesized in 1974 by the Pfizer pharmaceutical group. It was followed by HU-210 synthesized in 1988 by Mechoulam's group at the Hebrew University of Jerusalem.[3] Subsequently, John W. Huffman and his team at Clemson University in South Carolina focused on synthesizing CB1 and CB2 receptor agonists with analgesic properties. This work led to the creation of the JWH series, including JWH-018, which today comprises several hundred SCs.

While work on synthetic cannabinoids has greatly contributed to the understanding of the endocannabinoid system and opened therapeutic perspectives, notably for cancer pain or neurodegenerative diseases, it has also generated diverted development by clandestine laboratories as alternatives to phytocannabinoids.[4] In the mid-2000s, SCs began to appear in Europe as products called "Spice" and sold as "legal" cannabis substitutes. In 2008, German and Austrian forensic investigators first detected JWH-018 in a seizure product.

Subsequently, several cannabinoids were detected in herbal smoking blends also known as incense/air fresheners. These included "Spice Gold," "Spice Silver" and "Yucatan Fire," "Black Mamba," "K2," etc.[5,6] In recent years, alongside these smoking mixtures, new products, including e-liquids, have been sold on the underground market. Unbeknownst to users, synthetic cannabinoids have also been used to adulterate cannabidiol (CBD) and THC-based e-liquids, as well as other illicit drugs, such as opioids.[1,7,8] Another disturbing phenomenon is the recent adulteration of phytocannabinoids with synthetic cannabinoids. Typically, these adulterated products are plant material or resins with low THC content. In terms of appearance, smell and taste, these adulterated products would be very difficult to distinguish from "genuine" cannabis products and, as a result, users may be unaware that they are consuming synthetic cannabinoids. Most of these synthetic cannabinoids are manufactured in clandestine laboratories in Asia and then shipped to Europe and United States in bulk powder form. They are then sold at retail level, often mixed with crushed plants (*Turnera diffusa*, *Melissa officinalis*, phytocannabinoids, etc.) after dissolution with acetone or methanol, or in solid form (crystalline powder, leading to a non-homogenous mixture).[1] The mixture generally contains between 0.5% and 3% SC. Sometimes other substances are added, including vitamin E (tocopherol), whose role is not clearly defined, but which may act as an antioxidant.[9] The mixture is then dried and packaged. The products are sold on the Internet under the name of "legal highs" or "herbal blends" with the mention "not for human use," to circumvent controls, or in traditional outlets.[1] Currently, other preparations are available on the market: capsules, tablets, powders, or liquid-filled cartridges for e-cigarettes (e-liquids) sold as a new alternative to smoking cessation (Buddha-blue, C-Liquid, Herbal e-liquid, etc.).[10]

Chemical structure and nomenclature

With 209 identified on the drug market over the 13 years between January 1, 2008 and December 31, 2020, SCs constitute the largest group of new psychoactive substances monitored by the European Monitoring Centre for Drugs and Drug Addiction. Their chemical diversity and rapid emergence make this group of compounds a particular challenge in terms of detection and monitoring.[1,11]

Only a few SCs are structurally related to THC, the others belonging to different and varied chemical families.[10] There are seven main structural groups: naphthoylindoles, naphthylmethylindoles, naphthoylpyrroles, naphthylmethylindenes, phenylacetylindoles, cyclohexylphenols, and "classical cannabinoids" (e.g., HU-210, AM-906, AM-411, O-1184).[12] Finally, some of these compounds are chiral and can exist in two stereoisomeric forms.[10]

Colloquial and serial names

There are different and complex nomenclatures to identify SCs. In some cases, names were probably chosen by the manufacturers to facilitate their marketing.

The most striking examples are "AKB-48" and "2NE1," alternative names used for APINACA and APICA. "AKB-48" is the name of a popular Japanese girl band and "2NE1" is the name of a girl band from South Korea. To further complicate this situation, retailers use packaging that have their own names ("Pandora's box," "Black Mamba," etc.).[1,13]

Synthetic cannabinoids may have a "serial designation," derived from the initials of the institution, company, or researcher who synthesized them: "JWH" for John W. Huffmann, "AM" for Alexandros Makriyannis, "HU" for the Hebrew University of Jerusalem, "Win" for the Sterling-Winthrop company, and "CP" for the Carl Pfizer company.[1,2,13] This nomenclature does not consider the chemical structure of these molecules. For example, "AM" compounds include dibenzopyran, cyclohexyphenol, and indole derived structures in addition to synthetic analogues, JWH-series includes naphthoylindoles, phenylacetyl indoles, naphthylmethyl indoles, naphthylmethyl indene, and naphthoylpyrroles.[13]

Systematic chemical names

Each SC has a specific chemical name describing its structure, which could be abbreviated in alpha-numeric names. For example, APICA for N-(1-adamantyl)-1-pentyl-1H-indole-3-carboxamide, APINACA for N-(1-adamantyl)-1-pentyl-1H-indazole-3-carboxamide, or MDMD-CHMICA for methyl-(2S)-2-[(1-(cyclohexylmethyl)-1H-indol-3-yl)formamido]-3,3-dimethylbutanoate. Unfortunately, this nomenclature is complex, inaccessible to non-chemists and not suitable for research or clinical use. To standardize their nomenclature, EMCDDA proposed a classification based on their chemical structure which can be divided into four key structural components: the core, the tail, the linker, and the linked group. Assigning a code name to each component allows the chemical structure of the cannabinoid to be identified without the full chemical name. The chemical motif in each group is the assigned a unique code-letter and assembled into a name with the format of "Linked Group—TailCoreLinker" (Fig. 1). For example, the long chemical name N-(1-adamantyl)-1-pentylindole-3-carboxamide is abbreviated in A-PICA with this naming system, that has been adopted in reports of the World Health Organization and United Nations Office on Drugs and Crime.[1,13]

Epidemiology, patterns of use, and legal status

The success of synthetic cannabinoids is due to several reasons: major psychoactive effects, affordable price, purchase through the Internet, lack of toxicological screening procedures, and poor perception of the risk.[14]

Data on the prevalence of use of synthetic cannabinoids are based on population and subpopulation surveys. In population-based studies, the prevalence of current synthetic cannabinoid use is generally found to be less than 1% in Europe. Nevertheless, some subpopulations show higher prevalence. These

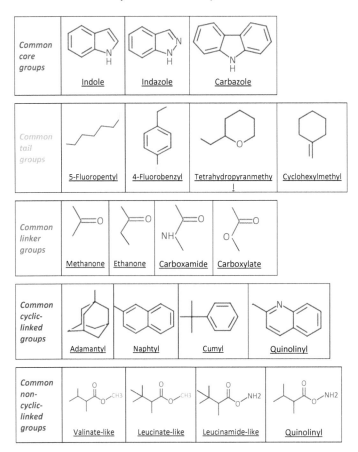

FIG. 1 Nomenclature syntax and structure of the common core, tail, linker, and linked groups.[1,13]

include marginalized people, individuals who are regularly tested for cannabis (prisoners, drivers, etc.), clubbers, recreational users, and those who experiment with new molecules ("psychonauts").[1,15,16] Among regular cannabis and tobacco users, SCs use appears to be relatively common, particularly to escape drug detection and/or to experience a cannabis-like euphoria.[17] Among pregnant women, cannabis use has risen in recent years.[18] Even if data regarding the SCs consumption during pregnancy are lacking, a similar trend may be expected.[19]

Because of the rapidity of the effect, the most common way to consume SC is by smoking ready-made or homemade "smoking mixtures," such as a cigarette ("joint") or via water pipe/bong, cigarette, blunt or pipe.[20] E-liquids containing SC can also be smoked in e-cigarettes. They can also be taken orally, but the onset of action is longer due to an extensive hepatic first-pass metabolism.[21] Rectal and intravenous absorption have also been described but seem to be rare.[1,20,22]

The following synthetic cannabinoids are included in the list of substances in Schedule II of the United Nations Convention on Psychotropic Substances of 1971 (INCB, 2020)[1]:

- Since 2015: AM-2201 (JWH-2201) and JWH-018 (AM-678)
- Since 2017: 5F-AKB-48 (5F-APINACA), MDMB-CHMICA, and XLR-11 (5F-UR-144)
- Since 2018: AB-CHMINACA, AB-PINACA, 5F-MDMB-PINACA (5F-ADB), AM-2201 carboxylate analogue quinolinyl derivative (5F-PB-22), and UR-144
- Since 2019: ADB-FUBINACA, AMB-FUBINACA (FUB-AMB), CUMYL-4CN-BINACA, ADB-CHMINACA, AB-FUBINACA, 5F-AMB (5F-AMB-PINACA), 5F-MDMD-PICA, and 4F-MDMD-BINACA
- Since 2021: MDMB-4en-PINACA and CUMYL-PeGACLONE

Pharmacology

While cannabis contains many cannabinoids, including THC, which is a potent CB1 receptor agonist, but also CBD, which modulates, through complex mechanisms, the effects of THC and leads to a reduction in the undesirable side effects of THC, SCs contain only CB1 and CB2 agonists and lack compounds such as CBD that may counteract psychoactive properties of CB1 agonists.[11,23,24] This could explain the very powerful psychotropic effects of these compounds.

Binding affinity

SCs bind to CB1 and CB2 receptors with varying degrees of affinity. Most competitive binding affinity studies with cannabinoid receptors have been conducted in rodent brain using tritium [3H]-labeled CP55,940 or WIN55,212-2.[20] The results were presented as the concentration of SC displacing 50% of the radiolabeled compound from the receptor (IC50), with Ki being lower as the affinity is higher. In general, the binding affinity of SC to CB1 and/or CB2 is compared to the binding affinity of THC, which is 40.7 and 36.4 nM for CB1 and CB2, respectively.[25] The majority of SCs have Ki in the range 1–10 nM or 10–100 nM for both CB1 and CB2 receptors. Some of them present higher affinity and lower Ki than THC; for example, JWH-210 (naphthoylindole group), acts as a potent cannabinoid agonist with both the CB1 and CB2 receptors, with Ki values of 0.46 nM at CB1 and 0.69 nM at CB2.[26] The most potent agonist currently available is HU-210 with Ki values of 0.06 nM at CB1 and 0.52 nM at CB2.[27]

SC receptor binding affinities and selectivity may vary according to their functional group substitution.[28,29] In vitro or in vivo functional assays, which evaluate the drug's efficacy and potency, are needed to determine a drug's intrinsic activity as a receptor ligand, i.e., full partial, or inverse agonist, or neutral antagonist.[20] CB1 agonist activation down regulates adenylyl cyclase and decreases cellular cAMP levels, which triggers a cascade of reactions that affects cellular signaling and neurotransmitter inhibition, including acetylcholine, dopamine, noradrenaline, and glutamine, and γ-aminobutyric acid (GABA), while CB2 activation reduced cellular cAMP levels.[30–32] The majority of SCs are unspecific CB1/CB2 ligands; only a few of them are CB2 selective such as AM-1221 and A-836,339.[33,34] Generally, the agonists show little selectivity between the CB1 and CB2 receptors, while the antagonist compounds are highly selective.[35]

Pharmacokinetics and metabolism

As previously described, smoking is the main administration form as peak blood concentrations reach very quickly, then being distributed to other organs such as the brain.[20] After oral absorption, the effects are delayed due to the digestive process. In addition, a hepatic first-pass effect reduces the bioavailability of the products.[36] Due to their highly lipophilic nature, they generally have high volumes of distribution and tend to accumulate in adipose tissue in case of repeated consumption.[36] SCs undergo extensive hepatic metabolism, even if other organs, such as lung, intestine, and kidney are involved.[37] Phase I reactions, catalyzed by multiple CYPs, mainly 2C9 and 1A2, result in mono- or tri-hydroxylated, carboxylated, or N-dealkylated metabolites.[37,38] The mono-hydroxylated metabolites, most abundant, are strong agonists of CB1 and/or CB2, prolonging, and potentiating the activity of the parent drug.[39] Metabolites then undergo glucuronidation and/or sulfonation by UDP-glucuronyltransferases 1A1, 1A9, and 2B7. They are eliminated predominantly by renal route.[40]

Clinical aspects

In general, studies show that SCs induce stronger and faster effects than THC. The duration of the effects is also shorter, although some cannabinoids such as HU-210 presents a duration of action nearly five times longer than THC.[11]

The composition of SCs by producers is continuously modified to avoid detection and regulation.[41] In addition, several synthetic cannabinoids can be simultaneously added to the herbaceous mixture.[6] This results in significant variations in potency, efficacy, duration of action, and toxicity. Some users report a pleasant feeling of euphoria, joy, talkativeness, and relaxation, while others describe agitation, perceptual disturbances (fractal, geometric patterns flash of colors), altered sense of time, and mild cognitive impairments.[41,42] SC psychoactive effects may be more intense in individuals with any/minimal levels of previous exposure to cannabis.[43]

Many users reported also hot flashes, mydriasis, burning eyes, xerostomia, and tachycardia. Severe complications are mainly cardiac, neurologic, and psychiatric. Indeed, SC intake is associated with a 30-fold higher risk of seeking emergency room as compared with traditional cannabis.[44]

Acute cardiac symptoms are relatively common with SCs. They include tachy- or bradycardia, arrhythmias, cardiomyopathy, myocardial infarction, prolonged QT interval and torsade de pointe, cardiac arrest, and sudden death.[45] Acute respiratory depression with hypoxia, hypercapnia, and respiratory acidosis is also described.[46–48] Strokes and seizures are other severe effects occurring with SCs.[49,50] The use of SCs can also lead a variety of gastro-intestinal, such as emesis, abdominal pain, persistent cannabinoid hyperemesis syndrome, and renal disturbances with acute renal failure, hypokalemia, tubular necrosis.[48] Finally, insulin resistance and metabolic acidosis are also described.[51]

Several studies demonstrate an association between SCs repeated consumption and cognitive impairments in a wide range of domains including working memory, long-term memory, attention difficulties, executive and visuospatial functions as well as cognitive flexibility. These impairments were found to be significantly greater than in individuals with cannabis use disorder and healthy controls.[42,52,53] According to Livny et al., working memory impairment is associated with structural and functional deficits in several brain regions including the middle frontal gyrus, frontal orbital gyrus, inferior frontal gyrus, insula, anterior cingulate cortex, and precuneus.[54] Moreover, there is strong evidence that repeated exposure to SCs is associated with higher ratings of depression and anxiety compared to the group of recreational cannabis users and non-users.[55] High dosages appear to correlate with significant levels of anxiety as well as an increased risk of "bad trips" characterized by paranoid feelings, altered experience of self, and feelings of living in different/parallel realities.[56,57]

Presumably due to their high potency and the absence of CBD in preparations, single, or repeated use of SCs is associated with an increased risk of developing schizophrenic spectrum disorders, including perceptual alterations, depersonalization, dissociation, delusions, auditory and visual hallucinations, paranoid delusions, disorganized behavior and speech, catatonia, agitation, and suicidal thoughts.[6,24] Negative symptoms include blunted affect, emotional withdrawal, psychomotor retardation, and lack of spontaneity.[58] These disorders may occur in healthy patients but may also exacerbate symptoms present in patients with psychotic disorders. Interestingly, Cohen et al. show that chronic SC users differ from natural cannabis users and non-users on dimensions of specific personality traits and schizotypy.[59]

The abuse potential of these molecules is poorly studied. Nevertheless, several cases of addictive behavior with SCs have been described.[60] According to Tai and Fantegrossi, this abuse potential is related to the highly potent CB1 agonist activity.[61] Different studies using behavioral tasks, including intravenous self-administration (IVSA), conditioned place preference (CPP), drug discrimination techniques (DD) and intracranial self-stimulation (ICSS), were performed in rodents with different SCs (JWH series, HU-210, AM-2201, etc.). Results confirm that SCs effects are mostly mediated by CB1 as they can be prevented by pretreatment with the CB1 antagonist/inverse agonist AM-251.[14]

Recently, Hur et al. confirmed that AM-1248, CB-13, and PB-22 have reinforcing effects, reinforcement-enhancing effects, and impulsive effects in rodents.[62] Withdrawal symptoms in humans were reported to be like, but more severe than, those reported during cannabis cessation (disturbed sleep, anxiety, craving, nausea, cramps, chills).[63] Zimmermann et al. reported a severe withdrawal syndrome after stopping gold spice, including craving, agitation, nightmares, tachycardia, and hypertension.[64] Nacca et al. reported severe anxiety, sweating, chills, craving, and anorexia 6 days after stopping SCs.[65]

Several deaths have been described after SCs' exposure, either on their own or in combination with recreational drugs, including other NPS such as synthetic cathinones, or with prescription drugs.[66–68] Autopsies revealed different pathological findings, such as acute kidney injury, fulminant liver failure, or pulmonary oedema.[69–71]

Synthetic cannabinoids and fertility

The toxicity of synthetic cannabinoids on human reproduction is not fully understood and appreciated. However, because cannabis impacts human fertility, synthetic cannabinoids might be expected to interfere with cannabinoid signaling and thus compromise pregnancy, especially since many synthetic cannabinoids have much higher affinity and potency to bind to cannabinoid receptors.

During pregnancy, a tight balance between proliferation, differentiation, and apoptosis of trophoblast cells is required for a proper placental development. The endocannabinoid system has a key role in these physiological processes, including placentation, decidualization, and implantation. Phytocannabinoids are known to interfere with trophoblast turnover, preventing trophoblast cell death and differentiation, and are implicated in intrauterine growth retardation, low birth weight, pre-term labor, and neurodevelopment disturbances.[72–76] The mechanisms of action are not fully documented. To the best of our knowledge, THC impairs the secretion of pituitary hormones, such as luteinizing hormone (LH), follicle stimulating hormone (FSH), and sex steroid hormones[77]; THC can also impact endometrial stromal cells differentiation and placenta development.[78]

Data concerning SCs are very scarce, partly because the use of these molecules is recent, but the possibility that SCs adversely impact pregnancy outcome should not be neglected, as many of them have much higher affinity to cannabinoid receptors than the phytocannabinoids.

Studies on the effects of SCs in placental cytotrophoblast cells may help to better understand the impact of cannabinoids consumption in pregnancy. Almada et al. demonstrated, in BeWo cell line, a human placental cytotrophoblast cell model, that WIN-55,212 induce apoptosis of cytotrophoblast cells that are required for syncytium formation.[79] This could be mediated by a disruption of mitochondrial membrane potential and activation of caspases-9 and -3/-7, independently of reactive oxygen species (ROS) production. Interestingly, these effects were prevented by pre-incubation with a selective CB1 antagonist (AM281), while CB2 antagonist was not able to counteract it. With the same model, Almada et al. explored the effects of THC and of the SCBs JWH-018, JWH-122, and UR-144.[80] All the cannabinoids caused a significant decrease in cell viability, even if this effect was only detected for the highest concentrations of THC. JWH-018 and JWH-122 increased reactive oxygen species production, and UR-144 and JWH-122 caused loss of mitochondrial membrane potential. All these compounds were able to induce caspase-9 activation. The involvement of apoptotic pathways was confirmed through the significant increase in caspase-3/-7 activities. For UR-144, this effect was reversed by the CB1 antagonist AM281, for JWH-018 and THC this effect was mediated by both cannabinoid receptors CB1 and CB2, and finally, it was cannabinoid receptor-independent for JWH-122. Although the mechanisms are complex and not fully understood, these studies confirm that SCs decrease the viability of cytotrophoblast cells.

Endometrial stromal cells also play an important role in decidualization. Fonseca et al. studied the impact of SCBs JWH-122, UR-144 and WIN55,212-2 (WIN) on endometrial stromal cells by using a telomerase-immortalized human endometrial stromal cell line (St-T1b), and primary human decidual fibroblasts (HdF).[81] They demonstrated that JWH-122 and UR-144 induce prompt reactive nitrogen and oxygen species formation as well as endoplasmic reticulum (ER) stress without reduction in cell viability, as this cellular stress is then compensated by the increase in reduced/oxidized glutathione ratio. In contrast, WIN induces ER stress, mitochondrial dysfunction, and cell apoptosis. The addition of the CB1 antagonist AM281 significantly reduces the effects of WIN on cell viability. In conclusion, although they may act through different mechanisms and potencies, the studied cannabinoids have the potential to disrupt gestational fundamental events.

Synthetic cannabinoids and the developing brain

As described by Giua et al. cannabis in utero exposure may have strong neurodevelopmental consequences (cf. Chapter 8). Information concerning the effects of SCs on the developing brain are scarce. Nevertheless, some animal and in vitro studies have been carried out on this subject.

Psychoyos et al. exposed chick embryos to O-2545-HCL (0.035–0.35 mg/mL) at gastrulation and tested them for morphological defects at stages equivalent to 9–14 somites.[82] In embryos treated with low dose of O-2545-HCl, the neural folds fail to elevate and to fuse, a phenotype comparable to exencephaly in rodent systems. In those treated with moderate dose of O-2545-HCl, the brain is poorly segmented into forebrain, midbrain, and hindbrain primordia, in a phenotype comparable to anencephaly in rodents. These studies also reveal that O-2545-HCl also affects gene expression in the developing CNS. These defects resulted from the downregulation of Pax6 expression within the nascent neural tube, which hindered the neural tube's ability to close along the anteroposterior axis of the embryo. Gilbert et al. observed, after administration of 0.0625, 0.125, 0.25, 0.5, 1.0, and 2.0 mg/kg CP-55,940 in mice on gestational day 8, major craniofacies and/or eyes malformations in all treated groups.[83] Malformations included lateral and median facial clefts, cleft palate, microphthalmia, iridial coloboma, anophthalmia, exencephaly, holoprosencephaly, and cortical dysplasia. Mereu et al. reported that 40-day-old and 80-day-old rat pups exhibited lack of memory retention and hyperactivity after prenatal exposure to a nontoxic dose of WIN 55,212-2.[84] These effects were attributed to a significant reduction in hippocampal K+-mediated glutamate release. Del Arco et al. exposed Wistar rats to 1, 5, and 25 µg/kg HU-210 during gestation and lactation and observed a decreased responsiveness of the hypothalamo-pituitary-adrenal axis.[85] More recently, Carvahlo et al. investigate whether chronic systemic exposure to WIN 55,212-2 causes morphological changes in the structure of dendrites and dendritic spines in adolescent and adult male rats' pyramidal neurons in the medial pre-frontal cortex and medium spiny neurons in the nucleus accumbens.[86] While no structural changes were observed in WIN 55,212-2-treated adolescent compared to control, exposure to WIN 55,212-2 significantly increased dendritic length, spine density, and the number of dendritic branches in pyramidal neurons in the pre-frontal cortex of adult subjects when compared to control and adolescent subjects. In nucleus accumbens, exposure to WIN 55,212-2 significantly decreased dendritic length and number of branches in adult rat subjects while no changes were observed in the adolescent groups. In contrast, spine density was significantly decreased in both the adult and adolescent groups.

In vitro studies were also conducted. Kim and Thayer demonstrated that Win-55,212-22 blocked formation of new synapses in rat hippocampal neurons obtained from 17-d embryos.[87] This inhibition was stereoselective and was reversed by a selective CB1 receptor antagonist. Jiang et al. demonstrate that HU-210 promotes proliferation, but not differentiation, of cultured embryonic hippocampal stem cells likely via a sequential activation of CB1 receptors, G(i/o) proteins, and ERK signaling.[88] Oudin et al. demonstrated that the CB2 agonist JWH-133 enhanced neuroblast migration in the subventricular zone. Ferreira et al. described in neurosphere cultures of subventricular and hippocampal dentate gyrus from early postnatal Sprague-Dawley rats, cell proliferation, and differentiation after CB1 activation by win-55,212-2.[89]

In conclusion, synthetic cannabinoids act as CB1 receptor super agonists and, as such, can lead to acute and chronic complications. Due to the ubiquitous distribution of CB1 and CB2 receptors, multiple organs may be affected. Exposure during pregnancy can lead to obstetrical complications and developmental abnormalities.

The multiplicity and rapid emergence of these molecules on the black market makes the study of their effects particularly complex. The monitoring of these molecules is a major public health issue, as is information and prevention.

References

1. Synthetic cannabinoids in Europe – a review. *European Monitoring Center for Drugs and Drug Addiction* 2021. https://www.emcdda.europa.eu/system/files/publications/14035/Synthetic-cannabinoids-in-Europe-EMCDDA-technical-report.pdf/.
2. Papaseit E, Pérez-Mañà C, Pérez-Acevedo AP, et al. Cannabinoids: from pot to lab. *Int J Med Sci* 2018;**15**(12):1286–95. https://doi.org/10.7150/ijms.27087.
3. De Luca MA, Fattore L. Therapeutic use of synthetic cannabinoids: still an open issue? *Clin Ther* 2018;**40**(9):1457–66. https://doi.org/10.1016/j.clinthera.2018.08.002.
4. Zou S, Kumar U. Cannabinoid receptors and the endocannabinoid system: signaling and function in the central nervous system. *Int J Mol Sci* 2018;**19**(3):833–56. https://doi.org/10.3390/ijms19030833.
5. Hudson S, Ramsey J. The emergence and analysis of synthetic cannabinoids. *Drug Test Anal* 2011;**3**(7–8):466–78. https://doi.org/10.1002/dta.268.
6. Fattore L, Fratta W. Beyond THC: the new generation of cannabinoid designer drugs. *Front Behav Neurosci* 2011;**21**(5):60. https://doi.org/10.3389/fnbeh.2011.00060. eCollection 2011.
7. Gurley BJ, Murphy TP, Gul W, et al. Content versus label claims in cannabidiol (CBD)-containing products obtained from commercial outlets in the state of Mississippi. *Diet Suppl* 2020;**17**(5):599–607. https://doi.org/10.1080/19390211.2020.1766634.
8. Ti L, Tobias S, Maghsoudi N, et al. Detection of synthetic cannabinoid adulteration in the unregulated drug supply in three Canadian settings. *Drug Alcohol Rev* 2021;**40**(4):580–5. https://doi.org/10.1111/dar.13237.
9. Cottencin O, Rolland B, Karila L. New designer drugs (synthetic cannabinoids and synthetic cathinones): review of literature. *Curr Pharm Des* 2014;**20**(25):4106–11. https://doi.org/10.2174/13816128113199990622.
10. Debruyne D, Le Boisselier R. Emerging drugs of abuse: current perspectives on synthetic cannabinoids. *Subst Abus Rehabil* 2015;**6**:113–29. https://doi.org/10.2147/SAR.S73586.

11. Alves VL, Gonçalves JL, Aguiar J, et al. The synthetic cannabinoids phenomenon: from structure to toxicological properties. A review. *Crit Rev Toxicol* 2020;**50**(5):359–82. https://doi.org/10.1080/10408444.2020.1762539.
12. Karila L, Benyamina A, Blecha L, et al. The synthetic cannabinoids phenomenon. *Curr Pharm Des* 2016;**22**(42):6420–5. https://doi.org/10.2174/1381612822666160919093450.
13. Potts AJ, Cano C, Thomas SHL, Hill SL. Synthetic cannabinoid receptor agonists: classification and nomenclature. *Clin Toxicol (Phila)* 2020;**58**(2):82–98. https://doi.org/10.1080/15563650.2019.1661425.
14. Pintori N, Loi B, Mereu M. Synthetic cannabinoids: the hidden side of Spice drugs. *Behav Pharmacol* 2017;**28**(6):409–19. https://doi.org/10.1097/FBP.0000000000000323.
15. Campbell S, Poole R. Editorial. Disorderly street users of novel psychoactive substances: what might help? *Crim Behav Ment Health* 2020;**30**(2–3):53–8. https://doi.org/10.1002/cbm.2146.
16. Gray P, Ralphs R, Williams L. The use of synthetic cannabinoid receptor agonists (SCRAs) within the homeless population: motivations, harms and the implications for developing an appropriate response. *Addict Res Theory* 2021;**29**(1):1–10. https://doi.org/10.1080/16066359.2020.1730820.
17. Gunderson EW, Haughey HM, Ait-Daoud N, et al. A survey of synthetic cannabinoid consumption by current cannabis users. *Subst Abuse* 2014;**35**(2):35–41. https://doi.org/10.1080/08897077.2013.846288.
18. Young-Wolff KC, Tucker LY, Alexeeff S, et al. Trends in self-reported and biochemically tested marijuana use among pregnant females in California from 2009–2016. *JAMA* 2017;**318**(24):2490–1. https://doi.org/10.1001/jama.2017.17225.
19. García-González J, de Quadros B, Havelange W, et al. Behavioral effects of developmental exposure to JWH-018 in wild-type and disrupted in schizophrenia 1 (disc1) mutant zebrafish. *Biomolecules* 2021;**11**(2):319. https://doi.org/10.3390/biom11020319.
20. Castaneto MS, Gorelick DA, Desrosiers NA, et al. Synthetic cannabinoids: epidemiology, pharmacodynamics, and clinical implications. *Drug Alcohol Depend* 2014;**144**:12–41. https://doi.org/10.1016/j.drugalcdep.2014.08.005.
21. Obafemi AL, Kleinschmidt K, Goto C, Fout D. Cluster of acute toxicity from ingestion of synthetic cannabinoid-laced brownies. *J Med Toxicol* 2015;**44**(4):426–9. https://doi.org/10.1007/s13181-015-0482-z.
22. Phillips J, Lim F, Hsu R. The emerging threat of synthetic cannabinoids. *Nurs Manage* 2017;**48**(3):22–30. https://doi.org/10.1097/01.NUMA.0000512504.16830.b6.
23. García-Gutiérrez MS, Navarrete F, Gasparyan A, et al. Cannabidiol: a potential new alternative for the treatment of anxiety, depression, and psychotic disorders. *Biomolecules* 2020;**10**(11):1575–609. https://doi.org/10.3390/biom10111575.
24. Altintas M, Inanc L, Oruc GA, et al. Clinical characteristics of synthetic cannabinoid-induced psychosis in relation to schizophrenia: a single-center cross-sectional analysis of concurrently hospitalized patients. *Neuropsychiatr Dis Treat* 2016;**12**:1893–900. https://doi.org/10.2147/NDT.S107622.
25. Showalter VM, Compton DR, Martin BR, Abood ME. Evaluation of binding in a transfected cell line expressing a peripheral cannabinoid receptor (CB2): identification of cannabinoid receptor subtype selective ligands. *J Pharmacol Exp Ther* 1996;**278**(3):989–99.
26. Huffman JW, Zengin G, Wu MJ, et al. Structure-activity relationships for 1-alkyl-3-(1-naphthoyl)indoles at the cannabinoid CB(1) and CB(2) receptors: steric and electronic effects of naphthoyl substituents. New highly selective CB(2) receptor agonists. *Bioorg Med Chem* 2005;**13**(1):89–112. https://doi.org/10.1016/j.bmc.2004.09.050.

27. Gurney SM, Scott KS, Kacinko SL, et al. Pharmacology, toxicology, and adverse effects of synthetic cannabinoid drugs. *Forensic Sci Rev* 2014;**26**(1):53–78.
28. Aung MM, Griffin G, Huffman JW, et al. Influence of the N-1 alkyl chain length of cannabimimetic indoles upon CB1 and CB2 receptor binding. *Drug Alcohol Depend* 2000;**60**: 133–40. https://doi.org/10.1016/s0376-8716(99)00152-0.
29. Wiley JL, Marusich JA, Huffman JW. Moving around the molecule: relationship between chemical structure and in vivo activity of synthetic cannabinoids. *Life Sci* 2014;**97**(1):55–63. https://doi.org/10.1016/j.lfs.2013.09.011.
30. Mechoulam R, Parker LA. The endocannabinoid system and the brain. *Annu Rev Psychol* 2013;**64**:21–47. https://doi.org/10.1146/annurev-psych-113011-143739.
31. Pertwee RG. Receptors and channels targeted by synthetic cannabinoid receptor agonists and antagonists. *Curr Med Chem* 2010;**17**:1360–81. https://doi.org/10.2174/092986710790980050.
32. Slipetz DM, O'Neill GP, Favreau L, et al. Activation of the human peripheral cannabinoid receptor results in inhibition of adenylyl cyclase. *Mol Pharmacol* 1995;**48**(2):352–61.
33. Makriyannis A, Deng H. *Cannabimimetic indole derivatives*; 2007. WO patent 200128557. Granted 2001-06-07.
34. McGaraughty S, Chu KL, Dart MJ, et al. A CB(2) receptor agonist, A-836339, modulates wide dynamic range neuronal activity in neuropathic rats: contributions of spinal and peripheral CB(2) receptors. *Neuroscience* 2009;**158**(4):1652–61. https://doi.org/10.1016/j.neuroscience.2008.11.015.
35. Console-Bram L, Marcu J, Abood ME. Cannabinoid receptors: nomenclature and pharmacological principles. *Prog Neuro-Psychopharmacol Biol Psychiatry* 2012;**38**(1):4–15. https://doi.org/10.1016/j.pnpbp.2012.02.009.
36. *Synthetic cannabinoids in herbal products*. UNODC; 2011. https://www.unodc.org/documents/scientific/Synthetic_Cannabinoids.pdf/.
37. Diao X, Huestis MA. New synthetic cannabinoids metabolism and strategies to best identify optimal markers metabolites. *Front Chem* 2019;**7**:109–23. https://doi.org/10.3389/fchem.2019.00109.
38. Zendulka O, Dovrtělová G, Nosková K, et al. Cannabinoids and cytochrome P450 interactions. *Curr Drug Metab* 2016;**17**(3):206–26. https://doi.org/10.2174/1389200217666151210142051.
39. Rajasekaran M, Brents LK, Franks LN, et al. Human metabolites of synthetic cannabinoids JWH-018 and JWH-073 bind with high affinity and act as potent agonists at cannabinoid type-2 receptors. *Toxicol Appl Pharmacol* 2013;**269**(2):100–8. https://doi.org/10.1016/j.taap.2013.03.012.
40. Patton AL, Seely KA, Yarbrough AL, et al. Altered metabolism of synthetic cannabinoid JWH-018 by human cytochrome P450 2C9 and variants. *Biochem Biophys Res Commun* 2018;**498**(3):597–602. https://doi.org/10.1016/j.bbrc.2018.03.028.
41. Cooper ZD. Adverse effects of synthetic cannabinoids: management of acute toxicity and withdrawal. *Curr Psychiatry Rep* 2016;**18**(5):52–71. https://doi.org/10.1007/s11920-016-0694-1.
42. Cohen K, Weinstein A. The effects of cannabinoids on executive functions: evidence from cannabis and synthetic cannabinoids—a systematic review. *Brain Sci* 2018;**8**(3):40–59. https://doi.org/10.3390/brainsci8030040.
43. Hermanns-Clausen M, Kneisel S, Szabo B, Auwärter V. Acute toxicity due to the confirmed consumption of synthetic cannabinoids: clinical and laboratory findings. *Addiction* 2013;**108**(3):534–44. https://doi.org/10.1111/j.1360-0443.2012.04078.x.
44. Winstock A, Lynskey M, Borschmann R, Waldron J. Risk of emergency medical treatment following consumption of cannabis or synthetic cannabinoids in a large global sample. *J Psychopharmacol* 2015;**29**(6):698–703. https://doi.org/10.1177/0269881115574493.

45. Drummer OH, Gerostamoulos D, Woodford NW. Cannabis as a cause of death: a review. *Forensic Sci Int* 2019;**298**:298–306. https://doi.org/10.1016/j.forsciint.2019.03.007.
46. Jinwala FN, Gupta M. Synthetic cannabis and respiratory depression. *J Child Adolesc Psychopharmacol* 2012;**22**(6):459–62. https://doi.org/10.1089/cap.2011.0122.
47. Alon MH, Saint-Fleur MO. Synthetic cannabinoid induced acute respiratory depression: case series and literature review. *Respir Med Case Rep* 2017;**22**:137–41. https://doi.org/10.1016/j.rmcr.2017.07.011.
48. Bukke VN, Archana M, Villani R, et al. Pharmacological and toxicological effects of phytocannabinoids and recreational synthetic cannabinoids: increasing risk of public health. *Pharmaceuticals (Basel)* 2021;**14**(10):965–92. https://doi.org/10.3390/ph14100965.
49. Pacher P, Steffens S, Haskó G, et al. Cardiovascular effects of marijuana and synthetic cannabinoids: the good, the bad, and the ugly. *Nat Rev Cardiol* 2018;**15**(3):151–66. https://doi.org/10.1038/nrcardio.2017.130.
50. Gounder K, Dunuwille J, Dunne J, et al. The other side of the leaf: seizures associated with synthetic cannabinoid use. *Epilepsy Behav* 2020;**104**. https://doi.org/10.1016/j.yebeh.2020.106901, 106901.
51. Tournebize J, Gibaja V, Kahn JP. Acute effects of synthetic cannabinoids: update 2015. *Subst Abus* 2017;**38**(3):344–66. https://doi.org/10.1080/08897077.2016.1219438.
52. Cengel HY, Bozkurt M, Evren C, et al. Evaluation of cognitive functions in individuals with synthetic cannabinoid use disorder and comparison to individuals with cannabis use disorder. *Psychiatry Res* 2018;**262**:46–54. https://doi.org/10.1016/j.psychres.2018.01.046.
53. Cohen K, Mama Y, Rosca P, et al. Chronic use of synthetic cannabinoids is associated with impairment in working memory and mental flexibility. *Front Psychiatry* 2020;**11**:602. https://doi.org/10.3389/fpsyt.2020.00602.
54. Livny A, Cohen K, Tik N, et al. The effects of synthetic cannabinoids (SCs) on brain structure and function. *Eur Neuropsychopharmacol* 2018;**28**(9):1047–57. https://doi.org/10.1016/j.euroneuro.2018.07.095.
55. Cohen K, Kapitány-Fövény M, Mama Y, et al. The effects of synthetic cannabinoids on executive functions. *Psychopharmacology* 2017;**234**(7):1121–34. https://doi.org/10.1007/s00213-017-4546-4.
56. Soussan C, Kjellgren A. The flip side of 'Spice': the adverse effects of synthetic cannabinoids as discussed on a Swedish Internet forum. *NAD Nord Stud Alcohol Drugs* 2014;**31**(2):207–19. https://doi.org/10.2478/nsad-2014-006.
57. Bilgrei OR. From 'herbal highs' to the 'heroin of cannabis': exploring the evolving discourse on synthetic cannabinoid use in a Norwegian Internet drug forum. *Int J Drug Policy* 2016;**29**:1–8. https://doi.org/10.1016/j.drugpo.2016.01.011.
58. Radhakrishnan R, Wilkinson ST, D'Souza DC. Gone to pot – a review of the association between cannabis and psychosis. *Front Psychiatry* 2014;**5**:54. https://doi.org/10.3389/fpsyt.2014.00054.
59. Cohen K, Rosenzweig S, Rosca P, et al. Personality traits and psychotic proneness among chronic synthetic cannabinoid users. *Front Psychiatry* 2020;**11**:355. https://doi.org/10.3389/fpsyt.2020.00355.
60. Inci R, Kelekci KH, Oguz N, et al. Dermatological aspects of synthetic cannabinoid addiction. *Cutan Ocul Toxicol* 2017;**36**(2):125–31. https://doi.org/10.3109/15569527.2016.1169541.
61. Tai S, Fantegrossi WE. Synthetic cannabinoids: pharmacology, behavioral effects, and abuse potential. *Curr Addict Rep* 2014;**1**(2):129–36. https://doi.org/10.1007/s40429-014-0014-y.

62. Hur KH, Ma SX, Lee BR, et al. Abuse potential of synthetic cannabinoids: AM-1248, CB-13, and PB-22. *Biomol Ther (Seoul)* 2021;**29**(4):384–91. https://doi.org/10.4062/biomolther.2020.212.
63. Andrabi S, Greene S, Moukaddam N, Li B. New drugs of abuse and withdrawal syndromes. *Emerg Med Clin North Am* 2015;**33**(4):779–95. https://doi.org/10.1016/j.emc.2015.07.006.
64. Zimmermann US, Winkelmann PR, Pilhatsch M, et al. Withdrawal phenomena and dependence syndrome after the consumption of "spice gold". *Dtsch Arztebl Int* 2009;**106**(27):464–7. https://doi.org/10.3238/arztebl.2009.0464.
65. Nacca N, Vatti D, Sullivan R, et al. The synthetic cannabinoid withdrawal syndrome. *J Addict Med* 2013;**7**(4):296–8. https://doi.org/10.1097/ADM.0b013e31828e1881.
66. Fujita Y, Koeda A, Fujino Y, et al. Clinical and toxicological findings of acute intoxication with synthetic cannabinoids and cathinones. *Acute Med Surg* 2015;**3**(3):230–6. https://doi.org/10.1002/ams2.182.
67. Kovács K, Kereszty É, Berkecz R, et al. Fatal intoxication of a regular drug user following N-ethyl-hexedrone and ADB-FUBINACA consumption. *J Forensic Legal Med* 2019;**65**: 92–100. https://doi.org/10.1016/j.jflm.2019.04.012.
68. Martinotti G, Santacroce R, Papanti D, et al. Synthetic cannabinoids: psychopharmacology, clinical aspects, psychotic onset. *CNS Neurol Disord Drug Targets* 2017;**16**(5):567–75. https://doi.org/10.2174/1871527316666170413101839.
69. Shanks KG, Winston D, Heidingsfelder J, Behonick G. Case reports of synthetic cannabinoid XLR-11 associated fatalities. *Forensic Sci Int* 2015;**252**:e6–9. https://doi.org/10.1016/j.forsciint.2015.04.021.
70. Shanks KG, Clark W, Behonick G. Death associated with the use of the synthetic cannabinoid ADB-FUBINACA. *J Anal Toxicol* 2016;**40**(3):236–9. https://doi.org/10.1093/jat/bkv142.
71. Shanks KG, Behonick GS. Death after use of the synthetic cannabinoid 5F-AMB. *Forensic Sci Int* 2016;**262**:e21–4. https://doi.org/10.1016/j.forsciint.2016.03.004.
72. Costa MA, Fonseca BM, Marques F, et al. The psychoactive compound of Cannabis sativa, delta(9)-tetrahydrocannabinol (THC) inhibits the human trophoblast cell turnover. *Toxicology* 2015;**334**:94–103. https://doi.org/10.1016/j.tox.2015.06.005.
73. Fergusson DM, Horwood LJ, Northstone K, et al. Maternal use of cannabis and pregnancy outcome. *BJOG* 2002;**109**(1):21–7. https://doi.org/10.1111/j.1471-0528.2002.01020.x.
74. Schneider M. Cannabis use in pregnancy and early life and its consequences: animal models. *Eur Arch Psychiatry Clin Neurosci* 2009;**259**(7):383–93. https://doi.org/10.1007/s00406-009-0026-0.
75. Jaques SC, Kingsbury A, Henshcke P, et al. Cannabis, the pregnant woman and her child: weeding out the myths. *J Perinatol* 2014;**34**(6):417–24. https://doi.org/10.1038/jp.2013.180.
76. Castel P, Simon P, Barbier M, et al. Focus on the endocannabinoid system and the reprotoxicity of marijuana in female users. *Gynecol Obstet Fertil Senol* 2020;**48**(4):384–92. https://doi.org/10.1016/j.gofs.2020.01.024.
77. Park B, McPartland JM, Glass M. Cannabis, cannabinoids and reproduction. *Prostaglandins Leukot Essent Fatty Acids* 2004;**70**(2):189–97. https://doi.org/10.1016/j.plefa.2003.04.007.
78. Almada M, Amaral C, Diniz-da-Costa M, Correia-da-Silva G, Teixeira NA, Fonseca BM. The endocannabinoid anandamide impairs in vitro decidualization of human cells. *Reproduction* 2016;**152**(4):351–61. https://doi.org/10.1530/REP-16-0364.
79. Almada M, Costa L, Fonseca BM, et al. The synthetic cannabinoid WIN-55,212 induced-apoptosis in cytotrophoblasts cells by a mechanism dependent on CB1 receptor. *Toxicology* 2017;**385**:67–73. https://doi.org/10.1016/j.tox.2017.04.013.

80. Almada M, Alves P, Fonseca BM, et al. Synthetic cannabinoids JWH-018, JWH-122, UR-144 and the phytocannabinoid THC activate apoptosis in placental cells. *Toxicol Lett* 2020;**319**: 129–37. https://doi.org/10.1016/j.toxlet.2019.11.004.
81. Fonseca BM, Fernandes R, Almada M, et al. Synthetic cannabinoids and endometrial stromal cell fate: dissimilar effects of JWH-122, UR-144 and WIN55,212-2. *Toxicology* 2019;**413**:40–7. https://doi.org/10.1016/j.tox.2018.11.006.
82. Psychoyos D, Hungund B, Cooper T, Finnell RH. A cannabinoid analogue of Delta9-tetrahydrocannabinol disrupts neural development in chick. *Birth Defects Res B Dev Reprod Toxicol* 2008;**83**(5):477–88. https://doi.org/10.1002/bdrb.20166.
83. Gilbert MT, Sulik KK, Fish EW, et al. Dose-dependent teratogenicity of the synthetic cannabinoid CP-55,940 in mice. *Neurotoxicol Teratol* 2016;**58**:15–22. https://doi.org/10.1016/j.ntt.2015.12.004.
84. Mereu G, Fa M, Ferraro L, et al. Prenatal exposure to a cannabinoid agonist produces memory deficits linked to dysfunction in hippocampal long-term potentiation and glutamate release. *Proc Natl Acad Sci U S A* 2003;**100**(8):4915–20. https://doi.org/10.1073/pnas.0537849100.
85. Del Arco I, Muñoz R, Rodríguez De Fonseca F, et al. Maternal exposure to the synthetic cannabinoid HU-210: effects on the endocrine and immune systems of the adult male offspring. *Neuroimmunomodulation* 2000;**7**(1):16–26. https://doi.org/10.1159/000026416.
86. Carvalho AF, Reyes BA, Ramalhosa F, et al. Repeated administration of a synthetic cannabinoid receptor agonist differentially affects cortical and accumbal neuronal morphology in adolescent and adult rats. *Brain Struct Funct* 2016;**221**(1):407–19. https://doi.org/10.1007/s00429-014-0914-6.
87. Kim DJ, Thayer SA. Activation of CB1 cannabinoid receptors inhibits neurotransmitter release from identified synaptic sites in rat hippocampal cultures. *Brain Res* 2000;**852**(2): 398–405. https://doi.org/10.1016/s0006-8993(99)02210-6.
88. Jiang W, Zhang Y, Xiao L, et al. Cannabinoids promote embryonic and adult hippocampus neurogenesis and produce anxiolytic- and antidepressant-like effects. *J Clin Invest* 2005;**115** (11):3104–16. https://doi.org/10.1172/JCI25509.
89. Oudin MJ, Gajendra S, Williams G, et al. Endocannabinoids regulate the migration of subventricular zone-derived neuroblasts in the postnatal brain. *J Neurosci* 2011;**31**(11): 4000–11. https://doi.org/10.1523/JNEUROSCI.5483-10.2011.

Chapter 13

Prenatal THC exposure interferes with the neurodevelopmental role of endocannabinoid signaling

Ismael Galve-Roperh[a,b], Adán de Salas-Quiroga[a,b,c], Samuel Simón Sánchez[a,b], and Manuel Guzmán[a,b]

[a]School of Biology and Instituto Universitario de Investigación Neuroquímica (IUIN), Complutense University, Madrid, Spain, [b]Centro de Investigación Biomédica en Red Enfermedades Neurodegenerativas (CIBERNED) and Instituto Ramón y Cajal de Investigaciones Sanitarias (IRYCIS), Madrid, Spain, [c]Champalimaud Centre for the Unknown, Lisboa, Portugal

Abbreviations

AEA	anandamide
ASD	autism spectrum disorder
CCK	cholecystokinin
CSMN	corticospinal motor neuron
DAGL	diacylglycerol lipase
ECB	endocannabinoid
ERK	extracellular regulated kinase
ES	embryonic stem
FCD	focal cortical dysplasia
hiPSC	human-induced pluripotent stem cell
MAGL	monoacylglycerol lipase
NDD	neurodevelopmental disorders
NP	neural progenitor
PCE	prenatal cannabinoid exposure
PI3K	phosphatidylinositol 3-kinase
VZ/SVZ	ventricular and subventricular zone
2AG	2-arachidonoylglycerol

Introduction

The contribution of the endocannabinoid (ECB) system (ECBs, their metabolizing enzymes, and CB_1 and CB_2 receptors) to brain development has been the subject of intense research in the last decades. Many studies have addressed the neurodevelopmental consequences of ECB signaling manipulation by using different genetic paradigms, conditional knockout mouse models and silencing strategies for various elements of the ECB system.[1,2] Other studies have also used administration of exogenous cannabinoids [i.e., plant-derived Δ^9-tetrahydrocannabinol (THC) and synthetic receptor agonists as WIN 55,212-2, CP-55,940, or others] during different developmental windows to investigate the consequences of prenatal and perinatal cannabinoid exposure (PCE).[3,4] In this chapter, we will focus particularly on PCE to THC, as a CB_1 receptor targeting molecule, and its impact on the neurodevelopmental role of the ECB system. Other plant-derived cannabinoids of interest, such as cannabidiol, have been less investigated and may involve, at least in part, ECB system-unrelated mechanisms of action that will not be discussed herein. The development of novel stem cell-derived methodologies, like human induced-pluripotent stem cell (hiPSC)-derived neuronal differentiation paradigms, and most recently, three-dimensional cerebral organoids, have emerged as powerful approaches to investigate the neurodevelopmental functions of the ECB system, as well as the impact of PCE in the immature human brain. The available experimental evidence reveals that the consequences of PCE in adulthood are due to its interference with CB_1 receptor signaling in different developing neural cell populations, which in turn control crucial developmental processes (proliferation, differentiation, migration, and maturation) responsible for neuronal network assembly and brain development.

The ECB system in neurogenic niches

The expression and function of ECB signaling elements in embryonic forebrain development has been extensively investigated. These studies have demonstrated a regulatory role of CB_1 receptors both in proliferating neural progenitor (NP) cells and in developing neurons, controlling their differentiation, migration and axonal pathfinding[1,4] (Fig. 1A). CB_1 receptor ablation interferes with NP cell proliferation in the embryonic ventricular and subventricular zones (VZ/SVZ).[5] CB_1 receptors control the transition from apical to basal progenitor cells by regulating the homeodomain transcription factor Pax6 activity and its proneurogenic action via Tbr2 (Eomes).[6] Noteworthy, the proliferative role of CB_1 receptors in embryonic NPs is conserved in the adult hippocampal and SVZ neurogenic niches[5,7–9] (Fig. 1B). Selective ablation of CB_1 receptors in neural stem cells interferes with NP proliferation, decreasing neurogenesis and affecting dendritic maturation. The existence of a NP-derived ECB tone involved in cell proliferation[5] has been confirmed in diacylglycerol lipase

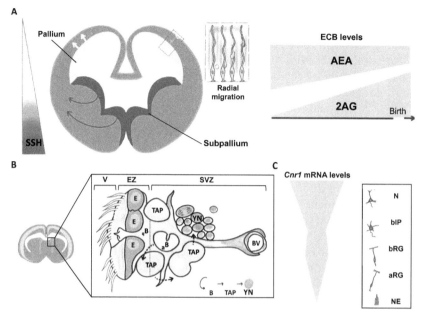

FIG. 1 Expression of the ECB system elements during forebrain embryonic development and in adult neurogenic areas. (A) The ECB system is expressed and functional in embryonic development, regulating neurogenesis of projection neurons (pallium) and interneuron development (subpallium). Changes of ECB levels during prenatal developmental stages. (B) The expression of the ECB system in embryonic brain is conserved in the adult neurogenic areas: hippocampus and subventricular zone. (C) Gene expression analyses demonstrate the dynamic regulation of CB_1 receptor levels in different neural cell populations (progenitors, neuroblasts and neurons). See main text for details and references.

(DAGL) α knockout mice, as well as by the specific inducible deletion of *Dagla* in astrocytes located in neurogenic niches.[10,11] Recently, a role for ECB signaling via CB_1 receptors in Muller glia transition to progenitor cells upon retinal injury has been reported.[12] In addition, SVZ neural stem cell-derived ECB tone has been proposed to act as a neuroprotective signal against brain excitotoxicity injury.[13] The involvement of ECB signaling in tuning adult neurogenesis may have important implications not only in cognitive, anxiety and depression processes regulated by newborn hippocampal and SVZ-derived neurons, but also in brain injury and neurodegenerative diseases.[14,15]

Despite the fact that the functional role of CB_1 receptors in NPs has been demonstrated in many studies,[5,16,17] their expression levels in progenitor cells in culture are low, and possibly they are even lower in vivo.[18] Hence, the cell-autonomous role of CB_1 receptors in NP proliferation in vivo is still a matter of debate. Transcriptomic analyses of *Cnr1* mRNA levels reveal a coincident pattern of expression in mouse and human cells. These data indicate that human CB_1 receptor expression occurs in a gradient from apical radial glial progenitors (aRG) basal RG neurons[19], and also point to lower CB_1 expression levels in human neurons. The existence of higher levels of CB_1 receptor expression in basal RG cells is in agreement with its role in the transition from Pax6$^+$ apical RG to basal Tbr2$^+$ NPs.[6] Using mouse embryonic stem (ES) cell-derived models of pyramidal neuron differentiation, CB_1 receptor levels have been shown to increase along differentiation,[20] similarly to what is observed in vivo.[21,22] In addition, dynamic changes in ECB levels are observed. Decreased anandamide (AEA) levels occur at later stages of prenatal development, while increased 2-arachidonoylglycerol (2AG) levels become predominant (Fig. 1C). Hence, a dynamic regulation of the ECB tone and CB_1 receptors is required for the appropriate control of the different NP cell populations and neuronal differentiation of projection neurons. Likewise, the ab hydrolase domain containing 4 (ABHD4) enzyme, which is involved in AEA synthesis, has been shown to participate in the elimination of detached cells during cortical development. Hence, ABHD4 levels are downregulated in postnatal brain to prevent anoikis of delaminated neuroblasts.[23] Signal transduction studies have delineated the mechanism of action of CB_1 receptors in the embryonic developing brain. CB_1 receptors in NPs regulate cell proliferation and identity via phosphatidylinositol 3-kinase (PI3K)/Akt and ERK signaling that activate mammalian target of rapamycin complex 1 (mTORC1), which, in turn, controls the activity of the transcription factor Pax6 and its proneurogenic action via Tbr2[6,24] (Fig. 2A). Also, CB_1 receptor-induced PI3K/Akt

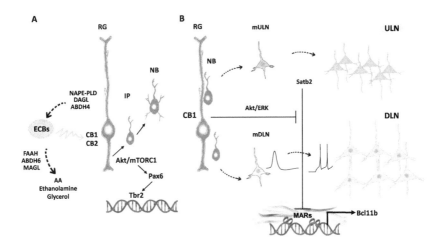

FIG. 2 Neural cell fate regulation of projection neurons by ECB signaling in prenatal development. (A) The ECB system regulates neural progenitor cell proliferation and neurogenesis. CB_1 receptors in NPs regulate cell proliferation and the transition from apical to basal intermediate progenitors via PI3K/Akt/mTORC1 signaling and regulating the transcription factors Pax6 and Tbr2. (B) In post-mitotic neurons, CB_1 receptors favor deep-layer projection neuron development by regulating the neurogenic program controlled by Bcl11b and Satb2.

signaling regulates Brca1, a member of the E3 ubiquitin ligase 3 family that also exerts transcriptional regulation, and is involved in neurite outgrowth.[24] In addition, CB_1 receptor-mediated PI3K/Akt signaling induces GSK3β and β-catenin activation that contribute to promote NP cell proliferation.[25] ECB signaling in NPs synergizes with EGF receptor and other growth factor-evoked mechanisms, acting as integrating signal modulators.[26] As discussed below, specific CB_1 receptor signaling pathways are regulated in differentiating post-mitotic neurons, and are responsible for controlling neuronal migration, differentiation, and axonal growth regulation.

Role of ECB signaling in pyramidal neuron development

Prenatal ECB signaling regulates the development of different neuronal populations, being glutamatergic pyramidal neurons, a subpopulation of basket cell GABAergic interneurons and dopaminergic neurons the most largely studied and affected by PCE.[3,4] CB_1 receptor signaling regulates long-range axon projection neuron development,[27–29] as well as short-range connectivity by influencing the morphogenesis of cholecystokinin (CCK)-expressing basket cell interneurons.[30] Studies with complete deletion and conditional CB_1 receptor loss of function in telencephalic glutamatergic neurons have demonstrated a role for the ECB system in promoting deep-layer corticothalamic and corticospinal motor neuron (CSMN) generation.[27–29] In corticothalamic neuron development, which is coordinated with thalamocortical axon growth by the "hand-shake model," appropriate subcellular distribution of CB_1 receptors, receptor membrane exposure and internalization, and the existence of a delicate temporal and spatial regulation of 2AG levels, have been reported.[29,31]

CB_1 receptors act post-mitotically in promoting the differentiation and maturation of deep layer pyramidal neurons by regulating the activity of Bcl11b and Satb2 transcription factors[20,28,32] (Fig. 2B). CB_1 receptor signaling favors *Bcl11b* transcription by alleviating Satb2-mediated repression via MAR sequences, and this allows the balanced expression of netrin1-mediated growth cone signaling receptors Dcc (deleted in colorectal carcinoma) and Unc5C,[33] hence contributing to the development of subcortical projection neurons. Proteomic studies in THC-administered mice revealed that CB_1 receptors, via JNK signaling, modulate the levels of the microtubule regulator protein superior cervical ganglion 10, controlling axonal morphology and corticofugal development.[34] The ECB system also influences corticofugal neuron development by controlling oligodendrocyte-derived Slit2, that acts on neuronal Robo1 receptors.[35] The role of ECB signaling in CSMN development has been confirmed in vitro by using ES-derived neuronal differentiation and hiPSC-derived cerebral organoid models.[20] Increased 2AG levels induced by the monoacylglycerol lipase (MAGL) inhibitor JZL-184, or the presence of THC during neuronal differentiation, promoted the generation of CSMNs and favored their maturation and acquisition of neuronal activity.

CB$_1$ receptor regulation of axonal growth and dendritic arborization relies on its ability to regulate actomyosin network dynamics. CB$_1$ receptors interact with members of the WAVE1 complex (Wiskott-Aldrich syndrome protein-family verprolin-homologous protein 1) and the Rho GTPase Rac1 in controlling actin nucleation and dynamics.[36] In addition, CB$_1$ receptors, in a cell context-dependent manner, can couple to G proteins other than G$_{i/o}$. Hence, CB$_1$ signaling can occur via G$_{12/13}$ and G$_z$, which regulates RhoA and downstream ROCK that controls cytoskeletal dynamics and axon growth.[37] CB$_1$ receptor regulation of axon growth relies on kinesin-1-dependent transport.[38] Thus, in the absence of kinesin-1, corticofugal axonal growth is resistant to CB$_1$ regulation, and kinesin-1 deficient mice reproduce the corticofugal axonal alterations observed in CB$_1$ receptor-deficient mice.[27,29] In retinal neurons, CB$_1$ receptors regulate axonal growth cone guidance, via crosstalk with netrin-1 receptor Dcc, controlling its membrane exposure in a PKA-dependent manner.[39] The role of CB$_1$ receptors in CSMN development and differentiation is mediated at least in part by ERK and Akt signaling, but independently of cAMP levels and mTORC1 activity.[20] In line with these findings, ERK signaling is known to be essential for the development and acquisition of neuronal excitability of Bcl11b$^+$ layer V neurons.[40] In vitro studies of neurite outgrowth regulation show that CB$_1$ receptors control the ERK cascade by acting at different levels, and can influence the duration (i.e., Rap1 vs. Ras and B-Raf vs. Raf1 involvement), tune the activity of small GTPase proteins, GTP exchanging factors and GTPase-activating proteins, and releasing cAMP mediated-ERK inhibition.[41,42] Recently, super-resolution microscopy analyses by STORM revealed that CB$_1$ receptor localization in membrane periodic skeleton structures is necessary for ERK activation via tyrosine kinase receptor transactivation.[43]

In addition to regulate NP proliferation and differentiation, the ECB system participates in various forms of neuronal migration. During embryonic development, CB$_1$ receptor signaling controls the radial migration of newborn cortical pyramidal neurons (Fig. 3A) and the tangential migration of GABAergic interneurons.[44–46] Moreover, in the postnatal brain, CB$_1$ and CB$_2$ receptors, engaged by 2AG, regulate the migration of SVZ-derived neuroblasts through the rostral migratory stream.[47] ECBs, by acting through CB$_1$ receptors, exert a promigratory function by ensuring RhoA proteasomal degradation in migrating newborn pyramidal neurons.[46] RhoA downregulation is required for neuronal migration out of the VZ/SVZ and hence its levels are tightly controlled, among other mechanisms, by the KCTD13/CUL3 ubiquitin ligase, a prevalent gene associated to autism spectrum disorder (ASD) and schizophrenia.[48] Importantly, RhoA signaling is also crucial in netrin-1/DCC mediated axon guidance decisions[49] and therefore RhoA may act as a signaling integration hub downstream of CB$_1$ receptors in regulating neural cell fate, migration, and morphogenesis.

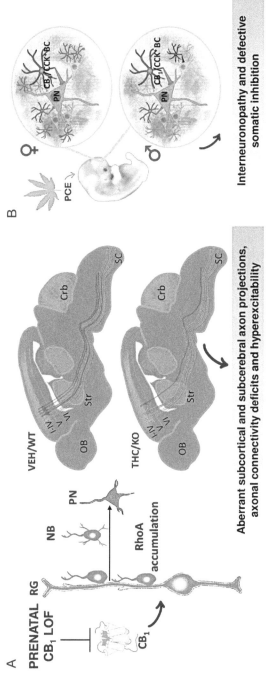

FIG. 3 Prenatal CB_1 receptor regulates neuronal migration and CCK^+ basket cell interneuron development. (A) Altered cortical laminarization owing to radial migration interference and aberrant long-range subcerebral axon projections are observed upon conditional embryonic CB_1 receptor loss of function. (B) PCE induces CB_1/CCK^+ basket cell interneuronopathy and defective somatic inhibition. Sex-dependent phenotypic differences to PCE have been highlighted for various neuronal populations.

Role of the ECB system in GABAergic interneuron development

Within the adult neocortex and hippocampus, CB_1 is largely enriched in GABAergic interneurons, particularly in CCK^+ basket cells and a subset of calbindin D28K-positive interneurons,[50,51] whereas low but consistent expression is detected in other subtypes and in projection neurons.[52] During mouse GABAergic interneuron development, CB_1 receptor-immunoreactive cells are detected by E12.5 in the subpial area of the ganglionic eminences, and are later enriched in the migratory streams at the marginal zone and preplate of the dorsal telencephalon, enroute to their final destinations in the hippocampus and neocortex.[53,54] CB_1 receptors control sequential steps of GABAergic interneuron ontogeny, as their tangential migration and positioning, morphogenesis and wiring.[30,44,45] AEA has been shown to regulate interneuron migration and morphogenesis by CB_1 receptor-dependent transactivation of TrkB signaling.[44] Likewise, activation of CB_1 receptors by AEA or WIN 55,212-2 promotes growth cone collapse and neurite retraction by driving the ERK and RhoA signaling cascades. Accordingly, conditional GABA-CB_1 knockout mice display impaired target selection in cortical interneurons.[30] Another study showed that deletion of CB_1 receptors reduces parvalbumin (PV) levels in fast spiking interneurons, alters dendritic distribution of dopamine D2 receptors, and affects mitochondrial numbers in PV-expressing interneurons in the medial prefrontal cortex (PFC), with potential implications for neuropsychiatric disorders.[55] Therefore, both genetic and pharmacologic disturbances of the ECB system result in defects in interneuron synaptic integration and excitatory/inhibitory (E/I) imbalance with a wide range of functional consequences. In addition to the abovementioned studies, there is abundant literature describing molecular, electrophysiological, and behavioral alterations relying on GABAergic interneuron dysfunction after systemic or conditional genetic manipulation of the ECB system[56] (i.e., selective ablation of *Cnr1* in interneurons, its conditional expression rescue[57–59] or *Dagla* deletion[60]). However, none of these studies have identified whether the deficits dependent on CB_1 dysfunction occur during GABAergic interneuron development or are caused by the lack of the receptor for proper adult brain function. Investigating the consequences of embryonically restricted administration of THC (or other CB_1 agonists), combined with the use of genetic models, is essential to elucidate the molecular and cellular mechanisms governed by the ECB system in regulating interneuron development and function (see following section). Interestingly, a recent study found different CB_1 receptor-interacting proteins in early developing interneurons and late, mature interneurons,[61] pointing to a functional role of CB_1 signaling in interneuron development.

Long-lasting consequences of prenatal THC exposure

The intrauterine developmental period is very sensitive to changes in endogenous regulatory mechanisms and to environmentally-induced prenatal stress.

Deregulated neurodevelopmental signaling pathways induce changes in NP proliferation, neuronal differentiation, synaptic connectivity, and neuronal network assembly, which constitute the ethiopathological substrates responsible for neurological and psychiatric neurodevelopmental disorders (NDD).[62] It is hence not surprising that PCE influences a variety of neurological and neuropsychiatric functions in the offspring, including E/I balance and neuronal synchrony, motor function, plasticity of the limbic reward system and cognition.[3,4] PCE impacts different aspects of motor function in adulthood, ranging from a hyperactive phenotype to specific changes in skilled motor function. Whereas it is difficult to elucidate the neurobiological mechanisms responsible of PCE consequences, the neuronal populations involved in some functions have been underscored. Hence, CB_1 receptor-driven deep-layer CSMN development alterations are responsible for skilled motor activity deficits observed in THC-treated mice.[3] The contribution of cerebellar alterations induced by prenatal THC exposure to motor function remains largely unexplored, and is a plausible hypothesis.[63] In this regard, CB_1 receptors regulate cerebellar secondary fissure formation.[64] Acute interference with ECB signaling, as induced by PCE and long-term plasticity changes (i.e., CB_1 receptor reorganization and/or desensitization, changes in ECB-metabolic enzymes, etc.), impacts different developmental processes (neural cell generation, differentiation, connectivity, and maturation) according to the precise developmental period affected. Thus, at prenatal stages, neuronal cell generation and differentiation are primarily affected by PCE,[2] while at perinatal stages neuronal maturation, connectivity,[3,4] and glial cell development[65–67] are the primarily affected neurodevelopmental processes. Once synaptic neuronal activity takes place, altered CB_1 receptor function is responsible for changes in neurotransmitter release and fine-tunes neuronal activity.[68] Alteration of levels and homeostasis of neurotransmitters (GABA, glutamate, serotonin, dopamine, and others), as induced by PCE, which are important modulators of NPs in mammalian cortical development,[19] likely contribute to the emotional, cognitive, and motor function alterations found in adulthood.[69]

Projection neurons

Prenatal pharmacological manipulation of CB_1 receptor signaling by administration of THC, WIN 55,212-2 or the CB_1 inverse agonist SR-141716, impact on long-range projection neuron development, deep-layer neuronal differentiation[27,32,34,70] and acquisition of pyramidal intrinsic excitability.[20,71] Importantly, despite the fact that THC and WIN 55,212-2 are agonists at CB_1 receptors, sustained administration of both compounds can result in attenuated embryonic CB_1 signaling induced by receptor desensitization and downregulation.[32] In addition, exogenous cannabinoids disrupt the proper balance of spatiotemporally restricted ECB cues and can induce changes in the levels of the enzymes responsible for ECB metabolism.[34,72] Consequently, PCE leads to a transient loss of function of CB_1 signaling and mimics some of the molecular, cellular, or functional traits observed upon CB_1 genetic deletion. Thus, in mice, PCE results in fewer CSMNs, impaired skilled motor function and increased seizure susceptibility, a similar phenotype to that found in CB_1-deficient littermates. Notably, conditional rescue of CB_1 receptor expression in different neuronal populations allowed to unequivocally identify the cellular substrates of PCE and its impact on the functional deficits observed.[32] In rats, sex-dependent mechanistic differences were observed in the PFC after PCE, and social interaction deficits were also sex-dimorphic, with males primarily affected.[71] Perinatal THC administration via lactation or WIN 55,212-2 administration induced deficits in social discrimination accompanied by changes in social exploratory behavior that are accompanied by loss of ECB-LTD, LTP, and augmented mGlu2/3 LTD.[73,74] Several studies have addressed potential strategies to rescue the consequences of PCE. Administration of an mGluR5 positive allosteric modulator or increasing ECB levels induced by the FAAH inhibitor URB597, can revert some of the functional alterations in the adult rat PFC induced by THC due to changes in pyramidal neuronal excitability and maturation.[71] Surprisingly, sex-dependent mechanistic differences were observed and TRPV1, but not CB_1 receptors, were involved in altered PFC long-term depression (LTD). Others have rescued PCE consequences by interfering with CB_1 receptor regulation (see below).

Dopaminergic neurons

Early studies focused on PCE in the corticolimbic system and found that PCE induces dopamine alterations.[75] Both prenatal[76,77] and postnatal THC administration[78] affect dopaminergic function (see Chapter 3). PCE decreased dopamine receptor D2R expression in cortical and subcortical areas, as a likely consequence of the hyperdopaminergic phenotype in the ventral tegmental area, increasing THC sensitivity on adolescence. Again, a sex-dependent phenotype was evident, with only male offspring being affected.[79] Pregnenolone, a neurosteroid that acts as a negative allosteric modulator of CB_1 receptors, including those located in the mitochondria,[80] rescues synaptic defects and normalizes dopaminergic activity and behavior in the PCE offspring.[77] These results highlight the potential of PCE to impact sensitive populations and sensitize the adult brain to psychotic states by inducing plastic adaptations of the mesolimbic dopaminergic system.

GABAergic interneurons

The impact of PCE in GABAergic interneuron development and function has been extensively investigated. Prenatal exposure to THC leads to a miswiring of CB_1/CCK^+ basket cells in the hippocampus and a decrease in the number of perisomatic boutons around pyramidal cell somata in CA1, with corresponding defects in LTD and presynaptic neurotransmitter release in adult mice[34] (Fig. 3B). Furthermore, prenatal THC or WIN 55,212-2 administration causes a selective decrease in the total number and dendritic complexity of CB_1/CCK^+ interneurons in the hippocampus, since other interneuron cell populations seem not affected by PCE. Accordingly, CB_1/CCK^+ interneuron-dependent feedforward and feedback inhibition is compromised, and adult mice exhibit altered social behavior.[81] A subsequent study demonstrated that the reported PCE-induced CB_1/CCK^+ deficiency also alters hippocampal oscillations, increases brain excitability, and preferentially impairs spatial memory performance. Importantly, a striking sexual dimorphism was evident, and only the male offspring prenatally exposed to THC exhibited CB_1/CCK^+ interneuronopathy and its corresponding functional deficits.[82] Recently, prenatal WIN 55,212-2 or THC administration were found to delay the expression of the K^+/Cl^- transporters potassium-chloride cotransporter 2 (KCC2) and sodium-potassium-chloride transporter (NKCC1), and hence the GABA excitatory to inhibitory transition.[83]

Importantly, most of the abovementioned pharmacological studies allow to identify developmentally restricted processes controlled by the ECB system, but in most cases fail to prove CB_1 receptor-dependency and to discriminate between direct impact (i.e., cell-autonomous GABAergic interneuron development) or indirect, non-cell-autonomous actions that could, in turn, alter interneuron development. Cortical projection neuron populations control the recruitment and layering of cortical interneurons, and thus the final assembly of local inhibitory circuits.[84] Projection neurons express different trophic factors and cell adhesion molecules involved in the selective subtype distribution of various cortical interneuron subtypes. For instance, different isoforms of neuregulin 1 expressed by cortical pyramidal neurons control the distribution and development of cortical interneurons, which differentially express its receptor ErbB4.[85,86] Likewise, dystroglycan expressed by pyramidal neurons has been recently shown to selectively control the development of CB_1/CCK^+ interneurons.[87] In addition, programmed cell death and survival are key processes in GABAergic interneuron circuit refinement and largely depend on the activity of pyramidal neurons during early postnatal development.[88,89] Considering the role of the ECB system in controlling major aspects of cortical projection neuron development and excitability,[1,4] it is tempting to speculate that PCE-dependent alterations in pyramidal neuron subpopulations could indirectly contribute to the defects observed in GABAergic interneuron development.[56] The use of conditional CB_1 receptor knockout mice, lacking the receptor in specific neuronal populations,[90,91] in conjunction with temporally restricted THC exposure, has been crucial to decipher the physiological role of the ECB system in the development of specific cell populations of the nervous system. Conditional CB_1 receptor ablation in telencephalic pyramidal neurons or in forebrain GABAergic neurons has confirmed that the sex-dimorphic CB_1/CCK^+ interneuronopathy and spatial memory impairment induced by PCE is a cell-autonomous effect.[82] Also, the unambiguous identification of the cellular targets of prenatal THC and its corresponding functional consequences has benefited from Cre recombinase-mediated, lineage-specific, CB_1 expression-rescue strategy in a CB_1-null background.[92] Thus, embryonic exposure to THC elicits defects in CSMN specification and skilled motor function that depend entirely on CB_1 signaling in cortical projection neurons.[32] In contrast, PCE also leads to E/I imbalance and increased seizure susceptibility that relies on embryonic CB_1 dysfunction in both glutamatergic pyramidal neurons and forebrain GABAergic interneurons. Intriguingly, the E/I imbalance found after PCE exhibits a sex-dependent effect, with only males being affected, exclusively when results from developmental CB_1 signaling disruption in GABAergic interneurons. In contrast, when the E/I imbalance emerges from altered embryonic CB_1 signaling in pyramidal neurons, both sexes appeared equally affected.[82] Hence, time-restricted manipulation of the ECB system by PCE in genetically modified mouse models has proven to constitute a valuable experimental approach to assess the cell autonomous role of CB_1 signaling in the development of different neuronal populations and brain areas.

Functional consequences of PCE

PCE induces functional and behavioral changes in the progeny at various levels. PCE-induced developmental alterations ensues, in most studies, with different types of cognitive and motor function alterations, E/I imbalance, and hyperexcitability. Higher variability is observed in the analyses of neuropsychiatric-related behaviors, as discussed below. According to the brain area affected by embryonic ECB system dysfunction (i.e., PFC, motor cortex, hippocampus, cerebellum, amygdala, etc.), different brain function abnormalities and behavioral alterations can be expected. Appropriate PFC activity is essential for cognitive function and social behavior. PCE interferes with cognitive function in part by altered cortical glutamatergic, GABAergic, dopaminergic, opioidergic, and serotonergic neurotransmission[69] (see Chapter 2). Deregulated trajectories of neuronal development induced by PCE impact brain's oscillatory activity,[82,93] which is essential for cognitive functions but also contributes to epilepsy and neurological diseases.[94] PCE-induced hippocampal CB_1/CCK^+ interneuronopathy constitutes one of the neurobiological substrates involved in oscillatory alterations.[82] CB_1/CCK^+ cells are required for experience-dependent plasticity[95] and spatial information coding.[96] In addition, reduction of perisomatic CA1 pyramidal cell inhibition from CCK-expressing basket cells is associated with seizures and network hyperexcitability.[97] The contribution of impaired development, migration, and function of interneurons is an etiopathogenic mechanism for epileptic encephalopathies, but also for cognitive and behavioral alterations of neuropsychiatric disorders,[98,99] and thus constitute a promising area of research. Also, PCE-induced hippocampal CB_1/CCK^+ interneuronopathy can exert unstudied additional consequences, as CB_1 receptors regulate GABAergic transmission to granule cells and hilar mossy cells, which in turn regulate adult hippocampal neurogenesis by influencing the direct and indirect pathways and astrocyte-mediated glutamatergic signaling.[100,101] Hence, it is likely that hippocampal neurogenesis-dependent mechanisms can contribute to the cognitive and emotional deficits induced by PCE.

Contribution of prenatal ECB signaling alterations to neurodevelopmental disorders

The neuromodulatory role of the ECB system in the adult brain provides many mechanistic explanations for the contribution of cannabinoid signaling disruption to alterations of the E/I balance, hence contributing to seizure severity and epilepsy.[102,103] In addition, the neurodevelopmental role of the ECB system prompts the question of whether epileptic syndromes and neuropsychiatric disorders of developmental origin may be originated or influenced by aberrant ECB system function. A common signaling mechanism altered in neurodevelopmental diseases is the mTORC1 pathway, typically regulated upstream by PI3K/Akt signaling, among many other additional mechanisms. The so-called mTORopathies, driven by mTORC1 overactivation, cover a wide range of diseases with a prominent epileptogenic phenotype [tuberous sclerosis, focal cortical dysplasia (FCD), etc.], but also ASD (fragile X syndrome, etc.) and intellectual disability.[104,105] In this respect, CB_1 receptors are coupled to PI3K/mTORC1 activation in neural cells and this mechanism mediates cannabinoid responses including cognitive function, brain excitability and neuroprotection.[6,106,107] Deregulated prenatal ECB signaling via CB_1 receptors contributes to hyperexcitability and epilepsy by interfering with pyramidal neurogenesis and migration, as well as interneuron development and function.[108] CB_1 receptors are overexpressed in type II FCD and organotypic cultures derived from palliative resections, and their pharmacological blockade by SR141716 attenuates mTORC1 overactivation.[109]

A correct neuronal fate balance between early-generated deep layer projection neurons and upper layer neurons is crucial to prevent the etiopathology of neurological and psychiatric NDDs, including schizophrenia, autism and intellectual disability,[110–112] but also epilepsy.[105] *SATB2*, a master regulator of upper layer neuron development, is a genetic risk locus for schizophrenia, mental retardation and educational attainment.[113] Patients with mutations or deletions within the *SATB2* locus experience severe learning difficulties and profound mental retardation, known as *SATB2*-associated syndrome.[114] In animal models, pharmacologically induced overproduction of upper cortical layer projection neurons recapitulates the development of autism-related traits, and correction of upper/deep layer projection neuron unbalance restores some of the behavioral autistic traits.[115] Thus, deregulated embryonic activity of the ECB system can contribute mechanistically to NDDs or, on the other hand, its manipulation during permissive developmental windows for cannabinoid-induced plasticity may counteract some of the molecular and cellular mechanisms involved in NDDs, and rescue some of their pathological phenotypes.

In this regard, CB_1 receptor signaling manipulation is under investigation as a potential disease-modifying strategy in preclinical research for fragile X and Down syndromes among other neuropsychiatric diseases.[116] Pharmacological enhancement of 2AG levels with JZL184 rescues synaptic and behavioral deficits in a fragile X and Down syndrome mouse models[117] (Lysenko et al., 2014; Thomazeau et al., 2014), and increasing AEA tone with URB597 rescues phenotypic traits of valproic acid-induced model of autism.[118] Alternatively, CB_1 receptor antagonism during developmental sensitive periods has been shown to rescue neuronal and behavioral deficits in mouse models of fragile X and Down syndromes.[119,120]

Prenatal cannabinoid exposure in human-based models

Great efforts are currently dedicated to determine the consequences of PCE in humans, and improved analytical analyses aim to discriminate from confounding social, economic, and cultural factors inherent to longitudinal studies.[121] The contribution of lifetime cannabinoid consumption and the risk of developing schizophrenia and psychosis have been intensively studied, and a number of genetic risk factors identified.[122] Genome-wide association studies show that PCE associates with preterm birth and higher frequency of autism, but indicate less power correlation with intellectual disability and learning disorders.[123] Importantly, modeling PCE can now move forward beyond the use of animal models to unravel the relation between PCE and psychiatric NDDs. The use of novel stem cell-derived neuron generation models opens new opportunities to understand the neurodevelopmental consequences of PCE,[3,4] as well as to unravel the contribution of the ECB system to NDDs that result in severe neurological alterations (refractory epilepsy)[102,103] and neuropsychiatric disorders.[116] Transcriptomic studies on hiPS-derived neurons exposed to THC revealed altered expression of genes involved in neurodevelopment, glutamatergic synaptic function, and mitochondrial activity. Exposure to THC induces a gene expression pattern that associates with neuropsychiatric conditions, particularly more evident for autism and intellectual disability.[124] Others have found that THC administration during rat adolescence induces transcriptomic changes of gene expression more related to schizophrenia than to ASD and intellectual disability.[125] THC exposure interferes with pyramidal neuron maturation in the PFC and regulates gene networks involved in the regulation of actin dynamics and dendritic spines. THC also induced the downregulation of several histone-binding genes including MECP2, a Rett syndrome causing gene. In this regard, the importance of ECB signaling and the impact of PCE in epigenetic regulation of gene expression and cross-generational inheritance of drug-induced long-term consequences constitutes an arising field to understand the neurodevelopmental consequences of PCE[4] (see Chapter 8).

The interference of developmental CB exposure with mitochondrial activity and dynamics, as a potential common mechanism of THC-induced neuronal aberrant plasticity and function, appears a reiterative finding. In hiPSCs-derived neurons, THC exposure induces alterations of mitochondrial genes and glutamatergic signaling.[124] Adolescence THC exposure interferes with mitochondrial function and membrane integrity.[126] Whether cannabinoid interference with mitochondrial dynamics and neuroenergetic metabolism is mediated via mitochondrial CB_1 receptors[91] remain to be elucidated.

Studies using hiPS cell-derived cerebral organoids have shown that chronic THC treatment decreases neuronal activity and neurite outgrowth,[93] whereas THC exposure and increased 2AG levels promote pyramidal neuronal maturation via CB_1 receptors in ES-derived projection neurons and cerebral organoids.[20] The involvement of CB_1 receptors in projection neuron and GABAergic interneuron development has been confirmed in cerebral organoids by other studies. Treatment with SR141716 increases excitatory synaptogenesis.[127] In a different neuronal lineage model, hiPS cell-derived dopaminergic neurons generated in the presence of THC and AEA were shown to display dual effects, with low cannabinoid concentrations exerting a positive effect on neuronal activity and high cannabinoid concentrations impairing activity.[128]

Many studies approach the investigation of PCE with WIN 55,212-2, HU-210 or other ligands. However, it is evident that synthetic cannabinoids induce stronger impact or at least exert different consequences than THC and other plant-derived cannabinoids. Plant-derived cannabinoids possess important differences in solubility, potency, metabolism and therefore, pharmacokinetics and pharmacodynamics. Hence, when modeling human cannabinoid exposure in laboratory animals and experimental models, it is important to keep in mind the intrinsic differences between the molecules employed. Moreover, distinct cannabinoid ligands contribute differentially to biased CB_1 receptor signaling and can target additional receptors and binding proteins.[129] In vitro studies and the use of organoid cultures have shown that high concentrations of WIN 55,212-2 induce severe neurotoxic actions,[130,131] which are not observed when THC is used. Exposure to new synthetic cannabinoids (the so-called "spice" compounds, like THJ-018 and EG-018) during neuronal differentiation in a hiPSC-derived model, induced premature neuronal and glial differentiation and affected voltage-dependent calcium channel neuronal activity.[132] Hence, the impact of PCE in neurogenesis and circuit development is different according to the pharmacological manipulation paradigm used. Prolonged administration of high cannabinoid concentrations or potent synthetic cannabinoid derivatives promote CB_1 receptor desensitization and loss of function,[32,93,133] and regulate yet unknown off-targets but likely not canonical CB receptors.[130,132] Different neuronal cell fate and functional consequences become evident with increased CB_1 signaling within physiological range, i.e., by the use of ECB metabolism regulators.[20]

Conclusions

PCE to THC, or other CB_1 receptor-interacting molecules, exerts long-lasting consequences by interfering with the neurodevelopmental role of the ECB system. According to the developmental time window of exposure, the frequency, the molecular composition and concentration of cannabinoid preparations, different alterations of neuronal populations and circuit establishment can occur. Hence, THC administration, either by engaging or, oppositely, interfering with CB_1 receptor signaling (desensitization/tolerance/ligand competition), influence the developmental processes regulated by the ECB system. Experimental evidence demonstrates that prenatal THC exposure interferes with pyramidal neuronal generation, differentiation, and long-range axon connectivity. In addition, GABAergic interneuron development, synaptic integration, and CB_1/CCK^+ basket cell function are regulated by the ECB system. Subcortically, developmental exposure to THC leads to a hyperdopaminergic state with important implications in drug addiction susceptibility. PCE induces, in general, subtle neurodevelopmental changes, however showing sex-dependent differences, with males being primarily affected. Overall, PCE leads to an unbalanced development of neuronal subpopulations that ultimately evokes synaptic and synchrony alterations that translate into alterations of spatial memory, cognitive function, E/I balance and emotional behavior homeostasis in adulthood. Neuronal cell reprogramming, synaptic and circuit reorganization as a consequence of PCE are of relevance as potential ethiopathological mechanisms of NDDs and open the window for their potential therapeutic applications. Finally, understanding with fidelity the consequences of PCE is essential for rational policy making in current times of expanding use of cannabis-based extracts for therapeutic and recreational uses.

References

1. Galve-Roperh I, Chiurchiù V, Díaz-Alonso J, Bari M, Guzmán M, Maccarrone M. Cannabinoid receptor signaling in progenitor/stem cell proliferation and differentiation. *Prog Lipid Res* 2013;**52**(4):633–50. https://doi.org/10.1016/j.plipres.2013.05.004.
2. Maccarrone M, Guzmán M, Mackie K, Doherty P, Harkany T. Programming of neural cells by (endo)cannabinoids: from physiological rules to emerging therapies. *Nat Rev Neurosci* 2014;**15**(12):786–801. https://doi.org/10.1038/nrn3846.
3. Scheyer AF, Melis M, Trezza V, Manzoni OJJ. Consequences of perinatal cannabis exposure. *Trends Neurosci* 2019;1–14. https://doi.org/10.1016/j.tins.2019.08.010.
4. Bara A, Ferland JMN, Rompala G, Szutorisz H, Hurd YL. Cannabis and synaptic reprogramming of the developing brain. *Nat Rev Neurosci* 2021;**22**(7):423–38. https://doi.org/10.1038/s41583-021-00465-5.
5. Aguado T, Monory K, Palazuelos J, et al. The endocannabinoid system drives neural progenitor proliferation. *FASEB J* 2005;**19**(12):1704–6. https://doi.org/10.1096/fj.05-3995fje.
6. Díaz-Alonso J, Aguado T, de Salas-Quiroga A, Ortega Z, Guzmán M, Galve-Roperh I. CB1 cannabinoid receptor-dependent activation of mTORC1/Pax6 signaling drives Tbr2 expression and basal progenitor expansion in the developing mouse cortex. *Cereb Cortex* 2015;**25**(9):2395–408. https://doi.org/10.1093/cercor/bhu039.

7. Aguado T, Romero E, Monory K, et al. The CB1 cannabinoid receptor mediates excitotoxicity-induced neural progenitor proliferation and neurogenesis (Journal of Biological Chemistry (2007) 282 (23892–23898)). *J Biol Chem* 2008;**283**(9):5971.
8. Zimmermann T, Maroso M, Beer A, et al. Neural stem cell lineage-specific cannabinoid type-1 receptor regulates neurogenesis and plasticity in the adult mouse hippocampus. *Cereb Cortex* 2018;**28**(12):4454–71. https://doi.org/10.1093/cercor/bhy258.
9. Rodrigues RS, Ribeiro FF, Ferreira F, Vaz SH, Sebastião AM, Xapelli S. Interaction between cannabinoid type 1 and type 2 receptors in the modulation of subventricular zone and dentate gyrus neurogenesis. *Front Pharmacol* 2017;**8**(Aug):516. https://doi.org/10.3389/fphar.2017.00516.
10. Gao Y, Vasilyev DV, Goncalves MB, et al. Loss of retrograde endocannabinoid signaling and reduced adult neurogenesis in diacylglycerol lipase knock-out mice. *J Neurosci* 2010;**30**(6):2017–24. https://doi.org/10.1523/JNEUROSCI.5693-09.2010.
11. Schuele L-L, Schuermann B, Bilkei-Gorzo A, Gorgzadeh S, Zimmer A, Leidmaa E. *Autocrine regulation of adult neurogenesis by the endocannabinoid 2-arachidonoylglycerol (2-AG)*; 2021.
12. Campbell WA, Blum S, Blackshaw S, Fischer AJ. Cannabinoid signaling promotes the de-differentiation and proliferation of Müller glia-derived progenitor cells. *Glia* 2021;**2**(June):1–19. https://doi.org/10.1002/glia.24056.
13. Butti E, Bacigaluppi M, Rossi S, et al. Subventricular zone neural progenitors protect striatal neurons from glutamatergic excitotoxicity. *Brain* 2012;**135**(11):3320–35. https://doi.org/10.1093/brain/aws194.
14. De Oliveira RW, Oliveira CL, Guimarães FS, Campos AC. Cannabinoid signalling in embryonic and adult neurogenesis: Possible implications for psychiatric and neurological disorders. *Acta Neuropsychiatr* 2019;**31**(1):1–16. https://doi.org/10.1017/neu.2018.11.
15. Prenderville JA, Kelly ÁM, Downer EJ. The role of cannabinoids in adult neurogenesis. *Br J Pharmacol* 2015;**172**(16):3950–63. https://doi.org/10.1111/bph.13186.
16. Xapelli S, Agasse F, Sardà-Arroyo L, et al. Activation of type 1 cannabinoid receptor (CB1R) promotes neurogenesis in murine subventricular zone cell cultures. *PLoS One* 2013;**8**(5). https://doi.org/10.1371/journal.pone.0063529, e63529.
17. Compagnucci C, Di Siena S, Bustamante MB, et al. Type-1 (CB1) cannabinoid receptor promotes neuronal differentiation and maturation of neural stem cells. *PLoS One* 2013;**8**(1). https://doi.org/10.1371/journal.pone.0054271, e54271.
18. Morozov YM, Mackie K, Rakic P. Cannabinoid Type 1 receptor is undetectable in rodent and primate cerebral neural stem cells but participates in radial neuronal. *Migration* 2020;1–19.
19. Xing L, Huttner WB. Neurotransmitters as modulators of neural progenitor cell proliferation during mammalian neocortex development. *Front Cell Dev Biol* 2020;**8**(May). https://doi.org/10.3389/fcell.2020.00391.
20. Paraíso-Luna J, Aguareles J, Martín R, et al. Endocannabinoid signalling in stem cells and cerebral organoids drives differentiation to deep layer projection neurons via CB 1 receptors. *Development* 2020. https://doi.org/10.1242/dev.192161.
21. Vitalis T, Lainé J, Simon A, Roland A, Leterrier C, Lenkei Z. The type 1 cannabinoid receptor is highly expressed in embryonic cortical projection neurons and negatively regulates neurite growth in vitro. *Eur J Neurosci* 2008;**28**(9):1705–18. https://doi.org/10.1111/j.1460-9568.2008.06484.x.
22. Begbie J, Doherty P, Graham A. Cannabinoid receptor, CB1, expression follows neuronal differentiation in the early chick embryo. *J Anat* 2004;**205**(3):213–8. https://doi.org/10.1111/j.0021-8782.2004.00325.x.

23] László Z.I., Lele Z., Zöldi M., et al. ABHD4-dependent developmental anoikis safeguards the embryonic brain. Nat Commun 2020;11(1):1-16. doi:https://doi.org/10.1038/s41467-020-18175-4.
24. Bromberg KD, Ma'ayan A, Neves SR, Iyengar R. Design logic of a cannabinoid receptor signaling network that triggers neurite outgrowth. *Science* 2008;**320**(5878):903–9. https://doi.org/10.1126/science.1152662.
25. Trazzi S, Steger M, Mitrugno VM, Bartesaghi R, Ciani E. CB 1 cannabinoid receptors increase neuronal precursor proliferation through AKT/glycogen synthase kinase-3 β/β-catenin signaling. *J Biol Chem* 2010;**285**(13):10098–109. https://doi.org/10.1074/jbc.M109.043711.
26. Sütterlin P, Williams EJ, Chambers D, et al. The molecular basis of the cooperation between EGF, FGF and eCB receptors in the regulation of neural stem cell function. *Mol Cell Neurosci* 2013;**52**:20–30. https://doi.org/10.1016/j.mcn.2012.10.006.
27. Mulder J, Aguado T, Keimpema E, et al. Endocannabinoid signaling controls pyramidal cell specification and long-range axon patterning. *Proc Natl Acad Sci U S A* 2008;**105**(25):8760–5. https://doi.org/10.1073/pnas.0803545105.
28. Díaz-Alonso J, Aguado T, Wu C-S, et al. The CB(1) cannabinoid receptor drives corticospinal motor neuron differentiation through the Ctip2/Satb2 transcriptional regulation axis. *J Neurosci* 2012;**32**(47):16651–65. https://doi.org/10.1523/JNEUROSCI.0681-12.2012.
29. Wu C-S, Zhu J, Wager-Miller J, et al. Requirement of cannabinoid CB(1) receptors in cortical pyramidal neurons for appropriate development of corticothalamic and thalamocortical projections. *Eur J Neurosci* 2010;**32**(5):693–706. https://doi.org/10.1111/j.1460-9568.2010.07337.x.
30. Berghuis P, Rajnicek AM, Morozov YM, et al. Hardwiring the brain: endocannabinoids shape neuronal connectivity. *Science (80-)* 2007;**316**(5828):1212–6. https://doi.org/10.1126/science.1137406.
31. Keimpema E, Barabas K, Morozov YM, et al. Differential subcellular recruitment of monoacylglycerol lipase generates spatial specificity of 2-arachidonoyl glycerol signaling during axonal pathfinding. *J Neurosci* 2010;**30**(42):13992–4007. https://doi.org/10.1523/JNEUROSCI.2126-10.2010.
32. de Salas-Quiroga A, Díaz-Alonso J, García-Rincón D, et al. Prenatal exposure to cannabinoids evokes long-lasting functional alterations by targeting CB 1 receptors on developing cortical neurons. *Proc Natl Acad Sci* 2015;**112**(44):13693–8. https://doi.org/10.1073/pnas.1514962112.
33. Srivatsa S, Parthasarathy S, Britanova O, et al. Unc5C and DCC act downstream of Ctip2 and Satb2 and contribute to corpus callosum formation. *Nat Commun* 2014;**5**:3708. https://doi.org/10.1038/ncomms4708.
34. Tortoriello G, Morris CV, Alpar A, et al. Miswiring the brain: Δ^9-tetrahydrocannabinol disrupts cortical development by inducing an SCG10/stathmin-2 degradation pathway. *EMBO J* 2014;**33**(7):668–85. https://doi.org/10.1002/embj.201386035.
35. Alpár A, Tortoriello G, Calvigioni D, et al. Endocannabinoids modulate cortical development by configuring Slit2/Robo1 signalling. *Nat Commun* 2014;**5**:4421. https://doi.org/10.1038/ncomms5421.
36. Njoo C, Agarwal N, Lutz B, Kuner R. The cannabinoid receptor CB1 interacts with the WAVE1 complex and plays a role in actin dynamics and structural plasticity in neurons. *PLoS Biol* 2015;**13**(10). https://doi.org/10.1371/journal.pbio.1002286, e1002286.
37. Roland AB, Ricobaraza A, Carrel D, et al. Cannabinoid-induced actomyosin contractility shapes neuronal morphology and growth. *elife* 2014;**3**. https://doi.org/10.7554/eLife.03159, e03159.

38. Saez TMM, Bessone IF, Rodriguez MS, et al. Kinesin-1-mediated axonal transport of CB1 receptors is required for cannabinoid-dependent axonal growth and guidance. *Development* 2020;**147**(8):1–14. https://doi.org/10.1242/dev.184069.
39. Argaw A, Duff G, Zabouri N, et al. Concerted action of CB1 cannabinoid receptor and deleted in colorectal cancer in axon guidance. *J Neurosci* 2011;**31**(4):1489–99. https://doi.org/10.1523/JNEUROSCI.4134-09.2011.
40. Xing L, Larsen RS, Bjorklund GR, et al. Layer specific and general requirements for ERK/MAPK signaling in the developing neocortex. *elife* 2016;**5**(February). https://doi.org/10.7554/eLife.11123.
41. Rueda D. The endocannabinoid anandamide inhibits neuronal progenitor cell differentiation through attenuation of the Rap1/B-Raf/ERK pathway. *J Biol Chem* 2002;**277**(48):46645–50. https://doi.org/10.1074/jbc.M206590200.
42. Jordan JD, He JC, Eungdamrong NJ, et al. Cannabinoid receptor-induced neurite outgrowth is mediated by Rap1 activation through Gαo/i-triggered proteasomal degradation of Rap1GAPII. *J Biol Chem* 2005;**280**(12):11413–21. https://doi.org/10.1074/jbc.M411521200.
43. Zhou R, Han B, Xia C, Zhuang X. Membrane-associated periodic skeleton is a signaling platform for RTK transactivation in neurons. *Science (80-)* 2019;**365**(6456):929–34. https://doi.org/10.1126/science.aaw5937.
44. Berghuis P, Dobszay MB, Wang X, et al. Endocannabinoids regulate interneuron migration and morphogenesis by transactivating the TrkB receptor. *Proc Natl Acad Sci U S A* 2005;**102**(52):19115–20. https://doi.org/10.1073/pnas.0509494102.
45. Saez TMM, Aronne MP, Caltana L, Brusco AH. Prenatal exposure to the CB1 and CB2 cannabinoid receptor agonist WIN 55,212-2 alters migration of early-born glutamatergic neurons and GABAergic interneurons in the rat cerebral cortex. *J Neurochem* 2014;**129**(4):637–48. https://doi.org/10.1111/jnc.12636.
46. Díaz-Alonso J, De Salas-Quiroga A, Paraíso-Luna J, et al. Loss of cannabinoid CB1 receptors induces cortical migration malformations and increases seizure susceptibility. *Cereb Cortex* 2017;**27**(11):5303–17. https://doi.org/10.1093/cercor/bhw309.
47. Oudin MJ, Gajendra S, Williams G, Hobbs C, Lalli G, Doherty P. Endocannabinoids regulate the migration of subventricular zone-derived neuroblasts in the postnatal brain. *J Neurosci* 2011;**31**(11):4000–11. https://doi.org/10.1523/JNEUROSCI.5483-10.2011.
48. Escamilla CO, Filonova I, Walker AK, et al. Kctd13 deletion reduces synaptic transmission via increased RhoA. *Nature* 2017;**551**(7679):227–31. https://doi.org/10.1038/nature24470.
49. Bonanomi D, Valenza F, Chivatakarn O, et al. p190RhoGAP filters competing signals to resolve axon guidance conflicts. *Neuron* 2019;**102**(3):602–620.e9. https://doi.org/10.1016/j.neuron.2019.02.034.
50. Marsicano G, Lutz B. Expression of the cannabinoid receptor CB1 in distinct neuronal subpopulations in the adult mouse forebrain. *Eur J Neurosci* 1999;**11**(12):4213–25. https://doi.org/10.1046/j.1460-9568.1999.00847.x.
51. Katona I, Sperlágh B, Sík A, et al. Presynaptically located CB1 cannabinoid receptors regulate GABA release from axon terminals of specific hippocampal interneurons. *J Neurosci* 1999;**19**(11):4544–58. http://www.ncbi.nlm.nih.gov/pubmed/10341254. [Accessed 15 March 2019].
52. Hill EL, Gallopin T, Férézou I, et al. Functional CB1 receptors are broadly expressed in neocortical GABAergic and glutamatergic neurons. *J Neurophysiol* 2007;**97**(4):2580–9. https://doi.org/10.1152/JN.00603.2006.
53. Morozov YM, Torii M, Rakic P. Origin, early commitment, migratory routes, and destination of cannabinoid type 1 receptor-containing interneurons. *Cereb Cortex* 2009;**19**(Suppl 1):i78–89. https://doi.org/10.1093/cercor/bhp028.

54. Antypa M, Faux C, Eichele G, Parnavelas JG, Andrews WD. Differential gene expression in migratory streams of cortical interneurons. *Eur J Neurosci* 2011;**34**(10):1584–94. https://doi.org/10.1111/j.1460-9568.2011.07896.x.
55. Fitzgerald ML, Chan J, Mackie K, Lupica CR, Pickel VM. Altered dendritic distribution of dopamine D2 receptors and reduction in mitochondrial number in parvalbumin-containing interneurons in the medial prefrontal cortex of cannabinoid-1 (CB1) receptor knockout mice. *J Comp Neurol* 2012;**520**(17):4013–31. https://doi.org/10.1002/cne.23141.
56. Song CG, Kang X, Yang F, et al. Endocannabinoid system in the neurodevelopment of GABAergic interneurons: Implications for neurological and psychiatric disorders. *Rev Neurosci* 2021. https://doi.org/10.1515/revneuro-2020-0134.
57. von Rüden EL, Jafari M, Bogdanovic RM, Wotjak CT, Potschka H. Analysis in conditional cannabinoid 1 receptor-knockout mice reveals neuronal subpopulation-specific effects on epileptogenesis in the kindling paradigm. *Neurobiol Dis* 2015;**73**:334–47. https://doi.org/10.1016/j.nbd.2014.08.001.
58. Remmers F, Lange MD, Hamann M, Ruehle S, Pape HC, Lutz B. Addressing sufficiency of the CB1 receptor for endocannabinoid-mediated functions through conditional genetic rescue in forebrain GABAergic neurons. *Brain Struct Funct* 2017;1–22. https://doi.org/10.1007/s00429-017-1411-5.
59. Sales-Carbonell C, Rueda-Orozco PE, Soria-Gómez E, Buzsáki G, Marsicano G, Robbe D. Striatal GABAergic and cortical glutamatergic neurons mediate contrasting effects of cannabinoids on cortical network synchrony. *Proc Natl Acad Sci U S A* 2013;**110**(2):719–24. https://doi.org/10.1073/pnas.1217144110.
60. Jenniches I, Ternes S, Albayram O, et al. Anxiety, stress, and fear response in mice with reduced endocannabinoid levels. *Biol Psychiatr* 2016;**79**(10):858–68. https://doi.org/10.1016/J.BIOPSYCH.2015.03.033.
61. Molina-Holgado E, Paniagua-Torija B, Arevalo-Martin A, et al. Cannabinoid REceptor 1 associates to different molecular complexes during GABAergic neuron maturation. *J Neurochem* 2021;(January):1–17. https://doi.org/10.1111/jnc.15381.
62. Sahin M, Sur M. Genes, circuits, and precision therapies for autism and related neurodevelopmental disorders. *Science (80-)* 2015;**350**(6263). https://doi.org/10.1126/science.aab3897.
63. Pinky PD, Majrashi M, Fujihashi A, et al. Effects of prenatal synthetic cannabinoid exposure on the cerebellum of adolescent rat offspring. *Heliyon* 2021;**7**(4). https://doi.org/10.1016/j.heliyon.2021.e06730, e06730.
64. Pestilli F, et al. Deep learning and cross-species analysis identify the influences of sex and cannabinoid signaling on cerebellar vermis morphology. *bioRxiv* 2021;**3**(2):6.
65. Aguado T, Palazuelos J, Monory K, Stella N, Cravatt B, Beat Lutz GM, et al. The endocannabinoid system promotes astroglial differentiation by acting on neural progenitor cells. *J Neurosci* 2006;**26**(5):1551–61. https://doi.org/10.1523/JNEUROSCI.3101-05.2006.
66. Molina-Holgado E, Vela JM, Arévalo-Martín A, et al. Cannabinoids promote oligodendrocyte progenitor survival: Involvement of cannabinoid receptors and phosphatidylinositol-3 kinase/Akt signaling. *J Neurosci* 2002;**22**(22):9742–53. https://doi.org/10.1523/jneurosci.22-22-09742.2002.
67. Huerga-Gómez A, Aguado T, Sánchez-de la Torre A, et al. Δ^9-Tetrahydrocannabinol promotes oligodendrocyte development and CNS myelination in vivo. *Glia* 2020;(May):532–45. https://doi.org/10.1002/glia.23911.
68. Castillo PE, Younts TJ, Chávez AE, Hashimotodani Y. Endocannabinoid signaling and synaptic function. *Neuron* 2012;**76**(1):70–81. https://doi.org/10.1016/j.neuron.2012.09.020.
69. Pinky PD, Bloemer J, Smith WD, et al. Prenatal cannabinoid exposure and altered neurotransmission. *Neuropharmacology* 2019;**149**(January):181–94. https://doi.org/10.1016/j.neuropharm.2019.02.018.

70. Alpár A, Tortoriello G, Calvigioni D, et al. Endocannabinoids modulate cortical development by configuring Slit2/Robo1 signalling. *Nat Commun* 2014;**5**:4421. https://doi.org/10.1038/ncomms5421.
71. Manduca A, Bernabeu A, et al. Sex-dependent effects of in utero cannabinoid exposure on cortical function. *elife* 2018;**1-31**. https://doi.org/10.7554/eLife.36234.
72. Castelli MP, Paola Piras A, D'Agostino A, et al. Dysregulation of the endogenous cannabinoid system in adult rats prenatally treated with the cannabinoid agonist WIN 55,212-2. *Eur J Pharmacol* 2007;**573**(1-3):11–9. https://doi.org/10.1016/j.ejphar.2007.06.047.
73. Scheyer AF, Borsoi M, Pelissier-Alicot A-L, Manzoni OJJ. Maternal exposure to the cannabinoid agonist WIN 55,12,2 during lactation induces lasting behavioral and synaptic alterations in the rat adult offspring of both sexes. *eNeuro* 2020;**7**(5). https://doi.org/10.1523/ENEURO.0144-20.2020.
74. Scheyer AF, Borsoi M, Pelissier-Alicot A-L, Manzoni OJJ. Perinatal THC exposure via lactation induces lasting alterations to social behavior and prefrontal cortex function in rats at adulthood. *Neuropsychopharmacology* 2020;**45**(11):1826–33. https://doi.org/10.1038/s41386-020-0716-x.
75. de Fonseca FR, Cebeira M, Fernández-Ruiz JJ, Navarro M, Ramos JA. Effects of pre- and perinatal exposure to hashish extracts on the ontogeny of brain dopaminergic neurons. *Neuroscience* 1991;**43**(2-3):713–23.
76. Wang X, Dow-Edwards D, Anderson V, Minkoff H, Hurd YL. In utero marijuana exposure associated with abnormal amygdala dopamine D2 gene expression in the human fetus. *Biol Psychiatry* 2004;**56**(12):909–15. https://doi.org/10.1016/j.biopsych.2004.10.015.
77. Frau R, Miczán V, Traccis F, et al. Prenatal THC exposure produces a hyperdopaminergic phenotype rescued by pregnenolone. *Nat Neurosci* 2019;**22**(12):1975–85. https://doi.org/10.1038/s41593-019-0512-2.
78. Renard J, Rosen LG, Loureiro M, et al. Adolescent cannabinoid exposure induces a persistent sub-cortical hyper-dopaminergic state and associated molecular adaptations in the prefrontal cortex. *Cereb Cortex* 2016;(February):bhv335. https://doi.org/10.1093/cercor/bhv335.
79. Sagheddu C, Traccis F, Serra V, Congiu M, Frau R, Cheer JF, et al. Mesolimbic dopamine dysregulation as a signature of information processing deficits imposed by prenatal THC exposure. *Prog Neuropsychopharmacol Biol Psychiatry* 2021;**105**:110128. https://doi.org/10.1016/j.pnpbp.2020.110128.
80. Vallée M, Vitiello S, Bellocchio L, et al. Pregnenolone can protect the brain from cannabis intoxication. *Science (80-)* 2014;**343**(6166):94–8. https://doi.org/10.1126/science.1243985.
81. Vargish GA, Pelkey KA, Yuan X, et al. Persistent inhibitory circuit defects and disrupted social behaviour following in utero exogenous cannabinoid exposure. *Mol Psychiatr* 2016;(January):1–12. https://doi.org/10.1038/mp.2016.17.
82. de Salas-Quiroga A, García-Rincón D, Gómez-Domínguez D, et al. Long-term hippocampal interneuronopathy drives sex-dimorphic spatial memory impairment induced by prenatal THC exposure. *Neuropsychopharmacology* 2020;(August):1–10. https://doi.org/10.1038/s41386-020-0621-3.
83. Scheyer AF, Borsoi M, Wager-Miller J, et al. Cannabinoid exposure via lactation in rats disrupts perinatal programming of the gamma-aminobutyric acid trajectory and select early-life behaviors. *Biol Psychiatr* 2020;**87**(7):666–77. https://doi.org/10.1016/j.biopsych.2019.08.023.
84. Lodato S, Rouaux C, Quast KB, et al. Excitatory projection neuron subtypes control the distribution of local inhibitory interneurons in the cerebral cortex. *Neuron* 2011;**69**(4):763–79. https://doi.org/10.1016/j.neuron.2011.01.015.

85. Fazzari P, Paternain AV, Valiente M, et al. Control of cortical GABA circuitry development by Nrg1 and ErbB4 signalling. *Nature* 2010;**464**(7293):1376–80. https://doi.org/10.1038/nature08928.
86. Flames N, Long JE, Garratt AN, et al. Short- and long-range attraction of cortical GABAergic interneurons by neuregulin-1. *Neuron* 2004;**44**(2):251–61. https://doi.org/10.1016/j.neuron.2004.09.028.
87. Miller DS, Wright KM. Neuronal Dystroglycan regulates postnatal development of CCK/cannabinoid receptor-1 interneurons. *Neural Dev* 2021;**091027**(158). https://doi.org/10.1101/2021.04.26.441492.
88. Southwell DG, Paredes MF, Galvao RP, et al. Intrinsically determined cell death of developing cortical interneurons. *Nature* 2012;**491**(7422):109–13. https://doi.org/10.1038/nature11523.
89. Wong FK, Bercsenyi K, Sreenivasan V, Portalés A, Fernández-Otero M, Marín O. Pyramidal cell regulation of interneuron survival sculpts cortical networks. *Nature* 2018;**557**(7707):668–73. https://doi.org/10.1038/s41586-018-0139-6.
90. Monory K, Blaudzun H, Massa F, et al. Genetic dissection of behavioural and autonomic effects of Δ^9-tetrahydrocannabinol in mice. *PLoS Biol* 2007;**5**(10):2354–68. https://doi.org/10.1371/journal.pbio.0050269.
91. Hebert-Chatelain E, Desprez T, Serrat R, et al. A cannabinoid link between mitochondria and memory. *Nature* 2016;**539**(7630):555–9. https://doi.org/10.1038/nature20127.
92. Ruehle S, Remmers F, Romo-Parra H, et al. Cannabinoid CB1 receptor in dorsal telencephalic glutamatergic neurons: distinctive sufficiency for hippocampus-dependent and amygdala-dependent synaptic and behavioral functions. *J Neurosci* 2013;**33**(25):10264–77. https://doi.org/10.1523/JNEUROSCI.4171-12.2013.
93. Ao Z, Cai H, Havert DJ, et al. One-stop microfluidic assembly of human brain organoids to model prenatal cannabis exposure. *Anal Chem* 2020. https://doi.org/10.1101/2020.01.15.908483.
94. Menendez De La Prida L, Staba RJ, Dian JA. Conundrums of high-frequency oscillations (80–800 Hz) in the epileptic brain. *J Clin Neurophysiol* 2015;**32**(3):207–19. https://doi.org/10.1097/WNP.0000000000000150.
95. Feng T, Alicea C, Pham V, Kirk A, Pieraut S. Experience-dependent inhibitory plasticity is mediated by CCK+ basket cells in the developing dentate gyrus. *J Neurosci* 2021;**41**(21):4607–19. https://doi.org/10.1523/jneurosci.1207-20.2021.
96. del Pino I, Brotons-Mas JR, Marques-Smith A, et al. Abnormal wiring of CCK+ basket cells disrupts spatial information coding. *Nat Neurosci* 2017;**20**(6):784–92. https://doi.org/10.1038/nn.4544.
97. Wyeth MS, Zhang N, Mody I, Houser CR. Selective reduction of cholecystokinin-positive basket cell innervation in a model of temporal lobe epilepsy. *J Neurosci* 2010;**30**(26):8993–9006. https://doi.org/10.1523/JNEUROSCI.1183-10.2010.
98. Tang X, Jaenisch R, Sur M. The role of GABAergic signalling in neurodevelopmental disorders. *Nat Rev Neurosci* 2021;**22**(5):290–307. https://doi.org/10.1038/s41583-021-00443-x.
99. Katsarou A-M, Moshé SL, Galanopoulou AS. Interneuronopathies and their role in early life epilepsies and neurodevelopmental disorders. *Epilepsia Open* 2017;**2**(3):284–306. https://doi.org/10.1002/epi4.12062.
100. Asrican B, Wooten J, Li YD, et al. Neuropeptides modulate local astrocytes to regulate adult hippocampal neural stem cells. *Neuron* 2020;**108**(2):349–366.e6. https://doi.org/10.1016/j.neuron.2020.07.039.
101. Yeh CY, Asrican B, Moss J, et al. Mossy cells control adult neural stem cell quiescence and maintenance through a dynamic balance between direct and indirect pathways. *Neuron* 2018;**99**(3):493–510.e4. https://doi.org/10.1016/j.neuron.2018.07.010.

102. Soltesz I, Alger BE, Kano M, et al. Weeding out bad waves: towards selective cannabinoid circuit control in epilepsy. *Nat Rev Neurosci* 2015;**16**(5):264–77. https://doi.org/10.1038/nrn3937.
103. Cristino L, Bisogno T, Di Marzo V. Cannabinoids and the expanded endocannabinoid system in neurological disorders. *Nat Rev Neurol* 2020;**16**(1):9–29. https://doi.org/10.1038/s41582-019-0284-z.
104. Parenti I, Rabaneda LG, Schoen H, Novarino G. Neurodevelopmental disorders: from genetics to functional pathways. *Trends Neurosci* 2020;**43**(8):608–21. https://doi.org/10.1016/j.tins.2020.05.004.
105. Nguyen LH, Bordey A. Convergent and divergent mechanisms of epileptogenesis in mTORopathies. *Front Neuroanat* 2021;**15**(April). https://doi.org/10.3389/fnana.2021.664695.
106. Puighermanal E, Busquets-Garcia A, Maldonado R, Ozaita A. Cellular and intracellular mechanisms involved in the cognitive impairment of cannabinoids. *Philos Trans R Soc B Biol Sci* 2012;**367**(1607):3254–63. https://doi.org/10.1098/rstb.2011.0384.
107. Blázquez C, Chiarlone A, Bellocchio L, et al. The CB_1 cannabinoid receptor signals striatal neuroprotection via a PI3K/Akt/mTORC1/BDNF pathway. *Cell Death Differ* 2015;**22**(10):1618–29. https://doi.org/10.1038/cdd.2015.11.
108. Bara A, Ferland JMN, Rompala G, Szutorisz H, Hurd YL. Cannabis and synaptic reprogramming of the developing brain. *Nat Rev Neurosci* 2021;**22**(7):423–38. https://doi.org/10.1038/s41583-021-00465-5.
109. García-Rincón D, Díaz-alonso J, Paraíso-luna J, et al. contribution of altered endocannabinoid system to overactive mTORC1 signaling in focal cortical dysplasia. *Front Pharmacol* 2019;**9**(January):1–13. https://doi.org/10.3389/fphar.2018.01508.
110. Carpentier PA, Haditsch U, Braun AE, et al. Stereotypical alterations in cortical patterning are associated with maternal illness-induced placental dysfunction. *J Neurosci* 2013;**33**(43):16874–88. https://doi.org/10.1523/JNEUROSCI.4654-12.2013.
111. Cera I, Whitton L, Donohoe G, Morris DW, Dechant G, Apostolova G. Genes encoding SATB2-interacting proteins in adult cerebral cortex contribute to human cognitive ability. *PLoS Genet* 2019;**15**(2):1–21. https://doi.org/10.1371/journal.pgen.1007890.
112. Willsey AJ, Sanders SJ, Li M, et al. Coexpression networks implicate human midfetal deep cortical projection neurons in the pathogenesis of autism. *Cell* 2013;**155**(5):997. https://doi.org/10.1016/j.cell.2013.10.020.
113. Whitton L, Apostolova G, Rieder D, et al. Genes regulated by SATB2 during neurodevelopment contribute to schizophrenia and educational attainment. *PLoS Genet* 2018;**14**(7):1–20. https://doi.org/10.1371/journal.pgen.1007515.
114. Zarate YA, Fish JL. SATB2-associated syndrome: Mechanisms, phenotype, and practical recommendations. *Am J Med Genet A* 2017;**173**(2):327–37. https://doi.org/10.1002/ajmg.a.38022.
115. Pucilowska J, Vithayathil J, Tavares EJ, Kelly C, Karlo JC, Landreth GE. The 16p11.2 deletion mouse model of autism exhibits altered cortical progenitor proliferation and brain cytoarchitecture linked to the ERK MAPK pathway. *J Neurosci* 2015;**35**(7):3190–200. https://doi.org/10.1523/JNEUROSCI.4864-13.2015.
116. Fernández-Ruiz J, Galve-Roperh I, Sagredo O, Guzmán M. Possible therapeutic applications of cannabis in the neuropsychopharmacology field. *Eur Neuropsychopharmacol* 2020;1–18. https://doi.org/10.1016/j.euroneuro.2020.01.013.
117. Jung K-M, Sepers M, Henstridge CM, et al. Uncoupling of the endocannabinoid signalling complex in a mouse model of fragile X syndrome. *Nat Commun* 2012;**3**:1080. https://doi.org/10.1038/ncomms2045.
118. Servadio M, Melancia F, Manduca A, et al. Targeting anandamide metabolism rescues core and associated autistic-like symptoms in rats prenatally exposed to valproic acid. *Transl Psychiatry* 2016;**6**(9). https://doi.org/10.1038/tp.2016.182, e902.

119. Busquets-Garcia A, Gomis-Gonzalez M, Guegan T, et al. Targeting the endocannabinoid system in the treatment of fragile X syndrome. *Nat Med* 2013;**19**(5):603–7. https://doi.org/10.1038/nm.3127.
120. Navarro-Romero A, Vázquez-Oliver A, Gomis-González M, et al. Cannabinoid type-1 receptor blockade restores neurological phenotypes in two models for Down syndrome. *Neurobiol Dis* 2019;**125**(October):92–106. https://doi.org/10.1016/j.nbd.2019.01.014.
121. Hurd YL. Cannabis and the developing brain challenge risk perception. *J Clin Invest* 2020;**140**(8):3947–9. https://doi.org/10.1172/JCI139051.
122. Pasman JA, Verweij KJH, Gerring Z, et al. GWAS of lifetime cannabis use reveals new risk loci, genetic overlap with psychiatric traits, and a Causal Influence of Schizophrenia. *Nat Neurosci* 2018;**21**(September). https://doi.org/10.1038/s41593-018-0206-1.
123. Corsi DJ, Donelle J, Sucha E, et al. Maternal cannabis use in pregnancy and child neurodevelopmental outcomes. *Nat Med* 2020;**26**(10):1536–40. https://doi.org/10.1038/s41591-020-1002-5.
124. Guennewig B, Bitar M, Obiorah I, et al. THC exposure of human iPSC neurons impacts genes associated with neuropsychiatric disorders. *Transl Psychiatr* 2018;**8**(1). https://doi.org/10.1038/s41398-018-0137-3.
125. Miller ML, Chadwick B, Dickstein DL, et al. Adolescent exposure to Δ^9-tetrahydrocannabinol alters the transcriptional trajectory and dendritic architecture of prefrontal pyramidal neurons. *Mol Psychiatr* 2018. https://doi.org/10.1038/s41380-018-0243-x.
126. Beiersdorf J, Hevesi Z, Calvigioni D, et al. Adverse effects of Δ^9-tetrahydrocannabinol on neuronal bioenergetics during postnatal development. *JCI Insight* 2020;**5**(23). https://doi.org/10.1172/jci.insight.135418.
127. Papariello A, Taylor D, Soderstrom K, Litwa K. CB1 antagonism increases excitatory synaptogenesis in a cortical spheroid model of fetal brain development. *Sci Rep* 2021;**11**(1):1–17. https://doi.org/10.1038/s41598-021-88750-2.
128. Stanslowsky N, Jahn K, Venneri A, et al. Functional effects of cannabinoids during dopaminergic specification of human neural precursors derived from induced pluripotent stem cells. *Addict Biol* 2016. https://doi.org/10.1111/adb.12394.
129. Manning JJ, Green HM, Glass M, Finlay DB. Pharmacological selection of cannabinoid receptor effectors: signalling, allosteric modulation and bias. *Neuropharmacology* 2021;**193**. https://doi.org/10.1016/j.neuropharm.2021.108611.
130. Notaras M, Lodhi A, Barrio-Alonso E, et al. Neurodevelopmental signatures of narcotic and neuropsychiatric risk factors in 3D human-derived forebrain organoids. *Mol Psychiatr* 2021; (June). https://doi.org/10.1038/s41380-021-01189-9.
131. Shabani M, Divsalar K, Janahmadi M. Destructive effects of prenatal WIN 55212-2 exposure on central nervous system of neonatal rats. *Addict Heal* 2012;**4**(1–2):9–19. https://pubmed.ncbi.nlm.nih.gov/24494131/. [Accessed 13 July 2021].
132. Miranda CC, Barata T, Vaz SH, Ferreira C, Quintas A, Bekman EP. hiPSC-based model of prenatal exposure to cannabinoids: effect on neuronal differentiation. *Front Mol Neurosci* 2020;**13**(July):1–11. https://doi.org/10.3389/fnmol.2020.00119.
133. Shum C, Dutan L, Annuario E, et al. Δ^9-tetrahydrocannabinol and 2-AG decreases neurite outgrowth and differentially affects ERK1/2 and Akt signaling in hiPSC-derived cortical neurons. *Mol Cell Neurosci* 2020;**103**. https://doi.org/10.1016/j.mcn.2019.103463.

Chapter 14

Cannabis effects on the adolescent brain

Kateryna Murlanova[a], Yuto Hasegawa[b], Atsushi Kamiya[b], and Mikhail V. Pletnikov[a]
[a]Department of Physiology and Biophysics, Jacobs School of Medicine and Biomedical Sciences, State University of New York at Buffalo, Buffalo, NY, United States, [b]Department of Psychiatry and Behavioral Sciences, Johns Hopkins University School of Medicine, Baltimore, MD, United States

Introduction

Human brain development is far from complete at the time of birth, and maturation continues throughout childhood and into adolescence, a time of particularly notable morphological and functional transformations in the brain. The high degree of neuroplasticity that occurs in this period renders the adolescent brain vulnerable to environmental factors. During adolescence, the endogenous cannabinoid (eCB) system that fine-tunes dynamic changes in mesocorticolimbic brain structures is maturing and the expression of cannabinoid receptors peaks. These receptors also respond to the main psychoactive component of cannabis, delta-9-tetrahydrocannabinol (Δ^9-THC). Therefore, exposure to Δ^9-THC during adolescence may substantially alter neurodevelopmental trajectories, resulting in a cascade of neurochemical and neurostructural aberrations.

Chronic cannabis exposure during adolescence can result in persistent deficits in cognitive domains such as attention, memory, and processing speed. Cannabis use during adolescence is also linked to an increased risk for psychiatric disorders, including psychosis (schizophrenia), depression, anxiety, and substance use disorders, later in life. Notably, not all cannabis users exhibit these long-lasting behavioral and cognitive impairments, suggesting there is a genetic vulnerability, i.e., a gene-environment relationship for cannabis sensitivity.

Unfortunately, little is known about the mechanisms of individual susceptibilities to the adverse effects of cannabis use in adolescence. The molecular mechanisms of gene and environment interactions differ across cell types, and recent studies have only begun to identify the molecular cascades activated by Δ^9-THC in a cell-type-specific manner. In this chapter, we review these interactions and their contributions to cannabis sensitivity and to the development of long-lasting behavioral abnormalities. We also lay out the known cell-type-specific mechanisms of the susceptibilities to the adverse effects of cannabis and discuss the proinflammatory signaling pathways involved in Δ^9-THC-induced behavioral impairments. Finally, we highlight new avenues to study the vulnerability to adverse effects of Δ^9-THC exposure, specifically, changes in brain cell energetics and the insights gleaned from studies in humans and animal models.

Adolescence is a critical period of brain development

Morphological and functional transformations in the brain during adolescence

Adolescence is the phase of life between late childhood and adulthood that begins with the onset of puberty (i.e., hormonal fluctuations and physical changes). This is a critical period in biological and psychosocial development during which neurobiological factors and experiences interact to shape normative brain development and alter behavior. This critical period is marked by autonomous exploration of the environment, a peak in sensation-seeking behavior, and sexual maturation, i.e., novel and increasingly complex experiences that are psychologically, socially, and cognitively demanding.[1]

The understanding of the developmental changes in the human adolescent brain has expanded rapidly, aided by longitudinal studies and neuroimaging tools, such as magnetic resonance imaging (MRI), digital MRI, functional MRI (fMRI), magnetic resonance spectroscopy, magnetoencephalography, positron emission tomography, single-photon emission computed tomography, and infrared techniques.[2-4] However, causal relationships and the precise morphological and molecular underpinnings of the observed developmental changes are difficult to dissect with human imaging and are more amenable to study using animal models of adolescence.

The adolescent brain undergoes both progressive and regressive transformations that are regionally specific and serve to refine functional connectivity. Multiple MRI studies have documented the changes in cortical structure, specifically, changes in gray matter (GM) volume and the parameters that contribute to volume—surface area and thickness[5,6]—with most regions following inverted-U-shaped developmental trajectories. Cortical GM volume rapidly increases during the first years of life, peaks in early childhood,[7] and then decreases throughout adolescence toward adulthood.[6,8-10] The timing of this is brain region dependent, with different structures maturing at different times and rates.[11,12] The folding of the cerebral cortex is also impacted, evidenced by a decrease in gyrification during adolescence.[13,14]

The maturational event most consistently linked to developmental cortical "thinning" is the reduction of synaptic density known as "synaptic pruning." Pruning during adolescence is highly specific and can be pronounced, resulting in a loss of nearly 50% of the synaptic connections in some regions but minimal decline in others.[15] The sequence in which the cortex matures, as found through structural and functional MRI scans, follows the same course as cognitive development. GM pruning begins in the primary sensorimotor cortices and spreads rostrally to association areas. GM loss occurs latest in the dorsolateral pre-frontal cortex (PFC).[16] Synaptic elimination during adolescence likely involves fine-tuning of the excitatory/inhibitory balance at the neuronal and network levels such that excitatory synapses are selectively reduced while inhibitory synapses are spared.[17] Because synapses are energetically costly, synaptic loss may contribute to the increases in brain efficiency detected during adolescence as a decline in brain energy utilization in humans and other species.[15] Furthermore, microglia and astrocytes play key roles in regulating synaptic elimination during adolescent development through direct or indirect phagocytosis of supernumerary synapses.[18] Thus, maturation during adolescence is accompanied by a substantial increase in astrocyte-synaptic colocalization, indicative of increased interactions between the two.[19]

The thinning and pruning described above are counteracted by neurogenesis and the establishment of new synaptic connections, which occur at higher rates than in adulthood.[20,21] However, the generation of modest amounts of new neurons is restricted to a few brain regions, for example, the hippocampus, and is thought to be important for some forms of learning and processing of stress-inducing stimuli.[22,23] Neurons generated during adolescence may also contribute to emotional and reward circuitry.[24] As the circuits become efficient and reliable, the synaptic architecture is stabilized by perineural nets that form around cell bodies and proximal dendrites.[1,25]

In parallel with GM contraction, white matter (WM) expansion and fiber optimization are observed during adolescence.[26–28] These co-occurring phenomena are attributed to developmental exuberance,[29] which refers to the overproduction and subsequent selection of neural connections. The expanding WM surface area corresponds to increases in the volume and density of WM.[30–32] These changes can be observed as increases in fractional anisotropy (index of the orientational coherence) and decreases in diffusivity (index of the magnitude of water diffusion) with an MRI technique known as diffusion tensor imaging.[33–35] The WM changes during adolescence are generally attributed to increases in the size and bundling of axons as well as increases in myelin volume.[36–38] The myelin sheath that enhances the speed and reliability of electrical transmission along axons is formed by oligodendrocytes, which also metabolically support axon function.[39] During adolescence, the speed and efficiency of information flow across relatively distant regions is accelerated, and the adaptive changes in WM optimize information transfer within and across neural networks to enhance cognitive processing.[15,36] However, myelination prevents future branching of neural circuits and, together with perineural nets, restricts plasticity and closes the adolescent critical period window.[1]

Altogether, the cortical thinning, synaptic pruning, and myelination as well as apoptotic[40] and epigenetic[41,42] processes lead to the maturation and remodeling of neural circuits. In this way, the increase in and persistence of sex steroids (estrogens, progestogens, and androgens) during adolescence modulate the neuroplasticity and restructuring of the adolescent brain.[42] Notably, GM and WM development is sexually dimorphic,[8,34,35,43] with less of a linear increase in WM volume and decrease in GM volume with age in females than in males.[44,45]

Imaging studies also reveal morphological sex differences in cortical and limbic structures.[46] However, the reports of sex-specific patterns in regional brain changes across adolescence are not always consistent,[12,47–49] and differences likely depend on age and stage of pubertal development.[44] Both males and females show non-linear changes with age, but these changes diverge between the sexes, which may contribute to the observed discrepancies. For example, the volumes of the hippocampus, amygdala, caudate, pallidum, and thalamus were found to be larger in boys throughout puberty and adolescence when corrected for total intracranial volume.[50,51] Nevertheless, numerous animal and human studies provide evidence that adolescent neurodevelopment is differentially triggered, modulated, and programmed by pubertal sex hormones.[52–54]

The sexual dimorphisms found are not only structural but also functional and, as a result, behavioral. One of the circuits undergoing significant sex-dependent changes during adolescence is known as the mesolimbic dopaminergic "reward" circuitry.[55,56] Although both adolescent males and females are highly vulnerable to substance use, males are more likely to use drugs. Functional neuroimaging studies have found greater activation in reward-relevant regions, particularly, the frontal lobe, striatum, and amygdala, in adolescent males than in females during reward-related tasks.[57] In line with this, boys also show heightened brain activation in the nucleus accumbens (NAc) during reward receipt compared to that in girls.[58]

Notable sex-dependent interplay between neurons, microglia, and steroid hormones may underlie the sexually dimorphic brain circuitry and behavior that emergence in adolescence. For example, microglia and complement-mediated immune signaling may participate in a sex-specific manner in developmental changes in brain circuitry and in reward behavior in adolescence.[59] In rodents, microglia shape NAc development by eliminating dopamine D1 receptors in males but not in females during adolescence.[60]

Sex differences in brain structure and function during adolescence may be linked to sex differences in the sensitivity to environmental factors.[44] Indeed, the developmental trajectories (both structural and functional) in the brain are guided by the complex interplay of many genetic and environmental factors.[27] The ongoing Adolescent Brain Cognitive Development℠ (ABCD) study, the largest study to date in the United States assessing brain development, aims to identify individual developmental trajectories (e.g., brain, cognitive, emotional, and academic) by using neuroimaging, cognitive, biospecimen, behavioral, and environmental measures, and youth and parent self-report metrics to examine more than 10,000 youths.[3,61] Such a comprehensive approach is expected to increase our understanding of environmental, genetic, social, and other biological factors, including substance use, that affect brain and cognitive development and can enhance or disrupt a young person's life trajectory.

Vulnerability to environmental factors during adolescence

The brain is undergoing extensive neuroanatomical and functional reorganization during adolescence and is exquisitely sensitive to factors in the environment. Adolescent experiences thus profoundly influence brain development, with long-lasting effects on behavior, cognition, and emotion. Individuals are more responsive to adverse environmental inputs such as social stress and substance use during adolescence than they are at other developmental stages.[62–69]

Adolescence is a time of increased exploration and risk taking, e.g., engaging in unsafe sexual behavior and dangerous driving and experimenting with drug use.[55] Substance (tobacco, cannabis, alcohol, cocaine, etc.) use is often initiated during adolescence, and >80% of drug users began using during adolescence.[70] These tendencies are thought to result from the elevated reward value coupled with underdeveloped inhibitory control and, thus, a hypersensitivity to reward.[56]

Studies in humans and laboratory animals generally support the notion that adolescents are more sensitive to reward than adults.[71–73] The reward circuitry comprises dopaminergic neurons in the ventral tegmental area (VTA) that project to the NAc, which is part of the ventral striatum.[74] During adolescence, dopamine D1 and D2 receptor binding in striatum and NAc peaks,[75] dopaminergic neurons release more dopamine in response to environmental or pharmacological stimulation,[76] and the striatum is hyperresponsive to reward, as shown in fMRI studies.[77,78] Therefore, the mesolimbic dopamine system is considered to be in a state of overdrive during adolescence, with increased sensitivity to environmental factors and implications for adolescent psychopathology.[79,80]

At the neurochemical level, a variety of neurotransmitter systems, including monoaminergic, GABAergic, and glutamatergic systems, undergo major developmental changes in the brain and have a higher sensitivity during adolescence.[81–84] For example, remodeling of GABAergic functionality in the PFC refines the excitatory-inhibitory balance, which is essential for the acquisition of adaptive adult behaviors and cognitive processing.[85] Additionally, glutamate-guided pruning in the frontal cortex steers the changes in corticolimbic connectivity during adolescence.[17]

During adolescence, connections between different networks of functionally related regions become stronger via intensification of within-network connections, particularly those linking more distant network regions.[15] There is an "imbalance" in functional development across brain regions, with development occurring earlier in posterior regions and anterior regions progressing later, leading to underdeveloped connections between midbrain corticolimbic (reward) and frontal (inhibitory) region circuits.[86,87] This imbalance may account for typical adolescent behavior patterns, such as engaging in risky behaviors (sensation seeking and impulsivity) and substance abuse.[65,88–90] Exposure to environmental factors, especially psychotropic drugs, can profoundly affect cortical circuit and corticolimbic network development, making adolescents highly sensitive to drug-related developmental disturbances.[68,91,92]

However, individuals are varied in their sensitivity to environmental triggers, with some more affected than others. The degree to which individuals "tune" to the environment may reflect gene expression, stress reactivity, and structural and functional neural characteristics that are context sensitive and reactive to specific environmental cues.[66] Multi-hit hypotheses accounting for the combination of environmental exposure, genetic risk, and lifestyle factors in the etiology of neurological disorders are increasingly used to explain individual susceptibility. For example, chronic adolescent cannabis use is commonly associated with altered brain development and poor cognitive performance and, when coupled with certain genetic polymorphisms (a "two-hit" hypothesis), appears to increase the risk of delayed-onset psychosis and schizophrenia.[93]

Endogenous cannabinoid (eCB) system in the adolescent brain

eCB system

The eCB system is a widespread neuromodulatory system comprising cannabinoid receptors (CB_1 and CB_2), endogenous ligands (primarily N-arachidonoylethanolamine, or anandamide [AEA], and 2-arachidonoylglycerol [2-AG]), and the enzymes responsible for the metabolism of the eCBs, including fatty acid amide hydrolase (FAAH) and monoacylglycerol lipase (MAGL).[94] Dynamic interactions between these eCB system components, which reach peak expression during adolescence, are important in adolescent brain development, synaptic plasticity, and the homeostatic maintenance of behavioral processes.[95–97]

Both CB_1 (encoded by the *CNR1* gene) and CB_2 (encoded by the *CNR2* gene) receptors are G-protein-coupled receptors, which, when activated, inhibit adenylyl cyclases and certain voltage-dependent calcium channels and activate several mitogen-activated protein kinases that regulate potassium channels, with some variation depending on the cell type.[94,98–100] Thus, CB_1 and/or CB_2 receptor activation has different consequences on cellular physiology, including synaptic function, gene transcription, cell motility, etc.[98,101,102]

eCBs mediate short- and long-term plasticity at both excitatory and inhibitory synapses via retrograde signaling.[100] In most cases, signaling starts with the de novo production of 2-AG in response to an increase in intracellular Ca^{2+} concentrations and/or activation of $G_{q/11}$-coupled receptors. 2-AG is then released into and traverses the extracellular space, via a mechanism not yet fully elucidated, and arrives at the pre-synaptic terminal where it binds to CB_1 receptors. Activated CB_1 receptors inhibit neurotransmitter release in two ways: first, by suppressing voltage-gated Ca^{2+} channels, which reduces pre-synaptic Ca^{2+} influx, and second, by inhibiting adenylyl cyclase and the subsequent cAMP/PKA pathway, which is involved in long-term depression (LTD).[100,103–105] For example, CB_1 receptor activation decreases pre-synaptic GABA release, thereby depressing GABAergic inhibitory control of post-synaptic neurons, resulting in excitation.[106,107] The eCB signal is terminated when 2-AG is degraded by MAGL, which is expressed in select synaptic terminals as well as glial cells.[108] Thus, the eCB system is well placed to modulate neuronal activity by providing negative feedback to filter and select afferent inputs.[109]

There is also evidence that eCBs signal in a non-retrograde or autocrine manner, in which they modulate neural function and synaptic transmission by engaging transient receptor potential vanilloid 1 (TRPV1) ion channel and CB_1 receptors on the post-synaptic cell.[106] Furthermore, eCBs can signal via glial cells to indirectly modulate pre-synaptic or post-synaptic function.[110,111] For example, CB_1 receptors are expressed by astrocytes[112,113] to influence glutamatergic neurotransmission.[114,115] CB_1 receptors are abundantly expressed by multiple neuronal populations throughout the brain, with a notable presence on glutamatergic and GABAergic neurons,[116,117] primarily the axon terminals of GABAergic neurons in the VTA and hippocampus.[118–121] CB_2 receptors are pre-dominantly expressed by peripheral immune cells[122,123] but also by glial cells and vascular elements[124–127] in the PFC, hippocampus, and basal ganglia and by VTA dopamine neurons.[128–131] The activation of neuronal CB_2 receptors, which are mainly expressed on the post-synaptic cell body, hyperpolarizes the membrane potential and thus is inhibitory.[131–133]

Maturation of the eCB system

The dynamic changes in the cortico- and mesolimbic structures that occur during adolescence[134,135] are fine-tuned by the eCB system[136,137] to regulate reward-, stress-, and anxiety-associated behaviors.[97,138,139] Numerous studies in rodents demonstrate that CB_1 receptor expression, particularly in cortical regions, striatum, cerebellum, and hippocampus, peaks in adolescence.[97,140–142] The increased expression corresponds to increases in CB_1 receptor binding in these areas in male and female animals at approximately the onset of puberty.[143] By contrast, CB_2 receptor expression in PFC and hippocampus is lower in adolescent animals than in adult animals, with asymptote values occurring around mid-adolescence.[130]

How CB_1 and CB_2 receptor expression changes throughout human brain development is less clear. A microarray analysis of post-mortem samples showed reduced expression of CB_1 receptors in dorsolateral PFC, particularly in cortical layer II, during adolescence compared to that during early post-natal life, which continued to decline until young adulthood.[144] However, the circumstances relating to death and variable post-mortem intervals are limitations of using human brain material, which may obscure the major impact of age on the expression of components of the eCB system. A study of non-human primates showed that CB_1 receptor mRNA expression in dorsolateral PFC was highest at birth, markedly decreased during the first three post-natal months, and then did not change during adolescence and adulthood.[145] The same study showed that CB_1 receptor immunoreactivity in the GM robustly increased during the perinatal period and achieved adult levels by 1 week after birth. However, a marked increase in CB_1 receptor-immunoreactive axons in layer IV of PFC was observed during adolescence. Refinements in the laminar distribution of CB_1-receptor-immunoreactive axons suggest that CB_1 receptors may play a role in the development of cognitive functions reliant on the dorsolateral PFC.

2-AG and AEA are the primary eCBs (endogenous lipids) in the brain that affect behavior. The intrinsic efficacies of these differ: 2-AG is a high-efficacy agonist of both CB_1 and CB_2 receptors, whereas AEA is a low-efficacy agonist of CB_1 receptors and a very-low-efficacy agonist of CB_2 receptors.[146] The levels of 2-AG and AEA change throughout adolescence.[97] In rodents, levels of these eCBs progressively increase from early adolescence to adulthood, with fluctuations in expression consistent with the temporal development of corticolimbic structures.[140,147] AEA levels in the hypothalamus peak immediately before the onset of puberty in female rats,[148] indicating the possible involvement of the eCB system in the timing of puberty.

Both 2-AG and AEA are synthesized from omega-6 polyunsaturated fatty acid and contain arachidonic acid in their structure. Arachidonic acid and docosahexaenoic acid make up $\sim 20\%$ of the fatty acids in the mammalian brain.[149] Components of arachidonic and docosahexaenoic acid pathways are coordinately expressed and underlie the cascade of interactions in membrane synthesis during neurodevelopment, neuroplasticity, and neurotransmission,[150,151] including those involved with WM development during adolescence.[152,153] The concentrations of arachidonic and docosahexaenoic acids remain relatively constant during adolescence,[154] but a deficiency in these is linked to developmental disorders, such as attention-deficit/hyperactivity disorder in adolescence.[155]

AEA is largely produced from N-arachidonoyl phosphatidyl ethanol (NAPE), which is hydrolyzed by NAPE-preferring phospholipase D (NAPE-PLD), whereas 2-AG is produced from 2-arachidonoyl-containing phospholipids, primarily phosphatidyl inositol bis-phosphate, which is hydrolyzed by phospholipase C-β; the resulting diacylglycerol is primarily hydrolyzed by diacylglycerol lipase (DAGL)α.[94] NAPE-PLD levels progressively increase in human cortex throughout development, reaching maximal levels in adolescence that are maintained during adulthood.[144] Similarly, there is a marked increase in NAPE-PLD expression and activity during adolescence in rodents.[156] Expression of cortical DAGLα peaks during adolescence, with a subsequent decrease later in life.[144]

2-AG and AEA are hydrolyzed by serine hydrolases. MAGL and α/β-hydrolase domain 6 and 12 (ABHD6 and -12) enzymes degrade 2-AG.[157] MAGL is located pre-synaptically, whereas ABHD6 is mainly found in dendrites, suggesting that the two different 2-AG-degrading enzymes have different functions.[99] MAGL levels decline in adolescence, while ABHD6 levels increase compared to that in early post-natal life,[144] suggesting that 2-AG degradation is anatomically balanced in pre- and post-synaptic compartments. AEA is degraded by the enzyme FAAH.[94,158] In rodents, FAAH activity during adolescence fluctuates in a reciprocal fashion to changes in AEA levels.[147] From early life until adolescence, whole-brain FAAH activity increases linearly and remains at a high level during adulthood.[156] Similarly, FAAH expression in human PFC increases steadily after infancy, peaking in adulthood.[144] These patterns indicate increased levels of regulation corresponding to AEA availability during adolescence. Given the rises in FAAH and NAPE-PLD expression, AEA turnover is likely higher in adolescence than earlier in life. The variations in the regional and temporal expression of components of the eCB system, which reflect the enhanced eCB system plasticity during adolescence, may regulate the window during this developmental period when corticolimbic circuitry is particularly sensitive to perturbations of eCB signaling.

The eCB system and sex hormones closely and bidirectionally interact.[159,160] For example, pregnenolone, the precursor to all steroid hormones, which increases during adolescence,[161,162] controls CB receptor activation.[163,164] Notably, there are sex differences in the dynamic expression of the eCB system during adolescence. For example, cannabinoid receptor levels peak earlier in female rodents than in males, and females have a higher striatal cannabinoid receptor density during adolescence.[165] Another rodent study demonstrated that CB_1 receptor levels in caudate putamen increase in males but decrease in females from adolescence to adulthood, whereas expression in the basolateral amygdala decreases only in males.[166] Such interactions may provide a biological basis for sex differences in the development of the eCB system during pubertal maturation as well sex-specific alterations following exposure to cannabinoid agonists.[167–170]

Cannabis and the adolescent brain

Cannabis is the most-used illicit drug worldwide, particularly among adolescents.[171,172] Δ^9-THC, one of the four major and most-researched compounds of the cannabis plant, is the primary active constituent responsible for the drug's psychoactive effects.[173] Δ^9-THC interacts with the orthosteric sites of CB_1 and

CB$_2$ receptors, resulting in partial agonism.[99,174] The very high binding affinity of Δ^9-THC to the CB$_1$ receptor ($K_i = 10$ nM) mediates its psychomimetic properties,[175,176] whereas binding to CB$_2$ receptors ($K_i = 24$ nM) produces immunosuppressive effects.[177,178]

Δ^9-THC does not simply mimic 2-AG and AEA[94,179] but rather evokes a complex interplay. By itself, Δ^9-THC acts as a partial agonist, but it also can antagonize or attenuate the actions of the endogenous eCBs.[180] The resulting effect of Δ^9-THC thus depends on the context of eCB stimulation of the cannabinoid system as well as the CB receptor densities and signaling pathways.[94]

As stated above, the eCB system contributes to maturation-related neural reorganization[97] and is itself highly plastic during adolescence. As a result, the adolescent brain, which is "under reconstruction," is particularly vulnerable to the effects of cannabis use and abuse.

Developmental effects of cannabis

Morphological changes in the adolescent brain

Cannabis exposure during adolescence changes brain morphology, though the specific patterns of changes in GM and WM macrostructure, cortical/subcortical volume, and cortical thickness based on imaging are inconsistent.[92] The macro- and microstructural changes to the adolescent brain in response to cannabis use differ according to brain region, age, and sex and are dependent on the extent and frequency of use.[181–183] Furthermore, the effects of alcohol and nicotine, which are often used concurrently, on brain morphology may obscure the effects of cannabis.[184,185] Moreover, the composition of cannabis over the last 20 years has changed, with increased Δ^9-THC contents in newer cultivated strains,[186] complicating data comparisons between earlier and more recent studies. Thus, more-complex longitudinal research with better-defined groups is needed to understand how cannabis affects central nervous system (CNS) maturation in adolescents and how enduring these effects are.

The developmental reduction in GM appears to be greater in adolescent cannabis users than in age-matched non-users.[143,187,188] MRI evaluations also show that the percentage of GM relative to the whole-brain volume is lower in early-onset users (before age 17) than in late-onset users.[189] Because the eCB system influences synaptic pruning,[190] this influence may be exacerbated by the additional stimulation from Δ^9-THC during adolescence. Further alterations in GM structure were not observed in those who continued use during adulthood, indicating that the structural effects occur pre-dominantly during adolescence,[191] possibly as a result of altered pruning.

Adolescent chronic users exhibit a greater loss of GM in the medial temporal cortex, parahippocampus, insula, and orbitofrontal cortex,[180] which was linked to reductions in both GM thickness and density.[183] There are several reports that cannabis use in adolescence reduces cortical thickness,[187,192] a metric that captures the cumulative effects of environmental exposures on neuropil (i.e., dendrites and glial cells) and capillary densities.[192] Notably, brain regions with high MAGL expression appear to be the most vulnerable to cannabis-related cortical thinning.[183]

However, there is also evidence that cannabis use in late adolescence increases cortical thickness and volume.[193–195] The Avon Longitudinal Study of Parents and Children demonstrated a trend for greater cortical thickness in male adolescents with fewer than five instances of cannabis use than in cannabis-naive controls.[192] In adolescents reporting only one or two instances of cannabis use, GM volume was greater in regions enriched with CB_1 receptors and *CNR1* gene expression, such as large medial temporal clusters incorporating the amygdala, hippocampus, and striatum, and extending into the left PFC.[196] Significantly greater GM volume was also observed in the lingual gyri, posterior cingulate, and cerebellum.[196] Increased GM density and shape differences were also reported in the left NAc extending to sub-callosal cortex, hypothalamus, sub-lenticular extended amygdala, and left amygdala in young adult recreational users.[193] Furthermore, hippocampal enlargement and asymmetry have been observed in cannabis-exposed adolescents.[197] These findings suggest that cannabis exposure, even at extremely low levels, alters the neural matrix of core reward structures, consistent with pre-clinical studies of changes in dendritic density and gliogenesis after exposure to cannabinoid agonists.[198,199] Even a single dose of Δ^9-THC transiently eliminated eCB-mediated LTD in the NAc and hippocampus in adolescent mice.[200] A delay in LTD could interrupt maturation-related neural pruning and preserve GM.[17]

Although CB_1 receptors are pre-dominantly found on neurons, they are also expressed by myelinating glial cells and thought to contribute to structural connectivity.[201–203] Accordingly, adolescent cannabis use appears to alter WM volume,[189,204] reduce WM structural integrity (e.g., decreased fractional anisotropy and diffusivity),[183,205,206] and alter connectivity in a number of regions, including connections between the frontal lobes and temporal and occipital regions as well as the left inferior longitudinal fasciculus.[92,207,208] Adolescent cannabis exposure was found to increase cerebral blood flow in WM fiber tracts, indicating poorer structural integrity, which is essential for cortical connectivity in reward circuitry and optimal cognitive functioning.[209] Adolescent cannabis users also exhibit reduced cerebral blood flow in PFC, insula, and temporal regions.[210]

Volumetric analyses demonstrated a larger cerebellar volume in adolescent cannabis users[211] but no differences in volume or shape of the hippocampus, amygdala, or NAc compared to that in non-users.[212,213] By contrast, a voxel-based morphometry study found that frequency of cannabis use was related to smaller hippocampal volumes in heavy users, and the severity of cannabis dependence was associated with smaller amygdala volumes.[214] Thus, differential patterns of structural changes suggest that alterations to the brain depend on the amount of cannabis use and the extent of abuse.

Functional changes in the adolescent brain

In contrast to the discrepancies for structural alterations in cortical and subcortical regions,[215] there is consistency in the functional changes in the brains of adolescent cannabis users, with a trend toward increased activation.[92] The resting state activity in the right parietal cortex and right PFC is higher in adolescent cannabis users than in non-users.[216] Adolescent cannabis users exhibit reduced connectivity between anterior cingulate cortex and dorsolateral PFC,[217] reduced interhemispheric connectivity in cerebellum and superior frontal gyri, and increased interhemispheric connectivity in supramarginal gyri.[216] Functional alterations in adolescent cannabis users are also evidenced by reduced sub-cortical global myo-inositol/creatine ratios and WM myo-inositol levels[218] as well as reduced glutamate, N-acetyl aspartate, creatine, and myo-inositol levels in the anterior cingulate,[219] which hint at neuronal and microglial toxicity.[204]

Most published fMRI studies in adolescent cannabis users are focused on working memory, reward, and emotional processing tasks.[215] Adolescent cannabis users show poorer verbal working memory performance, with elevated activity in pre-frontal and posterior parietal regions[220] and a failure to deactivate the hippocampus.[221] With regard to spatial working memory, adolescent cannabis users' exhibit heightened dorsolateral PFC activation and greater inferior and middle frontal deactivation than non-users despite similar task performance,[222] suggesting differences in their strategic approaches to the task. The age at first cannabis exposure in adolescents was associated with reduced blood-oxygen-level-dependent activation of the posterior parietal cortex, which correlated with longer reaction times in working memory tasks.[223] Moreover, a meta-analysis of fMRI studies revealed that the activation in inferior parietal gyrus and putamen in adolescents who used cannabis increased independently of specific cognitive processes.[224] During anticipatory stages of reward in a monetary incentive delay task (reward processing task), adolescent cannabis users exhibit striatal hyperactivity,[225] which may signify overly sensitive motivational brain circuitry and a weakened ability to disengage this circuit during non-rewarding events, which in turn, could drive risk-seeking behavior in adolescents. By contrast, positron emission tomography studies show there is a blunting of striatal dopamine release and a reduced dopamine synthesis capacity in adolescent cannabis users.[226] This might suggest lower striatal responsivity to reward, which relates to the reported apathy in chronic cannabis users.[227] Meanwhile, adolescent cannabis users exhibit greater bilateral amygdala activation in response to signals of threat (viewing an angry face rather than a neutral face during fMRI scanning) in an emotional processing task.[68]

Cannabis use among adolescents can also interfere with brain-derived neurotrophic factor (BDNF) signaling, which is critical for brain development, cognitive processing, and neuroplasticity. Adolescents demonstrate increased plasma BDNF levels during moderate cannabis use.[228] Because cannabinoids can transactivate BDNF receptors, cannabis use during adolescence could alter the plasticity involved in the acquisition of new skills and the ability to adapt and result in the "learning" of addiction-related behaviors.[229,230]

Behavioral effects of chronic cannabis use during adolescence
Insights from human studies

Accumulating evidence from human studies suggests that exposure to cannabinoids during adolescence has short- and long-term consequences on cognition and behavior. For example, cannabinoid exposure in adolescents results in acute, transient, and dose-related deficits in attention (sustained, divided, and selective), signal detection, allocation of attention, sensory gating, learning (verbal and associative), memory (working, spatial working, and procedural), time estimation, distance estimation, impulsivity, reaction time, information processing speed, and tracking accuracy.[231] Most of these effects peak ~15–45 min following exposure and decrease afterwards. A recent longitudinal study in adolescent cannabis users also showed that cannabis exposure impairs immediate, but not delayed, recall in an episodic memory task as well as decision-making.[232]

A meta-analysis of 69 cross-sectional studies of 2152 cannabis users and 6575 comparison participants showed that cognitive functioning is impaired by frequent cannabis use.[233] However, these impairments were attributed to residual effects from acute use or withdrawal, because the deficits diminished after abstinence periods of more than 72 h.[233] Verbal learning in adolescent cannabis users was shown to improve beginning 1 week into a period of abstinence.[234] Although abstinence was also shown to attenuate the IQ deficit associated with cannabis use in adolescents, the deficit persists nonetheless.[235] Attentional and psychomotor deficits also seem to persist in chronic users after termination of use.[236]

Chronic, heavy cannabis use during adolescence has been shown to impair memory, attention, working memory, executive function, and intelligence.[217,237] Notably, the impairments in general executive functioning and IQ with chronic use are greater in adolescents than in adults.[238] Furthermore, adults show a greater decrease in craving and impulsivity post-cannabis intoxication than adolescents.[238] Adolescent cannabis users also have a slower learning trajectory (paired-associate verbal learning task), which was linked to disrupted medial temporal and midbrain function.[239]

In a longitudinal cohort study, adolescent cannabis use (before age 17 years) was identified as a potential risk factor for hypomania in early adulthood,[240] and the nature of this association suggests a potential causal link. Adolescent cannabis use is associated with an approximately threefold increased risk for the onset of manic symptoms and worsens manic symptoms in those with pre-existing bipolar disorder.[241] Cannabis use may be linked to emotional dysregulation, schizophrenia-related psychosis, and increased vulnerability to certain classes of other addictive drugs.[242–244]

An estimated 35%–75% of adolescents with cannabis use disorder experience cannabis withdrawal syndrome during attempts to reduce or abstain from use.[245] Cannabis withdrawal syndrome includes changes in mood, anxiety, irritability, sensation of restlessness, appetite changes, sleep disturbance, and cannabis craving and is recognized in the DSM-5.[246] Because of the extended half-life of Δ^9-THC (27–57 h) and its metabolites (up to 5 days), symptoms of withdrawal, particularly cannabis craving and sleep disturbances, may persist for several weeks and/or possibly months after reducing or discontinuing cannabis use.[247] There is also an evidence that long-term/heavy cannabis use can lead to addiction. Approximately 1 in 11 people who experienced cannabis will become dependent in their lifetime, but this risk is almost doubled if use starts in adolescence.[248]

The findings from human studies reveal uncertainty regarding the persistence of cannabis-associated cognitive deficits and psychiatric conditions after abstinence.[249] Nevertheless, there appears to be a pattern in which there is greater harm in general when cannabis is used regularly than when used infrequently, when used daily as opposed to once in a while, and when used in greater amounts as opposed to smaller amounts.[92] Adolescent cannabis research suffers from difficulties in evaluating cannabis exposure history, poor control over potential sub-acute effects, and heterogeneity in cognitive measures and sample composition.[250] More research, especially with large-scale prospective studies and well-defined groups, is crucial to better understand the effects of adolescent cannabis use on behavior and to explore other contributors, such as genetic and environmental factors.

Insights from animal studies

Adolescent cannabis exposure has been studied extensively in animal models. These models similarly demonstrate more-pronounced cannabinoid effects in adolescents than in adults. To study the long-term effects of adolescent cannabinoid exposure in animals, three different CB_1 receptor agonists are generally used: Δ^9-THC and the synthetic cannabinoids WIN 55,212-2 (WIN) and CP 55,940 (CP). CP and WIN are full CB_1 receptor agonists, whereas Δ^9-THC is a partial agonist.[251] Substantial evidence from animal research supports that cannabis exposure during adolescence causes long-term or possibly permanent cognitive impairments. Repeated exposure of adolescent rhesus monkeys to Δ^9-THC impaired their performance in spatial, but not object, working memory tasks.[252] Repeated Δ^9-THC exposures did not result in tolerance to these effects, suggesting that adolescents remain susceptible to the acute cognitive deficits caused by cannabis regardless of their history of use.

A common pre-clinical model of adolescent cannabis exposure involves chronically treating an adolescent rodent with a cannabinoid, such as

Δ^9-THC or synthetic cannabimimetic compounds. Adolescence in rodents can be sub-divided into early adolescence (beginning around PND28), middle adolescence (beginning around post-natal day 38), and late adolescence (beginning around post-natal day 49); full sexual maturity is attained at post-natal day 60. Chronic adolescent cannabinoid treatment is associated with short- and long-term executive function impairment,[253] including working memory deficits[254,255] and impaired object recognition.[256–258] Furthermore, pre-clinical evidence suggests that chronic adolescent cannabis exposure increases the risk for neuropsychiatric conditions later in life.[251]

Adolescent Δ^9-THC exposure in female rats results in cognitive deficits and a depressive/psychotic phenotype in adulthood, whereas adult rats treated with Δ^9-THC show no impairments.[259] The same work showed that adolescent, but not adult, Δ^9-THC exposure alters selective histone modifications (mainly trimethylation of lysine 9 on histone 3 [H3K9me3]), impacting the expression of genes closely associated with synaptic plasticity. However, another study found that repeated exposure of rats to either cannabis smoke or Δ^9-THC during adolescence did not result in adverse affective or cognitive outcomes during adulthood.[260] Self-administration of WIN by adolescent rats also did not result in long-term cognitive dysfunction.[261] Moreover, male adolescent rats that self-administered high doses of Δ^9-THC showed enhanced working memory performance.[262] Similarly, adolescent rats exposed to cannabis smoke performed better on a delayed response working memory task in adulthood than clean-air controls.[263] Differences in drug exposure (Δ^9-THC, synthetic CB_1 agonists, and cannabis smoke) and washout length may account for these discrepancies. Nevertheless, these findings suggest that neural substrates of cognition are sensitive to CB_1 agonists.

The effects of adolescent cannabinoid exposure on anxiety behaviors in adult animals are even less consistent, with reports of anxiogenic-like[255,264,265] and anxiolytic-like[167] effects or no effects.[266] By contrast, Δ^9-THC or WIN exposure during adolescence more reliably induces depression-like behaviors, observed as decreased sucrose preference (i.e., anhedonia-like effects) and increased immobility times in the forced swim test in adult rats, reflecting their altered stress-coping strategy.[251,267]

Given that adolescence is associated with increased stress exposure,[268] individual differences in stress coping may contribute to the elevated risk of cannabis-induced psychopathology. For example, there could be a link between innate stress sensitivity and response to Δ^9-THC. Different responses to high (15 mg/kg) and low (1.5 mg/kg) doses of Δ^9-THC were found between two selectively bred strains of mice, i.e., socially submissive and stress susceptible (Sub) mice and socially dominant stress resilient (Dom) mice. The repeated low dose of Δ^9-THC lessened depressive-like behavior in Sub-mice exposed to forced swim, an acute stressor. The high dose of Δ^9-THC produced anxiogenic effect in Dom mice in the same setting.[269] Sub-mice developed strong aversion to the high dose of Δ^9-THC as measured by Conditioned Place Preference test, whereas Dom mice displayed no place preference or aversion. Despite Dom and Sub-mice were exposed to Δ^9-THC at their late adolescence, the effects persisted long after the cessation of drug exposure.[269]

A variety of long-lasting reward-related effects of adolescent cannabinoid exposure have been observed.[270] Adolescent WIN exposure alters the motivation for rewards (i.e., induces differences in incentive salience attribution in a sign/goal-tracking paradigm) and increases food palatability.[271] Adolescent cannabis consumption is also associated with enhanced intake and sensitivity to opiate drugs later in life[169,272] as well as cross-sensitization to cocaine.[273] Rats exposed to Δ^9-THC during adolescence self-administer more WIN and heroin in adulthood and exhibit a heightened vulnerability to stress-induced relapse of heroin seeking than those without a history of Δ^9-THC exposure.[274,275] Exposure to WIN during adolescence also prolongs immobility times in a tail suspension test during cocaine withdrawal in adulthood.[276] These works link adolescent cannabinoid exposure to susceptibility to addiction and reward-related consequences later in adulthood. It is important to note that from behavioral and developmental perspectives, many of the shared root causes (e.g., genetic pre-disposition, trauma, unstable psychiatric symptoms, thrill seeking, impulsivity, delay discounting, and environmental exposures) that increase an individual's likelihood of using cannabis may also increase the likelihood of using drugs of abuse.[277]

Cannabinoid exposure during adolescence was also shown to induce psychosis-like signs in rodents.[251] For example, chronic Δ^9-THC or WIN exposure during middle or late adolescence, but not during adulthood, induces persistent deficits in pre-pulse inhibition,[278,279] an operational measure of sensorimotor gating that is deficient in patients with schizophrenia.[280] These deficits were associated with hyperdopaminergic activity in the VTA.[264,265] Chronic cannabinoid exposure may also lead to abnormal spontaneous locomotor activity, observed as either hyperactivity[279,281] or hypoactivity,[282] although locomotor effects are not always observed for adolescent animals treated with Δ^9-THC or CP.[264] However, adolescent Δ^9-THC exposure consistently augments locomotor responses to psychomotor stimulants such as amphetamine or to phencyclidine in adulthood.[283–285]

There is some evidence for sex differences in the effects of adolescent Δ^9-THC. For example, female rodents are more likely to show cognitive deficits[259] and an altered emotional profile[266] following adolescent Δ^9-THC exposure. By contrast, voluntary oral consumption of Δ^9-THC by adolescent male rats impairs Pavlovian reward-predictive cue behaviors, consistent with a male-specific loss of CB_1 receptor-expressing vesicular glutamate transporter 1 synaptic terminals in the VTA.[286] Thus, Δ^9-THC consumption in adolescence may modify emotional circuits in female rodents but alter the sensitivity to rewarding stimuli in males. Silva et al. reported that pre-pubertal male and female rats are more sensitive to the anxiolytic effects of Δ^9-THC than post-pubertal rats.[287] The authors also noted that the two sexes do not reach puberty at the same time; thus, the testing of both sexes at the same age may contribute to inconsistencies in sex differences across studies. Furthermore, male and female rats metabolize Δ^9-THC differently. Adolescent female rats have higher levels of the active metabolite 11-OH-THC than their male conspecifics, particularly after repeated Δ^9-THC administration.[288] Thus, more research is needed to determine whether the sex differences in the effects of cannabinoids on cognition, motivation, anxiety, etc. reported to date are consistent across a broader range of ages, drug doses, drug metabolism rates, and behavioral tests in animals. Moreover, studies are needed to determine whether there are sex differences in the effects of cannabis and cannabinoids on these behavioral domains in humans.

Neurobiological mechanisms underlying cannabinoid-induced behavioral alterations: Insights from animal studies

Pre-clinical studies have advanced the understanding of how adolescent cannabis use affects brain function. For example, animal models have been used to establish causal relationships and thus better elucidate the neurobiological mechanisms underlying the cannabis-triggered processes. Adolescent rodent models confirmed that chronic cannabinoid exposure alters the eCB system, neurotransmission, dendritic architecture, synaptic activity, neural circuits, and neural and glial cellular processes.[289–291]

Long-term changes in the eCB system

It is postulated that the adolescent eCB system is vulnerable to long-term changes produced by exogenous cannabinoids, which in turn modulate excitatory and inhibitory synaptic plasticity. Repeated Δ^9-THC treatment in adolescent rats causes widespread CB_1 receptor downregulation (reduced number of receptors) and desensitization (attenuated receptor-mediated G-protein activity) in the PFC, striatum, hypothalamus, hippocampus, periaqueductal gray, ventral midbrain, and cerebellum.[292] CB_1 receptor desensitization/downregulation disrupts the eCB system, and inappropriate modulation of this highly adapting system during adolescence is likely to perturb CNS maturation, leading to profound consequences later in life.

Intermittent administration of Δ^9-THC to adolescent rats increases AEA levels in the NAc.[140] Because AEA is an important regulator of synaptic plasticity, the subsequent increased activation of CB_1 receptors on excitatory presynaptic terminals inhibits glutamate release[293] and thus disturbs glutamate function in the NAc. Moreover, eCB transmission was also disturbed by altering the normal ratios of AEA and 2-AG in the NAc and PFC.[140] AEA and 2-AG have a homeostatic relationship, and AEA in adult striatum inhibits 2-AG synthesis by regulating TRPV1 channel in GABAergic neurons.[294] Because AEA reduces glutamate inputs to striatal neurons,[295] changes in AEA levels will impact their excitability and indirectly interfere with inhibitory inputs controlled by 2-AG. Mice that were chronically administered WIN during adolescence similarly had elevated AEA levels in the hippocampus as adults, which promoted DNA hypermethylation at the intragenic region of the intracellular signaling modulator *Rgs7* and a decrease in its transcription.[296] Regulator of G-protein signaling 7 (RGS7) controls synaptic plasticity by negatively regulating neurotransmitter signaling via G protein-coupled receptors,[297] and a deficiency of RGS7 disrupts learning and memory.[296,298] Chronic administration of WIN to adolescent mice also increased the hippocampal expression of MAGL and FAAH genes, thereby reducing eCB signaling (via increases in eCB uptake and degradation).[278] Furthermore, inhibition of AEA hydrolysis during adolescence (by administration of the FAAH inhibitor URB597) reduced CB_1 receptor expression in caudate putamen, NAc, VTA, and hippocampus.[299] Altogether, these findings demonstrate that chronic administration of exogenous cannabinoids alters the regulated maintenance of eCB signaling.

Long-term changes in neurotransmission

Synthetic cannabinoids and Δ^9-THC disrupt the eCB system's regulatory control of GABA and glutamate neurotransmitter release.[300] Presynaptic CB_1 receptors inhibit the release of excitatory neurotransmitters in local GABAergic circuits in the developing PFC. Excessive activation of CB_1 receptors during adolescence via repeated exposure to exogenous cannabinoids significantly reduces the glutamatergic drive needed for the functional maturation of these GABAergic interneurons, resulting in prefrontal GABA hypofunction.[301] This reduces the inhibitory tone to cortical pyramidal cells and consequently decreases the synchronization of the prefrontal network.[302] The reduced inhibitory tone in PFC is evidenced by reduced expression of the GABAergic marker glutamate decarboxylase and hyperexcitability of pyramidal neurons, which show increased spontaneous neuron bursting and firing rates and potentiated high-gamma-power oscillatory activity.[264,265] Interestingly, female adolescent rats exposed to Δ^9-THC show decreased PFC GABAergic tone,[285] whereas male rats demonstrate increased hippocampal glutamatergic tone due to altered expression of synaptic receptor units.[303]

Cannabinoid agonists administered during adolescence also disinhibit nigrostriatal dopamine transmission by decreasing the GABAergic input from the ventral pallidum. Thus, adolescent rats treated with WIN had greater dopamine release in the dorsolateral striatum in adulthood, with decreased extracellular GABA levels in substantia nigra and increased activity of dopaminergic neurons, which was reversed by administering bicuculline, a $GABA_A$ receptor antagonist, into the ventral pallidum.[304] Hyperdopaminergic neurons are also present in the VTA of adult rats exposed to cannabinoids during adolescence.[264,265]

Acute adolescent Δ^9-THC administration also increases dopamine levels in NAc shell in rats,[305] which is more evident during mid- and late-adolescence than in adulthood.[306] The increased sensitivity of mesolimbic dopaminergic circuitry to Δ^9-THC in adolescence is consistent with the postulated gateway role of cannabis toward the use and abuse of other illicit drugs and progression to addiction later in life. For example, rats exposed to Δ^9-THC during adolescence exhibit increased self-administration of, but blunted striatal dopamine responses to, CB_1 receptor agonists in adulthood.[274] Furthermore, adolescent Δ^9-THC exposure also upregulates expression of the gene encoding the endogenous opioid proenkephalin (*Penk*) in the NAc,[307] which increases heroin self-administration.[308]

The eCB system also modulates serotonergic systems. It was proposed that heteromers of CB_1 receptors and serotonergic 2A receptors (5-HT2ARs) mediate the cognitive deficits produced by Δ^9-THC in mice.[309] Furthermore, adolescent Δ^9-THC exposure leads to the overactivation of prohallucinogenic 5-HT2AR signaling via the Akt/mTOR pathway,[310] which has been implicated in Δ^9-THC-induced memory impairments.[311] Rapamycin (macrolide compound, immunosuppressive agent) prevents the Δ^9-THC-induced activation of the Akt/mTOR pathway and restores 5-HT2AR signaling.[310] Rats that were administered WIN during adolescence exhibit decreased firing rates of 5-HT neurons in the dorsal raphe nucleus and increased norepinephrine levels in locus coeruleus in adulthood, paralleled by signs of anhedonia and anxiety.[312] Similarly, low-dose chronic Δ^9-THC exposure during adolescence reduces spontaneous firing of dorsal raphe 5-HT neurons.[267]

Chronic adolescent exposure to Δ^9-THC also increases the amount of fibers in parietal cortex expressing serotonin transporter in male rats.[313] This increase in serotonin transporter expression may reflect a reorganization of serotoninergic fibers as a result of the decrease in 5-HT levels, because CB_1 receptor activation inhibits 5-HT release.[109] Exposure to escalating doses of Δ^9-THC during adolescence reduces the 5-HT turnover rate and increases serotonin transporter levels, perhaps to compensate, in the hippocampus of male (but not female) mice.[314] The male Δ^9-THC-treated rats also had lower levels of BDNF in the PFC and exhibited spatial-related cognitive deficits.[314] However, other studies have reported increased brain levels of BDNF in response to adolescent or adult cannabinoid exposure.[315–318]

Overall, the changes described above could be interpreted as neuroadaptations and suggest interactions between adolescent cannabis use, BDNF, and psychotic disorders, even though the mechanisms involved remain unclear.

Long-term neuronal changes

Exposure to Δ^9-THC in adolescence can alter the dendritic architecture of prefrontal pyramidal neurons, with allostatic atrophy of dendrites as well as premature pruning of dendritic spines. These alterations, including spine loss, persist through adulthood.[319,320] The loss of stubby spines after adolescent Δ^9-THC exposure may reduce the capacity for developmental plasticity and refinement of neural circuits. The impact of Δ^9-THC on pruning mainly impacts the glutamatergic system in the PFC, resulting in increased levels of postsynaptic density protein 95 and the NMDA receptor GluN2A sub-unit in mid-adolescence. Moreover, animals exposed to Δ^9-THC during adolescence have higher amounts of GluN2B in adulthood than unexposed controls. Furthermore, the expression of the GluA1 sub-unit of AMPA receptors increases with exposure to Δ^9-THC during adolescence.[320] These alterations to the glutamatergic system during adolescent development result in deficits in eCB-LTD in the limbic part of the medial PFC in adulthood.[320]

Repeated administration of WIN during adolescence similarly increases GluA1 and post-synaptic density protein 95 expression as well as BDNF-TrkB signaling (elevated levels of BDNF protein and its specific receptor p-TrkB/TrkB ratio) in the NAc; repeated administration of TrkB antagonist ANA-12 attenuates this effect.[283] BDNF-TrkB signaling is implicated in the pathophysiology of drug addiction, including both depressive and psychotic symptoms.[321] Therefore, TrkB antagonism represents a potential preventive strategy in young adults after repeated use of cannabis during adolescence.

Δ^9-THC has short-term and protracted effects on the transcriptome of prelimbic pyramidal neurons in adolescent rats, with different sets of differentially expressed genes noted at 24 h and 2 weeks after exposure.[319] Notably, the gene networks affected include those involved in regulating the actin dynamics at excitatory synapses and dendritic spines. The top differentially expressed genes include *Dstn*, *Pacsin1*, and *Bap1*,[319] which are also dysregulated in mood and neurodegenerative disorders.[322–324] Moreover, genes associated with epigenetic modifications are strongly linked to those important for synaptic plasticity in Δ^9-THC-exposed animals, such as *Kmt2a*, the methylase responsible for trimethylation of lysine 4 on histone 3,[319] which are factors essential for neurogenesis[325,326] and synaptic plasticity[327,328] and are highly implicated in cellular processes associated with neurodevelopment and psychiatric disorders.[329–331] Other evidence for an epigenetic-driven molecular mechanism for adolescent-specific Δ^9-THC sensitivity is the increase in H3K9me3 in the PFC of female rats exposed to Δ^9-THC during adolescence. Accordingly, levels of the histone methyl transferase Suv39H1 are also elevated. Notably, pharmacological inhibition of Suv39H1 during adolescent Δ^9-THC exposure prevents long-lasting Δ^9-THC-induced behavioral (cognitive) deficits.[259]

Proteomics of PFC synaptosomes have also been used to study long-lasting effects of adolescent Δ^9-THC exposure, revealing decreased levels of proteins belonging mainly to mitochondria and energetic metabolism pathways.[289] In particular, cytochrome bc_1 complex sub-unit 1 and sub-unit 2 and ATP synthase α and β sub-units were less abundant following Δ^9-THC exposure in adolescence. It should be noted here that cannabinoids can activate CB_1 receptors on mitochondrial membranes (mtCB_1 receptors), resulting in altered mitochondrial activity and decreased cell respiration.[332] Synaptic mitochondria are important for maintaining and regulating neurotransmission[333]; thus, it is not surprising that reduced cell respiration in response to mtCB$_1$ receptor activation exerts negative effects on synaptic activity. Furthermore, Δ^9-THC is lipophilic and can accumulate in membranes, possibly altering membrane fluidity in intracellular organelles, including mitochondria.[334]

Long-term glial changes

In delineating the potential mechanisms involved in cannabis-induced changes, glial cells should also be considered. Astrocytes,[335] oligodendrocytes,[336] neural precursor cells,[337] and microglial cells[127] all express CB$_1$ receptors. CB$_2$ receptors are expressed by neural precursors[337] and microglia,[127] in which expression is increased under inflammatory conditions.[338] Furthermore, exogenous cannabinoids can modulate the activation of microglia[339] and astrocytes.[340] Indeed, researchers have begun examining the changes in glial cells and impairments of neuron-glia communication in animal models following cannabinoid administration in adolescence.

In addition to providing support to neurons and contributing to circuit remodeling and plasticity, microglia and astroglia are the principal immunoresponsive cells in the CNS, transiently upregulating inflammatory processes to protect against environmental insults.[341] These cells produce and release cytokines, prostanoids, free radicals, proinflammatory mediators, histocompatibility complex II, etc.[342,343] Drugs of abuse, including cannabis, can lead to the activation of microglia and astrocytes through signaling at innate immune receptors, which in turn influences neuronal function through secretion of inflammatory factors or synapse remodeling. Microglia and astroglia also express Toll-like receptors, which trigger transcription factor NF-κB and the production and release of inflammatory mediators.[344] The NF-kB pathway is upregulated in astrocytes as well as other cell types following cannabinoid exposure.[345,346]

Chronic Δ^9-THC exposure during adolescence has long-term effects on the proportion of reactive microglia (according to Iba1 expression) in PFC[347] and hippocampus,[313] suggesting that glial reactivity is permanently altered. For example, adolescent Δ^9-THC exposure induces a persistent neuroinflammatory state within the PFC of adult female rats, characterized by increases in proinflammatory markers (tumor necrosis factor α [TNF-α], inducible nitric oxide synthase [iNOS], and cyclooxygenase 2 [COX-2]) and a decrease in antiinflammatory interleukin 10.[347] This neuroinflammatory state is associated with a downregulation of CB$_1$ receptors on neuronal cells and an upregulation of CB$_2$ receptors on microglia. Notably, ibudilast, a non-selective phosphodiesterase inhibitor that suppresses glial cell activation,[348] prevents these changes, thereby minimizing the behavioral impairments that emerge in adulthood in response to adolescent Δ^9-THC exposure.[347] Chronic Δ^9-THC exposure during adolescence also impacts reactive microglia in the hippocampus,[313] increasing the percentage of reactive microglial cells in male rats but decreasing the percentage in females.[313] Of note, basal microglial reactivity differs between adolescent males and females, with females harboring more microglia with the reactive phenotype.[349] Thus, the effect of adolescent Δ^9-THC exposure on microglia reactivity differs according to the baseline reactivity. In agreement with this, chronic WIN administration to young rats reduces the number of activated microglia when coadministered with the proinflammatory agent lipopolysaccharide.[350]

Emerging evidence suggests microglia help shape neuronal circuits during the adolescent developmental window by interacting with neuronal and non-neuronal elements, thereby directly or indirectly impacting neuronal plasticity and function.[351] The effect of adolescent Δ^9-THC exposure on microglia likely impacts their role in synaptic pruning, thereby altering the neurodevelopmental trajectory.[352] The findings outlined above suggest that Δ^9-THC exposure has the potential to disrupt the bidirectional interaction between microglia and neurons in the adolescent brain and thus compromise microglia-synapse cross talk in the adult CNS.

Cannabinoid exposure similarly affects astrocytes, as chronic treatment with Δ^9-THC during adolescence increases glial fibrillar acidic protein immunoreactivity in the hippocampus in adult rats,[303,313] and this effect is more prominent in males than in females. The activation of astrocytes corresponds to a neuroinflammatory state[303] resembling that found regarding microglial activation in the PFC of female rats described above,[347] demonstrating region- and sex-specific differences in glial activation in response to cannabinoid exposure during adolescence.

The activation of astrocytes by cannabinoids during development impacts glutamatergic neurotransmission, because astrocytes regulate excitatory synapse formation as well as the surface expression and activity of glutamate NMDA and AMPA receptors.[353] Cannabinoids bind to CB_1 receptors on hippocampal astrocytes to release glutamate, which stimulates GluN2B-containing NMDA receptors and triggers AMPA receptor internalization at CA3-CA1 synapses, resulting in cannabinoid-induced LTD.[110] Prolonged activation of CB_1 receptors from repeated exposures to cannabinoids then reduces long-term potentiation in the hippocampus, mediated by CB_1R-[G$\beta\gamma$-Akt-ERK/MAPK-NF-κB]-COX-2 signaling.[354] Specifically, Δ^9-THC increases the activity and expression of COX-2, an inducible enzyme that converts arachidonic acid to prostanoids, and the production of prostaglandin E2, which may contribute to extracellular glutamate accumulation. Moreover, pharmacological or genetic inhibition of COX-2 blocked the Δ^9-THC-induced downregulation and internalization of glutamate receptors, the alteration of dendritic spine density on hippocampal neurons, and prevented the impairment of hippocampal long-term synaptic plasticity and associated cognitive deficits.[354]

More recent studies have examined the contributions from mtCB$_1$ receptors in astroglia,[355,356] the activation of which reduces the phosphorylation of the mitochondrial complex I sub-unit NDUFS4, which decreases the stability and activity of complex I. This leads to a reduction in astroglial reactive oxygen species and glycolytic production of lactate through the hypoxia-inducible factor 1 pathway, resulting in neuronal redox stress and, ultimately, social behavioral impairments.[356] However, these events have not been studied in the adolescent brain following chronic cannabinoid exposure. Thus, astroglial bioenergetics and astroglia–neuron communication are key research targets that are of paramount importance for understanding both brain physiology and pathology.

CB$_1$ receptors are also expressed by oligodendrocytes, the glial cells responsible for generating the myelin sheaths that support neurons and ensure rapid and reliable conduction of nerve pulses. Chronic cannabis exposure during adolescence may result in apoptosis of oligodendrocyte progenitors and alter WM development.[357] Although the underlying pathophysiological mechanisms are understudied, cannabinoids may disrupt myelination, which is not complete until late adolescence.[336] Δ^9-THC activation of CB$_1$ receptors on oligodendrocytes suppresses depolarization-induced Ca^{2+} influx in these cells, which may affect axonal recognition and the initiation of myelin wrapping.[358] Indeed, early cannabis use has been linked to WM microstructure abnormalities, and chronic cannabinoid exposure downregulates the expression of myelin basic protein and myelin proteolipid protein.[359,360] This downregulated expression persists more than 2 weeks after Δ^9-THC exposure, but only in rats exposed during adolescence,[361] supporting the notion that adolescence is a critical period regarding the impact of cannabis on brain development.

Altogether, these findings demonstrate that cannabis use during adolescence impacts glial cells as well as neurons to influence neural plasticity and behavioral outcomes. The observation that behavioral deficits induced by cannabinoid exposure in pre-clinical studies are attenuated with certain pharmacological agents (e.g., inhibitors of COX-2 and phosphodiesterase) presents a promising treatment strategy to promote detoxification after heavy adolescent cannabis abuse or potentially control side effects induced by cannabis-based therapies.

Developmental variability in cannabinoid pharmacokinetics

Δ^9-THC is highly lipophilic and quickly crosses the blood-brain barrier in accordance with the route of administration.[362] Δ^9-THC is also readily transformed to 11-OH-THC (psychotropic metabolite), which in turn is metabolized to the inactive metabolite THC-COOH.[363,364] Most Δ^9-THC biotransformation takes place in the liver.[365] 11-OH-THC enters the brain more rapidly and efficiently than Δ^9-THC, and thus further contributes to the psychoactive effects of the Δ^9-THC.[366] The presence of age-dependent adjustments in the distribution, biotransformation, and elimination of exogenous cannabinoids is an important consideration, particularly because the eCB system itself is modified during adolescence, as discussed earlier.

The pharmacokinetic properties of Δ^9-THC, including peak drug concentrations in plasma and brain, brain-to-plasma ratio, and CYP_{450}-mediated metabolism, were compared between adolescent and adult mice.[367] The peak plasma concentrations of Δ^9-THC, 11-OH-THC, and THC-COOH were all substantially greater in adolescent mice than in adult mice, which could be attributable to progressive diversion toward white adipose tissue, which accumulates in adulthood. However, adolescent mice eliminated Δ^9-THC at a faster rate than adults, demonstrated by the shorter plasma half-life and the faster conversion of Δ^9-THC to THC-COOH in the younger mice.[367] This pattern is consistent with that of CYP_{450} expression in the developing liver.[368] Conversely, adolescent mice had 40%–60% lower brain concentrations and brain-to-plasma ratios of Δ^9-THC; however, the concentrations of the THC-COOH metabolite in the brain were up to 70% higher in adolescents than in adults.[367] Thus, adolescent mice exposed to Δ^9-THC experience lower peak brain concentrations of the non-metabolized drug but higher brain concentrations of its metabolites than adults. Possible explanations for these age-related differences include changes in the transport of Δ^9-THC metabolites across the blood-brain barrier and/or the local metabolism of Δ^9-THC within the CNS.

There are also sex-dependent differences in the Δ^9-THC metabolism in adolescent rats. Specifically, brain levels of 11-OH-THC are significantly higher in adolescent females than in males after both acute and repeated exposure to Δ^9-THC.[288] Because 11-OH-THC retains the psychotropic properties of Δ^9-THC,[369,370] the high levels of the metabolite may potentiate the effects of Δ^9-THC exposure, which would account for the increased sensitivity of adolescent females to some of the effects of cannabinoids.[371]

Genetic vulnerability to cannabis

The underlying mechanisms of individual vulnerability to cannabis remain unknown. Substantial evidence implicates genetic factors contributing to association between early cannabis use and an increased risk of developing schizophrenia, cannabis use disorder and cognitive impairments.[95,372,373] The link between chronic adolescent cannabis consumption and psychotic symptoms in humans was attributed to variation in *CNR1*,[374,375] *COMT*,[376,377] *FAAH*,[204,378] or *AKT1*.[379,380] These studies are consistent with the "two-hit" hypothesis, which posits that both genetics and environmental factors, and their interaction contribute to individual risks of psychiatric disorders.

Genetic underpinnings of cannabis use phenotypes are described in detail by Thorpe et al.[381] For example, genetic variants in *CNR1* have been reported to attribute to the development of cannabis dependence in adolescence[382] as well as increased risk of schizophrenia.[375] In line with this, heavy cannabis consumption by individuals with the certain *CNR1* genotypes was found to be associated with decreased volume of the WM and cognitive impairment.[374] It should be noted that several studies have failed to find a link between polymorphisms in the *CNR1* gene and increased risk of schizophrenia.[383,384]

Rodent studies have also demonstrated interaction between mutations in several genes and adolescent cannabis exposure. *Comt* knockout mice exposed to Δ^9-THC during adolescence exhibited long-term impairments in exploration, spatial working memory, and anxiety,[385] as well as deficits in pre-pulse inhibition, sociability, and social novelty preference.[386] Adolescent Δ^9-THC treatment of *Comt* knockout mice decreased size of dopaminergic and GABAergic cells in the VTA and PFC, respectively, and increased expression of CB_1 receptor in the hippocampus in adult mice.[387]

Using a knock-in mouse model, it was shown that *FAAH* C385A polymorphism enhances the mesolimbic dopamine circuitry projecting from the VTA to the NAc and alters CB_1 receptor levels on inhibitory and excitatory terminals in the VTA. These developmental changes collectively increased vulnerability of adolescent mice with *FAAH* C385A polymorphism to Δ^9-THC preference that persisted into adulthood.[388]

Surprisingly, single gene mutations can also have some protective effects. For example, the *Neuregulin 1* (*Nrg1*) mutation has been shown to attenuate some negative effects of Δ^9-THC exposure during adolescence. NRG1 is involved in the regulation of the expression and activation of neurotransmitter receptors, synaptogenesis and was associated with an increased susceptibility to schizophrenia.[389] In contrast to adolescent, adult *Nrg1* mutant mice appeared to be more sensitive to the behavioral effects of acute or chronic cannabis exposure, i.e., enhancement of pre-pulse inhibition, hypolocomotive and anxiogenic effects.[390,391] *Nrg1* mutant mice chronically exposed to Δ^9-THC during adolescence displayed a short-term increase in pre-pulse inhibition followed by a rapid normalization 1 week after Δ^9-THC administration. *Nrg1* mice were more resistant than wild-type animals to the negative effect of adolescent Δ^9-THC on social behavior and less susceptible to induction of anxiety-like behavior by acute Δ^9-THC.[392] Unexposed *Nrg1* mice had reduced brain density of 5-HT2A and CB_1 receptors density in the brain, and exposure to Δ^9-THC increased the NMDA receptor density in the hippocampus in these mutants.[392]

The molecular mechanisms of gene-environment interaction might differ across cell types; therefore, recent studies have begun to study molecular cascades activated by cannabinoids with a cell type-specific resolution. The disrupted-in-schizophrenia 1 (*DISC1*) gene is a scaffolding protein involved in neurodevelopment, neuro-signaling, synaptic functioning and is implicated in schizophrenia.[393,394] Perturbation of DISC1 expression in astrocytes, but not neurons, worsened the effects of adolescent Δ^9-THC exposure on recognition memory in adult mice,[346,395] suggesting that the same risk factor in different brain cells leads to different behavioral outcomes in mice treated with cannabinoid. It was suggested that DISC1 and Δ^9-THC converged and caused synergistic activation of the proinflammatory NF-κB-COX2 pathway in astrocytes leading to secretion of astrocyte glutamate and dysfunction of GABAergic neurons in the hippocampus.[346]

Conclusion and future directions

Recent studies have convincingly demonstrated adverse effects of adolescent exposure to cannabis on neurobehavioral development. Both human and preclinical research shows that adolescent use of cannabis, particularly with high content of Δ^9-THC, exerts long-lasting effects on brain maturation and is strongly associated with cognitive impairment in adulthood. Although adverse effects of cannabis have been strongly linked to general vulnerability of adolescent brain to environmental insults, emerging evidence also points to genetic pre-disposition that could exacerbate harmful effects of cannabis in susceptible individuals. However, the underlying mechanisms of gene-environment interplay remains poorly understood.

We propose that the future research on uncovering the underpinning of adverse effects of cannabis neurobehavioral development be focused on several critical directions. Given that ubiquitous expression of cannabinoid receptors on non-neuronal brain cells and peripheral tissues, the role of non-neuronal eCB signaling in mediating effects of Δ^9-THC and other potent ingredients of cannabis calls for more investigations. Most recent research revealed that cannabinoid receptors are expressed on mitochondria and may be involved in mediating detrimental effects of cannabis by affecting energy metabolism.[355,356] Future studies will determine what types of brains cells could particularly be involved in mitochondria-dependent effects during adolescence. Further, there are only few studies that tried to address the molecular mechanisms of gene-environment interaction in the context of adolescent cannabis exposure in patients with cognitive dysfunction and/or major psychiatric disorders. Future investigations will explore these complex pathogenic interactions using model organisms with various genetic lesions, including multiple point mutations, copy number variants or rare and highly penetrant mutations. The impressive progress in the development of in vivo technologies provides the opportunities to characterize the effects of cannabis in a longitudinal fashion to uncover the trajectories of abnormal brain maturations and/or potentially compensatory changes during chronic exposure to cannabis. These directions could help determine potential therapeutic targets to ameliorate pathogenic alterations and/or stimulate processes to fight back effects of cannabis on the maturing brain.

References

1. Larsen B, Luna B. Adolescence as a neurobiological critical period for the development of higher-order cognition. *Neurosci Biobehav Rev* 2018;**94**:179–95.
2. Ernst M, Mueller SC. The adolescent brain: insights from functional neuroimaging research. *Dev Neurobiol* 2008;**68**(6):729–43.

3. Hagler Jr DJ, Hatton S, Cornejo MD, Makowski C, Fair DA, Dick AS, et al. Image processing and analysis methods for the adolescent brain cognitive development study. *NeuroImage* 2019;**202**:116091.
4. Morris AS, Squeglia LM, Jacobus J, Silk JS. Adolescent brain development: implications for understanding risk and resilience processes through neuroimaging research. *J Res Adolesc* 2018;**28**(1):4–9.
5. Forde NJ, Ronan L, Zwiers MP, Schweren LJS, Alexander-Bloch AF, Franke B, et al. Healthy cortical development through adolescence and early adulthood. *Brain Struct Funct* 2017;**222**(8):3653–63.
6. Vijayakumar N, Allen NB, Youssef G, Dennison M, Yücel M, Simmons JG, et al. Brain development during adolescence: a mixed-longitudinal investigation of cortical thickness, surface area, and volume. *Hum Brain Mapp* 2016;**37**(6):2027–38.
7. Gilmore JH, Knickmeyer RC, Gao W. Imaging structural and functional brain development in early childhood. *Nat Rev Neurosci* 2018;**19**(3):123–37.
8. Gennatas ED, Avants BB, Wolf DH, Satterthwaite TD, Ruparel K, Ciric R, et al. Age-related effects and sex differences in gray matter density, volume, mass, and cortical thickness from childhood to young adulthood. *J Neurosci Off J Soc Neurosci* 2017;**37**(20):5065–73.
9. Giedd JN, Raznahan A, Alexander-Bloch A, Schmitt E, Gogtay N, Rapoport JL. Child psychiatry branch of the National Institute of Mental Health longitudinal structural magnetic resonance imaging study of human brain development. *Neuropsychopharmacology* 2015;**40**(1):43–9.
10. Tamnes CK, Herting MM, Goddings A-L, Meuwese R, Blakemore S-J, Dahl RE, et al. Development of the cerebral cortex across adolescence: a multisample study of inter-related longitudinal changes in cortical volume, surface area, and thickness. *J Neurosci* 2017;**37**(12):3402.
11. Narvacan K, Treit S, Camicioli R, Martin W, Beaulieu C. Evolution of deep gray matter volume across the human lifespan. *Hum Brain Mapp* 2017;**38**(8):3771–90.
12. Wierenga L, Langen M, Ambrosino S, van Dijk S, Oranje B, Durston S. Typical development of basal ganglia, hippocampus, amygdala and cerebellum from age 7 to 24. *NeuroImage* 2014;**96**:67–72.
13. Cao B, Mwangi B, Passos IC, Wu M-J, Keser Z, Zunta-Soares GB, et al. Lifespan gyrification trajectories of human brain in healthy individuals and patients with major psychiatric disorders. *Sci Rep* 2017;**7**(1):511.
14. Klein D, Rotarska-Jagiela A, Genc E, Sritharan S, Mohr H, Roux F, et al. Adolescent brain maturation and cortical folding: evidence for reductions in gyrification. *PLoS One* 2014;**9**(1), e84914.
15. Spear LP. Adolescent neurodevelopment. *J Adolesc Health* 2013;**52**(2 Suppl 2):S7–S13.
16. Gogtay N, Giedd JN, Lusk L, Hayashi KM, Greenstein D, Vaituzis AC, et al. Dynamic mapping of human cortical development during childhood through early adulthood. *Proc Natl Acad Sci USA* 2004;**101**(21):8174–9. https://doi.org/10.1073/pnas.0402680101.
17. Selemon LD. A role for synaptic plasticity in the adolescent development of executive function. *Transl Psychiatr* 2013;**3**(3):e238.
18. Fossati G, Matteoli M, Menna E. Astrocytic factors controlling synaptogenesis: a team play. *Cells* 2020;**9**(10):2173.
19. Testen A, Ali M, Sexton HG, Hodges S, Dubester K, Reissner KJ, et al. Region-specific differences in morphometric features and synaptic colocalization of astrocytes during development. *Neuroscience* 2019;**400**:98–109.
20. He J, Crews FT. Neurogenesis decreases during brain maturation from adolescence to adulthood. *Pharmacol Biochem Behav* 2007;**86**(2):327–33.

21. Kozareva DA, Cryan JF, Nolan YM. Born this way: hippocampal neurogenesis across the lifespan. *Aging Cell* 2019;**18**(5), e13007.
22. Curlik DM, DiFeo G, Shors TJ. Preparing for adulthood: thousands upon thousands of new cells are born in the hippocampus during puberty, and most survive with effortful learning. *Front Neurosci* 2014;**8**:70.
23. O'Leary JD, Hoban AE, Murphy A, O'Leary OF, Cryan JF, Nolan YM. Differential effects of adolescent and adult-initiated exercise on cognition and hippocampal neurogenesis. *Hippocampus* 2019;**29**(4):352–65.
24. Kirshenbaum GS, Lieberman SR, Briner TJ, Leonardo ED, Dranovsky A. Adolescent but not adult-born neurons are critical for susceptibility to chronic social defeat. *Front Behav Neurosci* 2014;**8**:289.
25. Drzewiecki CM, Willing J, Juraska JM. Influences of age and pubertal status on number and intensity of perineuronal nets in the rat medial prefrontal cortex. *Brain Struct Funct* 2020;**225**(8):2495–507.
26. Bray S, Krongold M, Cooper C, Lebel C. Synergistic effects of age on patterns of white and gray matter volume across childhood and adolescence. *eNeuro* 2015;**2**(4), ENEURO.0003-0015.2015.
27. Foulkes L, Blakemore S-J. Studying individual differences in human adolescent brain development. *Nat Neurosci* 2018;**21**(3):315–23.
28. Lynch KM, Cabeen RP, Toga AW, Clark KA. Magnitude and timing of major white matter tract maturation from infancy through adolescence with NODDI. *NeuroImage* 2020;**212**:116672.
29. Innocenti GM, Price DJ. Exuberance in the development of cortical networks. *Nat Rev Neurosci* 2005;**6**(12):955–65.
30. Brouwer RM, Mandl RC, Schnack HG, van Soelen IL, van Baal GC, Peper JS, et al. White matter development in early puberty: a longitudinal volumetric and diffusion tensor imaging twin study. *PLoS One* 2012;**7**(4), e32316.
31. Genc S, Malpas CB, Ball G, Silk TJ, Seal ML. Age, sex, and puberty related development of the corpus callosum: a multi-technique diffusion MRI study. *Brain Struct Funct* 2018;**223**(6):2753–65.
32. Giorgio A, Watkins KE, Chadwick M, James S, Winmill L, Douaud G, et al. Longitudinal changes in grey and white matter during adolescence. *NeuroImage* 2010;**49**(1):94–103.
33. Schmithorst VJ, Yuan W. White matter development during adolescence as shown by diffusion MRI. *Brain Cogn* 2010;**72**(1):16–25.
34. Geeraert BL, Lebel RM, Lebel C. A multiparametric analysis of white matter maturation during late childhood and adolescence. *Hum Brain Mapp* 2019;**40**(15):4345–56.
35. Tamnes CK, Roalf DR, Goddings A-L, Lebel C. Diffusion MRI of white matter microstructure development in childhood and adolescence: methods, challenges and progress. *Dev Cogn Neurosci* 2018;**33**:161–75.
36. Bells S, Lefebvre J, Longoni G, Narayanan S, Arnold DL, Yeh EA, et al. White matter plasticity and maturation in human cognition. *Glia* 2019;**67**(11):2020–37.
37. Paus T. Growth of white matter in the adolescent brain: myelin or axon? *Brain Cogn* 2010;**72**(1):26–35.
38. Whitaker KJ, Vértes PE, Romero-Garcia R, Váša F, Moutoussis M, Prabhu G, et al. Adolescence is associated with genomically patterned consolidation of the hubs of the human brain connectome. *Proc Natl Acad Sci U S A* 2016;**113**(32):9105–10.
39. Simons M, Nave K-A. Oligodendrocytes: myelination and axonal support. *Cold Spring Harb Perspect Biol* 2015;**8**(1):a020479.

40. Yamaguchi Y, Miura M. Programmed cell death in neurodevelopment. *Dev Cell* 2015; **32**(4):478–90.
41. Morrison KE, Rodgers AB, Morgan CP, Bale TL. Epigenetic mechanisms in pubertal brain maturation. *Neuroscience* 2014;**264**:17–24.
42. Vigil P, Del Río JP, Carrera B, ArÁnguiz FC, Rioseco H, Cortés ME. Influence of sex steroid hormones on the adolescent brain and behavior: an update. *Linacre Q* 2016;**83**(3):308–29.
43. Lenroot RK, Gogtay N, Greenstein DK, Wells EM, Wallace GL, Clasen LS, et al. Sexual dimorphism of brain developmental trajectories during childhood and adolescence. *NeuroImage* 2007;**36**(4):1065–73.
44. Kaczkurkin AN, Raznahan A, Satterthwaite TD. Sex differences in the developing brain: insights from multimodal neuroimaging. *Neuropsychopharmacology* 2019;**44**(1):71–85.
45. Ladouceur CD, Peper JS, Crone EA, Dahl RE. White matter development in adolescence: the influence of puberty and implications for affective disorders. *Dev Cogn Neurosci* 2012; **2**(1):36–54.
46. Lenroot RK, Giedd JN. Sex differences in the adolescent brain. *Brain Cogn* 2010;**72**(1):46–55.
47. Goddings AL, Mills KL, Clasen LS, Giedd JN, Viner RM, Blakemore SJ. The influence of puberty on subcortical brain development. *NeuroImage* 2014;**88**:242–51.
48. Koolschijn PCMP, Crone EA. Sex differences and structural brain maturation from childhood to early adulthood. *Dev Cogn Neurosci* 2013;**5**:106–18.
49. Satterthwaite TD, Vandekar S, Wolf DH, Ruparel K, Roalf DR, Jackson C, et al. Sex differences in the effect of puberty on hippocampal morphology. *J Am Acad Child Adolesc Psychiatr* 2014;**53**(3):341–350.e341.
50. Herting MM, Johnson C, Mills KL, Vijayakumar N, Dennison M, Liu C, et al. Development of subcortical volumes across adolescence in males and females: a multisample study of longitudinal changes. *NeuroImage* 2018;**172**:194–205.
51. Wierenga LM, Bos MGN, Schreuders E, Vd Kamp F, Peper JS, Tamnes CK, et al. Unraveling age, puberty and testosterone effects on subcortical brain development across adolescence. *Psychoneuroendocrinology* 2018;**91**:105–14.
52. Gur RE, Gur RC. Sex differences in brain and behavior in adolescence: findings from the Philadelphia neurodevelopmental cohort. *Neurosci Biobehav Rev* 2016;**70**:159–70.
53. Kight KE, McCarthy MM. Androgens and the developing hippocampus. *Biol Sex Differ* 2020;**11**(1):30.
54. Sisk CL. Development: pubertal hormones meet the adolescent brain. *Curr Biol* 2017;**27**(14): R706–8.
55. Fuhrmann D, Knoll LJ, Blakemore SJ. Adolescence as a sensitive period of brain development. *Trends Cogn Sci* 2015;**19**(10):558–66.
56. Walker DM, Bell MR, Flores C, Gulley JM, Willing J, Paul MJ. Adolescence and reward: making sense of neural and behavioral changes amid the chaos. *J Neurosci* 2017;**37**(45): 10855–66.
57. Hammerslag LR, Gulley JM. Sex differences in behavior and neural development and their role in adolescent vulnerability to substance use. *Behav Brain Res* 2016;**298**(Pt A):15–26.
58. Alarcón G, Cservenka A, Nagel BJ. Adolescent neural response to reward is related to participant sex and task motivation. *Brain Cogn* 2017;**111**:51–62.
59. Thion MS, Ginhoux F, Garel S. Microglia and early brain development: an intimate journey. *Science* 2018;**362**(6411):185–9.
60. Kopec AM, Smith CJ, Ayre NR, Sweat SC, Bilbo SD. Microglial dopamine receptor elimination defines sex-specific nucleus accumbens development and social behavior in adolescent rats. *Nat Commun* 2018;**9**(1):3769.

61. Karcher NR, Barch DM. The ABCD study: understanding the development of risk for mental and physical health outcomes. *Neuropsychopharmacology* 2021;**46**(1):131–42.
62. Eiland L, Romeo RD. Stress and the developing adolescent brain. *Neuroscience* 2013;**249**:162–71.
63. Heffernan AL, Hare DJ. Tracing environmental exposure from neurodevelopment to neurodegeneration. *Trends Neurosci* 2018;**41**(8):496–501.
64. Jordan CJ, Andersen SL. Sensitive periods of substance abuse: early risk for the transition to dependence. *Dev Cogn Neurosci* 2017;**25**:29–44.
65. Nock NL, Minnes S, Alberts JL. Neurobiology of substance use in adolescents and potential therapeutic effects of exercise for prevention and treatment of substance use disorders. *Birth Defects Res* 2017;**109**(20):1711–29.
66. Schriber RA, Guyer AE. Adolescent neurobiological susceptibility to social context. *Dev Cogn Neurosci* 2016;**19**:1–18.
67. Spear LP. Heightened stress responsivity and emotional reactivity during pubertal maturation: implications for psychopathology. *Dev Psychopathol* 2009;**21**(1):87–97.
68. Spechler PA, Orr CA, Chaarani B, Kan K-J, Mackey S, Morton A, et al. Cannabis use in early adolescence: evidence of amygdala hypersensitivity to signals of threat. *Dev Cogn Neurosci* 2015;**16**:63–70.
69. Yohn NL, Blendy JA. Adolescent chronic unpredictable stress exposure is a sensitive window for long-term changes in adult behavior in mice. *Neuropsychopharmacology* 2017;**42**(8):1670–8.
70. LeNoue SR, Riggs PD. Substance abuse prevention. *Child Adolesc Psychiatr Clin N Am* 2016;**25**(2):297–305.
71. Altikulaç S, Bos MGN, Foulkes L, Crone EA, van Hoorn J. Age and gender effects in sensitivity to social rewards in adolescents and young adults. *Front Behav Neurosci* 2019;**13**(171).
72. Peeters M, Oldehinkel T, Vollebergh W. Behavioral control and reward sensitivity in adolescents' risk taking behavior: a longitudinal TRAILS study. *Front Psychol* 2017;**8**:231.
73. Simon NW, Moghaddam B. Neural processing of reward in adolescent rodents. *Dev Cogn Neurosci* 2015;**11**:145–54.
74. Russo SJ, Nestler EJ. The brain reward circuitry in mood disorders. *Nat Rev Neurosci* 2013;**14**(9):609–25.
75. Galvan A. Adolescent development of the reward system. *Front Hum Neurosci* 2010;**4**:6.
76. Laviola G, Pascucci T, Pieretti S. Striatal dopamine sensitization to d-amphetamine in periadolescent but not in adult rats. *Pharmacol Biochem Behav* 2001;**68**(1):115–24.
77. Galvan A, Hare TA, Parra CE, Penn J, Voss H, Glover G, et al. Earlier development of the accumbens relative to orbitofrontal cortex might underlie risk-taking behavior in adolescents. *J Neurosci* 2006;**26**(25):6885–92.
78. Van Leijenhorst L, Zanolie K, Van Meel CS, Westenberg PM, Rombouts SA, Crone EA. What motivates the adolescent? Brain regions mediating reward sensitivity across adolescence. *Cereb Cortex* 2010;**20**(1):61–9.
79. Richards JS, Arias Vásquez A, von Rhein D, van der Meer D, Franke B, Hoekstra PJ, et al. Adolescent behavioral and neural reward sensitivity: a test of the differential susceptibility theory. *Transl Psychiatr* 2016;**6**(4):e771.
80. Telzer EH. Dopaminergic reward sensitivity can promote adolescent health: a new perspective on the mechanism of ventral striatum activation. *Dev Cogn Neurosci* 2016;**17**:57–67.
81. Gleich T, Lorenz RC, Pöhland L, Raufelder D, Deserno L, Beck A, et al. Frontal glutamate and reward processing in adolescence and adulthood. *Brain Struct Funct* 2015;**220**(6):3087–99.

82. Kilb W. Development of the GABAergic system from birth to adolescence. *Neuroscientist* 2012;**18**(6):613–30.
83. Pitzer M. The development of monoaminergic neurotransmitter systems in childhood and adolescence. *Int J Dev Neurosci* 2019;**74**:49–55.
84. Spear LP. The adolescent brain and age-related behavioral manifestations. *Neurosci Biobehav Rev* 2000;**24**(4):417–63.
85. Caballero A, Tseng KY. GABAergic function as a limiting factor for prefrontal maturation during adolescence. *Trends Neurosci* 2016;**39**(7):441–8.
86. Dow-Edwards D, MacMaster FP, Peterson BS, Niesink R, Andersen S, Braams BR. Experience during adolescence shapes brain development: from synapses and networks to normal and pathological behavior. *Neurotoxicol Teratol* 2019;**76**:106834.
87. Ernst M. The triadic model perspective for the study of adolescent motivated behavior. *Brain Cogn* 2014;**89**:104–11.
88. Miech RA, Johnston L, O'Malley P, Bachman J, Schulenberg JE, Patrick M. *Monitoring the future national survey results on drug use, 1975–2017: volume I, secondary school students*. The National Institute on Drug Abuse at The National Institutes of Health; 2018.
89. Romer D, Reyna VF, Satterthwaite TD. Beyond stereotypes of adolescent risk taking: placing the adolescent brain in developmental context. *Dev Cogn Neurosci* 2017;**27**:19–34.
90. Willoughby T, Good M, Adachi PJC, Hamza C, Tavernier R. Examining the link between adolescent brain development and risk taking from a social–developmental perspective. *Brain Cogn* 2013;**83**(3):315–23.
91. Andersen SL. Trajectories of brain development: point of vulnerability or window of opportunity? *Neurosci Biobehav Rev* 2003;**27**(1):3–18.
92. Blest-Hopley G, Colizzi M, Giampietro V, Bhattacharyya S. Is the adolescent brain at greater vulnerability to the effects of cannabis? A narrative review of the evidence. *Front Psychiatr* 2020;**11**:859.
93. Volkow ND, Swanson JM, Evins AE, DeLisi LE, Meier MH, Gonzalez R, et al. Effects of cannabis use on human behavior, including cognition, motivation, and psychosis: a review. *JAMA Psychiatr* 2016;**73**(3):292–7.
94. Lu H-C, Mackie K. An introduction to the endogenous cannabinoid system. *Biol Psychiatr* 2016;**79**(7):516–25.
95. Hurd YL, Manzoni OJ, Pletnikov MV, Lee FS, Bhattacharyya S, Melis M. Cannabis and the developing brain: insights into its long-lasting effects. *J Neurosci* 2019;**39**(42):8250–8.
96. Mechoulam R, Parker LA. The endocannabinoid system and the brain. *Annu Rev Psychol* 2013;**64**:21–47.
97. Meyer HC, Lee FS, Gee DG. The role of the endocannabinoid system and genetic variation in adolescent brain development. *Neuropsychopharmacology* 2018;**43**(1):21–33.
98. Howlett AC, Barth F, Bonner TI, Cabral G, Casellas P, Devane WA, et al. International Union of Pharmacology. XXVII. Classification of cannabinoid receptors. *Pharmacol Rev* 2002;**54**(2):161–202.
99. Lu H-C, Mackie K. Review of the endocannabinoid system. *Biol Psychiatr* 2020;**6**:607–15.
100. Zou S, Kumar U. Cannabinoid receptors and the endocannabinoid system: signaling and function in the central nervous system. *Int J Mol Sci* 2018;**19**(3):833.
101. Howlett AC, Abood ME. CB(1) and CB(2) receptor pharmacology. *Adv Pharmacol* 2017;**80**:169–206.
102. Kano M. Control of synaptic function by endocannabinoid-mediated retrograde signaling. *Proc Jpn Acad Ser B Phys Biol Sci* 2014;**90**(7):235–50.

103. Augustin SM, Lovinger DM. Functional relevance of endocannabinoid-dependent synaptic plasticity in the central nervous system. *ACS Chem Neurosci* 2018;**9**(9):2146–61.
104. Kano M, Ohno-Shosaku T, Hashimotodani Y, Uchigashima M, Watanabe M. Endocannabinoid-mediated control of synaptic transmission. *Physiol Rev* 2009;**89**(1):309–80.
105. Ohno-Shosaku T, Kano M. Endocannabinoid-mediated retrograde modulation of synaptic transmission. *Curr Opin Neurobiol* 2014;**29**:1–8.
106. Castillo PE, Younts TJ, Chávez AE, Hashimotodani Y. Endocannabinoid signaling and synaptic function. *Neuron* 2012;**76**(1):70–81.
107. Musella A, Fresegna D, Rizzo FR, Gentile A, Bullitta S, De Vito F, et al. A novel crosstalk within the endocannabinoid system controls GABA transmission in the striatum. *Sci Rep* 2017;**7**(1):7363.
108. Murataeva N, Straiker A, Mackie K. Parsing the players: 2-arachidonoylglycerol synthesis and degradation in the CNS. *Br J Pharmacol* 2014;**171**(6):1379–91.
109. Peters KZ, Cheer JF, Tonini R. Modulating the neuromodulators: dopamine, serotonin, and the endocannabinoid system. *Trends Neurosci* 2021;**44**(6):464–77.
110. Han J, Kesner P, Metna-Laurent M, Duan T, Xu L, Georges F, et al. Acute cannabinoids impair working memory through astroglial CB1 receptor modulation of hippocampal LTD. *Cell* 2012;**148**(5):1039–50.
111. Stella N. Endocannabinoid signaling in microglial cells. *Neuropharmacology* 2009;**56**(Suppl 1):244–53.
112. Rasooli-Nejad S, Palygin O, Lalo U, Pankratov Y. Cannabinoid receptors contribute to astroglial Ca^{2+}-signalling and control of synaptic plasticity in the neocortex. *Philos Trans R Soc Lond Ser B Biol Sci* 2014;**369**(1654), 20140077.
113. Robin LM, Oliveira da Cruz JF, Langlais VC, Martin-Fernandez M, Metna-Laurent M, Busquets-Garcia A, et al. Astroglial CB(1) receptors determine synaptic D-serine availability to enable recognition memory. *Neuron* 2018;**98**(5). 935–944.e935.
114. Metna-Laurent M, Marsicano G. Rising stars: modulation of brain functions by astroglial type-1 cannabinoid receptors. *Glia* 2015;**63**(3):353–64.
115. Oliveira da Cruz JF, Robin LM, Drago F, Marsicano G, Metna-Laurent M. Astroglial type-1 cannabinoid receptor (CB1): a new player in the tripartite synapse. *Neuroscience* 2016;**323**:35–42.
116. Hill EL, Gallopin T, Férézou I, Cauli B, Rossier J, Schweitzer P, et al. Functional CB1 receptors are broadly expressed in neocortical GABAergic and glutamatergic neurons. *J Neurophysiol* 2007;**97**(4):2580–9.
117. Mackie K. Distribution of cannabinoid receptors in the central and peripheral nervous system. *Handb Exp Pharmacol* 2005;(168):299–325.
118. Albayram Ö, Passlick S, Bilkei-Gorzo A, Zimmer A, Steinhäuser C. Physiological impact of CB1 receptor expression by hippocampal GABAergic interneurons. *Pflugers Arch* 2016;**468**(4):727–37.
119. Fitzgerald ML, Shobin E, Pickel VM. Cannabinoid modulation of the dopaminergic circuitry: implications for limbic and striatal output. *Prog Neuro-Psychopharmacol Biol Psychiatr* 2012;**38**(1):21–9.
120. Loureiro M, Renard J, Zunder J, Laviolette SR. Hippocampal cannabinoid transmission modulates dopamine neuron activity: impact on rewarding memory formation and social interaction. *Neuropsychopharmacology* 2015;**40**(6):1436–47.
121. Szabo B, Siemes S, Wallmichrath I. Inhibition of GABAergic neurotransmission in the ventral tegmental area by cannabinoids. *Eur J Neurosci* 2002;**15**(12):2057–61.

122. McCoy KL. Interaction between cannabinoid system and toll-like receptors controls inflammation. *Mediat Inflamm* 2016;**2016**, 5831315.
123. Turcotte C, Blanchet M-R, Laviolette M, Flamand N. The CB(2) receptor and its role as a regulator of inflammation. *Cell Mol Life Sci* 2016;**73**(23):4449–70.
124. Atwood BK, Mackie K. CB2: a cannabinoid receptor with an identity crisis. *Br J Pharmacol* 2010;**160**(3):467–79.
125. Cabral GA, Raborn ES, Griffin L, Dennis J, Marciano-Cabral F. CB2 receptors in the brain: role in central immune function. *Br J Pharmacol* 2008;**153**(2):240–51.
126. Ramirez SH, Haskó J, Skuba A, Fan S, Dykstra H, McCormick R, et al. Activation of cannabinoid receptor 2 attenuates leukocyte-endothelial cell interactions and blood-brain barrier dysfunction under inflammatory conditions. *J Neurosci* 2012;**32**(12):4004–16.
127. Stella N. Cannabinoid and cannabinoid-like receptors in microglia, astrocytes, and astrocytomas. *Glia* 2010;**58**(9):1017–30.
128. Chen D-J, Gao M, Gao F-F, Su Q-X, Wu J. Brain cannabinoid receptor 2: expression, function and modulation. *Acta Pharmacol Sin* 2017;**38**(3):312–6.
129. García MC, Cinquina V, Palomo-Garo C, Rábano A, Fernández-Ruiz J. Identification of CB_2 receptors in human nigral neurons that degenerate in Parkinson's disease. *Neurosci Lett* 2015;**587**:1–4.
130. García-Cabrerizo R, García-Fuster MJ. Opposite regulation of cannabinoid CB1 and CB2 receptors in the prefrontal cortex of rats treated with cocaine during adolescence. *Neurosci Lett* 2016;**615**:60–5.
131. Zhang HY, Gao M, Liu QR, Bi GH, Li X, Yang HJ, et al. Cannabinoid CB2 receptors modulate midbrain dopamine neuronal activity and dopamine-related behavior in mice. *Proc Natl Acad Sci U S A* 2014;**111**(46):E5007–15.
132. den Boon FS, Chameau P, Schaafsma-Zhao Q, van Aken W, Bari M, Oddi S, et al. Excitability of prefrontal cortical pyramidal neurons is modulated by activation of intracellular type-2 cannabinoid receptors. *Proc Natl Acad Sci U S A* 2012;**109**(9):3534–9.
133. Stempel AV, Stumpf A, Zhang HY, Özdoğan T, Pannasch U, Theis AK, et al. Cannabinoid type 2 receptors mediate a cell type-specific plasticity in the hippocampus. *Neuron* 2016;**90**(4):795–809.
134. Arain M, Haque M, Johal L, Mathur P, Nel W, Rais A, et al. Maturation of the adolescent brain. *Neuropsychiatr Dis Treat* 2013;**9**:449–61.
135. Gee DG, Bath KG, Johnson CM, Meyer HC, Murty VP, van den Bos W, et al. Neurocognitive development of motivated behavior: dynamic changes across childhood and adolescence. *J Neurosci* 2018;**38**(44):9433–45.
136. Atkinson DL, Abbott JK. Cannabinoids and the brain: the effects of endogenous and exogenous cannabinoids on brain systems and function. In: Compton MT, Manseau MW, editors. *The complex connection between cannabis and schizophrenia*. San Diego: Academic Press; 2018. p. 37–74.
137. Wenzel JM, Cheer JF. Endocannabinoid regulation of reward and reinforcement through interaction with dopamine and endogenous opioid signaling. *Neuropsychopharmacology* 2018;**43**(1):103–15.
138. Parsons LH, Hurd YL. Endocannabinoid signalling in reward and addiction. *Nat Rev Neurosci* 2015;**16**(10):579–94.
139. Solinas M, Yasar S, Goldberg SR. Endocannabinoid system involvement in brain reward processes related to drug abuse. *Pharmacol Res* 2007;**56**(5):393–405.
140. Ellgren M, Artmann A, Tkalych O, Gupta A, Hansen HS, Hansen SH, et al. Dynamic changes of the endogenous cannabinoid and opioid mesocorticolimbic systems during adolescence: THC effects. *Eur Neuropsychopharmacol* 2008;**18**(11):826–34.

141. Heng L, Beverley JA, Steiner H, Tseng KY. Differential developmental trajectories for CB1 cannabinoid receptor expression in limbic/associative and sensorimotor cortical areas. *Synapse* 2011;**65**(4):278–86.
142. Verdurand M, Nguyen V, Stark D, Zahra D, Gregoire M-C, Greguric I, et al. Comparison of cannabinoid CB(1) receptor binding in adolescent and adult rats: a positron emission tomography study using [F]MK-9470. *Int J Mol Imaging* 2011;**2011**:548123.
143. Schneider M. Puberty as a highly vulnerable developmental period for the consequences of cannabis exposure. *Addict Biol* 2008;**13**(2):253–63.
144. Long LE, Lind J, Webster M, Weickert CS. Developmental trajectory of the endocannabinoid system in human dorsolateral prefrontal cortex. *BMC Neurosci* 2012;**13**:87.
145. Eggan SM, Mizoguchi Y, Stoyak SR, Lewis DA. Development of cannabinoid 1 receptor protein and messenger RNA in monkey dorsolateral prefrontal cortex. *Cereb Cortex* 2010;**20**(5):1164–74.
146. Reggio PH. Endocannabinoid binding to the cannabinoid receptors: what is known and what remains unknown. *Curr Med Chem* 2010;**17**(14):1468–86.
147. Lee TT, Hill MN, Hillard CJ, Gorzalka BB. Temporal changes in N-acylethanolamine content and metabolism throughout the peri-adolescent period. *Synapse* 2013;**67**(1):4–10.
148. Wenger T, Gerendai I, Fezza F, González S, Bisogno T, Fernandez-Ruiz J, et al. The hypothalamic levels of the endocannabinoid, anandamide, peak immediately before the onset of puberty in female rats. *Life Sci* 2002;**70**(12):1407–14.
149. Rapoport SI. Arachidonic acid and the brain. *J Nutr* 2008;**138**(12):2515–20.
150. Belkind-Gerson J, Carreón-Rodríguez A, Contreras-Ochoa CO, Estrada-Mondaca S, Parra-Cabrera MS. Fatty acids and neurodevelopment. *J Pediatr Gastroenterol Nutr* 2008;**47**(Suppl 1):S7–9.
151. Ryan VH, Primiani CT, Rao JS, Ahn K, Rapoport SI, Blanchard H. Coordination of gene expression of arachidonic and docosahexaenoic acid cascade enzymes during human brain development and aging. *PLoS One* 2014;**9**(6), e100858.
152. McNamara RK, Schurdak JD, Asch RH, Peters BD, Lindquist DM. Deficits in docosahexaenoic acid accrual during adolescence reduce rat forebrain white matter microstructural integrity: an in vivo diffusion tensor imaging study. *Dev Neurosci* 2018;**40**(1):84–92.
153. McNamara RK, Szeszko PR, Smesny S, Ikuta T, DeRosse P, Vaz FM, et al. Polyunsaturated fatty acid biostatus, phospholipase A(2) activity and brain white matter microstructure across adolescence. *Neuroscience* 2017;**343**:423–33.
154. Miller LR, Jorgensen MJ, Kaplan JR, Seeds MC, Rahbar E, Morgan TM, et al. Alterations in levels and ratios of n-3 and n-6 polyunsaturated fatty acids in the temporal cortex and liver of vervet monkeys from birth to early adulthood. *Physiol Behav* 2016;**156**:71–8.
155. Chen JR, Hsu SF, Hsu CD, Hwang LH, Yang SC. Dietary patterns and blood fatty acid composition in children with attention-deficit hyperactivity disorder in Taiwan. *J Nutr Biochem* 2004;**15**(8):467–72.
156. Morishita J, Okamoto Y, Tsuboi K, Ueno M, Sakamoto H, Maekawa N, et al. Regional distribution and age-dependent expression of N-acylphosphatidylethanolamine-hydrolyzing phospholipase D in rat brain. *J Neurochem* 2005;**94**(3):753–62.
157. Chanda PK, Gao Y, Mark L, Btesh J, Strassle BW, Lu P, et al. Monoacylglycerol lipase activity is a critical modulator of the tone and integrity of the endocannabinoid system. *Mol Pharmacol* 2010;**78**(6):996–1003.
158. Dainese E, Oddi S, Simonetti M, Sabatucci A, Angelucci CB, Ballone A, et al. The endocannabinoid hydrolase FAAH is an allosteric enzyme. *Sci Rep* 2020;**10**(1):2292.

159. Gorzalka BB, Dang SS. Minireview: endocannabinoids and gonadal hormones: bidirectional interactions in physiology and behavior. *Endocrinology* 2012;**153**(3):1016–24.
160. Struik D, Sanna F, Fattore L. The modulating role of sex and anabolic-androgenic steroid hormones in cannabinoid sensitivity. *Front Behav Neurosci* 2018;**12**:249.
161. Apter D. Serum steroids and pituitary hormones in female puberty: a partly longitudinal study. *Clin Endocrinol* 1980;**12**(2):107–20.
162. Hill M, Lukác D, Lapcík O, Sulcová J, Hampl R, Pouzar V, et al. Age relationships and sex differences in serum levels of pregnenolone and 17-hydroxypregnenolone in healthy subjects. *Clin Chem Lab Med* 1999;**37**(4):439–47.
163. Busquets-Garcia A, Soria-Gómez E, Redon B, Mackenbach Y, Vallée M, Chaouloff F, et al. Pregnenolone blocks cannabinoid-induced acute psychotic-like states in mice. *Mol Psychiatr* 2017;**22**(11):1594–603.
164. Vallée M, Vitiello S, Bellocchio L, Hébert-Chatelain E, Monlezun S, Martin-Garcia E, et al. Pregnenolone can protect the brain from cannabis intoxication. *Science* 2014; **343**(6166):94–8.
165. Rodríguez de Fonseca F, Ramos JA, Bonnin A, Fernández-Ruiz JJ. Presence of cannabinoid binding sites in the brain from early postnatal ages. *Neuroreport* 1993;**4**(2):135–8. https://doi.org/10.1097/00001756-199302000-00005.
166. Vangopoulou C, Bourmpoula MT, Koupourtidou C, Giompres P, Stamatakis A, Kouvelas ED, et al. Effects of an early life experience on rat brain cannabinoid receptors in adolescence and adulthood. *IBRO Rep* 2018;**5**:1–9.
167. Biscaia M, Marín S, Fernández B, Marco EM, Rubio M, Guaza C, et al. Chronic treatment with CP 55,940 during the peri-adolescent period differentially affects the behavioural responses of male and female rats in adulthood. *Psychopharmacology* 2003;**170**(3):301–8.
168. Borsoi M, Manduca A, Bara A, Lassalle O, Pelissier-Alicot A-L, Manzoni OJ. Sex differences in the behavioral and synaptic consequences of a single in vivo exposure to the synthetic cannabimimetic WIN55,212-2 at puberty and adulthood. *Front Behav Neurosci* 2019;**13**:23.
169. Nguyen JD, Creehan KM, Kerr TM, Taffe MA. Lasting effects of repeated Δ(9)-tetrahydrocannabinol vapour inhalation during adolescence in male and female rats. *Br J Pharmacol* 2020;**177**(1):188–203.
170. Schepis TS, Desai RA, Cavallo DA, Smith AE, McFetridge A, Liss TB, et al. Gender differences in adolescent marijuana use and associated psychosocial characteristics. *J Addict Med* 2011;**5**(1):65–73.
171. Albaugh MD, Ottino-Gonzalez J, Sidwell A, Lepage C, Juliano A, Owens MM, et al. Association of cannabis use during adolescence with neurodevelopment. *JAMA Psychiatr* 2021; **78**(9):1–11.
172. Wilson J, Freeman TP, Mackie CJ. Effects of increasing cannabis potency on adolescent health. *Lancet Child Adolesc Health* 2019;**3**(2):121–8.
173. Atakan Z. Cannabis, a complex plant: different compounds and different effects on individuals. *Ther Adv Psychopharmacol* 2012;**2**(6):241–54.
174. Pertwee RG. The diverse CB1 and CB2 receptor pharmacology of three plant cannabinoids: delta9-tetrahydrocannabinol, cannabidiol and delta9-tetrahydrocannabivarin. *Br J Pharmacol* 2008;**153**(2):199–215.
175. Fakhoury M. Role of the endocannabinoid system in the pathophysiology of schizophrenia. *Mol Neurobiol* 2017;**54**(1):768–78.
176. Shahbazi F, Grandi V, Banerjee A, Trant JF. Cannabinoids and cannabinoid receptors: the story so far. *iScience* 2020;**23**(7), 101301.

177. Basu S, Dittel BN. Unraveling the complexities of cannabinoid receptor 2 (CB2) immune regulation in health and disease. *Immunol Res* 2011;**51**(1):26–38.
178. Cabral GA, Jamerson M. Marijuana use and brain immune mechanisms. In: Cui C, Shurtleff D, Harris RA, editors. *International review of neurobiology*. Vol. 118. Academic Press; 2014. p. 199–230.
179. Baggelaar MP, Maccarrone M, van der Stelt M. 2-Arachidonoylglycerol: a signaling lipid with manifold actions in the brain. *Prog Lipid Res* 2018;**71**:1–17.
180. Dhein S. Different effects of cannabis abuse on adolescent and adult brain. *Pharmacology* 2020;**105**(11−12):609–17.
181. Bloomfield MAP, Hindocha C, Green SF, Wall MB, Lees R, Petrilli K, et al. The neuropsychopharmacology of cannabis: a review of human imaging studies. *Pharmacol Ther* 2019;**195**:132–61.
182. Jacobus J, Tapert SF. Effects of cannabis on the adolescent brain. *Curr Pharm Des* 2014;**20**(13):2186–93.
183. Manza P, Yuan K, Shokri-Kojori E, Tomasi D, Volkow ND. Brain structural changes in cannabis dependence: association with MAGL. *Mol Psychiatry* 2020;**25**(12):3256–66.
184. Karoly HC, Ross JM, Ellingson JM, Feldstein Ewing SW. Exploring cannabis and alcohol co-use in adolescents: a narrative review of the evidence. *J Dual Diagn* 2020;**16**(1):58–74.
185. Subramaniam P, McGlade E, Yurgelun-Todd D. Comorbid cannabis and tobacco use in adolescents and adults. *Curr Addict Rep* 2016;**3**(2):182–8.
186. Freeman TP, Groshkova T, Cunningham A, Sedefov R, Griffiths P, Lynskey MT. Increasing potency and price of cannabis in Europe, 2006-16. *Addiction* 2019;**114**(6):1015–23.
187. Battistella G, Fornari E, Annoni J-M, Chtioui H, Dao K, Fabritius M, et al. Long-term effects of cannabis on brain structure. *Neuropsychopharmacology* 2014;**39**(9):2041–8.
188. Cohen M, Rasser PE, Peck G, Carr VJ, Ward PB, Thompson PM, et al. Cerebellar grey-matter deficits, cannabis use and first-episode schizophrenia in adolescents and young adults. *Int J Neuropsychopharmacol* 2012;**15**(3):297–307.
189. Wilson W, Mathew R, Turkington T, Hawk T, Coleman RE, Provenzale J. Brain morphological changes and early marijuana use: a magnetic resonance and positron emission tomography study. *J Addict Dis* 2000;**19**(1):1–22.
190. Ligresti A, De Petrocellis L, Di Marzo V. From phytocannabinoids to cannabinoid receptors and endocannabinoids: pleiotropic physiological and pathological roles through complex pharmacology. *Physiol Rev* 2016;**96**(4):1593–659.
191. Koenders L, Cousijn J, Vingerhoets WA, van den Brink W, Wiers RW, Meijer CJ, et al. Grey matter changes associated with heavy cannabis use: a longitudinal sMRI study. *PLoS One* 2016;**11**(5), e0152482.
192. French L, Gray C, Leonard G, Perron M, Pike GB, Richer L, et al. Early cannabis use, polygenic risk score for schizophrenia and brain maturation in adolescence. *JAMA Psychiatr* 2015;**72**(10):1002–11.
193. Gilman JM, Kuster JK, Lee S, Lee MJ, Kim BW, Makris N, et al. Cannabis use is quantitatively associated with nucleus Accumbens and amygdala abnormalities in young adult recreational users. *J Neurosci* 2014;**34**(16):5529.
194. Jacobus J, Squeglia LM, Meruelo AD, Castro N, Brumback T, Giedd JN, et al. Cortical thickness in adolescent marijuana and alcohol users: a three-year prospective study from adolescence to young adulthood. *Dev Cogn Neurosci* 2015;**16**:101–9.
195. Lopez-Larson MP, Bogorodzki P, Rogowska J, McGlade E, King JB, Terry J, et al. Altered prefrontal and insular cortical thickness in adolescent marijuana users. *Behav Brain Res* 2011;**220**(1):164–72.

196. Orr C, Spechler P, Cao Z, Albaugh M, Chaarani B, Mackey S, et al. Grey matter volume differences associated with extremely low levels of cannabis use in adolescence. *J Neurosci* 2019;**39**(10):1817.
197. Medina KL, Schweinsburg AD, Cohen-Zion M, Nagel BJ, Tapert SF. Effects of alcohol and combined marijuana and alcohol use during adolescence on hippocampal volume and asymmetry. *Neurotoxicol Teratol* 2007;**29**(1):141–52.
198. Bortolato M, Bini V, Frau R, Devoto P, Pardu A, Fan Y, et al. Juvenile cannabinoid treatment induces frontostriatal gliogenesis in Lewis rats. *Eur Neuropsychopharmacol* 2014;**24**(6):974–85.
199. Gilbert MT, Soderstrom K. Late-postnatal cannabinoid exposure persistently elevates dendritic spine densities in area X and HVC song regions of zebra finch telencephalon. *Brain Res* 2011;**1405**:23–30.
200. Mato S, Chevaleyre V, Robbe D, Pazos A, Castillo PE, Manzoni OJ. A single in-vivo exposure to delta 9THC blocks endocannabinoid-mediated synaptic plasticity. *Nat Neurosci* 2004;**7**(6):585–6.
201. Ilyasov AA, Milligan CE, Pharr EP, Howlett AC. The endocannabinoid system and oligodendrocytes in health and disease. *Front Neurosci* 2018;**12**:733.
202. Moldrich G, Wenger T. Localization of the CB1 cannabinoid receptor in the rat brain. An immunohistochemical study. *Peptides* 2000;**21**(11):1735–42.
203. Molina-Holgado E, Vela JM, Arévalo-Martín A, Almazán G, Molina-Holgado F, Borrell J, et al. Cannabinoids promote oligodendrocyte progenitor survival: involvement of cannabinoid receptors and phosphatidylinositol-3 kinase/Akt signaling. *J Neurosci* 2002;**22**(22):9742–53.
204. Lisdahl KM, Wright NE, Medina-Kirchner C, Maple KE, Shollenbarger S. Considering cannabis: the effects of regular cannabis use on neurocognition in adolescents and young adults. *Curr Addict Rep* 2014;**1**(2):144–56.
205. Arnone D, Barrick TR, Chengappa S, Mackay CE, Clark CA, Abou-Saleh MT. Corpus callosum damage in heavy marijuana use: preliminary evidence from diffusion tensor tractography and tract-based spatial statistics. *NeuroImage* 2008;**41**(3):1067–74.
206. Orr JM, Paschall CJ, Banich MT. Recreational marijuana use impacts white matter integrity and subcortical (but not cortical) morphometry. *NeuroImage Clin* 2016;**12**:47–56.
207. Ashtari M, Cervellione K, Cottone J, Ardekani BA, Sevy S, Kumra S. Diffusion abnormalities in adolescents and young adults with a history of heavy cannabis use. *J Psychiatr Res* 2009;**43**(3):189–204.
208. Epstein KA, Kumra S. White matter fractional anisotropy over two time points in early onset schizophrenia and adolescent cannabis use disorder: a naturalistic diffusion tensor imaging study. *Psychiatr Res* 2015;**232**(1):34–41.
209. Courtney KE, Baca R, Doran N, Jacobson A, Liu TT, Jacobus J. The effects of nicotine and cannabis co-use during adolescence and young adulthood on white matter cerebral blood flow estimates. *Psychopharmacology* 2020;**237**(12):3615–24.
210. Jacobus J, Goldenberg D, Wierenga CE, Tolentino NJ, Liu TT, Tapert SF. Altered cerebral blood flow and neurocognitive correlates in adolescent cannabis users. *Psychopharmacology* 2012;**222**(4):675–84.
211. Medina KL, Nagel BJ, Tapert SF. Abnormal cerebellar morphometry in abstinent adolescent marijuana users. *Psychiatr Res* 2010;**182**(2):152–9.
212. Koenders L, Lorenzetti V, de Haan L, Suo C, Vingerhoets W, van den Brink W, et al. Longitudinal study of hippocampal volumes in heavy cannabis users. *J Psychopharmacol* 2017;**31**(8):1027–34.

213. Weiland BJ, Thayer RE, Depue BE, Sabbineni A, Bryan AD, Hutchison KE. Daily marijuana use is not associated with brain morphometric measures in adolescents or adults. *J Neurosci Off J Soc Neurosci* 2015;**35**(4):1505–12.
214. Cousijn J, Wiers RW, Ridderinkhof KR, van den Brink W, Veltman DJ, Goudriaan AE. Grey matter alterations associated with cannabis use: results of a VBM study in heavy cannabis users and healthy controls. *NeuroImage* 2012;**59**(4):3845–51.
215. Silveri MM, Dager AD, Cohen-Gilbert JE, Sneider JT. Neurobiological signatures associated with alcohol and drug use in the human adolescent brain. *Neurosci Biobehav Rev* 2016;**70**:244–59.
216. Orr C, Morioka R, Behan B, Datwani S, Doucet M, Ivanovic J, et al. Altered resting-state connectivity in adolescent cannabis users. *Am J Drug Alcohol Abuse* 2013;**39**(6):372–81.
217. Camchong J, Lim KO, Kumra S. Adverse effects of cannabis on adolescent brain development: a longitudinal study. *Cereb Cortex* 2017;**27**(3):1922–30.
218. Silveri MM, Jensen JE, Rosso IM, Sneider JT, Yurgelun-Todd DA. Preliminary evidence for white matter metabolite differences in marijuana-dependent young men using 2D J-resolved magnetic resonance spectroscopic imaging at 4 Tesla. *Psychiatr Res* 2011;**191**(3):201–11.
219. Prescot AP, Locatelli AE, Renshaw PF, Yurgelun-Todd DA. Neurochemical alterations in adolescent chronic marijuana smokers: a proton MRS study. *NeuroImage* 2011;**57**(1):69–75.
220. Jager G, Block RI, Luijten M, Ramsey NF. Cannabis use and memory brain function in adolescent boys: a cross-sectional multicenter functional magnetic resonance imaging study. *J Am Acad Child Adolesc Psychiatr* 2010;**49**(6):561–72 [572.e561-563].
221. Jacobsen LK, Mencl WE, Westerveld M, Pugh KR. Impact of cannabis use on brain function in adolescents. *Ann N Y Acad Sci* 2004;**1021**:384–90.
222. Schweinsburg AD, Nagel BJ, Schweinsburg BC, Park A, Theilmann RJ, Tapert SF. Abstinent adolescent marijuana users show altered fMRI response during spatial working memory. *Psychiatr Res* 2008;**163**(1):40–51.
223. Tervo-Clemmens B, Simmonds D, Calabro FJ, Day NL, Richardson GA, Luna B. Adolescent cannabis use and brain systems supporting adult working memory encoding, maintenance, and retrieval. *NeuroImage* 2018;**169**:496–509.
224. Blest-Hopley G, Giampietro V, Bhattacharyya S. Residual effects of cannabis use in adolescent and adult brains—a meta-analysis of fMRI studies. *Neurosci Biobehav Rev* 2018;**88**:26–41.
225. Jager G, Block RI, Luijten M, Ramsey NF. Tentative evidence for striatal hyperactivity in adolescent cannabis-using boys: a cross-sectional multicenter fMRI study. *J Psychoactive Drugs* 2013;**45**(2):156–67.
226. Ernst M, Luciana M. Neuroimaging of the dopamine/reward system in adolescent drug use. *CNS Spectr* 2015;**20**(4):427–41.
227. Bloomfield MA, Morgan CJ, Kapur S, Curran HV, Howes OD. The link between dopamine function and apathy in cannabis users: an [18F]-DOPA PET imaging study. *Psychopharmacology* 2014;**231**(11):2251–9.
228. Miguez MJ, Chan W, Espinoza L, Tarter R, Perez C. Marijuana use among adolescents is associated with deleterious alterations in mature BDNF. *AIMS Public Health* 2019;**6**(1):4–14.
229. Leal G, Bramham CR, Duarte CB. BDNF and hippocampal synaptic plasticity. *Vitam Horm* 2017;**104**:153–95.
230. Verheij MM, Vendruscolo LF, Caffino L, Giannotti G, Cazorla M, Fumagalli F, et al. Systemic delivery of a brain-penetrant TrkB antagonist reduces cocaine self-administration and normalizes TrkB Signaling in the nucleus accumbens and prefrontal cortex. *J Neurosci* 2016;**36**(31):8149–59.

231. Ganesh S, Vidya KL, Rashid AA, Singh J, D'Souza DC. Revisiting the consequences of adolescent cannabinoid exposure through the lens of the endocannabinoid system. *Curr Addict Rep* 2018;**5**(4):418–27.
232. Duperrouzel JC, Hawes SW, Lopez-Quintero C, Pacheco-Colón I, Coxe S, Hayes T, et al. Adolescent cannabis use and its associations with decision-making and episodic memory: preliminary results from a longitudinal study. *Neuropsychology* 2019;**33**(5):701–10.
233. Scott JC, Slomiak ST, Jones JD, Rosen AFG, Moore TM, Gur RC. Association of cannabis with cognitive functioning in adolescents and young adults: a systematic review and meta-analysis. *JAMA Psychiatr* 2018;**75**(6):585–95.
234. Schuster RM, Gilman J, Schoenfeld D, Evenden J, Hareli M, Ulysse C, et al. One month of cannabis abstinence in adolescents and young adults is associated with improved memory. *J Clin Psychiatr* 2018;**79**(6). 17m11977.
235. Meier MH, Caspi A, Ambler A, Harrington H, Houts R, Keefe RSE, et al. Persistent cannabis users show neuropsychological decline from childhood to midlife. *Proc Natl Acad Sci* 2012;**109**(40):E2657–64.
236. Curran HV, Freeman TP, Mokrysz C, Lewis DA, Morgan CJA, Parsons LH. Keep off the grass? Cannabis, cognition and addiction. *Nat Rev Neurosci* 2016;**17**(5):293–306.
237. Broyd SJ, van Hell HH, Beale C, Yücel M, Solowij N. Acute and chronic effects of cannabinoids on human cognition—a systematic review. *Biol Psychiatry* 2016;**79**(7):557–67.
238. Gorey C, Kuhns L, Smaragdi E, Kroon E, Cousijn J. Age-related differences in the impact of cannabis use on the brain and cognition: a systematic review. *Eur Arch Psychiatr Clin Neurosci* 2019;**269**(1):37–58.
239. Blest-Hopley G, O'Neill A, Wilson R, Giampietro V, Bhattacharyya S. Disrupted parahippocampal and midbrain function underlie slower verbal learning in adolescent-onset regular cannabis use. *Psychopharmacology* 2019;**238**:1315–31.
240. Marwaha S, Winsper C, Bebbington P, Smith D. Cannabis use and hypomania in young people: a prospective analysis. *Schizophr Bull* 2018;**44**(6):1267–74.
241. Gibbs M, Winsper C, Marwaha S, Gilbert E, Broome M, Singh SP. Cannabis use and mania symptoms: a systematic review and meta-analysis. *J Affect Disord* 2015;**171**:39–47.
242. Connor JP, Stjepanović D, Le Foll B, Hoch E, Budney AJ, Hall WD. Cannabis use and cannabis use disorder. *Nat Rev Dis Primers* 2021;**7**(1):16.
243. Kiburi SK, Molebatsi K, Ntlantsana V, Lynskey MT. Cannabis use in adolescence and risk of psychosis: are there factors that moderate this relationship? A systematic review and meta-analysis. *Subst Abus* 2021;1–25.
244. Mustonen A, Niemelä S, Nordström T, Murray GK, Mäki P, Jääskeläinen E, et al. Adolescent cannabis use, baseline prodromal symptoms and the risk of psychosis. *Br J Psychiatr* 2018;**212**(4):227–33.
245. Greene MC, Kelly JF. The prevalence of cannabis withdrawal and its influence on adolescents' treatment response and outcomes: a 12-month prospective investigation. *J Addict Med* 2014;**8**(5):359–67.
246. American Psychiatric Association. *Diagnostic and statistical manual of mental disorders*. Washington, DC: American Psychiatric Association; 2013.
247. Simpson AK, Magid V. Cannabis use disorder in adolescence. *Child Adolesc Psychiatr Clin N Am* 2016;**25**(3):431–43.
248. Krebs M-O, Kebir O, Jay TM. Exposure to cannabinoids can lead to persistent cognitive and psychiatric disorders. *Eur J Pain* 2019;**23**(7):1225–33.
249. Mashhoon Y, Sagar KA, Gruber SA. Cannabis use and consequences. *Pediatr Clin N Am* 2019;**66**(6):1075–86.

250. Kroon E, Kuhns L, Cousijn J. The short-term and long-term effects of cannabis on cognition: recent advances in the field. *Curr Opin Psychol* 2021;**38**:49–55.
251. Renard J, Rushlow WJ, Laviolette SR. What can rats tell us about adolescent cannabis exposure? Insights from preclinical research. *Can J Psychiatr* 2016;**61**(6):328–34.
252. Verrico CD, Gu H, Peterson ML, Sampson AR, Lewis DA. Repeated Δ^9-tetrahydrocannabinol exposure in adolescent monkeys: persistent effects selective for spatial working memory. *Am J Psychiatr* 2014;**171**(4):416–25.
253. Cohen K, Weinstein A. The effects of cannabinoids on executive functions: evidence from cannabis and synthetic cannabinoids—a systematic review. *Brain Sci* 2018;**8**(3):40.
254. Chen H-T, Mackie K. Adolescent $\Delta(9)$-tetrahydrocannabinol exposure selectively impairs working memory but not several other mPFC-mediated behaviors. *Front Psychiatr* 2020;**11**:576214.
255. Iemolo A, Montilla-Perez P, Nguyen J, Risbrough VB, Taffe MA, Telese F. Reelin deficiency contributes to long-term behavioral abnormalities induced by chronic adolescent exposure to $\Delta 9$-tetrahydrocannabinol in mice. *Neuropharmacology* 2021;**187**:108495.
256. Kasten CR, Zhang Y, Boehm SL. Acute and long-term effects of Δ^9-tetrahydrocannabinol on object recognition and anxiety-like activity are age- and strain-dependent in mice. *Pharmacol Biochem Behav* 2017;**163**:9–19.
257. Kevin RC, Wood KE, Stuart J, Mitchell AJ, Moir M, Banister SD, et al. Acute and residual effects in adolescent rats resulting from exposure to the novel synthetic cannabinoids AB-PINACA and AB-FUBINACA. *J Psychopharmacol* 2017;**31**(6):757–69.
258. Murphy M, Mills S, Winstone J, Leishman E, Wager-Miller J, Bradshaw H, et al. Chronic adolescent $\Delta(9)$-tetrahydrocannabinol treatment of male mice leads to long-term cognitive and behavioral dysfunction, which are prevented by concurrent cannabidiol treatment. *Cannabis Cannabinoid Res* 2017;**2**(1):235–46.
259. Prini P, Rusconi F, Zamberletti E, Gabaglio M, Penna F, Fasano M, et al. Adolescent THC exposure in female rats leads to cognitive deficits through a mechanism involving chromatin modifications in the prefrontal cortex. *J Psychiatr Neurosci* 2018;**43**(2):87–101.
260. Bruijnzeel AW, Knight P, Panunzio S, Xue S, Bruner MM, Wall SC, et al. Effects in rats of adolescent exposure to cannabis smoke or THC on emotional behavior and cognitive function in adulthood. *Psychopharmacology* 2019;**236**(9):2773–84.
261. Kirschmann EK, Pollock MW, Nagarajan V, Torregrossa MM. Effects of adolescent cannabinoid self-administration in rats on addiction-related Behaviors and working memory. *Neuropsychopharmacology* 2017;**42**(5):989–1000.
262. Stringfield SJ, Torregrossa MM. Intravenous self-administration of delta-9-THC in adolescent rats produces long-lasting alterations in behavior and receptor protein expression. *Psychopharmacology* 2021;**238**(1):305–19.
263. Hernandez CM, Orsini CA, Blaes SL, Bizon JL, Febo M, Bruijnzeel AW, et al. Effects of repeated adolescent exposure to cannabis smoke on cognitive outcomes in adulthood. *J Psychopharmacol* 2020;, 269881120965931.
264. Renard J, Rosen LG, Loureiro M, De Oliveira C, Schmid S, Rushlow WJ, et al. Adolescent cannabinoid exposure induces a persistent sub-cortical hyper-dopaminergic state and associated molecular adaptations in the prefrontal cortex. *Cereb Cortex* 2017;**27**(2):1297–310.
265. Renard J, Szkudlarek HJ, Kramar CP, Jobson CEL, Moura K, Rushlow WJ, et al. Adolescent THC exposure causes enduring prefrontal cortical disruption of GABAergic inhibition and dysregulation of sub-cortical dopamine function. *Sci Rep* 2017;**7**(1):11420.
266. Rubino T, Vigano D, Realini N, Guidali C, Braida D, Capurro V, et al. Chronic $\Delta 9$-tetrahydrocannabinol during adolescence provokes sex-dependent changes in the

emotional profile in adult rats: behavioral and biochemical correlates. *Neuropsychopharmacology* 2008;**33**(11):2760–71.
267. De Gregorio D, Dean Conway J, Canul M-L, Posa L, Bambico FR, Gobbi G. Effects of chronic exposure to low-dose delta-9-tetrahydrocannabinol in adolescence and adulthood on serotonin/norepinephrine neurotransmission and emotional behavior. *Int J Neuropsychopharmacol* 2020;**23**(11):751–61.
268. Roberts AG, Lopez-Duran NL. Developmental influences on stress response systems: implications for psychopathology vulnerability in adolescence. *Compr Psychiatry* 2019;**88**:9–21.
269. Kardash T, Rodin D, Kirby M, Davis N, Koman I, Gorelick J, et al. Link between personality and response to THC exposure. *Behav Brain Res* 2020;**379**:112361.
270. Thorpe HHA, Hamidullah S, Jenkins BW, Khokhar JY. Adolescent neurodevelopment and substance use: receptor expression and behavioral consequences. *Pharmacol Ther* 2020;**206**:107431.
271. Schoch H, Huerta MY, Ruiz CM, Farrell MR, Jung KM, Huang JJ, et al. Adolescent cannabinoid exposure effects on natural reward seeking and learning in rats. *Psychopharmacology* 2018;**235**(1):121–34.
272. Hurd YL, Michaelides M, Miller ML, Jutras-Aswad D. Trajectory of adolescent cannabis use on addiction vulnerability. *Neuropharmacology* 2014;**76**(Pt B):416–24.
273. Scherma M, Qvist JS, Asok A, Huang SC, Masia P, Deidda M, et al. Cannabinoid exposure in rat adolescence reprograms the initial behavioral, molecular, and epigenetic response to cocaine. *Proc Natl Acad Sci U S A* 2020;**117**(18):9991–10002.
274. Scherma M, Dessì C, Muntoni AL, Lecca S, Satta V, Luchicchi A, et al. Adolescent Δ(9)-tetrahydrocannabinol exposure alters WIN 55,212-2 self-administration in adult rats. *Neuropsychopharmacology* 2016;**41**(5):1416–26.
275. Stopponi S, Soverchia L, Ubaldi M, Cippitelli A, Serpelloni G, Ciccocioppo R. Chronic THC during adolescence increases the vulnerability to stress-induced relapse to heroin seeking in adult rats. *Eur Neuropsychopharmacol* 2014;**24**(7):1037–45.
276. Aguilar MA, Ledesma JC, Rodríguez-Arias M, Penalva C, Manzanedo C, Miñarro J, et al. Adolescent exposure to the synthetic cannabinoid WIN 55212-2 modifies cocaine withdrawal symptoms in adult mice. *Int J Mol Sci* 2017;**18**(6):1326.
277. Williams AR. Cannabis as a gateway drug for opioid use disorder. *J Law Med Ethics* 2020;**48**(2):268–74.
278. Gleason KA, Birnbaum SG, Shukla A, Ghose S. Susceptibility of the adolescent brain to cannabinoids: long-term hippocampal effects and relevance to schizophrenia. *Transl Psychiatr* 2012;**2**(11):e199.
279. Wegener N, Koch M. Behavioural disturbances and altered Fos protein expression in adult rats after chronic pubertal cannabinoid treatment. *Brain Res* 2009;**1253**:81–91.
280. Mena A, Ruiz-Salas JC, Puentes A, Dorado I, Ruiz-Veguilla M, De la Casa LG. Reduced prepulse inhibition as a biomarker of schizophrenia. *Front Behav Neurosci* 2016;**10**:202.
281. Poulia N, Delis F, Brakatselos C, Polissidis A, Koutmani Y, Kokras N, et al. Detrimental effects of adolescent escalating low-dose Δ(9)-tetrahydrocannabinol leads to a specific biobehavioural profile in adult male rats. *Br J Pharmacol* 2021;**178**(7):1722–36.
282. Harte LC, Dow-Edwards D. Sexually dimorphic alterations in locomotion and reversal learning after adolescent tetrahydrocannabinol exposure in the rat. *Neurotoxicol Teratol* 2010;**32**(5):515–24.
283. Dong C, Tian Z, Zhang K, Chang L, Qu Y, Pu Y, et al. Increased BDNF-TrkB signaling in the nucleus accumbens plays a role in the risk for psychosis after cannabis exposure during adolescence. *Pharmacol Biochem Behav* 2019;**177**:61–8.

284. Gomes FV, Guimarães FS, Grace AA. Effects of pubertal cannabinoid administration on attentional set-shifting and dopaminergic hyper-responsivity in a developmental disruption model of schizophrenia. *Int J Neuropsychopharmacol* 2014;**18**(2), pyu018.
285. Zamberletti E, Beggiato S, Steardo Jr L, Prini P, Antonelli T, Ferraro L, et al. Alterations of prefrontal cortex GABAergic transmission in the complex psychotic-like phenotype induced by adolescent delta-9-tetrahydrocannabinol exposure in rats. *Neurobiol Dis* 2014;**63**:35–47.
286. Kruse LC, Cao JK, Viray K, Stella N, Clark JJ. Voluntary oral consumption of Δ^9-tetrahydrocannabinol by adolescent rats impairs reward-predictive cue behaviors in adulthood. *Neuropsychopharmacology* 2019;**44**(8):1406–14.
287. Silva L, Black R, Michaelides M, Hurd YL, Dow-Edwards D. Sex and age specific effects of delta-9-tetrahydrocannabinol during the periadolescent period in the rat: the unique susceptibility of the prepubescent animal. *Neurotoxicol Teratol* 2016;**58**:88–100. https://doi.org/10.1016/j.ntt.2016.02.005.
288. Wiley JL, Burston JJ. Sex differences in $\Delta(9)$-tetrahydrocannabinol metabolism and in vivo pharmacology following acute and repeated dosing in adolescent rats. *Neurosci Lett* 2014;**576**:51–5.
289. Rubino T, Realini N, Braida D, Alberio T, Capurro V, Viganò D, et al. The depressive phenotype induced in adult female rats by adolescent exposure to THC is associated with cognitive impairment and altered neuroplasticity in the prefrontal cortex. *Neurotox Res* 2009;**15**(4):291–302.
290. Rubino T, Realini N, Braida D, Guidi S, Capurro V, Viganò D, et al. Changes in hippocampal morphology and neuroplasticity induced by adolescent THC treatment are associated with cognitive impairment in adulthood. *Hippocampus* 2009;**19**(8):763–72.
291. Rubino T, Zamberletti E, Parolaro D. Adolescent exposure to cannabis as a risk factor for psychiatric disorders. *J Psychopharmacol* 2012;**26**(1):177–88.
292. Burston JJ, Wiley JL, Craig AA, Selley DE, Sim-Selley LJ. Regional enhancement of cannabinoid CB_1 receptor desensitization in female adolescent rats following repeated delta-tetrahydrocannabinol exposure. *Br J Pharmacol* 2010;**161**(1):103–12.
293. Hoffman AF, Lupica CR. Direct actions of cannabinoids on synaptic transmission in the nucleus accumbens: a comparison with opioids. *J Neurophysiol* 2001;**85**(1):72–83.
294. Maccarrone M, Rossi S, Bari M, De Chiara V, Fezza F, Musella A, et al. Anandamide inhibits metabolism and physiological actions of 2-arachidonoylglycerol in the striatum. *Nat Neurosci* 2008;**11**(2):152–9.
295. Gerdeman GL, Ronesi J, Lovinger DM. Postsynaptic endocannabinoid release is critical to long-term depression in the striatum. *Nat Neurosci* 2002;**5**(5):446–51.
296. Tomas-Roig J, Benito E, Agis-Balboa R, Piscitelli F, Hoyer-Fender S, Di Marzo V, et al. Chronic exposure to cannabinoids during adolescence causes long-lasting behavioral deficits in adult mice. *Addict Biol* 2017;**22**(6):1778–89.
297. Xie K, Martemyanov KA. Control of striatal signaling by g protein regulators. *Front Neuroanat* 2011;**5**:49.
298. Ostrovskaya O, Xie K, Masuho I, Fajardo-Serrano A, Lujan R, Wickman K, et al. RGS7/Gβ5/R7BP complex regulates synaptic plasticity and memory by modulating hippocampal GABABR-GIRK signaling. *eLife* 2014;**3**, e02053.
299. Marco EM, Rubino T, Adriani W, Viveros MP, Parolaro D, Laviola G. Long-term consequences of URB597 administration during adolescence on cannabinoid CB1 receptor binding in brain areas. *Brain Res* 2009;**1257**:25–31.
300. Bossong MG, Niesink RJM. Adolescent brain maturation, the endogenous cannabinoid system and the neurobiology of cannabis-induced schizophrenia. *Prog Neurobiol* 2010;**92**(3):370–85.

301. Cass DK, Flores-Barrera E, Thomases DR, Vital WF, Caballero A, Tseng KY. CB1 cannabinoid receptor stimulation during adolescence impairs the maturation of GABA function in the adult rat prefrontal cortex. *Mol Psychiatr* 2014;**19**(5):536–43.
302. Caballero A, Tseng KY. Association of cannabis use during adolescence, prefrontal CB1 receptor signaling, and schizophrenia. *Front Pharmacol* 2012;**3**:101.
303. Zamberletti E, Gabaglio M, Grilli M, Prini P, Catanese A, Pittaluga A, et al. Long-term hippocampal glutamate synapse and astrocyte dysfunctions underlying the altered phenotype induced by adolescent THC treatment in male rats. *Pharmacol Res* 2016;**111**:459–70.
304. Pérez-Valenzuela EJ, Andrés Coke ME, Grace AA, Fuentealba Evans JA. Adolescent exposure to WIN 55212-2 render the nigrostriatal dopaminergic pathway activated during adulthood. *Int J Neuropsychopharmacol* 2020;**23**(9):626–37.
305. Cadoni C, Simola N, Espa E, Fenu S, Di Chiara G. Strain dependence of adolescent cannabis influence on heroin reward and mesolimbic dopamine transmission in adult Lewis and Fischer 344 rats. *Addict Biol* 2015;**20**(1):132–42.
306. Corongiu S, Dessì C, Cadoni C. Adolescence versus adulthood: differences in basal mesolimbic and nigrostriatal dopamine transmission and response to drugs of abuse. *Addict Biol* 2020;**25**(1), e12721.
307. Ellgren M, Spano SM, Hurd YL. Adolescent cannabis exposure alters opiate intake and opioid limbic neuronal populations in adult rats. *Neuropsychopharmacology* 2007;**32**(3):607–15.
308. Tomasiewicz HC, Jacobs MM, Wilkinson MB, Wilson SP, Nestler EJ, Hurd YL. Proenkephalin mediates the enduring effects of adolescent cannabis exposure associated with adult opiate vulnerability. *Biol Psychiatry* 2012;**72**(10):803–10.
309. Galindo L, Moreno E, López-Armenta F, Guinart D, Cuenca-Royo A, Izquierdo-Serra M, et al. Cannabis users show enhanced expression of CB(1)-5HT(2A) receptor heteromers in olfactory neuroepithelium cells. *Mol Neurobiol* 2018;**55**(8):6347–61.
310. Ibarra-Lecue I, Mollinedo-Gajate I, Meana JJ, Callado LF, Diez-Alarcia R, Urigüen L. Chronic cannabis promotes pro-hallucinogenic signaling of 5-HT2A receptors through Akt/mTOR pathway. *Neuropsychopharmacology* 2018;**43**(10):2028–35.
311. Puighermanal E, Busquets-Garcia A, Gomis-González M, Marsicano G, Maldonado R, Ozaita A. Dissociation of the pharmacological effects of THC by mTOR blockade. *Neuropsychopharmacology* 2013;**38**(7):1334–43.
312. Bambico FR, Nguyen N-T, Katz N, Gobbi G. Chronic exposure to cannabinoids during adolescence but not during adulthood impairs emotional behaviour and monoaminergic neurotransmission. *Neurobiol Dis* 2010;**37**(3):641–55.
313. Lopez-Rodriguez AB, Llorente-Berzal A, Garcia-Segura LM, Viveros MP. Sex-dependent long-term effects of adolescent exposure to THC and/or MDMA on neuroinflammation and serotoninergic and cannabinoid systems in rats. *Br J Pharmacol* 2014;**171**(6):1435–47.
314. Poulia N, Delis F, Brakatselos C, Lekkas P, Kokras N, Dalla C, et al. Escalating low-dose Δ(9)-tetrahydrocannabinol exposure during adolescence induces differential behavioral and neurochemical effects in male and female adult rats. *Eur J Neurosci* 2020;**52**(1):2681–93.
315. Blázquez C, Chiarlone A, Bellocchio L, Resel E, Pruunsild P, García-Rincón D, et al. The CB1 cannabinoid receptor signals striatal neuroprotection via a PI3K/Akt/mTORC1/BDNF pathway. *Cell Death Differ* 2015;**22**(10):1618–29.
316. Butovsky E, Juknat A, Goncharov I, Elbaz J, Eilam R, Zangen A, et al. In vivo up-regulation of brain-derived neurotrophic factor in specific brain areas by chronic exposure to Δ^9-tetrahydrocannabinol. *J Neurochem* 2005;**93**(4):802–11.
317. Derkinderen P, Valjent E, Toutant M, Corvol JC, Enslen H, Ledent C, et al. Regulation of extracellular signal-regulated kinase by cannabinoids in hippocampus. *J Neurosci* 2003;**23**(6):2371–82.

318. Segal-Gavish H, Gazit N, Barhum Y, Ben-Zur T, Taler M, Hornfeld SH, et al. BDNF overexpression prevents cognitive deficit elicited by adolescent cannabis exposure and host susceptibility interaction. *Hum Mol Genet* 2017;**26**(13):2462–71.
319. Miller ML, Chadwick B, Dickstein DL, Purushothaman I, Egervari G, Rahman T, et al. Adolescent exposure to Δ(9)-tetrahydrocannabinol alters the transcriptional trajectory and dendritic architecture of prefrontal pyramidal neurons. *Mol Psychiatr* 2019;**24**(4):588–600.
320. Rubino T, Prini P, Piscitelli F, Zamberletti E, Trusel M, Melis M, et al. Adolescent exposure to THC in female rats disrupts developmental changes in the prefrontal cortex. *Neurobiol Dis* 2015;**73**:60–9.
321. Ren Q, Ma M, Yang C, Zhang JC, Yao W, Hashimoto K. BDNF–TrkB signaling in the nucleus accumbens shell of mice has key role in methamphetamine withdrawal symptoms. *Transl Psychiatr* 2015;**5**(10):e666.
322. Guan J, Cai JJ, Ji G, Sham PC. Commonality in dysregulated expression of gene sets in cortical brains of individuals with autism, schizophrenia, and bipolar disorder. *Transl Psychiatr* 2019;**9**(1):152.
323. Pennington K, Beasley CL, Dicker P, Fagan A, English J, Pariante CM, et al. Prominent synaptic and metabolic abnormalities revealed by proteomic analysis of the dorsolateral prefrontal cortex in schizophrenia and bipolar disorder. *Mol Psychiatr* 2008;**13**(12):1102–17.
324. Witt SH, Streit F, Jungkunz M, Frank J, Awasthi S, Reinbold CS, et al. Genome-wide association study of borderline personality disorder reveals genetic overlap with bipolar disorder, major depression and schizophrenia. *Transl Psychiatr* 2017;**7**(6):e1155.
325. Delgado RN, Mansky B, Ahanger SH, Lu C, Andersen RE, Dou Y, et al. Maintenance of neural stem cell positional identity by *mixed-lineage leukemia 1*. *Science* 2020;**368**(6486):48.
326. Lim DA, Huang YC, Swigut T, Mirick AL, Garcia-Verdugo JM, Wysocka J, et al. Chromatin remodelling factor Mll1 is essential for neurogenesis from postnatal neural stem cells. *Nature* 2009;**458**(7237):529–33.
327. Jakovcevski M, Ruan H, Shen EY, Dincer A, Javidfar B, Ma Q, et al. Neuronal Kmt2a/Mll1 histone methyltransferase is essential for prefrontal synaptic plasticity and working memory. *J Neurosci* 2015;**35**(13):5097–108.
328. Shen EY, Jiang Y, Javidfar B, Kassim B, Loh Y-HE, Ma Q, et al. Neuronal deletion of Kmt2a/Mll1 histone methyltransferase in ventral striatum is associated with defective spike-timing-dependent striatal synaptic plasticity, altered response to dopaminergic drugs, and increased anxiety. *Neuropsychopharmacology* 2016;**41**(13):3103–13.
329. Huang HS, Matevossian A, Whittle C, Kim SY, Schumacher A, Baker SP, et al. Prefrontal dysfunction in schizophrenia involves mixed-lineage leukemia 1-regulated histone methylation at GABAergic gene promoters. *J Neurosci* 2007;**27**(42):11254–62.
330. Shen E, Shulha H, Weng Z, Akbarian S. Regulation of histone H3K4 methylation in brain development and disease. *Philos Trans R Soc Lond Ser B Biol Sci* 2014;**369**:1652.
331. Sun H, Kennedy PJ, Nestler EJ. Epigenetics of the depressed brain: role of histone acetylation and methylation. *Neuropsychopharmacology* 2013;**38**(1):124–37.
332. Hebert-Chatelain E, Desprez T, Serrat R, Bellocchio L, Soria-Gomez E, Busquets-Garcia A, et al. A cannabinoid link between mitochondria and memory. *Nature* 2016;**539**(7630):555–9.
333. Rangaraju V, Calloway N, Ryan TA. Activity-driven local ATP synthesis is required for synaptic function. *Cell* 2014;**156**(4):825–35.
334. Beiersdorf J, Hevesi Z, Calvigioni D, Pyszkowski J, Romanov R, Szodorai E, et al. Adverse effects of Δ9-tetrahydrocannabinol on neuronal bioenergetics during postnatal development. *JCI Insight* 2020;**5**(23), e135418.

335. Scheller A, Kirchhoff F. Endocannabinoids and heterogeneity of glial cells in brain function. *Front Integr Neurosci* 2016;**10**:24.
336. Mato S, Victoria Sánchez-Gómez M, Matute C. Cannabidiol induces intracellular calcium elevation and cytotoxicity in oligodendrocytes. *Glia* 2010;**58**(14):1739–47.
337. Maccarrone M, Guzmán M, Mackie K, Doherty P, Harkany T. Programming of neural cells by (endo)cannabinoids: from physiological rules to emerging therapies. *Nat Rev Neurosci* 2014;**15**(12):786–801.
338. Cassano T, Calcagnini S, Pace L, De Marco F, Romano A, Gaetani S. Cannabinoid receptor 2 signaling in neurodegenerative disorders: from pathogenesis to a promising therapeutic target. *Front Neurosci* 2017;**11**:30.
339. Cutando L, Maldonado R, Ozaita A. Microglial activation and cannabis exposure. In: Preedy VR, editor. *Handbook of cannabis and related pathologies*. San Diego: Academic Press; 2017. p. 401–12 [Chapter 41].
340. Hablitz LM, Gunesch AN, Cravetchi O, Moldavan M, Allen CN. Cannabinoid signaling recruits astrocytes to modulate presynaptic function in the suprachiasmatic nucleus. *eNeuro* 2020;**7**(1), ENEURO.0081-19.2020.
341. Yang QQ, Zhou JW. Neuroinflammation in the central nervous system: symphony of glial cells. *Glia* 2019;**67**(6):1017–35.
342. Bachiller S, Jiménez-Ferrer I, Paulus A, Yang Y, Swanberg M, Deierborg T, et al. Microglia in neurological diseases: a road map to brain-disease dependent-inflammatory response. *Front Cell Neurosci* 2018;**12**:488.
343. Liddelow SA, Barres BA. Reactive astrocytes: production, function, and therapeutic potential. *Immunity* 2017;**46**(6):957–67.
344. Guerri C, Pascual M. Impact of neuroimmune activation induced by alcohol or drug abuse on adolescent brain development. *Int J Dev Neurosci* 2019;**77**:89–98.
345. Do Y, McKallip RJ, Nagarkatti M, Nagarkatti PS. Activation through cannabinoid receptors 1 and 2 on dendritic cells triggers NF-kappaB-dependent apoptosis: novel role for endogenous and exogenous cannabinoids in immunoregulation. *J Immunol* 2004;**173**(4):2373–82.
346. Jouroukhin Y, Zhu X, Shevelkin AV, Hasegawa Y, Abazyan B, Saito A, et al. Adolescent Delta(9)-tetrahydrocannabinol exposure and astrocyte-specific genetic vulnerability converge on nuclear factor-kappaB-cyclooxygenase-2 signaling to impair memory in adulthood. *Biol Psychiatry* 2019;**85**(11):891–903.
347. Zamberletti E, Gabaglio M, Prini P, Rubino T, Parolaro D. Cortical neuroinflammation contributes to long-term cognitive dysfunctions following adolescent delta-9-tetrahydrocannabinol treatment in female rats. *Eur Neuropsychopharmacol* 2015;**25**(12):2404–15.
348. Mizuno T, Kurotani T, Komatsu Y, Kawanokuchi J, Kato H, Mitsuma N, et al. Neuroprotective role of phosphodiesterase inhibitor ibudilast on neuronal cell death induced by activated microglia. *Neuropharmacology* 2004;**46**(3):404–11.
349. Han J, Fan Y, Zhou K, Blomgren K, Harris RA. Uncovering sex differences of rodent microglia. *J Neuroinflammat* 2021;**18**(1):74.
350. Marchalant Y, Rosi S, Wenk GL. Anti-inflammatory property of the cannabinoid agonist WIN-55212-2 in a rodent model of chronic brain inflammation. *Neuroscience* 2007;**144**(4):1516–22.
351. Paolicelli RC, Ferretti MT. Function and dysfunction of microglia during brain development: consequences for synapses and neural circuits. *Front Synaptic Neurosci* 2017;**9**:9.
352. Patel PK, Leathem LD, Currin DL, Karlsgodt KH. Adolescent neurodevelopment and vulnerability to psychosis. *Biol Psychiatry* 2021;**89**(2):184–93.
353. Hahn J, Wang X, Margeta M. Astrocytes increase the activity of synaptic GluN2B NMDA receptors. *Front Cell Neurosci* 2015;**9**:117.

354. Chen R, Zhang J, Fan N, Teng Z-Q, Wu Y, Yang H, et al. Δ9-THC-caused synaptic and memory impairments are mediated through COX-2 signaling. *Cell* 2013;**155**(5):1154–65.
355. Gutiérrez-Rodríguez A, Bonilla-Del Río I, Puente N, Gómez-Urquijo SM, Fontaine CJ, Egaña-Huguet J, et al. Localization of the cannabinoid type-1 receptor in subcellular astrocyte compartments of mutant mouse hippocampus. *Glia* 2018;**66**(7):1417–31.
356. Jimenez-Blasco D, Busquets-Garcia A, Hebert-Chatelain E, Serrat R, Vicente-Gutierrez C, Ioannidou C, et al. Glucose metabolism links astroglial mitochondria to cannabinoid effects. *Nature* 2020;**583**(7817):603–8.
357. Zalesky A, Solowij N, Yücel M, Lubman DI, Takagi M, Harding IH, et al. Effect of long-term cannabis use on axonal fibre connectivity. *Brain* 2012;**135**(7):2245–55.
358. Mato S, Alberdi E, Ledent C, Watanabe M, Matute C. CB1 cannabinoid receptor-dependent and -independent inhibition of depolarization-induced calcium influx in oligodendrocytes. *Glia* 2009;**57**(3):295–306.
359. Grigorenko E, Kittler J, Clayton C, Wallace D, Zhuang S, Bridges D, et al. Assessment of cannabinoid induced gene changes: tolerance and neuroprotection. *Chem Phys Lipids* 2002;**121**(1-2):257–66.
360. Kittler JT, Grigorenko EV, Clayton C, Zhuang S-Y, Bundey SC, Trower MM, et al. Large-scale analysis of gene expression changes during acute and chronic exposure to Δ^9-THC in rats. *Physiol Genomics* 2000;**3**(3):175–85.
361. Quinn HR, Matsumoto I, Callaghan PD, Long LE, Arnold JC, Gunasekaran N, et al. Adolescent rats find repeated Δ^9-THC less aversive than adult rats but display greater residual cognitive deficits and changes in hippocampal protein expression following exposure. *Neuropsychopharmacology* 2008;**33**(5):1113–26.
362. Calapai F, Cardia L, Sorbara EE, Navarra M, Gangemi S, Calapai G, et al. Cannabinoids, blood-brain barrier, and brain disposition. *Pharmaceutics* 2020;**12**(3):265.
363. Dinis-Oliveira RJ. Metabolomics of Δ^9-tetrahydrocannabinol: implications in toxicity. *Drug Metab Rev* 2016;**48**(1):80–7.
364. Sharma P, Murthy P, Bharath MMS. Chemistry, metabolism, and toxicology of cannabis: clinical implications. *Iran J Psychiatr* 2012;**7**(4):149–56.
365. Zhu J, Peltekian K. Cannabis and the liver: things you wanted to know but were afraid to ask. *Can Liver J* 2019;**2**:1–7.
366. Grotenhermen F. Pharmacokinetics and pharmacodynamics of cannabinoids. *Clin Pharmacokinet* 2003;**42**(4):327–60.
367. Torrens A, Vozella V, Huff H, McNeil B, Ahmed F, Ghidini A, et al. Comparative pharmacokinetics of Δ(9)-tetrahydrocannabinol in adolescent and adult male mice. *J Pharmacol Exp Ther* 2020;**374**(1):151–60.
368. Sadler NC, Nandhikonda P, Webb-Robertson B-J, Ansong C, Anderson LN, Smith JN, et al. Hepatic cytochrome P450 activity, abundance, and expression throughout human development. *Drug Metab Dispos* 2016;**44**(7):984.
369. Huestis MA. Pharmacokinetics and metabolism of the plant cannabinoids, Δ^9-tetrahydrocannibinol, cannabidiol and cannabinol. In: Pertwee RG, editor. *Cannabinoids*. Berlin, Heidelberg: Springer Berlin Heidelberg; 2005. p. 657–90.
370. Lucas CJ, Galettis P, Schneider J. The pharmacokinetics and the pharmacodynamics of cannabinoids. *Br J Clin Pharmacol* 2018;**84**(11):2477–82.
371. Craft RM, Marusich JA, Wiley JL. Sex differences in cannabinoid pharmacology: a reflection of differences in the endocannabinoid system? *Life Sci* 2013;**92**(8-9):476–81.
372. Renard J, Krebs M-O, Le Pen G, Jay TM. Long-term consequences of adolescent cannabinoid exposure in adult psychopathology. *Front Neurosci* 2014;**8**:361.

373. Saito A, Ballinger MDL, Pletnikov MV, Wong DF, Kamiya A. Endocannabinoid system: potential novel targets for treatment of schizophrenia. *Neurobiol Dis* 2013;**53**:10–7.
374. Ho BC, Wassink TH, Ziebell S, Andreasen NC. Cannabinoid receptor 1 gene polymorphisms and marijuana misuse interactions on white matter and cognitive deficits in schizophrenia. *Schizophr Res* 2011;**128**(1-3):66–75.
375. Tao R, Li C, Jaffe AE, Shin JH, Deep-Soboslay A, Yamin RE, et al. Cannabinoid receptor CNR1 expression and DNA methylation in human prefrontal cortex, hippocampus and caudate in brain development and schizophrenia. *Transl Psychiatr* 2020;**10**(1):158.
376. Caspi A, Moffitt TE, Cannon M, McClay J, Murray R, Harrington H, et al. Moderation of the effect of adolescent-onset cannabis use on adult psychosis by a functional polymorphism in the catechol-O-methyltransferase gene: longitudinal evidence of a gene X environment interaction. *Biol Psychiatry* 2005;**57**(10):1117–27.
377. Fatjó-Vilas M, Prats C, Fañanás L. COMT genotypes, cannabis use, and psychosis: gene-environment interaction evidence from human populations, and its methodological concerns. In: Preedy VR, editor. *Handbook of cannabis and related pathologies*. San Diego: Academic Press; 2017. p. e29–41 [Chapter 4].
378. Melroy-Greif WE, Wilhelmsen KC, Ehlers CL. Genetic variation in FAAH is associated with cannabis use disorders in a young adult sample of Mexican Americans. *Drug Alcohol Depend* 2016;**166**:249–53.
379. Di Forti M, Iyegbe C, Sallis H, Kolliakou A, Falcone MA, Paparelli A, et al. Confirmation that the AKT1 (rs2494732) genotype influences the risk of psychosis in cannabis users. *Biol Psychiatr* 2012;**72**(10):811–6.
380. Fatjó-Vilas M, Soler J, Ibáñez MI, Moya-Higueras J, Ortet G, Guardiola-Ripoll M, et al. The effect of the AKT1 gene and cannabis use on cognitive performance in healthy subjects. *J Psychopharmacol* 2020;**34**(9):990–8.
381. Thorpe HHA, Talhat MA, Khokhar JY. High genes: genetic underpinnings of cannabis use phenotypes. *Prog Neuro-Psychopharmacol Biol Psychiatry* 2021;**106**:110164.
382. Hartman CA, Hopfer CJ, Haberstick B, Rhee SH, Crowley TJ, Corley RP, et al. The association between cannabinoid receptor 1 gene (CNR1) and cannabis dependence symptoms in adolescents and young adults. *Drug Alcohol Depend* 2009;**104**(1-2):11–6.
383. Seifert J, Ossege S, Emrich HM, Schneider U, Stuhrmann M. No association of CNR1 gene variations with susceptibility to schizophrenia. *Neurosci Lett* 2007;**426**(1):29–33.
384. Tsai SJ, Wang YC, Hong CJ. Association study of a cannabinoid receptor gene (CNR1) polymorphism and schizophrenia. *Psychiatr Genet* 2000;**10**(3):149–51.
385. O'Tuathaigh CMP, Hryniewiecka M, Behan A, Tighe O, Coughlan C, Desbonnet L, et al. Chronic adolescent exposure to Δ-9-tetrahydrocannabinol in COMT mutant mice: impact on psychosis-related and other phenotypes. *Neuropsychopharmacology* 2010;**35**(11):2262–73.
386. O'Tuathaigh CM, Clarke G, Walsh J, Desbonnet L, Petit E, O'Leary C, et al. Genetic vs. pharmacological inactivation of COMT influences cannabinoid-induced expression of schizophrenia-related phenotypes. *Int J Neuropsychopharmacol* 2012;**15**(9):1331–42.
387. Behan ÁT, Hryniewiecka M, O'Tuathaigh CMP, Kinsella A, Cannon M, Karayiorgou M, et al. Chronic adolescent exposure to Delta-9-tetrahydrocannabinol in COMT mutant mice: impact on indices of dopaminergic, endocannabinoid and GABAergic pathways. *Neuropsychopharmacology* 2012;**37**(7):1773–83.
388. Burgdorf CE, Jing D, Yang R, Huang C, Hill MN, Mackie K, et al. Endocannabinoid genetic variation enhances vulnerability to THC reward in adolescent female mice. *Sci Adv* 2020;**6**(7), eaay1502.

389. Harrison PJ, Law AJ. Neuregulin 1 and schizophrenia: genetics, gene expression, and neurobiology. *Biol Psychiatr* 2006;**60**(2):132–40.
390. Boucher AA, Arnold JC, Duffy L, Schofield PR, Micheau J, Karl T. Heterozygous neuregulin 1 mice are more sensitive to the behavioural effects of delta-9-tetrahydrocannabinol. *Psychopharmacology* 2007;**192**(3):325–36.
391. Boucher AA, Hunt GE, Micheau J, Huang X, McGregor IS, Karl T, et al. The schizophrenia susceptibility gene neuregulin 1 modulates tolerance to the effects of cannabinoids. *Int J Neuropsychopharmacol* 2011;**14**(5):631–43.
392. Long LE, Chesworth R, Huang XF, McGregor IS, Arnold JC, Karl T. Transmembrane domain Nrg1 mutant mice show altered susceptibility to the neurobehavioural actions of repeated THC exposure in adolescence. *Int J Neuropsychopharmacol* 2013;**16**(1):163–75.
393. Johnstone M, Thomson PA, Hall J, McIntosh AM, Lawrie SM, Porteous DJ. DISC1 in schizophrenia: genetic mouse models and human genomic imaging. *Schizophr Bull* 2011;**37**(1):14–20.
394. Millar JK, Wilson-Annan JC, Anderson S, Christie S, Taylor MS, Semple CA, et al. Disruption of two novel genes by a translocation co-segregating with schizophrenia. *Hum Mol Genet* 2000;**9**(9):1415–23.
395. Ballinger MD, Saito A, Abazyan B, Taniguchi Y, Huang CH, Ito K, et al. Adolescent cannabis exposure interacts with mutant DISC1 to produce impaired adult emotional memory. *Neurobiol Dis* 2015;**82**:176–84.

Index

Note: Page numbers followed by *f* indicate figures *t* indicate tables and *b* indicate boxes.

A

Adolescent brain and cannabis
  behavioral effects of
    animal studies, 296–299
    human studies, 295–296
  CB1 and CB2 receptors, 291–292
  Δ^9-tetrahydrocannabinol (Δ^9-THC), 283, 291–292
  developmental effects of
    functional changes, 294–295
    morphological changes, 292–293
  endocannabinoid (ECB)
    2-arachidonoylglycerol (2-AG), 290–291
    anandamide (AEA), 290–291
    CB1 and CB2 receptors, 288–290
    components, 288
    diacylglycerol lipase (DAGL), 290–291
    fatty acid amide hydrolase (FAAH), 291
    GABAergic neurons, 288–289
    monoacylglycerol lipase (MAGL), 291
    N-arachidonoyl phosphatidyl ethanol-preferring phospholipase D (NAPE-PLD), 290–291
    retrograde signaling, 288–289
    and sex hormones, 291
  epigenetic mechanisms
    amygdala, 101–102
    behavioral brake, 101–102
    definition of, 100–101
    differentially expressed miRNAs, 102
    DNA methylation, 102–103
    expression of, 100–101
    histone methyltransferase SUV39H1, 101–102
    histone modification, 101–102
    maternal immune activation (MIA), 102
  genetic vulnerability
    CNR1 genotypes, 306–307
    disrupted-in-schizophrenia 1 (DISC1), 307–308
    neuregulin 1 (Nrg1), 307
  glial cells
    astrocytes, 95–96, 98–99
    microglia, 96–97, 99
    oligodendrocytes, 97, 99–100
  high degree of neuroplasticity, 283
  morphological and functional transformations, 284–287
  neurobiological mechanisms, long-term changes
    endocannabinoid (ECB) system, 299–300
    glial cells, 303–305
    neuron, 301–303
    neurotransmission, 300–301
  psychiatric disorders, 283
  vulnerability to environmental factors, 287–288
Adolescent cannabinoid exposure
  emotional processing, control of
    basolateral amygdala (BLA), 173–174, 175*f*
    mammalian pre-frontal cortex (PFC), 173–174
    schizophrenia-related psychopathology, 173
  neurodevelopmental effects, 179–180
  pre-frontal cortex (PFC), transmission in
    CB1R expression, 175–177, 176*f*
    clinical and pre-clinical evidence, 177
    ventral tegmental area (VTA), 174–175
  schizophrenia
    Akt signaling pathway, 185–187
    behavioral abnormalities, 190
    beta-catenin protein expression, 183–185
    biomarkers, 189–190
    DAergic control, 188–189
    excitatory *vs.* inhibitory control mechanisms, 188–189
    GABAergic control, 188–190
    glycogen-synthase kinase 3 (GSK-3) signaling pathway, 183–185
    mammalian target of rapamycin (mTOR) signaling pathways, 187–188
    mesolimbic dopamine (DA) system, 181–182
    P70S6K signaling pathways, 187–188

Adolescent cannabinoid exposure *(Continued)*
   paired-pulse inhibition (PPI) protocol, 181–182
   vulnerability, 180–181
   Wnt signaling pathway, 183–185
  translational rodent models
   advantages, 178
   cannabinoid receptor type 1 (CB1R), 178–179
AEA. *See* Anandamide (AEA)
2-AG. *See* 2-Arachidonoylglycerol (2-AG)
Anandamide (AEA)
  adolescent brain and cannabis, 290–291
  axonal growth, 157–158
  developing brain, 63–64, 63f
  enzymatic production and degradation, 153–155
  mature synapses, 152–153, 154–155f
  in neurogenic niches, 261–263
2-Arachidonoyl glycerol (2-AG)
  adolescent brain and cannabis, 290–291
  in developing brain, 62–63, 63f, 65
  enzymatic production and degradation, 153–155
  neurite outgrowth, 157–158
  plasma membrane precursors, 30
Artificial cannabinoids, 28–30
Astrocytes
  adolescent exposure, 98–99
  endocannabinoid (ECB) system, 95–96
Astrocytic progenitors (APs), 68–69

B

Behavioral consequences, cannabis exposure
  animal studies, during adolescence
   on anxiety behaviors, 297
   CB1 receptor agonists, 296
   Δ^9-tetrahydrocannabinol (Δ^9-THC), 296–297
   endocannabinoid (ECB) system, long-term changes, 299–300
   glial cells, long-term changes, 303–305
   innate stress sensitivity and response, 297–298
   neuron, long-term changes, 301–303
   neurotransmission, long-term changes, 300–301
   pre-clinical model, 296–297
   psychosis-like signs, 298
   sex differences, 298–299
  clinical studies

Adolescent Brain Cognitive Development (ABCD) study, 82–83
  autism spectrum disorder (ASD), 83
  during breastfeeding, 83–84
  Maternal Health Practices and Child Development Study (MHPCD), 81–82
  The Ottawa Prenatal Prospective Study (OPPS), 81–82
human studies, during adolescence
  attentional and psychomotor deficits, 295
  on cognition and behavior, 295
  hypomania, 295–296
  withdrawal syndrome, 296
rodent studies
  extrapolation, 84–86
  Hedgehog (HH) signaling, 86–87
  intrinsic limitations, 84
  during lactation, 87
  long-term depression (LTD), 84–86
  mesolimbic dopaminergic system, 86
  ultrasonic vocalizations (USVs), 84–86

C

Cannabidiol (CBD)
  in breast milk, 129–130
  consumption of, 150
  cytochrome P450 inhibition, 155
  drug administration, 6–7
  maternal cannabis exposure, 87–88
  neurogenesis, 157
  serotonin 2A receptors (5-HT2AR), 217–218
Cannabinoid
  adolescent exposure
   emotional processing, control of, 173–174, 175f
   long-term mesolimbic dopamine activity states, 181–182
   neurodevelopmental effects, 179–180
   pre-frontal cortex (PFC), transmission in, 174–177, 176f
   pre-frontal cortical GABAergic and glutamatergic functional balance, 188–190
   pre-frontal cortical regulation, 183–188, 184f, 186f
   schizophrenia vulnerability, 180–181
   translational rodent models, 178–179
  artificial form, 28–30
  hyperemesis syndrome (HES), 149–150
  phytocannabinoids, 1–2
Cannabis
  2-arachidonoyl glycerol (2-AG), 30

Index

adolescent brain
  behavioral effects of, 295–299
  brain anatomy and connectivity, 35–36
  cannabis use disorder (CUD), 36
  CB1 and CB2 receptors, 291–292
  cognitive aptitude and behavior, 34–35
  Δ^9-tetrahydrocannabinol (Δ^9-THC), 37, 283, 291–292
  developmental effects of, 292–295
  dopamine (DA) function, 37
  endocannabinoid (ECB), 283, 288–291
  genetic vulnerability, 306–308
  gray matter volume (GMV), 36
  high degree of neuroplasticity, 283
  morphological and functional transformations, 284–287
  neurobiological mechanisms, 299–306
  psychiatric disorders, 283
  vulnerability to environmental factors, 287–288
anandamide (AEA), 30
artificial cannabinoids, 28–30
cannabis light, 45–46
cannabis use disorder (CUD), 34
Δ^9-tetrahydrocannabinol (Δ^9-THC)
  adolescent brain development, 26
  cannabinoid hyperemesis syndrome (HES), 149–150
  dopaminergic system, 26
  e-cigarettes, 149–150
  effects of, 13
  fetus and newborns, 150–151
  molecular targets, 25–26, 25f
  phyto-cannabinoid bioactivity, 23–25, 24f
  primary psychotropic compound, 23–25
  properties of, 149–150
  psychotropic effects, 26–27
dopaminergic system, 26
endocannabinoid (ECB) system
  and brain development, 31, 32f
  enzymatic production and degradation, 153–155, 154–155f
  first cannabinoid receptor (CB1R), 152–153
  neuronal maturation, 33
  in neurons and glia, 28, 29f
  post-natal signaling, 159–161
  pre-natal signaling, 156–159
  second cannabinoid receptor (CB2R), 152–153
  transient receptor potential sub-family vanilloid type 1 ion channel (TRPV1), 152–153

flower cigarettes (joints) and edibles (cookies), 27–28, 27f
GPR55, 30–31
human studies, 33–37
legalization of, 27–28, 27f
maternal exposure
  animal models, 80
  behavioral consequences of, 80–87
  cannabidiol (CBD), 87–88
  challenges, 80, 87–88
  Δ^9-tetrahydrocannabinol (Δ^9-THC), 79
  during pregnancy, 79–80
medicinal uses, 149
monitory system, 48
paternal exposure, 87–88
prefrontal cortex (PFC), 31–33, 32f
during pregnancy
  central nervous system, effects on, 151
  cognitive function, 151
  Generation R, 152
  Stanford-Binet Intelligence Scale, 151–152
peri-natal exposure
  animal models, advantage of, 80
  challenges, 87–88
  cognitive deficits, in rats, 84–86
  pre-clinical studies, 87
prenatal exposure
  Δ^9-tetrahydrocannabinol (Δ^9-THC), 16–17
  DNA methylation, 16
  Drd2 gene, 16–17
  histone modifications, 16
  Ktm2a, 16–17
  molecular and epigenetic dysregulation, 18t, 19
  molecular reprogramming, 14–15
  neurodevelopmental effects, 16–17
  non-coding RNAs, 16
randomized control trials (RCTs), 33
ritualistic uses, 149
rodent studies
  additional systems-level and behavioral responses, 42–44
  adult mesocorticolimbic systems, 40–42
  cannabimimetic activities, 38, 39f
  cocaine, 41–42
  Δ^9-tetrahydrocannabinol (Δ^9-THC), 37–38, 44–45
  molecular and cellular responses, 38–40
  opiates, 41
  reward/addiction-like behaviors, 40–42

Cannabis *(Continued)*
   "tetrad response", 38
   schizophrenia
      adolescence and, 210–211
      CB1 receptor (CB1R), 214
      consumption, 197–198
      controlled studies, humans, 213
      glutamatergic and GABAergic terminals, 214
      negative effects, 214
      observational studies, 212–213
      reverse causation, 212
      risk factor for, 210–213
      serotonin 5-HT2A receptors, 215
      Swedish conscript survey, 210
   supraphysiological impact of, 13
   terpenoids (terpenes), 27
CBD. *See* Cannabidiol (CBD)

D

Δ^9-tetrahydrocannabinol (Δ^9-THC)
   adolescent brain
      and cannabis, 283, 291–292, 296–297
      development, 26
   adolescent exposure
      astrocytes, 98–99
      microglia, 99
      oligodendrocytes, 99–100
   in breast milk, 129–130
   cannabidiol (CBD)
      cannabinoid hyperemesis syndrome (HES), 149–150
      cannabinoid receptors (CBRs), 14
      consumption of, 150
      interactions with, 45–47
   cytochrome P450 metabolism, 133
   in developing brain, 65
   dopaminergic system, 26
   dose of drug, 7–8
   Drd2 gene, 16–17
   e-cigarettes, 149–150
   effects of, 13
   fetus and newborns, 150–151
   intra-peritoneal injections, 5
   maternal cannabis exposure, 79
   molecular targets, 25–26, 25*f*
   oral administration, 5
   phyto-cannabinoid bioactivity, 23–25, 24*f*
   prenatal exposure, 14–15
   pre-natal use, 6–7
   primary psychotropic compound, 23–25
   properties of, 149–150
   psychotropic effects, 26–27
   rodent studies, 37–38, 44–45
   transgenerational epigenetics, 132
Δ^9-THC. *See* Δ-tetrahydrocannabinol (Δ^9-THC)
Delta-9-tetrahydrocannabinol (Δ^9-THC). *See* Δ^9-tetrahydrocannabinol (Δ^9-THC)
Developmental cannabis exposure, 3*t*, 8
Developmental origins of health and disease (DOHaD) theory, 108, 111
DNA methylation, 16

E

ECB. *See* Endocannabinoid (ECB) system
ECS. *See* Endocannabinoid (ECB) system
Emotional salience processing
   CB1 receptor (CB1R), 172
   challenges, 172
   clinical and pre-clinical studies, 172
   disturbances in, 171–172
   dopamine (DA) pathway, 172
   endocannabinoid (ECB) system, 172
   mesocorticolimbic structures, 172
Endocannabinoid (ECB) system
   2-arachidonoyl glycerol (2-AG), 290–291
   adolescent playfulness, 65–66
   anandamide (AEA), 290–291
   CB1 and CB2 receptor, 288–290
   components, 288
   copulation, 70
   Δ^9-tetrahydrocannabinol (Δ^9-THC), effects of, 13
   developing brain
      2-arachidonoyl glycerol (2-AG), 62–63, 63*f*, 65
      anandamide (AEA), 63–64, 63*f*
      cannabis use, 64
      CB1 and CB2 receptors, 63–65
      components of, 64
      Δ^9-tetrahydrocannabinol (Δ^9-THC), 65
      fatty acid amide hydrolase (FAAH), 63–64
      prostaglandin synthesis, 62, 63*f*
      reproduction, role in, 62–63
      sexual differentiation, 61–62, 63*f*
   diacylglycerol lipase (DAGL), 290–291
   emotional salience processing, 172
   enzymatic production and degradation
      2-arachidonoyl glycerol (2-AG), 153
      anandamide (AEA), 153
      cytochrome P450 complex, 153–155
      "entourage" effect, 153–155

N-palmitoylethanolamine (PEA), 153–155
ω-6 poly-unsaturated fatty acids (PUFA), 153–155
fatty acid amide hydrolase (FAAH), 291
in female gametogenesis, 131
first cannabinoid receptor (CB1R), 152–153
GABAergic interneuron development, 266
GABAergic neurons, 288–289
glial cells
   astrocytes, 95–96, 98–99
   microglia, 96–97, 99
   oligodendrocytes, 97, 99–100
in male gametogenesis, 131
microglia phagocytic activity
   astrocytic progenitors (APs), 68–69
   complement system, 67–68
   features, 67–68
   medial extended amygdala (MeA), 68–69, 69f
   synthesizers, 69
monoacylglycerol lipase (MAGL), 291
N-arachidonoyl phosphatidyl ethanol-preferring phospholipase D (NAPE-PLD), 290–291
in neurogenic niches
   anandamide (AEA) levels, 261–263
   CB1 receptors, role of, 261–263
   diacylglycerol lipase (DAGL), 260–261
   expression and function of, 260–261, 261f
   neural cell fate regulation, 261–263, 262f
   neural progenitor (NP) cells, 260–261
neuropsychiatric disorders, 111–112, 112b
post-natal signaling
   early adolescence toward adulthood, 160–161
   neonatal findings, 160
   perinatal period, 159–160
pre-natal cannabinoid exposure (PCE)
   CB1 receptor, 266–267
   dopaminergic neurons, 268
   functional consequences of, 270–271
   GABAergic interneurons, 268–270
   in human-based models, 272–273
   neurodevelopmental disorders, 271–272
   projection neurons, 267–268
pre-natal signaling
   adolescent synapse refinement, 158
   dopaminergic circuitry, 159
   embryo implantation and growth, 156
   neurite outgrowth and target innervation, 157–158
   neurulation and neurogenesis, 156–157
   in pyramidal neuron development
      Bcl11b and Satb2 transcription factors, 262f, 263
      CB1 receptor signaling, 264, 265f
      RhoA signaling, 264
retrograde signaling, 288–289
role of
   embryo implantation and development, 132
   parental germline, 130–131
   synaptic programming, 133–134
second cannabinoid receptor (CB2R), 152–153
sex difference in playfulness, 66–67
and sex hormones, 291
signaling
   and brain development, 31, 32f
   neuronal maturation, 33
   in neurons and glia, 28, 29f
significance of, 67
social behavior network (SBN)
   hormone-sensitive, 61
   medial extended amygdala (MeA), 66, 68–69, 69f
   medial preoptic area (mPOA), 70–71
   nodes of, 60–61, 60f
synaptogenesis, 152–153, 154–155f, 156–157
transient receptor potential sub-family vanilloid type 1 ion channel (TRPV1), 152–153
Epigenetic mechanisms
   amygdala, 101–102
   behavioral brake, 101–102
   Δ^9-tetrahydrocannabinol (Δ^9-THC), 13
   definition of, 15, 100–101
   differentially expressed miRNAs, 102
   DNA methylation, 102–103
   endocannabinoid (ECB) system, 13
   expression of, 100–101
   histone methyltransferase SUV39H1, 101–102
   histone modification, 101–102
   histone modifications, 16
   maternal immune activation (MIA), 102
   non-coding RNAs, 16
   prenatal cannabis exposure (PCE)
      Δ^9-tetrahydrocannabinol (Δ^9-THC), 16–17
      DNA methylation, 16
      Drd2 gene, 16–17
      histone modifications, 16

Epigenetic mechanisms *(Continued)*
　Ktm2a, 16–17
　molecular and epigenetic dysregulation, 18*t*, 19
　molecular reprogramming, 14–15
　neurodevelopmental effects, 16–17
　non-coding RNAs, 16
Epigenetic reprogramming cohabitation, 130–131

F
First Episode Psychosis (FEP), 199

G
Glial cells
　astrocytes
　　adolescent exposure, 98–99
　　endocannabinoid (ECB) system, 95–96
　microglia
　　adolescent exposure, 99
　　endocannabinoid (ECB) system, 96–97
　oligodendrocytes
　　adolescent exposure, 99–100
　　endocannabinoid (ECB) system, 97

H
Hemp. *See* Cannabis
Histone modifications, 16

M
Mammalian brain
　addictive behaviors, 171–172
　DAergic neuronal activity, 172
　emotional salience processing
　　CB1 receptor (CB1R), 172
　　challenges, 172
　　clinical and pre-clinical studies, 172
　　disturbances in, 171–172
　　dopamine (DA) pathway, 172
　　endocannabinoid (ECB) system, 172
　　mesocorticolimbic structures, 172
　maturation of, 171
　neuropsychiatric disorders, 171–172
Mammalian target of rapamycin (mTOR) pathways
　alterations in, 222
　mTOR complex 1 (mTORC1) and mTOR complex 2 (mTORC2), 221
　regulation of, 219, 220*f*
　ribosomal protein S6 (rpS6), 218
　role in, 218
　signaling, 187–188, 221
Maternal cannabis exposure
　animal models, 80
　behavioral consequences of
　　clinical studies, 80–84
　　rodent studies, 84–87
　cannabidiol (CBD), 87–88
　challenges, 80, 87–88
　Δ^9-tetrahydrocannabinol (Δ^9-THC), 79
　during pregnancy, 79–80
Medial extended amygdala (MeA)
　androgen receptors (ARs), 66
　animal models, 66
　development of, 68–69, 69*f*
　vs. medial preoptic area (mPOA), 70
Medial preoptic area (mPOA), 60–61, 60*f*, 70–71
Microglia
　adolescent exposure, 99
　endocannabinoid (ECB) system, 96–97
　phagocytic activity
　　astrocytic progenitors (APs), 68–69
　　complement system, 67–68
　　features, 67–68
　　medial extended amygdala (MeA), 68–69, 69*f*
　　synthesizers, 69

N
Neurodevelopmental cannabinoid exposure
　cerebral organoids, 1–2
　drug administration
　　adolescent models, 4
　　dose of, 3*t*, 7–8
　　inhaled, 4–5
　　injection, 6
　　nature of, 6–7
　　oral, 5
　　pharmacokinetics, 2–4
　　pre-natal models, 2–4
　　vaping chamber, 5
　　voluntary consumption, 4–5
　phytocannabinoids, 1–2
　pre-clinical models, 2
Neurotransmission system
　dopamine, 203, 204*f*
　γ-aminobutyric acid (GABA), 204*f*, 206
　glutamate, 204*f*, 205–206
　serotonin, 203–205, 204*f*
Non-coding RNAs, 16

O

Oligodendrocytes
  adolescent exposure, 99–100
  endocannabinoid (ECB) system, 97

P

Paternal cannabis exposure, 87–88
PCE. *See* Pre-natal cannabis exposure (PCE)
Perinatal cannabis exposure
  animal models, advantage of, 80
  in breast milk, 129–130
  CB1R, expression of, 130
  challenges, 87–88
  cognitive deficits, in rats, 84–86
  early development, 130
  embryo implantation and development
    endocannabinoid (ECB) system, role of, 132
    maternal exocannabinoid exposure and repercussions, 133
  meconium analysis, 129
  parental germline
    endocannabinoid (ECB) system, role of, 130–131
    epigenetic reprogramming cohabitation, 130–131
    in female gametogenesis, 131
    in male gametogenesis, 131
    transgenerational effects, 132
  post-fecundation phase, 136–137
  pre-clinical studies, 87
  synaptic programming
    dopaminergic system, 136
    endocannabinoid (ECB) system, role of, 133–134
    GABAergic synapses, 135–136
    glutamatergic synapses, 134–135
    sex-specific neurodevelopmental perturbations, 134
Post-natal endocannabinoid signaling
  early adolescence toward adulthood, 160–161
  neonatal findings, 160
  perinatal period, 159–160
Pre-natal cannabis exposure (PCE).
    See also Maternal cannabis exposure
  animal models, advantage of, 80
  CB1 receptor, 266–267
  challenges, 87–88, 121–122
  cognitive deficits, in rats, 84–86
  developmental origins of health and disease (DOHaD) theory, 108

  DNA methylation, 16
  dopaminergic neurons, 268
  Drd2 gene, 16–17
  functional consequences of, 270–271
  GABAergic interneurons, 268–270
  histone modifications, 16
  in human-based models, 272–273
  Ktm2a, 16–17
  mental health
    children and adolescents, 107
    early detection interventions, 107–108
  mesolimbic dopamine system function
    environmental challenges, 116, 118
    female progeny, 117
    male progeny, 116–117
    neuronal circuits, 114
    pre frontal cortex (PFC), 117
    pre-pulse inhibition (PPI), 114
    sensorimotor gating, 114, 114*b*
    sex-specific alterations, 116–117
    ventral tegmental area (VTA), 113–115, 115*f*
  molecular and epigenetic dysregulation, 18*t*, 19
  molecular reprogramming, 14–15
  neurodevelopmental disorders, 16–17, 271–272
  neuropsychiatric disorders
    children and adolescents, 110
    developmental origins of health and disease (DOHaD) theory, 111
    endocannabinoid (ECB), 111–112, 112*b*
    endophenotypes, 110–111
    longitudinal human cohorts, 110
    metaplasticity, 112
    risk factors, 109–110
    synaptic plasticity, 112
  non-coding RNAs, 16
  pre-clinical studies, 87
  projection neurons, 267–268
  "two-hit" hypothesis, 108–109
  vulnerability measurement
    clinical staging model, 118–119
    fatty acid amide (FAAH), 121
    metabotropic glutamate receptor type 5 (mGluR5), 121
    pregnenolone, 120–121
    sensory information processing (SIP), 118–119, 119*f*
    sex-differential biological sensitivity, 119, 120*f*

Pre-natal endocannabinoid signaling
  adolescent synapse refinement, 158
  dopaminergic circuitry, 159
  embryo implantation and growth, 156
  neurite outgrowth and target innervation, 157–158
  neurulation and neurogenesis, 156–157
Pre-pulse inhibition (PPI), 114
Primary non-psychotropic compound.
  See Cannabidiol (CBD)
Primary psychotropic compound.
  See Δ^9-tetrahydrocannabinol (Δ^9-THC)
Psychosis, 198–199

R

Ribosomal protein S6 (rpS6), 218, 221
Rough-and-tumble play. *See* Social play

S

SBN. *See* Social behavior network (SBN)
Schizophrenia
  Akt pathway
    alterations in, 222
    protein kinase B, 219
    regulation of, 219, 220f
    ribosomal protein S6 (rpS6), 218
    role in, 218
    signaling, 185–187, 221
  behavioral abnormalities, 190
  beta-catenin protein expression, 183–185
  biomarkers, 189–190
  cannabis and
    adolescence and, 210–211
    CB1 receptor (CB1R), 214
    consumption, 197–198
    controlled studies, humans, 213
    glutamatergic and GABAergic terminals, 214
    negative effects, 214
    observational studies, 212–213
    reverse causation, 212
    risk factor for, 210–213
    serotonin 5-HT2A receptors, 215
    Swedish conscript survey, 210
  cognitive symptoms, 199
  comorbidity in
    anxiety disorders, 209
    clusters, 208
    depression symptoms, 209
    Epidemiologic Catchment Area (ECA) Study, 209
  DAergic control, 188–189
  environmental factors
    cannabis use, 208
    early childhood trauma and adversity, 208
    migration and minority group, 207
    obstetric complications, 207–208
    pre-natal stressful exposures, 207–208
    stress-vulnerability model, 206
    urbanicity, 207
  excitatory *vs.* inhibitory control mechanisms, 188–189
  First Episode Psychosis (FEP), 199
  functional brain imaging, 200–201
  GABAergic control, 188–190
  genetics
    copy number variations (CNV), 201
    genome-wide association studies (GWAS), 201–203
    Next Generation Sequencing (NGS), 201
    single-nucleotide polymorphisms (SNPs), 201
  glycogen-synthase kinase 3 (GSK-3) signaling pathway, 183–185
  late adolescence and early adulthood, 197–198
  mammalian target of rapamycin (mTOR) pathways
    alterations in, 222
    mTOR complex 1 (mTORC1) and mTOR complex 2 (mTORC2), 221
    regulation of, 219, 220f
    ribosomal protein S6 (rpS6), 218
    role in, 218
    signaling, 187–188, 221
  mesolimbic dopamine (DA) system, 181–182
  negative symptoms, 199
  neuroimaging, 200
  neurotransmission system
    dopamine, 203, 204f
    γ-aminobutyric acid (GABA), 204f, 206
    glutamate, 204f, 205–206
    serotonin, 203–205, 204f
  onset and progression of, 201, 202f
  P70S6K signaling pathways, 187–188
  paired-pulse inhibition (PPI) protocol, 181–182
  positive symptoms, 198–199
  prevalence of, 197–198
  prodromal period, 199
  ribosomal protein S6 (rpS6), 218, 221
  serotonin 2A receptors (5-HT2AR)
    alterations in, 216–217

and cannabinoids, 217–218
hallucinogenic and non-hallucinogenic drugs, 198
intracellular signaling, 216
stages of, 199, 199f
vulnerability, 180–181
Wnt signaling pathway, 183–185
Sensorimotor gating, 114, 114b
Social behavior network (SBN)
hormone-sensitive, 61
medial extended amygdala (MeA)
androgen receptors (ARs), 66
animal models, 66
development of, 68–69, 69f
vs. medial preoptic area (mPOA), 70
medial preoptic area (mPOA), 60–61, 60f, 70–71
nodes of, 60–61, 60f
Social play, 65–66
Synaptic programming
dopaminergic system, 136
endocannabinoid (ECB) system, role of, 133–134
GABAergic synapses, 135–136
glutamatergic synapses, 134–135
sex-specific neurodevelopmental perturbations, 134
Synthetic cannabinoids (SC)
advantages, 245
chemical structure and nomenclature
colloquial and serial names, 244–245
structural groups, 244
systematic chemical names, 245, 246f
clinical aspects
acute cardiac symptoms, 249
addictive behavior with, 250
cognitive impairments, 249
composition of, 249
schizophrenic spectrum disorders, 249–250
withdrawal symptoms, 250
cyclohexylphenol, 243
definition of, 243
developing brain and, 252–253
fertility and
CB1 and CB2 antagonist, 251
endometrial stromal cells, 251–252
in placental cytotrophoblast cells, 251
trophoblast turnover, 250–251
legal status, 247
patterns of use, 247
pharmacology
binding affinity, 247–248
pharmacokinetics and metabolism, 248
phytocannabinoids with, 243–244
prevalence of, 245–247
synthesis of, 243

T

THC. See Δ^9-tetrahydrocannabinol (Δ^9-THC)